AF686020

TREATISE ON INVERTEBRATE PALEONTOLOGY

Prepared under the Guidance of the
Joint Committee on Invertebrate Paleontology

Paleontological Society — *Society of Economic Paleontologists and Mineralogists* — *Palaeontographical Society*

Directed and Edited by
RAYMOND C. MOORE
Assisted by CHARLES W. PITRAT, DORIS E. N. ZELLER, and LAVON McCORMICK

Part W
MISCELLANEA

CONODONTS
CONOIDAL SHELLS OF UNCERTAIN AFFINITIES
WORMS
TRACE FOSSILS AND PROBLEMATICA

By W. H. HASS, WALTER HÄNTZSCHEL, D. W. FISHER, B. F. HOWELL, F. H. T. RHODES, K. J. MÜLLER, and R. C. MOORE

GEOLOGICAL SOCIETY OF AMERICA
and
UNIVERSITY OF KANSAS PRESS

1962

© 1962 BY THE UNIVERSITY OF KANSAS PRESS
AND THE GEOLOGICAL SOCIETY OF
AMERICA

ALL RIGHTS RESERVED

Library of Congress Catalogue Card
Number: 53-12913

Text Composed by
THE UNIVERSITY OF KANSAS PRESS
Lawrence, Kansas

Illustrations and Offset Lithography
MERIDEN GRAVURE COMPANY
Meriden, Connecticut

Binding
RUSSELL-RUTTER COMPANY
New York City

Address all communications to The Geological Society of America, 419 West 117 Street, New York 27, N.Y.

The *Treatise on Invertebrate Paleontology* has been made possible by (1) grants of funds from The Geological Society of America through the bequest of Richard Alexander Fullerton Penrose, Jr., for preparation of illustrations and partial defrayment of organizational expense and the United States National Science Foundation, in December, 1959, for completion of the *Treatise* project; (2) contribution of the knowledge and labor of specialists throughout the world, working in co-operation under sponsorship of The Palaeontographical Society, The Paleontological Society, and The Society of Economic Paleontologists and Mineralogists; and (3) acceptance by the University of Kansas Press of publication without cost to the Societies concerned and without any financial gain to the Press.

JOINT COMMITTEE ON INVERTEBRATE PALEONTOLOGY

RAYMOND C. MOORE, University of Kansas, *Chairman*

G. ARTHUR COOPER, United States National Museum, *Executive Member*

NORMAN D. NEWELL, Columbia University and American Museum of Natural History, *Executive Member*

C. J. STUBBLEFIELD, Geological Survey of Great Britain, *Executive Member*

Representing THE PALAEONTOGRAPHICAL SOCIETY

L. BAIRSTOW, British Museum (Natural History)
O. M. B. BULMAN, Sedgwick Museum, Cambridge University
C. P. CHATWIN, Geological Survey of Great Britain (retired)
L. R. COX, British Museum (Natural History)
T. NEVILLE GEORGE, University of Glasgow
H. L. HAWKINS, University of Reading
R. V. MELVILLE, Geological Survey of Great Britain
H. M. MUIR-WOOD, British Museum (Natural History)
C. J. STUBBLEFIELD, Geological Survey of Great Britain
H. DIGHTON THOMAS, British Museum (Natural History)

Representing THE PALEONTOLOGICAL SOCIETY

G. ARTHUR COOPER, United States National Museum
CARL O. DUNBAR, Yale University
B. F. HOWELL, Princeton University
A. K. MILLER, State University of Iowa
R. R. SHROCK, Massachusetts Institute of Technology
F. M. SWARTZ, Pennsylvania State University
H. E. VOKES, Tulane University
A. SCOTT WARTHIN, JR., Vassar College
J. W. WELLS, Cornell University
W. P. WOODRING, United States Geological Survey

Representing THE SOCIETY OF ECONOMIC PALEONTOLOGISTS AND MINERALOGISTS

CARL C. BRANSON, University of Oklahoma
J. WYATT DURHAM, University of California
DON L. FRIZZELL, Missouri School of Mines
H. V. HOWE, Louisiana State University
C. G. LALICKER, Wichita, Kansas
STUART A. LEVINSON, Humble Oil & Refining Company
N. D. NEWELL, Columbia University; American Museum of Natural History
F. W. ROLSHAUSEN, Humble Oil & Refining Company
H. B. STENZEL, Shell Development Company
J. M. WELLER, University of Chicago

GEOLOGICAL SOCIETY OF AMERICA

FREDERICK BETZ, JR., *Editor-in-Chief* AGNES CREAGH, *Managing Editor*

UNIVERSITY OF KANSAS PRESS

CLYDE K. HYDER, *Editor*

TREATISE ON INVERTEBRATE PALEONTOLOGY

Directed and Edited by

RAYMOND C. MOORE

Assisted by CHARLES W. PITRAT, DORIS E. N. ZELLER, and LAVON MCCORMICK

PARTS

The indicated Parts (excepting A and X) are to be published at whatever time each is ready. All may be assembled ultimately in bound volumes. In the following list, already published Parts are marked with a double asterisk (**) and those in press or nearing readiness for press are marked with a single asterisk (*). Each is cloth-bound with title in gold on the cover. Copies are available on orders sent to the Geological Society of America at 419 West 117th Street, New York 27, N.Y., at prices quoted, which very incompletely cover costs of producing and distributing them but on receipt of payment the Society will ship copies without additional charge to any address in the world.

The list of contributing authors is subject to change.

(A)—INTRODUCTION
B—PROTISTA 1 (chrysomonads, coccolithophorids, diatoms, etc.).
*C—PROTISTA 2 (foraminifers).
**D—PROTISTA 3 (radiolarians, tintinnines) (xii+195 p., 1050 figs.). $3.00.
**E—ARCHAEOCYATHA, PORIFERA (xviii + 122 p., 728 figs.). $3.00.
**F—COELENTERATA (xvii + 498 p., 2,700 figs.). $8.00.
**G—BRYOZOA (xii + 253 p., 2,000 figs.). $3.00.
H—BRACHIOPODA.
**I—MOLLUSCA 1 (chitons, scaphopods, gastropods) (xxiii+351 p., 1,732 figs.). $7.50.
J—MOLLUSCA 2 (gastropods).
*K—MOLLUSCA 3 (nautiloid cephalopods).
**L—MOLLUSCA 4 (ammonoid cephalopods) (xxii + 490 p., 3,800 figs.). $8.50.
M—MOLLUSCA 5 (dibranchiate cephalopods).
N—MOLLUSCA 6 (pelecypods).
**O—ARTHROPODA 1 (trilobitomorphs) (xix+560 p., 2880 figs.). $10.50.
**P—ARTHROPODA 2 (chelicerates, pycnogonids) (xvii + 181 p., 565 figs.). $3.50.
**Q—ARTHROPODA 3 (ostracodes) (xxiii+442 p., 3,476 figs.). $11.50.
R—ARTHROPODA 4 (branchiopods, cirripeds, malacostracans, myriapods, insects).
S—ECHINODERMATA 1 (cystoids, blastoids, edrioasteroids, etc.).
T—ECHINODERMATA 2 (crinoids).
U—ECHINODERMATA 3 (echinozoans, asterozoans).
**V—GRAPTOLITHINA (xvii + 101 p., 358 figs.). $3.00.
**W—MISCELLANEA (worms, conodonts, problematical fossils) (this volume).
(X)—ADDENDA (index).

CONTRIBUTING AUTHORS

AGER, D. V., Imperial College of Science and Technology, London, England.

AMSDEN, T. W., Oklahoma Geological Survey, Norman, Okla.

†ARKELL, W. J.

BAIRSTOW, LESLIE, British Museum (Natural History), London, Eng.

BARKER, R. WRIGHT, Shell Development Co., Houston, Tex.

†BASSLER, R. S.

BATTEN, ROGER, L., University of Wisconsin, Madison, Wisconsin.

BAYER, FREDERICK M., U.S. National Museum, Washington, D.C.

BEAVER, HAROLD H., Humble Oil & Refining Co., Houston, Texas.

BENSON, R. H., University of Kansas, Lawrence, Kans.

Berdan, Jean M., U.S. Geological Survey, Washington, D.C.

Biernat, Gertruda, Państwowe Wydawnictwo Naukowe, Warszawa, Polska.

Boardman, R. S., U.S. Geological Survey, Washington, D.C.

Bold, W. A. van den, Louisiana State University, Baton Rouge, La.

Boschma, H., Rijksmuseum van Natuurlijke Historie, Leiden, Netherlands.

Boucot, A. J., Massachusetts Institute of Technology, Cambridge, Mass.

Bowsher, Arthur L., Sinclair Oil & Gas Co., Tulsa, Okla.

Bramlette, M. N., Scripps Institution of Oceanography, La Jolla, California.

Branson, Carl C., University of Oklahoma, Norman, Okla.

Brown, D. A., Australian National University, Canberra, Austral.

Bulman, O. M. B., Sedgwick Museum, Cambridge University, Cambridge, Eng.

Campbell, Arthur S., St. Mary's College, St. Mary's College, Calif.

Carpenter, Frank M., Biological Laboratories, Harvard University, Cambridge, Mass.

Casey, Raymond, Geological Survey of Great Britain, London, Eng.

Caster, K. E., University of Cincinnati, Cincinnati, Ohio.

Chavan, André, L'École d'Anthropologie, Seyssel (Ain), France.

Cole, W. Storrs, Cornell University, Ithaca, N.Y.

Cooper, G. Arthur, U.S. National Museum, Washington, D.C.

Cox, L. R., British Museum (Natural History), London, Eng.

Curry, Dennis, Northwood, Middlesex, Eng.

†Davies, L. M.

Dechaseaux, C., Laboratoire de Paléontologie à la Sorbonne, Paris, France.

Dell, R. K., Dominion Museum, Wellington, N. Z.

Douglass, R. C., U.S. National Museum, Washington, D.C.

Durham, J. Wyatt, Museum of Paleontology, University of California, Berkeley, Calif.

Eames, F. E., British Petroleum Company, Middlesex, Eng.

Elliott, Graham F., Iraq Petroleum Company, London, Eng.

Emerson, W. K., American Museum of Natural History, New York, N.Y.

Erben, H. K., Universität Bonn, Bonn, West Ger.

Exline, Harriet, Rolla, Mo.

Fay, Robert O., Oklahoma Geological Survey, Norman, Okla.

Fell, H. Barraclough, Victoria University of Wellington, Wellington, N.Z.

Fischer, Alfred G., Princeton University, Princeton, N.J.

Fisher, D. W., New York State Museum, Albany, N.Y.

Fleming, Charles A., New Zealand Geological Survey, Lower Hutt, N. Z.

Frizzell, Donald L., Missouri School of Mines, Rolla, Mo.

Furnish, William M., State University of Iowa, Iowa City, Iowa.

†Gardner, Julia

George, T. Neville, University of Glasgow, Glasgow, Scot.

Glaessner, M. F., University of Adelaide, Adelaide, S. Austral.

Glenister, B. F., State University of Iowa, Iowa City, Iowa.

Gordon, Mackenzie, Jr., U.S. Geological Survey, Washington, D.C.

Grant, Richard E., U.S. National Museum, Washington, D.C.

Haas, Fritz, Chicago Natural History Museum, Chicago, Ill.

Hanai, Tetsuro, University of Tokyo, Tokyo, Japan.

Hanna, G. Dallas, California Academy of Sciences, San Francisco, Calif.

Häntzschel, Walter, Geologisches Staatsinstitut, Hamburg, Ger.

Harrington, H. J., Tennessee Gas Transmission Company, Houston, Tex.

†Hass, Wilbert H.

Hatai, Kotora, Tohoku University, Sendai, Japan.

Hawkins, H. L., Reading University, Reading, Eng.

Hedgpeth, Joel, College of the Pacific, Dillon Beach, Calif.

Henningsmoen, Gunnar, Paleontologisk Museum, University of Oslo, Oslo, Norway.

Hertlein, L. G., California Academy of Sciences, San Francisco, Calif.

Hessland, Ivar, Geologiska Institutet, University of Stockholm, Stockholm, Swed.

Hessler, Robert R., Woods Hole Oceanographic Institution, Woods Hole, Mass.
Hill, Dorothy, University of Queensland, Brisbane, Austral.
Holthuis, L. B., Rijksmuseum van Natuurlijke Historie, Leiden, Netherlands.
Howe, Henry V., Louisiana State University, Baton Rouge, La.
Howell, B. F., Princeton University, Princeton, N.J.
Hyman, Libbie H., American Museum of Natural History, New York, N.Y.
Jaanusson, Valdar, Paleontological Inst., Uppsala, Sweden.
Jeletzky, J. A., Geological Survey of Canada, Ottawa, Can.
Kamptner, Erwin, Naturhistorisches Museum, Wien, Aus.
Keen, Myra, Stanford University, Stanford, Calif.
Kesling, Robert V., Paleontological Museum, University of Michigan, Ann Arbor, Mich.
Kier, Porter, U. S. National Museum, Washington, D.C.
†Knight, J. Brookes
Kummel, Bernhard, Museum of Comparative Zoölogy, Harvard College, Cambridge, Mass.
La Rocque, Aurèle, Ohio State University, Columbus, Ohio.
†Laubenfels, M. W. de
Lecompte, Marius, Université de Louvain, Louvain, Belg.
Leonard, A. Byron, University of Kansas, Lawrence, Kans.
Levinson, S. A., Humble Oil & Refining Company, Houston, Tex.
Lochman-Balk, Christina, New Mexico Institute of Mining and Geology, Socorro, N.Mex.
Loeblich, A. R., Jr., California Research Corporation, La Habra, Calif.
Lohman, Kenneth E., U.S. Geological Survey, Washington, D.C.
Lowenstam, Heinz A., California Institute of Technology, Pasadena, Calif.
Ludbrook, N. H., Department of Mines, Adelaide, S.Austral.
Marwick, J., New Zealand Geological Survey, Wellington, N.Z.
Melville, R. V., British Embassy, Paris, France.
Miller, Arthur K., State University of Iowa, Iowa City, Iowa.
Montanaro Gallitelli, Eugenia, Istituto di Geologia e Paleontologia, Università, Modena, Italy.
Moore, Raymond C., University of Kansas, Lawrence, Kans.
Morrison, J. P. E., U.S. National Museum, Washington, D.C.
Muir-Wood, Helen M., British Museum (Natural History), London, Eng.
Müller, Klaus J., Technische Universität Berlin, West-Berlin (Charlottenburg), Ger.
Newell, Norman D., American Museum of Natural History (and Columbia University), New York, N.Y.
Okulitch, Vladimir J., University of British Columbia, Vancouver, Can.
Olsson, Axel A., Coral Gables, Florida.
Palmer, Allison R., U.S. Geological Survey, Washington, D.C.
Palmer, Katherine Van Winkle, Paleontological Research Institution, Ithaca, N.Y.
Peck, Raymond E., University of Missouri, Columbia, Mo.
Petrunkevitch, Alexander, Osborn Zoological Laboratory, Yale University, New Haven, Conn.
Popenoe, W. P., University of California, Los Angeles, Calif.
Poulsen, Christian, Universitetets Mineralogisk-Geologiske Institut, København, Denm.
Powell, A. W. B., Auckland Institute and Museum, Auckland, N.Z.
Puri, Harbans, Florida State Geological Survey, Tallahassee, Fla.
Rasetti, Franco, Johns Hopkins University, Baltimore, Md.
Rasmussen, H. W., Universitetets Mineralogisk-Geologiske Institut, København, Denm.
Regnéll, Gerhard, Paleontologisk-Geologiska Institutionen, Lunds Universitet, Lund, Swed.
Rehder, Harald A., U.S. National Museum, Washington, D.C.
Reichel, M., Bernouillanum, Basel University, Basel, Switz.
Reyment, Richard A., University of Stockholm, Stockholm, Swed.
Rhodes, F. H. T., University of Wales, Swansea, Wales.
†Richter, Emma
†Richter, Rudolf

ROBERTSON, ROBERT, Academy of Natural Sciences, Philadelphia, Pa.
ROLFE, W. D. I., Museum of Comparative Zoölogy, Harvard College, Cambridge, Mass.
ROSEWATER, JOSEPH, U.S. National Museum, Washington, D.C.
ROWELL, A. J., Nottingham University, Nottingham, Eng.
SCHINDEWOLF, O. H. Geologisch-paläontologisches Institut, Universität Tübingen, Tübingen, Ger.
SCHMIDT, HERTA, Senckenbergische Naturforschende Gesellschaft, Frankfurt-a.-M., Ger.
SCOTT, HAROLD W., University of Illinois, Urbana, Ill.
SDZUY, KLAUS, Senckenbergische Naturforschende Gesellschaft, Frankfurt-a.-M., Ger.
SHAVER, ROBERT, Indiana University and Indiana Geological Survey, Bloomington, Ind.
SIEVERTS-DORECK, HERTHA, Stuttgart-Möhringen, Ger.
SINCLAIR, G. W., Geological Survey of Canada, Ottawa, Can.
SMITH, ALLYN G., California Academy of Sciences, San Francisco, Calif.
SOHN, I. G., U.S. Geological Survey, Washington, D.C.
SOOT-RYEN, T., Zoologisk Museum, Oslo, Nor.
†SPENCER, W. K.
†STAINBROOK, MERRILL A.
STEHLI, F. G., Western Reserve University, Cleveland, Ohio.
STENZEL, H. B., Shell Development Co., Houston, Tex.
STEPHENSON, LLOYD W., Dover, Ohio.
STØRMER, LEIF, Paleontologisk Institutt, Universitetet Oslo, Oslo, Nor.
STOUGH, JOAN, Humble Oil & Refining Company, Houston, Tex.
STRUVE, WOLFGANG, Senckenbergische Naturforschende Gesellschaft, Frankfurt-a.-M., Ger.
STUBBLEFIELD, C. J., Geological Survey and Museum, London, Eng.
STUMM, ERWIN C., Museum of Paleontology, University of Michigan, Ann Arbor, Mich.
SWEET, WALTER C., Ohio State University, Columbus, Ohio.
SWAIN, FREDERICK M., University of Minnesota, Minneapolis, Minn.
SYLVESTER-BRADLEY, P. C., University of Leicester, Leicester, Eng.
TAPPAN, HELEN, University of California, Los Angeles, Calif.
TASCH, PAUL, University of Wichita, Wichita, Kans.
TEICHERT, CURT, U.S. Geological Survey, Federal Center, Denver, Colo.
THOMPSON, M. L., Illinois State Geological Survey, Urbana, Ill.
THOMPSON, R. H., University of Kansas, Lawrence, Kans.
†TIEGS, O. W.
TRIPP, RONALD P., Seven Oaks, Kent, Eng.
TURNER, RUTH, Museum of Comparative Zoölogy, Harvard College, Cambridge, Mass.
UBAGHS, G., Université de Liège, Liège, Belg.
VOKES, H. E., Tulane University, New Orleans, La.
WAINWRIGHT, JOHN, Shell Development Co., Houston, Texas.
†WANNER, J.
WEIR, JOHN, University of Glasgow, Glasgow, Scot.
WELLER, J. MARVIN, University of Chicago, Chicago, Ill.
WELLS, JOHN W., Cornell University, Ithaca, N.Y.
WHITTINGTON, H. B., Museum of Comparative Zoölogy, Harvard College, Cambridge, Mass.
WILLIAMS, ALWYN, Queens University of Belfast, Belfast, N.Ire.
WILLS, L. J., Romsley, Eng.
†WITHERS, T. H.
WRIGHT, A. D., Queens University of Belfast, Belfast, N. Ire.
WRIGHT, C. W., London, Eng.
†WRIGLEY, ARTHUR
YOCHELSON, ELLIS L., U.S. Geological Survey, Washington, D.C.
YONGE, C. M., University of Glasgow, Glasgow, Scot.

†—Deceased.

EDITORIAL PREFACE

The aim of the *Treatise on Invertebrate Paleontology,* as originally conceived and consistently pursued, is to present the most comprehensive and authoritative, yet compact statement of knowledge concerning invertebrate fossil groups that can be formulated by collaboration of competent specialists in seeking to organize what has been learned of this subject up to the mid-point of the present century. Such work has value in providing a most useful summary of the collective results of multitudinous investigations and thus should constitute an indispensable text and reference book for all persons who wish to know about remains of invertebrate organisms preserved in rocks of the earth's crust. This applies to neozoologists as well as paleozoologists and to beginners in study of fossils as well as to thoroughly trained, long-experienced professional workers, including teachers, stratigraphical geologists, and individuals engaged in research on fossil invertebrates. The making of a reasonably complete inventory of present knowledge of invertebrate paleontology may be expected to yield needed foundation for future research and it is hoped that the *Treatise* will serve this end.

The *Treatise* is divided into parts which bear index letters, each except the initial and concluding ones being defined to include designated groups of invertebrates. The chief purpose of this arrangement is to provide for independence of the several parts as regards date of publication, because it is judged desirable to print and distribute each segment as soon as possible after it is ready for press. Pages in each part will bear the assigned index letter joined with numbers beginning with 1 and running consecutively to the end of the part. When the parts ultimately are assembled into volumes, no renumbering of pages and figures is required.

The outline of subjects to be treated in connection with each large group of invertebrates includes (1) description of morphological features, with special reference to hard parts, (2) ontogeny, (3) classification, (4) geological distribution, (5) evolutionary trends and phylogeny, and (6) systematic description of genera, subgenera, and higher taxonomic units. In general, paleoecological aspects of study are omitted or little emphasized because comprehensive treatment of this subject is given in the *Treatise on Marine Ecology and Paleoecology* (H. S. LADD, Editor, Geological Society of America, Memoir 67, 1957), prepared under auspices of a committee of the United States National Research Council. A selected list of references is furnished in each part of the *Treatise.*

Features of style in the taxonomic portions of this work have been fixed by the Editor with aid furnished by advice from the Joint Committee on Invertebrate Paleontology representing the societies which have undertaken to sponsor the *Treatise.* It is the Editor's responsibility to consult with authors and co-ordinate their work, seeing that manuscript properly incorporates features of adopted style. Especially he has been called on to formulate policies in respect to many questions of nomenclature and procedure. The subject of family and subfamily names is reviewed briefly in a following section of this preface, and features of *Treatise* style in generic descriptions are explained.

In December, 1959, the National Science Foundation of the United States, through its Division of Biological and Medical Sciences and the Program Director for Systematic Biology, made a grant in the amount of $210,000 for the purpose of aiding the completion of yet-unpublished volumes of the *Treatise.* Payment of this sum was provided to be made in installments distributed over a five-year period, with administration of disbursements handled by the University of Kansas. Expenditures planned are primarily for needed assistance to authors and may be arranged through approved institutions located anywhere. Important help for the Director-Editor of the *Treatise* has been made available from the grant, but no part of his stipend comes from it. Grateful acknowledgment to the Foundation is expressed on behalf of the societies sponsoring the *Treatise,* the University of Kansas, and innumerable individuals benefited by the *Treatise* project.

FORM OF ZOOLOGICAL NAMES

Many questions arise in connection with the form of zoological names. These include such matters as adherence to stipulations concerning Latin or Latinized nature of words accepted as zoological names, gender of generic and subgeneric names, nominative or adjectival form of specific names, required endings for some family-group names, and numerous others. Regulation extends to capitalization, treatment of particles belonging to modern patronymics, use of neo-Latin letters, and approved methods for converting diacritical marks. The magnitude and complexities of nomenclature problems surely are enough to warrant the complaint of those who hold that zoology is the study of animals rather than of names applied to them.

CLASSIFICATION OF ZOOLOGICAL NAMES

In accordance with the "Copenhagen Decisions on Zoological Nomenclature" (London, 135 p., 1953), zoological names may be classified usefully in various ways. The subject is summarized here with introduction of designations for some categories which the *Treatise* proposes to distinguish in systematic parts of the text for the purpose of giving readers comprehension of the nature of various names together with authorship and dates attributed to them.

CO-ORDINATE NAMES OF TAXA GROUPS

Five groups of different-rank taxonomic units (termed *taxa,* sing., *taxon*) are discriminated, within each of which names are treated as co-ordinate, being transferrable from one category to another without change of authorship or date. These are: (1) Species Group (subspecies, species); (2) Genus Group (subgenus, genus); (3) Family Group (tribe, subfamily, family, superfamily); (4) Order/Class Group (suborder, order, subclass, class); and (5) Phylum Group (subphylum, phylum). In the first 3 of these groups, but not others, the author of the first-published valid name for any taxon is held to be the author of all other taxa in the group which are based on the same nominate type and the date of publication for purposes of priority is that of the first-published name. Thus, if author A in 1800 introduces the family name X-idae to include 3 genera, one of which is *X-us;* and if author B in 1850 divides the 20 genera then included in X-idae into subfamilies called X-inae and Y-inae; and if author C in 1950 combines X-idae with other later-formed families to make a superfamily X-acea (or X-oidea, X-icae, etc.); the author of X-inae, X-idae and X-acea is A, 1800, under the Rules. Because taxonomic concepts introduced by authors B and C along with appropriate names surely are not attributable to author A, some means of recording responsibility of B and C are needed. This is discussed later in explaining proposed use of *"nom. transl."*

The co-ordinate status of zoological names belonging to the species group is stipulated in Art. 11 of the present Rules; genus group in Art. 6 of the present Rules; family group in paragraph 46 of the Copenhagen Decisions; order/class group and phylum group in paragraphs 65 and 66 of the Copenhagen Decisions.

ORIGINAL AND SUBSEQUENT FORMS OF NAMES

Zoological names may be classified according to form (spelling) given in original publication and employed by subsequent authors. In one group are names which are entirely identical in original and subsequent usage. Another group comprises names which include with the original subsequently published variants of one sort or another. In this second group, it is important to distinguish names which are inadvertent changes from those constituting intentional emendations, for they have quite different status in nomenclature. Also, among intentional emendations, some are acceptable and some quite unacceptable under the Rules.

VALID AND INVALID NAMES

Valid names. A valid zoological name is one that conforms to all mandatory provisions of the Rules (Copenhagen Decisions, p. 43-57) but names of this group are divisible into subgroups as follows: (1) *"inviolate names,"* which as originally published not only meet all mandatory requirements of the Rules but are not subject to any sort of alteration (most generic and subgeneric names); (2) *"perfect names,"* which as they appear in original publication (with or without precise duplication by subsequent authors) meet all mandatory requirements and need no correction of any kind but which nevertheless are legally alterable under present Rules (as in changing the form of ending of a published class/order-group name); (3) *"imperfect names,"* which as originally published and with or without subsequent duplication meet mandatory requirements but contain defects such as incorrect gender of an adjectival specific name (for example, *Spironema recta* instead of *Spironema rectum*) or incorrect stem or form of ending of a family-group name (for example, Spironemidae instead of Spironematidae); (4) *"transferred names,"* which are derived by valid emendation from either of the 2nd or 3rd subgroups or from a pre-existing transferred name (as illustrated by change of a family-group name from -inae to -idae or making of a superfamily name); (5) *"improved names,"* which include necessary as well as somewhat arbitrarily made emendations allowable under the Rules for taxonomic categories not now covered by regulations as to name form and alterations that are distinct from changes that distinguish the 4th subgroup (including names derived from the 2nd and 3rd subgroups and possibly some alterations of 4th subgroup names). In addition, some zoological names included among those recognized as valid are classifiable in special categories, while at the same time belonging to one or more of the above-listed subgroups. These chiefly include (7) *"substitute names,"* introduced to replace invalid names such as junior homonyms; and (8) *"conserved names,"* which are names that would have to be rejected by application of the Rules except for saving them in their original or an altered spelling by action of the International Commission on Zoological Nomenclature in exercising its plenary powers to this end. Whenever a name requires replacement, any individual may publish a "new name" for it and the first one so introduced has priority over any others; since newness is temporary and relative, the replacement designation is better called substitute name rather than new name. Whenever it is considered desirable to save for usage an otherwise necessarily rejectable name, an individual cannot by himself accomplish the preservation, except by unchallenged action taken in accordance with certain provisions of the Copenhagen Decisions; otherwise he must seek validation through ICZN.

It is useful for convenience and brevity of distinction in recording these subgroups of valid zoological names to introduce Latin designations, following the pattern of *nomen nudum, nomen novum,* etc. Accordingly, the subgroups are (1) *nomina inviolata* (sing., *nomen inviolatum,* abbr., *nom. inviol.*); (2) *nomina perfecta* (sing., *nomen perfectum,* abbr., *nom. perf.*); (3) *nomina imperfecta* (sing., *nomen imperfectum,* abbr., *nom. imperf.*); (4) *nomina translata* (sing., *nomen translatum,* abbr., *nom. transl.*); (5) *nomina correcta* (sing., *nomen correctum,* abbr., *nom. correct.*); (6) *nomina substituta* (sing., *nomen substitutum,* abbr., *nom. subst.*); (7) *nomina conservata* (sing., *nomen conservatum,* abbr., *nom. conserv.*).

Invalid names. Invalid zoological names consisting of originally published names that fail to comply with mandatory provisions of the Rules and consisting of inadvertent changes in spelling of names have no status in nomenclature. They are not available as replacement names and they do not preoccupy for purposes of the Law of Homonomy. In addition to *nomen nudum,* invalid names may be distinguished as follows: (1) *"denied names,"* which consist of originally published names (with or without subsequent duplication) that do not meet mandatory requirements of the Rules; (2) *"null names,"* which comprise unintentional alterations of names; and (3) *"vain or void names,"* which consist of invalid emendations of previously published valid or invalid names. Void names do have status in nomenclature, being classified as junior synonyms of valid names.

Proposed Latin designations for the indicated kinds of invalid names are as follows: (1) *nomina negata* (sing., *nomen negatum,* abbr., *nom. neg.*); (2) *nomina nulla* (sing., *nomen nullum,* abbr., *nom. null.*); (3) *nomina vana* (sing., *nomen vanum,* abbr., *nom. van.*). It is desirable in the *Treatise* to identify invalid names, particularly in view of the fact that many of these names (*nom. neg., nom null.*) have been considered incorrectly to be junior objective synonyms (like *nom. van.*), which have status in nomenclature.

SUMMARY OF NAME CLASSES

Partly because only in such publications as the *Treatise* is special attention to classes of zoological names called for and partly because new designations are now introduced as means of recording distinctions explicitly as well as compactly, a summary may be useful. In the following tabulation valid classes of names are indicated in boldface type, whereas invalid ones are printed in italics.

Definitions of Name Classes

nomen conservatum (nom. conserv.). Name otherwise unacceptable under application of the Rules which is made valid, either with original or altered spelling, through procedures specified by the Copenhagen Decisions or by action of ICZN exercising its plenary powers.

nomen correctum (nom. correct.). Name with intentionally altered spelling of sort required or allowable under the Rules but not dependent on transfer from one taxonomic category to another ("improved name"). (*See* Copenhagen Decisions, paragraphs 50, 71-2-a-i, 74, 75, 79, 80, 87, 101; in addition, change of endings for categories not now fixed by Rules.)

nomen imperfectum (nom. imperf.). Name that as originally published (with or without subsequent identical spelling) meets all mandatory requirements of the Rules but contains defect needing correction ("imperfect name"). (*See* Copenhagen Decisions, paragraphs 50-1-b, 71-1-b-i, 71-1-b-ii, 79, 80, 87, 101.)

nomen inviolatum (nom. inviol.). Name that as originally published meets all mandatory requirements of the Rules and also is uncorrectable or alterable in any way ("inviolate name"). (*See* Copenhagen Decisions, paragraphs 152, 153, 155-157).

nomen negatum (nom. neg.). Name that as originally published (with or without subsequent identical spelling) constitutes invalid original spelling and although possibly meeting all other mandatory requirements of the Rules, is not correctable to establish original authorship and date ("denied name"). (*See* Copenhagen Decisions, paragraph 71-1-b-iii.)

nomen nudum (nom. nud.). Name that as originally published (with or without subsequent identical spelling) fails to meet mandatory requirements of the Rules and having no status in nomenclature, is not correctable to establish original authorship and date ("naked name"). (*See* Copenhagen Decisions, paragraph 122.)

nomen nullum (nom. null.). Name consisting of an unintentional alteration in form (spelling) of a previously published name (either valid name, as *nom. inviol., nom. perf., nom. imperf., nom. transl.;* or invalid name, as *nom. neg., nom. nud., nom. van.,* or another *nom. null.*) ("null name"). (*See* Copenhagen Decisions, paragraphs 71-2-b, 73-4.)

nomen perfectum (nom. perf.). Name that as originally published meets all mandatory requirements of the Rules and needs no correction of any kind but which nevertheless is validly alterable ("perfect name").

nomen substitutum (nom. subst.). Replacement name published as substitute for an invalid name, such as a junior homonym (equivalent to "new name").

nomen translatum (nom. transl.). Name that is derived by valid emendation of a previously published name as result of transfer from one taxonomic category to another within the group to which it belongs ("transferred name").

nomen vanum (nom. van.). Name consisting of an invalid intentional change in form (spelling) from a previously published name, such invalid emendations having status in nomenclature as junior objective synonyms ("vain or void name"). (*See* Copenhagen Decisions, paragraphs 71-2-a-ii, 73-3.)

Except as specified otherwise, zoological names accepted in the *Treatise* may be understood to be classifiable either as *nomina inviolata* or *nomina perfecta* (omitting from notice *nomina correcta* among specific names) and these are not discriminated. Names which are not accepted for one reason or another include junior homonyms, a few senior synonyms classifiable as *nomina negata* or *nomina nuda,* and numerous junior synonyms which include both objective *(nomina vana)* and subjective (all classes of valid names) types; effort to classify the invalid names as completely as possible is intended.

NAME CHANGES IN RELATION TO GROUP CATEGORIES

SPECIFIC AND SUBSPECIFIC NAMES

Detailed consideration of valid emendation of specific and subspecific names is unnecessary here because it is well understood and relatively inconsequential. When the form of adjectival specific names is changed to obtain agreement with the gender of a generic name in transferring a species from one genus to another, it is never needful to label the changed name as a *nom. transl.* Likewise, transliteration of a letter accompanied by a diacritical mark in manner now called for by the Rules (as in changing originally published *bröggeri* to *broeggeri*) or elimination of a hyphen (as in changing originally published *cornu-oryx* to *cornuoryx* does not require "*nom. correct.*" with it. Revised provisions for emending specific and subspecific names are stated in the report on Copenhagen Decisions (p. 43-46, 51-57).

GENERIC AND SUBGENERIC NAMES

So rare are conditions warranting change of the originally published valid form of generic and subgeneric names that lengthy discussion may be omitted. Only elimination of diacritical marks of some names in this category seems to furnish basis for valid emendation. It is true that many changes of generic and subgeneric names have been published, but virtually all of these are either *nomina vana* or *nomina nulla*. Various names which formerly were classed as homonyms are not now, for two names that differ only by a single letter (or in original publication by presence or absence of a diacritical mark) are construed to be entirely distinct. Revised provisions for emendation of generic and subgeneric names also are given in the report on Copenhagen Decisions (p. 43-47).

Examples in use of classificatory designations for generic names as previously given are the following, which also illustrate designation of type species, as explained later.

Kurnatiophyllum THOMSON, 1875 [**K. concentricum;* SD GREGORY, 1917] [=*Kumatiophyllum* THOMSON, 1876 *(nom. null.); Cymatophyllum* THOMSON, 1901 *(nom. van.); Cymatiophyllum* LANG, SMITH & THOMAS, 1940 *(nom. van.)*].

Stichophyma POMEL, 1872 [**Manon turbinatum* RÖMER, 1841; SD RAUFF, 1893] [=*Stychophyma* VOSMAER, 1885 *(nom. null.); Sticophyma* MORET, 1924 *(nom. null.)*].

Stratophyllum SMYTH, 1933 [**S. tenue*] [=*Ethmoplax* SMYTH, 1939 (*nom. van. pro Stratophyllum); Stratiphyllum* LANG, SMITH & THOMAS, 1940 (*nom. van. pro Stratophyllum* SMYTH) (*non Stratiphyllum* SCHEFFEN, 1933)].

Placotelia OPPLIGER, 1907 [**Porostoma marconi* FROMENTEL, 1859; SD DELAUBENFELS, herein] [=*Plakotelia* OPPLIGER, 1907 *(nom. neg.)*].

Walcottella DELAUB., *nom. subst.,* 1955 [*pro Rhopalicus* SCHRAMM., 1936 (*non* FÖRSTER, 1856)].

Cyrtograptus CARRUTHERS, 1867 [*nom. correct.* LAPWORTH, 1873 (*pro Cyrtograpsus* CARRUTHERS, 1867), *nom. conserv.* proposed BULMAN, 1955 (ICZN pend.)]

FAMILY-GROUP NAMES; USE OF "*NOM. TRANSL.*"

The Rules now specify the form of endings only for subfamily (-inae) and family (-idae) but decisions of the Copenhagen Congress direct classification of all family-group assemblages (taxa) as co-ordinate, signifying that for purposes of priority a name published for a unit in any category and based on a particular type genus shall date from its original publication for a unit in any category, retaining this priority (and authorship) when the unit is treated as belonging to a lower or higher category. By exclusion of -inae and -idae, respectively reserved for subfamily and family, the endings of names used for tribes and superfamilies must be unspecified different letter combinations. These, if introduced subsequent to designation of a subfamily or family based on the same nominate genus, are *nomina translata,* as is also a subfamily that is elevated to family rank or a family reduced to subfamily rank. In the *Treatise* it is desirable to distinguish the valid emendation comprised in the changed ending of each transferred family group name by the abbreviation "*nom. transl.*" and record of the author and date belonging to this emendation. This is particularly important in the case of superfamilies, for it is the author who introduced this taxon that one wishes to know about rather than the author of the superfamily as defined by the Rules, for the latter is merely the individual who first defined some lower-rank family-group taxon that contains the nominate genus of the superfamily. The publication of the author containing introduction of the superfamily *nomen translatum* is likely to furnish the information on taxonomic considerations that support definition of the unit.

Examples of the use of "*nom. transl.*" are the following.

Subfamily STYLININAE d'Orbigny, 1851

[*nom. transl.* EDWARDS & HAIME, 1857 (*ex* Stylinidae D'ORBIGNY, 1851]

Superfamily ARCHAEOCTONOIDEA Petrunkevitch, 1949

[*nom. transl.* PETRUNKEVITCH, herein (*ex* Archaeoctonidae PETRUNKEVITCH, 1949)]

Superfamily CRIOCERATITACEAE Hyatt, 1900

[*nom. transl.* WRIGHT, 1952 (*ex* Crioceratitidae HYATT, 1900)]

FAMILY-GROUP NAMES; USE OF "*NOM. CORRECT.*"

Valid emendations classed as *nomina correcta* do not depend on transfer from one category of family-group units to another but most commonly involve correction of the stem of the nominate genus; in addition, they include somewhat arbitrarily chosen modification of ending for names of tribe or superfamily. Examples of the use of "*nom. correct.*" are the following.

Family STREPTELASMATIDAE Nicholson, 1889

[*nom. correct.* WEDEKIND, 1927 (*ex* Streptelasmidae NICHOLSON, 1889, *nom. imperf.*)]

Family PALAEOSCORPIIDAE Lehmann, 1944

[*nom. correct.* PETRUNKEVITCH, herein (*ex* Palaeoscorpionidae LEHMANN, 1944, *nom. imperf.*)]

Family AGLASPIDIDAE Miller, 1877

[*nom. correct.* STØRMER, herein (*ex* Aglaspidae MILLER, 1877, *nom. imperf.*)]

Superfamily AGARICIICAE Gray, 1847

[*nom. correct.* WELLS, herein (*ex* Agaricioidae VAUGHAN & WELLS, 1943, *nom. transl. ex* Agariciidae GRAY, 1847)]

FAMILY-GROUP NAMES; USE OF "*NOM. CONSERV.*"

It may happen that long-used family-group names are invalid under strict application of the Rules. In order to retain the otherwise invalid name, appeal to ICZN is needful. Examples of use of *nom. conserv.* in this connection, as cited in the *Treatise,* are the following.

Family ARIETITIDAE Hyatt, 1874

[*nom. correct.* HAUG, 1885 (*pro* Arietidae HYATT, 1875), *nom. conserv.* proposed ARKELL, 1955 (ICZN pend.)]

Family STEPHANOCERATIDAE Neumayr, 1875

[*nom. correct.* FISCHER, 1882 (*pro* Stephanoceratinen NEUMAYR, 1875, invalid vernacular name), *nom conserv.* proposed ARKELL, 1955 (ICZN pend.)]

FAMILY-GROUP NAMES; REPLACEMENTS

Family-group names are formed by adding letter combinations (prescribed for family and subfamily but not now for others) to the stem of the name belonging to genus (nominate genus) first chosen as type of the assemblage. The type genus need not be the oldest in terms of receiving its name and definition, but it must be the first-published as name-giver to a family-group taxon among all those included. Once fixed, the family-group name remains tied to the nominate genus even if its name is changed by reason of status as a junior homonym or junior synonym, either objective or subjective. According to the Copenhagen Decisions, the family-group name requires replacement only in the event that the nominate genus is found to be a junior homonym, and then a substitute family-group name is accepted if it is formed from the oldest available substitute name for the nominate genus. Authorship and date attributed to the replacement family-group name are determined by first publication of the changed family-group name.

The aim of family-group nomenclature is greatest possible stability and uniformity, just as in case of other zoological names. Experience indicates the wisdom of sustaining family-group names based on junior subjective synonyms if they have priority of publication, for opinions of different workers as to the synonymy of generic names founded on different type species may not agree and opinions of the same worker may alter from time to time. The retention similarly of first-published family-group names which are found to be based on junior objective synonyms is less clearly desirable, especially if a replacement name derived from the senior objective synonym has been recognized very long and widely. To displace a much-used family-group name based on the senior objective synonym by disinterring a forgotten and virtually unused family-group name based on a junior objective synonym because the latter happens to have priority of publication is unsettling. Conversely, a long-used family-group name founded on a junior objective synonym and having priority of publication is better continued in nomenclature than a replacement name based on the senior objective synonym. The Copenhagen Decisions (paragraph 45) take account of these considerations by providing a relatively simple procedure for fixing the desired choice in stabilizing family-group names. In conformance with this, the *Treatise* assigns to contributing authors responsibility for adopting provisions of the Copenhagen Decisions.

Replacement of a family-group name may be needed if the former nominate genus is transferred to another family-group. Then the first-published name-giver of a family-group assemblage in the remnant taxon is to be recognized in forming a replacement name.

FAMILY-GROUP NAMES; AUTHORSHIP AND DATE

All family-group taxa having names based on the same type genus are attributed to the author who first published the name for any of these assemblages, whether tribe, subfamily, or family (superfamily being almost inevitably a later-conceived taxon). Accordingly, if a family is divided into subfamilies or a subfamily into tribes, the name of no such subfamily or tribe can antedate the family name. Also, every family containing differentiated subfamilies must have a nominate *(sensu stricto)* subfamily, which is based on the same type genus as that for the family, and the author and date set down for the nominate subfamily invariably are identical with those of the family, without reference as to whether the author of the family or some subsequent author introduced subdivisions.

Changes in the form of family-group names of the sort constituting *nomina correcta,* as previously discussed, do not affect authorship and date of the taxon concerned, but in publications such as the *Treatise* it is desirable to record the authorship and date of the correction.

ORDER/CLASS-GROUP NAMES; USE OF "*NOM. CORRECT.*"

Because no stipulation concerning the form of order/class-group names is given yet by the Rules, emendation of all such names actually consists of arbitrarily devised changes in the form of endings. Nothing precludes substitution of a new name for an old one, but a change of this sort is not considered to be an emendation. Examples of the use of "*nom. correct.*" as applied to order/class-group names are the following.

Order DISPARIDA Moore & Laudon, 1943

[*nom. correct.* MOORE, 1952 (*ex* Disparata MOORE & LAUDON, 1943)]

Suborder FAVIINA Vaughan & Wells, 1943

[*nom. correct.* WELLS, herein (*ex* Faviida VAUGHAN & WELLS, 1943)]

Suborder FUNGIINA Verrill, 1865

[*nom. correct.* WELLS, herein (*ex* Fungiida DUNCAN, 1884, *ex* Fungacea VERRILL, 1865)]

TAXONOMIC EMENDATION

Emendation has two measurably distinct aspects as regards zoological nomenclature. These embrace (1) alteration of a name itself in various ways for various reasons, as has been reviewed, and (2) alteration of taxonomic scope or concept in application of a given zoological name, whatever its hierarchical rank. The latter type of emendation primarily concerns classification and inherently is not associated with change of name, whereas the other type introduces change of name without necessary expansion, restriction, or other modification in applying the name. Little attention generally has been paid to this distinction in spite of its significance.

Most zoologists, including paleozoologists, who have signified emendation of zoological names refer to what they consider a material change in application of the name such as may be expressed by an importantly altered diagnosis of the assemblage covered by the name. The abbreviation "*emend.*" then may accompany the name, with statement of the author and date of the emendation. On the other hand, a multitude of workers concerned with systematic zoology think that publication of "*emend.*" with a zoological name is valueless because more or less alteration of taxonomic sort is introduced whenever a subspecies, species, genus, or other assemblage of animals is incorporated under or removed from the coverage of a given zoological name. Inevitably associated with such classificatory expansions and restrictions is some degree of emendation affecting diagnosis. Granting this, still it is true that now and then somewhat radical revisions are put forward, generally with published statement of reasons for changing the application of a name. To erect a signpost at such points of most significant change is worth while, both as aid to subsequent workers in taking account of the altered nomenclatural usage and as indication that not-to-be-overlooked discussion may be found at a particular place in the literature. Authors of contributions to the *Treatise* are encouraged to include records of all specially noteworthy emendations of this nature, using the abbreviation "*emend.*" with the name to which it refers and citing the author and date of the emendation.

In Part G (Bryozoa) and Part D (Protista 3) of the *Treatise,* the abbreviation "*emend.*" is employed to record various sorts of name emendations, thus conflicting with usage of "*emend.*" for change in taxonomic application of a name without alteration of the name itself. This is objectionable. In Part E (Archaeocyatha, Porifera) and later-issued divisions of the *Treatise,* use of "*emend.*" is restricted to its customary sense, that is, significant alteration in taxonomic scope of a name such as calls for noteworthy modifications of a diagnosis. Other means of designating emendations that relate to form of a name are introduced.

STYLE IN GENERIC DESCRIPTIONS

CITATION OF TYPE SPECIES

The name of the type species of each genus and subgenus is given next following the generic name with its accompanying author and date, or after entries needed for definition of the name if it is involved in homonymy. The originally published combination of generic and trivial names for this species is cited, accompanied by an asterisk (*), with notation of the author and date of original publication. An exception in this procedure is made, however, if the species was first published in the same paper and by the same author as that containing definition of the genus which it serves as type; in such case, the initial letter of the generic name followed by the trivial name is given without repeating the name of the author and date, for this saves needed space. Examples of these 2 sorts of citations are as follows:

Diplotrypa NICHOLSON, 1879 [**Favosites petropolitanus* PANDER, 1830].

Chainodictyon FOERSTE, 1887 [**C. laxum*].

If the cited type species is a junior synonym of some other species, the name of this latter also is given, as follows:

Acervularia SCHWEIGGER, 1819 [**A. baltica* (=**Madrepora ananas* LINNÉ, 1758)].

It is judged desirable to record the manner of establishing the type species, whether by original designation or by subsequent designation, but various modes of original designation are not distinguished.

Fixation of type-species originally. The type-species of a genus or subgenus, according to provisions of the Rules, may be fixed in various ways originally (that is, in the publication containing first proposal of the generic name) or it may be fixed in specified ways subsequent to the original publication. Fixation of a type-species originally may be classified as automatic if the new genus was introduced for a single species (monotypy), or if the names of species referred to the genus are objectively synonymous. In addition, fixation of a type-species originally may be established in several ways by original designation, as by explicit statement given by an author, by use of *typus* or *typicus* as a new specific name, and by absolute tautonymy (e.g., *Mesolobus mesolobus*). According to convention adopted in the *Treatise,* the absence of indication as to the manner of fixing the type-species is to be understood as signifying fixation of the type-species in one way or another originally. Where an author wishes to specify the mode of original fixation, however, this may be done by such abbreviations as "M" (monotypy), "OS" (objective synonymy), and "OD" (original designation), the first- and last-mentioned being most common and the other very rare.

Fixation of type-species subsequently. The type species of many genera are not determinable from the publication in which the generic name was introduced and therefore such genera can acquire a type species only by some manner of subsequent designation. Most commonly this is established by publishing a statement naming as type species one of the species originally included in the genus, and in the *Treatise* fixation of the type species in this manner is indicated by the letters "SD" accompanied by the name of the subsequent author (who may be the same person as the original author) and the date of publishing the subsequent designation. Some genera, as first described and named, included no mentioned species and these necessarily lack a type species until a date subsequent to that of the original publication when one or more species are assigned to such a genus. If only a single species is thus assigned, it automatically becomes the type species and in the *Treatise* this subsequent monotypy is indicated by the letters "SM." Of course, the first publication containing assignment of species to the genus which originally lacked any included species is the one concerned in fixation of the type species, and if this named 2 or more species as belonging to the genus but did not designate a type species, then a later "SD" designation is necessary. Examples of the use of "SD" and "SM" as employed in the *Treatise* follow.

Hexagonaria GÜRICH, 1896 [**Cyathophyllum hexagonum* GOLDFUSS, 1826; SD LANG, SMITH & THOMAS, 1940].

Muriceides STUDER, 1887 [**M. fragilis* WRIGHT & STUDER, 1889; SM WRIGHT & STUDER, 1889].

Another mode of fixing the type-species of a genus that may be construed as a special sort of subsequent designation is action of the International Commission on Zoological Nomenclature using its plenary powers. Definition in this way may set aside application of the Rules so as to arrive at a decision considered to be in the best interest of continuity and stability of zoological nomenclature. When made, it is binding and commonly is cited in the *Treatise* by the letters "ICZN," accompanied by the date of announced decision and (generally) reference to the appropriate numbered Opinion.

HOMONYMS

Most generic names are distinct from all others and are indicated without ambiguity by citing their originally published spelling accompanied by name of the author and date of first publication. If the same generic name has been applied to 2 or more distinct taxonomic units, however, it is necessary to differentiate such homonyms, and this calls for distinction between junior homonyms and senior homonyms. Because a junior homonym is invalid, it must be replaced by some other name. For example, *Callopora* HALL, 1851, introduced for Paleozoic trepostome bryozoans, is invalid because GRAY in 1848 published the same name for Cretaceous-to-Recent cheilostome bryozoans, and BASSLER in 1911 introduced the new name *Hallopora* to replace HALL's homonym. The *Treatise* style of entry is:

Hallopora BASSLER, 1911 [*pro Callopora* HALL, 1851 (*non* GRAY, 1848)].

In like manner, a needed replacement generic name may be introduced in the *Treatise* (even though first publication of generic names otherwise in this work is avoided). The requirement that an exact bibliographic reference must be given for the replaced name commonly can be met in the *Treatise* by citing a publication recorded in the list of references, using its assigned index number, as shown in the following example.

Mysterium DELAUBENFELS, *nom. subst.* [*pro Mystrium* SCHRAMMEN, 1936 (ref. 40, p. 60) (*non* ROGER, 1862)] [**Mystrium porosum* SCHRAMMEN, 1936].

For some replaced homonyms, a footnote reference to the literature is necessary. A senior homonym is valid, and in so far as the *Treatise* is concerned, such names are handled according to whether the junior homonym belongs to the same major taxonomic division (class or phylum) as the senior homonym or to some other; in the former instance, the author and date of the junior homonym are cited as:

Diplophyllum HALL, 1851 [*non* SOSHKINA, 1939] [**D. caespitosum*].

Otherwise, no mention of the existence of a junior homonym generally is made.

Homonyms by misidentification. When an author uses a generic name for species not congeneric with the type-species, it is needful to record the misuse of the generic name, even though this is only determinable subjectively. In the *Treatise* homonyms by misidentification are cited in synonymies as illustrated in the following example.

Asmussia PACHT, 1849 [**A. membranacea*] [=*Posidonomya* PACHT, 1852 (*non* BRONN, 1834); *Estheria* JONES, 1856 (*non* ROBINEAU-DESVOIDY, 1830; *nec* RUEPPELL, 1837)]. . . .

Synonymic homonyms. An author sometimes publishes a generic name in two or more papers of different date, each of which indicates that the name is new. This is a bothersome source of errors for later workers who are unaware that a supposed first publication which they have in hand is not actually the original one. Although the names were separately published, they are identical and therefore definable as homonyms; at the same time they are absolute synonyms. For the guidance of all concerned, it seems desirable to record such names as synonymic homonyms and in the *Treatise* the junior one of these is indicated by the abbreviation "jr. syn. hom."

Identical family-group names not infrequently are published as new names by different authors, the author of the later-introduced name being ignorant of previous publication(s) by one or more other workers. In spite of differences in taxonomic concepts as indicated by diagnoses and grouping of genera and possibly in assigned rank, these family-group taxa are nomenclatural homonyms, based on the same type-genus, and they are also synonyms. Wherever encountered, such synonymic homonyms are distinguished in the *Treatise* as in dealing with generic names.

SYNONYMS

Citation of synonyms is given next following record of the type species and if 2 or more synonyms of differing date are recognized, these are arranged in chronological order. Objective synonyms are indicated by accompanying designation "(obj.)," others being understood to constitute subjective synonyms. Examples showing *Treatise* style in listing synonyms follow.

Calapoecia BILLINGS, 1865 [**C. anticostiensis;* SD LINDSTRÖM, 1883] [=*Columnopora* NICHOLSON, 1874; *Houghtonia* ROMINGER, 1876].

Staurocyclia HAECKEL, 1882 [**S. cruciata* HAECKEL, 1887] [=*Coccostaurus* HAECKEL, 1882 (obj.); *Phacostaurus* HAECKEL, 1887 (obj.)].

A synonym which also constitutes a homonym is recorded as follows:

Lyopora NICHOLSON & ETHERIDGE, 1878 [**Palaeopora? favosa* M'COY, 1850] [=*Liopora* LANG, SMITH & THOMAS, 1940 (*non* GIRTY, 1915)].

Some junior synonyms of either objective or subjective sort may take precedence desirably over senior synonyms wherever uniformity and continuity of nomenclature are served by retaining a widely used but technically rejectable name for a generic assemblage. This requires action of ICZN using its plenary powers to set aside the unwanted name and validate the wanted one, with placement of the concerned names on appropriate official lists. In the *Treatise* citation of such a conserved generic name is given in the manner shown by the following example.

Tetragraptus SALTER, 1863 [*nom. correct.* HALL, 1865 (*pro Tetragrapsus* SALTER, 1863), *nom. conserv.* proposed BULMAN, 1955, ICZN pend.] [**Fucoides serra* BRONGNIART, 1828 (=*Graptolithus bryonoides* HALL, 1858].

ABBREVIATIONS

Abbreviations used in this division of the *Treatise* are explained in the following alphabetically arranged list.

Abbreviations

abs., abstract
aff., affinis
Afr., Africa, -an
Ala., Alabama
Alb., Albian
Algonk., Algonkian
alter., alternate
Alta., Alberta
Am., America, -n
ant., anterior
Antarct., Antarctica
append., appendix
approx., approximately
Apt., Aptian
Arenig., Arenigian
Arg., Argentina
Ariz., Arizona
AsiaM., Asia Minor
auctt., auctorum
Aus., Austria
Austral., Australia
Balt., Baltic
Barton., Bartonian
B.C., British Columbia
Belg., Belgium, Belgique
Boh., Bohemia
Bol., Bolivia
Br.I., British Isles
Brit., Britain, British
C., Central
ca., circa
Calif., California
Cam., Cambrian
Campan., Campanian
Can., Canada
Caradoc., Caradocian

Carb., Carboniferous
Carib., Caribbean
Cenoz., Cenozoic
cf., confer (compare)
Chester., Chesteran
Cinc., Cincinnatian
cm., centimeter
Co., Company
Colo., Colorado
Colomb., Colombia
Conn., Connecticut
cosmop., cosmopolitan
Cret., Cretaceous
Czech., Czechoslovakia
D.C., District of Columbia
Dec., December
Denm., Denmark
Dev., Devonian
dia., diameter
diagram., diagrammatic
Dol., Dolomite
Downton., Downtonian
E., East
Ecuad., Ecuador
ed., edition, editor
e.g., exempli gratia (for example)
Eifel., Eifelian
emend., emendatus (-a)
Eng., England
Eoc., Eocene
Est., Estonia
et al., et alii
ety., etymology
Eu., Europe
F., Formation
fam., family
Famenn., Famennian
fig., figure, -s
Fin., Finland
Fr., France, French, Français, -e
Frasn., Frasnian
Ft., Fort
G.Brit., Great Britain
Geol., Geology, Geological, Geologische
Ger., Germany, German
Givet., Givetian
Gotl., Gotland, Gotlandian
Gr., Group
Greenl., Greenland
horiz., horizontal
hom., homonym
Hung., Hungary, Hungarica
I.(Is.), Island, -s
ICZN, International Commission on Zoological Nomenclature
i.e., id est (that is)
Ill., Illinois
illus., illustration, -s
Ind., Indiana
Ind.O., Indian Ocean
Internatl., International
Ire., Ireland
Ital., Italian
Jap., Japan
jr., junior
Jur., Jurassic
Kans., Kansas
Kinderhook., Kinderhookian
Ky., Kentucky
L., Lower
lat., lateral
Lias., Liassic
Llandeil., Llandeilian
Llandov., Llandoverian
Llanvirn., Llanvirnian
long., longitudinal
Ls., Limestone
Ludlov., Ludlovian
M., Middle
Maastricht., Maastrichtian
mag., magnification
Man., Manitoba
Mass., Massachusetts
Md., Maryland
Meramec., Meramecian
Mich., Michigan
Minn., Minnesota
Mio., Miocene
Miss., Mississippi, Mississippian
mm., millimeter
Mo., Missouri
mod., modified
Mont., Montana
n., new
N., North
N.Am., North America
Namur., Namurian
N.B., New Brunswick
NC., North Central
N.Car., North Carolina
NE., Northeast
Neb., Nebraska
Neog., Neogene
Neth., Netherlands
Nev., Nevada
Newf., Newfoundland
N.J., New Jersey
no., number
nom. correct., nomen correctum
nom. imperf., nomen imperfectum
nom. nov., nomen novum
nom. nud., nomen nudum
nom. null., nomen nullum
nom. subst., nomen substitutum
nom. transl., nomen translatum
nom. van., nomen vanum
Nor., Norway
NW., Northwest
N.Y., New York
N.Z., New Zealand
O., Ocean
obj., objective
Okla., Oklahoma
Oligo., Oligocene
Ont., Ontario
Ord., Ordovician
orient., orientation
Osag., Osagian
p., page, -s
Pa., Pennsylvania
Pac., Pacific
Pak., Pakistan
Paleoc., Paleocene
Paleog., Paleogene
Palest., Palestine
pend., pending
Penn., Pennsylvanian
Perm., Permian
Philip., Philippines
pl., plate, -s, plural
Pleist., Pleistocene
Plio., Pliocene
Pol., Poland
Port., Portugal
Portland., Portlandian
post., posterior
Precam., Precambrian
Prot., Proterozoic
pt., part, -s
Quat., Quaternary
Que., Quebec
Rec., Recent
reconstr., reconstructed, -ion
Rhaet., Rhaetian
Rotl., Rotliegende
Rupel., Rupelian
S., South, Sea
S.Am., South America
SC., South Central
schem., schematic
Scot., Scotland
SD, subsequent designation
S.Dak., South Dakota
SE., Southeast
sec., section, -s
Senon., Senonian
ser., series, serial
Sh., Shale
Shrops., Shropshire
Sib., Siberia
Sil., Silurian
simpl., simplified
sp., species
Sp., Spain
Spitz., Spitzbergen
s.s., sensu stricto
Ss., Sandstone
Staatsinst., Staatsinstitut
subsp., subspecies
suppl., supplement
SW., Southwest
Swed., Sweden, Swedish
Switz., Switzerland
Syst., System
tang., tangential
Tasm., Tasmania
tech., technical, technische
Tenn., Tennessee
Tert., Tertiary
Tex., Texas
Torton., Tortonian
transl., translated, translation
transv., transverse
Tremadoc., Tremadocian
Trias., Triassic

Turk., Turkey
Turon., Turonian
U., Upper
U.S., United States
USA, United States (America)
USSR, Union of Soviet Socialist Republics
v., volume, -s
Va., Virginia
var., variety
vert., vertical
Vict., Victoria
Vt., Vermont
W., West
Wenlock., Wenlockian
Wis., Wisconsin
W.Va., West Virginia
Wyo., Wyoming
Z., Zone
Zool., Zoology

REFERENCES TO LITERATURE

Each part of the *Treatise* is accompanied by a selected list of references to paleontological literature consisting primarily of recent and comprehensive monographs available but also including some older works recognized as outstanding in importance. The purpose of giving these references is to aid users of the *Treatise* in finding detailed descriptions and illustratios of morphological features of fossil groups, discussions of classifications and distribution, and especially citations of more or less voluminous literature. Generally speaking, publications listed in the *Treatise* are not original sources of information concerning taxonomic units of various rank but they tell the student where he may find them; otherwise it is necessary to turn to such aids as the *Zoological Record* or NEAVE's *Nomenclator Zoologicus.* References given in the *Treatise* are arranged alphabetically by authors and accompanied by index numbers which serve the purpose of permitting citation most concisely in various parts of the text; these citations of listed papers are enclosed invariably in parentheses and are distinguishable from dates because the index numbers comprise no more than 3 digits. Ordinarily, index numbers for literature references are given at the end of generic or family diagnoses.

The following is a statement of the full names of serial publications which are cited in abbreviated form in the *Treatise* lists of references. The information thus provided should be useful in library research work. The list is alphabetized according to the serial titles which were employed at the time of original publication. Those following it in brackets are those under which the publication may be found currently in the *Union List of Serials,* the United States Library of Congress listing, and most library card catalogues. The names of serials published in Cyrillic are transliterated; in the reference lists these titles, which may be abbreviated, are accompanied by transliterated authors' names and titles, with English translation of the title. The place of publication is added (if not included in the serial title).

List of Serial Publications

Académie des Sciences URSS, Comptes Rendus [Akademiya Nauk SSSR, Leningrad].
Académie Impériale des Sciences, St. Pétersbourg, Mémoires [Akademiya Nauk SSSR, Leningrad].
Académie Royale des Sciences, des Lettres et des Beaux-Arts de Belgique, Bulletin. Bruxelles.
Académie Tchèque des Sciences, Bulletin International, Classe des Sciences Mathématiques, Naturelles et de la Médecine [Česká Akademie véd a umění v Praze].
Academy of Natural Sciences of Philadelphia, Proceedings.
Accademia Pontaniana, Atti. Naples.
Acta Palaeontologica Polonica [Polska Akademia Nauk, Komitet Geologiczny]. Warszawa.
[K.[1]] Akademie der Wissenschaften, St. Petersburg [Akademiya Nauk SSSR, Leningrad].
Akademie der Wissenschaften und der Literatur, Mainz, mathematisch-naturwissenschaftliche Klasse, Abhandlungen.
Akademie der Wissenschaftlichen zu München, mathematische-physikalische Klasse, Sitzungsberichten.
Akademie der Wissenschaftlichen zu Wien, mathmatisch-naturwissenschaftliche Klasse, Denkschriften.
Akademiya Nauk SSSR, Leningrad, Doklady; Geologicheskikh Institut; Izvestiya, Seriya Biologicheskaya.
American Association of Petroleum Geologists, Bulletin. Tulsa, Okla.
American Geologist. Minneapolis, Minn.
American Journal of Conchology. Philadelphia, Pa.
American Journal of Science. New Haven, Conn.
American Midland Naturalist. Notre Dame, Ind.
Annals and Magazine of Natural History. London.
Annales d'Histoire Naturelle. Paris.
Archiv für Anatomie, Physiologie und Wissenschaftliche Medicin, Jahrgang. Leipzig.
Archiv für Naturgeschichte. Leipzig.
Archiv für Zoologie und Zootomie. Berlin.
Archives de Musée Teyler. Haarlem.
Arquivos do Museu Paranaense. Curitiba, Brazil.

[1] *K.* preceding a serial title stands for all forms meaning royal, imperial, e.g., Königliche, Kaiserliche, Kongelig, etc.

Beiträge zur naturkundlichen Forschung in Südwestdeutschland. Karlsruhe.
Beiträge zur Paläontologie und Geologie Österreich-Ungarns und des Orients. Wien.
[K.]Böhemische Gesellschaft der Wissenschaften, Sitzungsberichten; Abhandlungen. [České Společnosti Náuk, Prague.]
Buffalo Society of Natural Science, Bulletin. Buffalo, N.Y.
Bulletin of Zoological Nomenclature. London.
Cambridge Philosophical Society, Biological Reviews. Cambridge, Eng.
Canadian Alpine Journal. Winnipeg, Canada.
Canadian Naturalist. Montreal, Canada.
Carnegie Museum, Annals. Pittsburgh, Pa.
Challenger. Report on the Scientific Results of the Exploring Voyage of HMS Challenger. Zoology. Edinburgh.
[K.] Česká Společnost Náuk, Prague, Třída Mathematicko-Přírodovědecká, Věstník.
Cincinnati Society of Natural History, Journal.
Connecticut Geological and Natural History Survey, Bulletin. Hartford.
Dansk Geologisk Forening, Meddelelser. København.
Decheniana. Verhandlungen des Naturhistorischen Vereins der Rheinlande und Westfalens. Bonn.
Denison University Bulletin, Journal of the Scientific Laboratories. Granville, Ohio.
Deutsche geologische Gesellschaft, Zeitschrift. Berlin.
Edinburgh Geological Society, Transactions.
Ergebnisse und Fortschritte der Zoologie. Jena.
Experientia. Basel.
Fieldiana, Geology. Chicago.
Field Museum of Natural History, Publications, Geological Series. Chicago.
Földtani Közlöny [Magyaroni Földtani Társulat, Folyóirata]. Budapest.
Fortschritte der Geologie. Köln & Leipzig.
Fortschritte der Paläontologie. Berlin.
Freiberger Forschungshefte. Berlin.
[K.] Fysiografiska Sällskapet, Lund, Handlingar.
Geological Magazine. London & Hertford.
Geological Society of America, Bulletin; Special Paper; Memoir. New York.
Geological Society of China, Bulletin. Peiping.
Geological Society of London, Memoirs; Proceedings; Quarterly Journal; Transactions.
Geological Survey of Canada, Department of Mines and Resources, Mines and Geology Branch, Bulletin. Ottawa.
Geological Survey of Czechoslovakia, Sbornik. Paleontology. Praha.
Geological Survey of Great Britain and the Museum of Practical Geology, Memoirs. London.
Geological Survey of India, Memoirs [Palaeontologia Indica]. Calcutta.
Geological Survey of New Jersey, Paleontology. Trenton.
Geologie. Berlin.
[K.K.] Geologische Reichsanstalt, Wien, Jahrbuch.
Geologische Rundschau (Geologische Vereinigung). Stuttgart, Leipzig & Berlin.
Geologiska Föreningen, Stockholm, Förhandlingar.
Geologist's Association, Proceedings. London.
Helminthological Society of Washington, Proceedings. Washington, D.C.
Hessisches Landesamt für Bodenforschung, Notizblatt; Abhandlungen. Wiesbaden.
Illinois State Geological Survey, Report of Investigations. Urbana.
Instituto Geológico y Minero de España, Boletín; Memorias; Notas y Comunicaciones. Madrid.
International Geological Congress, 20th Session, Abstracts. Mexico City.
Journal of Geology. Chicago.
Journal of Paleontology. Tulsa, Okla.
Lund Universitet, Årsskrift.
Magazine of Zoology and Botany. Edinburgh & London.
Maryland Geological Survey. Baltimore.
Micropaleontology (American Museum of Natural History). New York.
Mitteilungen des Naturwissenschaftlichen Vereins für Steiermark. Graz, Austria.
Moskovskogo Obshchestvo Ispytateley Prirody (Société Impériale des Naturalistes de Moscou, Bulletin).
Musée royal d'Histoire naturelle de Belgique, Mémoires; Bulletin. Bruxelles.
Museo de Historia Natural de Mendoza, Revista.
Museo Libico Storia Naturale, Annali. Tripoli.
Muséum National d'Histoire Naturelle, Bulletin; Mémoires. Paris.
Natural History Society of New Brunswick, Bulletin. St. John.
Naturforschende Gesellschaft in Danzig, Schriften.
Naturhistorisk Tidsskrift. København.
The Nautilus. Philadelphia & Boston.
Neues Jahrbuch für Geologie und Paläontologie (Neues Jahrbuch für Mineralogie, Geologie, und Paläontologie), Beilage-Band. Stuttgart.
Neues Jahrbuch für Mineralogie (Neues Jahrbuch für Mineralogie, Geologie und Paläontologie), Abhandlungen; Beilage-Bande; Monatshefte. Stuttgart.
New Mexico State Bureau of Mines and Mineral Resources, Bulletin; Circular; Memoir. Socorro.
New York Academy of Science, Annals. New York.
New York State Geological Survey, Palaeontology of New York. Albany.
New York State Museum, Bulletin. New York.
New Zealand Journal of Geology and Geophysics. Wellington.
Norsk Geologisk Tidsskrift. Oslo.
[K.] Norske Videnskabers Selskab, Skrifter. Trondhjem, Norway.
Norske Videnskaps-Akadami, Oslo, Skrifter.
North Carolina Geological Survey, Report [now Department of Conservation and Development of the State of North Carolina, Division of Mineral Resources]. Raleigh.
Nova Acta Leopoldina, Neue Folge. [Nova Acta Academiae Caesareae Leopoldino-Carolinae Germanicae naturae curiosorum]. Halle.
Oklahoma Geology Notes. Norman.
Osnovy Paleontologiy, Spravochnik Dlya Paleontologov I Geologov SSSR. Yu. A. Orlov, ed. Akademiya Nauk SSSR. Moskva.
Österreichische Akademie der Wissenschaften, mathematisch-naturwissenschaftliche Klasse, Denkschriften. Wien.
Palaeobiologica. Wien.
Palaeontographia Italica. Pisa.

Palaeontographica, Abteilung A; Abteilung B. Stuttgart & Kassel.
Palaeontographica Bohemiae [Česká akademie věd a umění v Praze, Vydavá, Třída II]. Praha.
Paleontological Society of Russia, Journal. Moskva.
Paläontologische Zeitschrift. Berlin & Stuttgart.
Pan-American Geologist. Des Moines, Iowa.
[K.] Physikalisch-Oekonomische Gesellschaft zu Königsberg, Schriften.
Rivista Italiana de Paleontologia. Parma.
Royal Society of Canada, Transactions. Ottawa.
Royal Society of London, Philosophical Transactions, Series A; Series B.
Royal Society of South Australia, Transactions; Memoirs; Proceedings. Adelaide.
Royal Society of Victoria, Proceedings. Melbourne.
Russiche-Kaiserliche mineralogische Gesellschaft zu St. Petersburg, Verhandlungen [Vserossiyskoe Mineralogicheskoe Obshchestvo, Leningrad].
Schlesische Gesellschaft für vaterländische Kulture, Jahres-bericht. Breslau.
Schweizerische Palaeontologische Gesellschaft, Zurich, Abhandlungen.
Science. New York.
Sciences Naturelles, Annales. Paris.
Senckenbergiana Letheae* [Senkenbergische Naturforschende Gesellschaft Wissenschaftliche Mitteilungen] [*"Letheae" added to title, 1954]. Frankfurt am Main.
Senckenbergische Naturforschende Gesellschaft, Abhandlungen. Frankfurt am Main.
Service Géologique du Maroc, Division des Mines et de la Géologie, Notes. Rabat.
Serviço Geológico e Mineralógico do Brasil, Ministerio da Agricultura, Monographias; Boletim. Rio de Janeiro.
Smithsonian Institution, Harriman Alaska Series, Geology and Paleontology. Washington, D.C.
Smithsonian Miscellaneous Collections. Washington, D.C.
Societatis Regiae Scientarum Gottingensis, Commentationis.
Societa Geologica Italiana, Bolletine. Roma.
Société de Biologie, Hebdomadaires des Séances et Mémoires, Comptes Rendus. Paris.
Société Géologique de France, Bulletin; Mémoires. Paris.
Société d'Histoire Naturelle de Paris, Mémoires.
Société d'Histoire Naturelle de Toulouse, Bulletin.
Société Impériale des Naturalistes de Moscou, Bulletin. [Moskovskogo Obshchestvo Ispytateley Prirody].
Société Paléontologique de la Suisse, Mémoires. Zurich.
Société de Physique et d'Histoire Naturalle de Genève, Mémoires.
Société Vaudoise des Sciences Naturelles, Bulletin. Lausanne, Switz.
Société Zoologique de France, Bulletin. Paris.
Southern California Academy of Sciences, Bulletin. Los Angeles.
Svensk Faunistick Revy [1955, Zoologisk Revy]. Holmiae.
[K.] Svenska Vetenskapsakademien, Handlungen. Holmiae.
Sveriges Geologiska Undersökning, Årsbok; Afhandlingar. Holmiae.
Tartu Ülikooli Geoloogia-Instituudi Toimestused, Acta et Commentationes Universitatis Tartuensis (Dorpatensis). Tartu, Estonia.
Tohoku University, Science Reports, Second Series (Geology), Institute of Geology and Paleontology. Sendai, Japan.
U.S. Geological and Geographical Survey of the Territories, Annual Report. Washington, D.C.
U.S. Geological Survey, Bulletin; Professional Paper; Water-Supply Paper. Washington, D.C.
U.S. National Museum, Proceedings. Washington, D.C.
University of Iowa Studies in Natural History. Iowa City.
University of Kansas Paleontological Contributions. Lawrence.
University of Missouri Studies. Columbia.
University of Missouri School of Mines and Metallurgy, Technical Series. Bulletin. Rolla.
University of Texas, Bulletin. Austin.
Verein für vaterländische Naturkunde in Württemberg, Jahreshefte. Stuttgart.
Videnskabs-Salskabet, Khristiania, Fordhandlingar.
Vsesouiznyi Nauch-no Issledovatel'skiy Geologo-Razvedochoyi neftiano Institut [Paleontologicheskiy Sbornik]. Moskva.
Wagner Free Institute of Science of Philadelphia, Bulletin.
Zeitschrift für Geschiebeforschung. Berlin.
Zeitschrift für Mineralogie. Berlin.
Zeitschrift für Natur- und Heilkunde der K. Medicinisch-chirurgische Akademie St. Petersburg.
Zeitschrift für Wissenschlaftliche Zoologie. Leipzig.
Zoological Journal. London.
Zoologische Jahrbücher, Supplementen. Jena.
Zoologischer Anzeiger. Leipzig.

SOURCES OF ILLUSTRATIONS

At the end of figure captions an index number is given to supply record of the author of illustrations used in the *Treatise,* reference being made either (1) to publications cited in reference lists or (2) to the names of authors without indication of individual publications concerned. Previously unpublished illustrations are marked by the letter "n" (signifying "new") with the name of the author.

STRATIGRAPHIC DIVISIONS

Classification of rocks forming the geologic column as commonly cited in the *Treatise* in terms of units defined by concepts of time is reasonably uniform and firm throughout most of the world as regards major divisions (e.g., series, systems, and rocks representing eras) but it is variable and unfirm as regards smaller divisions (e.g., substages, stages, and subseries), which are provincial in application. Users of the *Treatise* have suggested the desirability of publishing reference lists showing the stratigraphic arrangement of at least the most commonly cited divisions. Accordingly, a tabulation of European and North American units, which broadly is applicable also to other continents, is given here.

Generally Recognized Divisions of Geologic Column

EUROPE	NORTH AMERICA
ROCKS OF CENOZOIC ERA	**ROCKS OF CENOZOIC ERA**
NEOGENE SYSTEM[1]	**NEOGENE SYSTEM**[1]
Pleistocene Series (including Recent)	Pleistocene Series (including Recent)
Pliocene Series	Pliocene Series
Miocene Series	Miocene Series
PALEOGENE SYSTEM	**PALEOGENE SYSTEM**
Oligocene Series	Oligocene Series
Eocene Series	Eocene Series
Paleocene Series	Paleocene Series
ROCKS OF MESOZOIC ERA	**ROCKS OF MESOZOIC ERA**
CRETACEOUS SYSTEM	**CRETACEOUS SYSTEM**
Upper Cretaceous Series	**Gulfian Series (Upper Cretaceous)**
Maastrichtian Stage[2]	Navarroan Stage
Campanian Stage[2]	Tayloran Stage
Santonian Stage[2]	Austinian Stage
Coniacian Stage[2]	
Turonian Stage	
Cenomanian Stage	Woodbinian (Tuscaloosan) Stage
	Comanchean Series (Lower Cretaceous)
	Washitan Stage
Lower Cretaceous Series	
Albian Stage	Fredericksburgian Stage
	Trinitian Stage
Aptian Stage	
	Coahuilan Series (Lower Cretaceous)
Barremian Stage[3]	Nuevoleonian Stage
Hauterivian Stage[3]	
Valanginian Stage[3]	Durangoan Stage
Berriasian Stage[3]	
JURASSIC SYSTEM	**JURASSIC SYSTEM**
Upper Jurassic Series	**Upper Jurassic Series**
Portlandian Stage[4]	Portlandian Stage
Kimmeridgian Stage	Kimmeridgian Stage
Oxfordian Stage	Oxfordian Stage
Middle Jurassic Series	**Middle Jurassic Series**
Callovian Stage	Callovian Stage
Bathonian Stage	Bathonian Stage
Bajocian Stage	Bajocian Stage

Lower Jurassic Series (Liassic)	**Lower Jurassic Series (Liassic)**
Toarcian Stage	Toarcian Stage
Pliensbachian Stage	Pliensbachian Stage
Sinemurian Stage	Sinemurian Stage
Hettangian Stage	Hettangian Stage
TRIASSIC SYSTEM	**TRIASSIC SYSTEM**
Upper Triassic Series	**Upper Triassic Series**
Rhaetian Stage[5]	(Not recognized)
Norian Stage	Norian Stage
Carnian Stage	Carnian Stage
Middle Triassic Series	**Middle Triassic Series**
Ladinian Stage	Ladinian Stage
Anisian Stage (Virglorian)	Anisian Stage
Lower Triassic Series	**Lower Triassic Series**
Scythian Series (Werfenian)	Scythian Stage
ROCKS OF PALEOZOIC ERA	**ROCKS OF PALEOZOIC ERA**
PERMIAN SYSTEM	**PERMIAN SYSTEM**
Upper Permian Series	**Upper Permian Series**
Tartarian Stage[6]	Ochoan Stage
Kazanian Stage[7]	Guadalupian Stage
Kungurian Stage	
Lower Permian Series	**Lower Permian Series**
Artinskian Stage[8]	Leonardian Stage
Sakmarian Stage	Wolfcampian Stage
CARBONIFEROUS SYSTEM	
Upper Carboniferous Series	**PENNSYLVANIAN SYSTEM**
Stephanian Stage	**Kawvian Series (Upper Pennsylvanian)**
	Virgilian Stage
	Missourian Stage
Westphalian Stage	**Oklan Series (Middle Pennsylvanian)**
	Desmoinesian Stage
	Bendian Stage
	Ardian Series (Lower Pennsylvanian)
	Morrowan Stage
	MISSISSIPPIAN SYSTEM
Namurian Stage	**Tennesseean Series (Upper Mississippian)**
	Chesteran Stage
Lower Carboniferous Series	
Viséan Stage	Meramecian Stage
	Waverlyan Series (Lower Mississippian)
Tournaisian Stage	Osagian Stage
Strunian Stage	Kinderhookian Stage
DEVONIAN SYSTEM	**DEVONIAN SYSTEM**
Upper Devonian Series	**Chautauquan Series (Upper Devonian)**
Famennian Stage	Conewangoan Stage
	Cassadagan Stage
Frasnian Stage	**Senecan Series (Upper Devonian)**
	Chemungian Stage
	Fingerlakesian Stage

Middle Devonian Series	**Erian Series (Middle Devonian)**
Givetian Stage	Taghanican Stage
	Tioughniogan Stage
Couvinian Stage	Cazenovian Stage
Lower Devonian Series	**Ulsterian Series (Lower Devonian)**
Coblenzian Stage	Onesquethawan Stage
	Deerparkian Stage
Gedinnian Stage	Helderbergian Stage
SILURIAN SYSTEM	**SILURIAN SYSTEM**
Upper Silurian Series	**Cayugan Series (Upper Silurian)**
Ludlovian Stage	Keyseran Stage
	Tonolowayan Stage
	Salinan Stage
Middle Silurian Series	**Niagaran Series (Middle Silurian)**
Wenlockian Stage	Lockportian Stage
	Cliftonian Stage
Llandoverian Stage (upper part)	Clintonian Stage
Lower Silurian Series	**Medinan Series (Lower Silurian)**
Llandoverian Stage (lower part)	Alexandrian Stage
ORDOVICIAN SYSTEM	**ORDOVICIAN SYSTEM**
Upper Ordovician Series	**Cincinnatian Series (Upper Ordovician)**
Ashgillian Stage	Richmondian Stage
Caradocian Stage (upper part)	Maysvillian Stage
	Edenian Stage
Middle Ordovician	**Champlainian Series (Middle Ordovician)**
Caradocian Stage (lower part)	Mohawkian Stage
	Trentonian Substage
	Blackriveran Substage
Llandeilian Stage	Chazyan Stage
Llanvirnian Stage	
Lower Ordovician Series	**Canadian Series (Lower Ordovician)**
Arenigian Stage	
Tremadocian Stage[9]	
CAMBRIAN SYSTEM	**CAMBRIAN SYSTEM**
Upper Cambrian Series	**Croixian Series (Upper Cambrian)**
	Trempealeauan Stage
	Franconian Stage
	Dresbachian Stage
Middle Cambrian Series	**Albertan Series (Middle Cambrian)**
Lower Cambrian Series	**Waucoban Series (Lower Cambrian)**
EOCAMBRIAN SYSTEM	**EOCAMBRIAN SYSTEM**
ROCKS OF PRECAMBRIAN AGE	**ROCKS OF PRECAMBRIAN AGE**

RAYMOND C. MOORE

[1] Considered by some to exclude post-Pliocene deposits.

[2] Classed as division of Senonian Subseries.

[3] Classed as division of Neocomian Subseries.

[4] Includes Purbeckian deposits.

[5] Interpreted as lowermost Jurassic in some areas.

[6] Includes some Lower Triassic and equivalent to upper Thuringian (Zechstein) deposits.

[7] Equivalent to lower Thuringian (Zechstein) deposits.

[8] Equivalent to upper Autunian and part of Rotliegend deposits.

[9] Classed as uppermost Cambrian by some geologists.

PART W

MISCELLANEA

By W. H. Hass, Walter Häntzschel, D. W. Fisher, B. F. Howell,

F. H. T. Rhodes, K. J. Müller, and R. C. Moore

CONTENTS

INTRODUCTION

By Raymond C. Moore

[University of Kansas]

The letter "W," assigned to this volume of the *Treatise,* indicates a position next to last in the planned sequence of units. This is explained readily by its intended content of "left-overs"—mostly groups of fossils set apart as unknown or very doubtful as to taxonomic affinities. Such a residuum might be expected to follow the publication of all other units, possibly with a miscellany of minor groups that for some reason had been omitted from already-issued volumes in which they would logically have found place. Obviously, the presumption expressed does not accord with fact, since several important divisions of the *Treatise* are yet unfinished. Readers may be reminded that an initial feature of this collaborative project was to publish each planned volume whenever it could be made ready for the press. No good end would be served by withholding the appearance of some completed unit while waiting on another, even though one or more of the latter might be much more generally sought after by paleontologists.

It is hardly appropriate for me to suggest a rating among fossil groups that undertakes to classify them according to what might be considered their relative usefulness or value. How could this be done unless the criteria for judgment were agreed upon, and even then would unanimity of opinions or approach to it serve any desirable purpose? Inquiry of this sort seems to be fruitless, but it is brought to notice again and again in dealing with parts of the fossil record. Among groups of organic remains or traces treated in Part W, surely the large number of described and named problematical fossils, including many which now are judged to be inorganic in origin (and thus not acceptable as evidence of any sort of animal or plant life), are at least significant assemblages. At the same time they are most difficult to interpret and to classify satisfactorily. Even so, they should not be excluded from consideration; in treating them as comprehensively and authoritatively as possible, Dr. Häntzschel has made a valuable contribution to paleontology in his section of this volume, especially because of the widely scattered nature of records in the literature and their many sorts of inadequacies.

A group of fossil remains now established as having exceptional value for stratigraphical correlations and age determinations of sedimentary deposits ranging from Cambrian to Triassic, inclusive, comprises the very widely distributed, highly varied, and locally very abundant conodonts. They are assigned to *Treatise* Part W because no yet-discovered evidence satisfactorily points to taxonomic placement. Certainly they are remains of animals, but whether belonging to invertebrates or vertebrates is unsettled.

It is a great regret for me to record the untimely death of Dr. WILBERT H. HASS, of the U.S. Geological Survey, who contributed the major article on conodonts prepared for the *Treatise*. This occurred on 30 November, 1959. He was a foremost American specialist in study of these fossils, benefited by approximately two decades of intensive studies of conodonts in the field and laboratory. His death brought to an end in mid-career his important series of contributions and what undoubtedly would have been a greater increase in knowledge of paleontology. Supplementary discussions of some aspects of conodont researches have been prepared by F. H. T. RHODES, of the University of Wales and KLAUS MÜLLER, of the Technische Universität, Berlin.

Another group of fossils that in some ways is comparable to the conodonts consists of remains termed scolecodonts. These are fairly well identified as the jaw parts of annelid worms. They are useful stratigraphic markers also, and, along with other remains of various sorts of worms preserved as fossils, are described by B. F. HOWELL.

Diverse sorts of narrowly conical shells classed as hyolithids, tentaculitids, and some others are probably molluscan groups. These are assigned to Part W and described by D. W. FISHER, State Paleontologist of New York. They include stratigraphically important genera and species, which are uncertainly classifiable taxonomically.

Finally, Part W includes a record of the problematical fossils, already mentioned, consisting of traces and impressions (so-called body fossils) prepared by W. HÄNTZSCHEL from widely scattered literature. Some of these fossils are stratigraphically useful in spite of uncertainty as to their biological placement.

CONODONTS

By Wilbert H. Hass

[U.S. Geological Survey, Washington, D.C. Publication authorized by the Director.]

CONTENTS

INTRODUCTION

NATURE OF CONODONTS

The term *Conodonten* was coined by Christian H. Pander (52) in 1856 for some toothlike and platelike microfossils discovered by him in the Paleozoic rocks of eastern Europe. These fossils comprise a monophyletic group of lamellar structures that range from a fraction of a millimeter to about 3 mm. in length. In their natural state, they are composed chiefly of calcium phosphate, are either amber or grayish-black, and are translucent to opaque; when weathered, many are friable and light gray. Conodonts have a world-wide distribution, but present knowledge is based chiefly upon European and North American faunas. They definitely range from the Lower Ordovician into the Upper Triassic, and recent finds indicate that they may range from the Upper Cambrian into the Upper Cretaceous. Chiefly as a result of investigations since 1926, conodonts have become an extremely useful tool of the stratigraphic paleontologist despite the fact that there has never been unanimity either on the zoological affinity of the animal that bore conodonts or on the function that was performed by these structures.

Presumably, the conodont-bearing animal was soft-bodied, bilaterally symmetrical, marine, and pelagic. These assumptions seem likely because conodonts and the basal plate to which some conodonts are still attached appear to be the only parts of these animals that have been preserved, because specimens with the same form occur as rights and lefts, and because conodonts are found associated with many kinds of marine fossils in all kinds of marine sedimentary rocks. Many answers have been given to the question "What are conodonts?" At one time or another they have been assigned to such different groups of organisms as the mollusks, worms, arthropods, primitive vertebrates, and fishes; and some or all of them have been considered to be spines, scales, dermal denticles, copulative claspers, gill-arch structures, mandibles, teeth and other ingestive aids, and supports for a tissue that covered them.

Many conodonts are good index fossils. They are durable, abundant, distinctive, widespread in their geographic distribution, and restricted in their stratigraphic range. Moreover, being minute, they are well suited for subsurface investigations; and being present in all kinds of marine sedimentary formations, they provide a dependable means of correlating lithologically different, bio-stratigraphic equivalents. Some formations, in which the more common kinds of fossils are either scarce or absent, abound in conodonts, and problems concerned with the age, faunal zonation, and correlation of such formations fall to a great degree within the province of the conodont specialist. The Devonian and Mississippian black shale sequence is a good example. Portions of this sequence occur throughout much of the interior of the United States and a part of Canada, and range from a featheredge to several thousand feet in thickness. The age and correlation of these black shales have been controversial subjects for many years, but through conodont studies it has now become possible to correlate certain parts of this sequence with formations in the North American standard Upper Devonian succession of New York and Pennsylvania, and other portions with formations in the lower part of the standard Mississippian succession of the middle Mississippi Valley area. Hence, through conodont studies, it will be possible eventually not only to solve the controversial black shale problem, but also to obtain much valuable information on the Late Devonian and Early Mississippian paleogeography of North America.

Some investigators have confused conodonts with scolecodonts—the jaw structures of polychaete worms. Scolecodonts, however, are morphologically distinct; some of their more obvious characteristics being that they are brittle, jet black, siliceo-chitinous, and insoluble in hydrochloric acid.

The writer thanks R. C. Moore and his staff for making editorial suggestions and for preparing the illustrations.

Some literature reaching the writer's desk after March 31, 1957, has not been considered in the preparation of this paper.

METHODS OF PREPARATION

Inasmuch as conodonts occur in all kinds of marine sedimentary rocks, the method used to prepare a collection for study depends not only upon the condition of the specimens, but also upon the nature of the matrix. Whole well-preserved specimens can be recovered from calcareous rocks with a 10 to 15 percent solution of acetic or formic acid. Acetic acid (CH_3COOH) is inexpensive but works so slowly that several weeks may be required to recover the conodonts in a 10- to 20-pound rock sample. The rate at which the reaction proceeds is governed to some degree by the amount of calcium acetate present in the solution. This salt is but slightly soluble in water and, by coating the undigested part of a sample, retards the chemical reaction, so that the solute must be replaced frequently. Formic acid ($HCOOH$) digests calcareous rocks rather quickly but, in the process, may alter some conodont specimens to a chalky white. Both acetic and formic acids have sharp disagreeable odors and should be used in a well-ventilated room. Some investigators prefer monochloracetic acid (CH_2ClOOH) because its calcium salt, being quite soluble in water, does not impede the chemical reaction. But monochloracetic acid must be used very carefully, for, on contact, it inflames and blisters the skin. Citric and tartaric acids have also been used to recover conodonts from calcareous rocks.

Specimens in an indurated noncalcareous rock are seldom recovered in good condition if removed from the enclosing matrix.

Crushing the matrix to pebble size will free some specimens but, as indicated above, most of those recovered will be fragments. Some siltstones and sandstones can be disintegrated either with an ultrasonic device or with a 10 to 15 percent aqueous solution of hydrogen peroxide. Rocks that disintegrate in water, with or without boiling, should be flushed until the water is no longer muddy.

Most conodonts will pass through a 16-mesh sieve and will be retained on the 150-mesh screen. As their specific gravity ranges from 2.84 to 3.10 (Ellison, 25), free specimens can be separated from grains of quartz (2.65-2.66), calcite (2.72), and several other constituents of sedimentary rocks by making a heavy mineral separation with a solution of bromoform and acetone. The specific gravity of this solution must fall between that of calcite (2.72) and the lightest of conodonts (2.84). This optimum can be achieved easily by placing a piece of calcite in acetone and adding bromoform until the mineral just floats. Free conodont specimens can also be concentrated with an isodynamic separator. This electromagnetic device is capable of separating mineral grains that have very slight differences of susceptibility, and can be used to advantage on collections containing numerous iron sulfide grains. With this device, the writer has been able to extract a 0.5 gram concentrate of conodonts from a heavy-mineral sample weighing over 225 grams.

Most conodonts in black shales cannot be removed in good condition from the enclosing matrix. Little can be done to prepare such material for study, though in some instances it is possible to expose one side of a specimen by flaking away the matrix with a sharply pointed needle. Excellent rubber replicas of conodonts can be made by filling their molds with a fast-setting, ammonia-soluble, latex compound. These replicas may eventually deteriorate, but for a few years, at least, each one retains all of the minute surface features of the specimen it represents.

Good prints of conodonts can be made by enlarging the photographic negatives of specimens taken with standard equipment at 10 magnifications. Specimens to be photographed should first be coated with a film of ammonium chloride. Stereographs have been used in only a few papers on conodonts; for the most part, this method of illustrating fossils appears to have a rather limited use in the study of conodonts.

The introduction to Branson & Mehl's (7-10) *Conodont Studies* contains many useful suggestions on the preparation and care of conodont samples and specimens.

MORPHOLOGY

TERMINOLOGY

Since the time of Pander most students have assumed that conodonts once functioned as ingestive aids, and, as a result, a descriptive terminology has been adopted that is highly suggestive of teeth and other mouth parts. There is, however, no reason for believing that the like-named parts of conodonts and of ingestive aids had similar origins or identical functions. Despite this, the terminology now used is adequate even though it has not been completely standardized. The morphological terms in use are listed below, and the parts of the conodont to which some of these terms refer are indicated in Figs. 1-4.

GLOSSARY OF MORPHOLOGICAL TERMS

[Terms of lesser importance are printed in italics.]

a-side. Same as anterior side; also has been used to refer to posterior side of platelike conodont.

aboral *(unten).* Toward underside of conodont.

aboral attachment scar. Same as pulp cavity, especially an expanded pulp cavity or one which is larger than a small-sized pit; also has been used to refer to that portion of aboral side to which the basal plate was attached.

aboral cavity. Same as pulp cavity.

aboral edge *(Aboralkante).* Sharp edge along mid-line of aboral side.

aboral extension. Portion of expanded base of main cusp extending below level of posterior bar.

aboral groove *(Basalfurche, Basisrinne).* Groove along mid-line of aboral side of conodont.

aboral margin. Trace of aboral side of unit in lateral view; also has been used to refer to aboral side.

aboral process. Same as linguiform process.

aboral projection. Same as anticusp; also has been used to refer to aboral extension.

aboral side *(Aboralrand).* Side onto which pulp

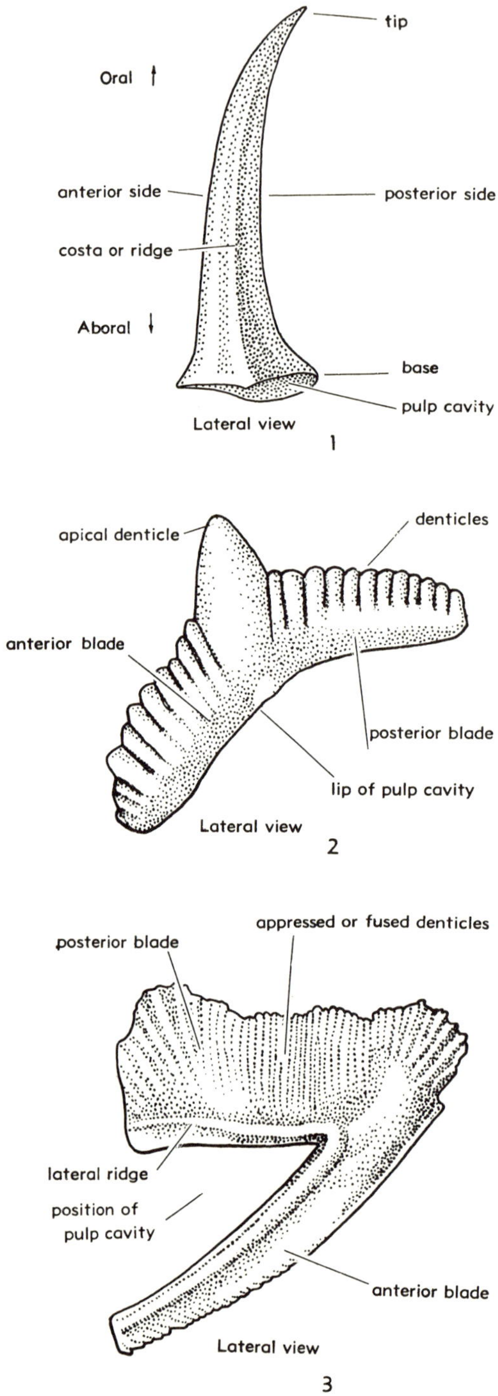

FIG. 1. Distacodontid and compound bladelike conodonts.——*1. Distacodus incurvus* (PANDER) (52).——*2. Ozarkodina typica* BRANSON & MEHL (7).——*3. Dinodus fragosus* (E. R. BRANSON) (Hass, n).

cavity opens; undersurface area to which basal plate is attached.

aboral surface. Same as aboral side.

aboro-lateral groove. Same as aboral groove.

accessory lobes. Nodose processes on posterior portion of platform located between transverse ridges and blade (*see* FAY, 27, p. 9).

anterior *(vorn).* Toward front end of conodont.

anterior arch. Arch located at anterior end of compound conodont.

anterior bar *(Vorderast, vorder Hälfte).* Bar located along antero-posterior mid-line and anterior to pulp cavity; also has been used to refer to blade of platelike conodont, even though this blade is actually located posterior (not anterior) to pulp cavity.

anterior blade. Blade located along antero-posterior mid-line and anterior to pulp cavity.

anterior curvature. Same as anterior blade, bar, or limb.

anterior deflection. Down-turned distal end of anterior blade, bar, or limb.

anterior denticles. Denticles of anterior blade, bar, or limb.

anterior edge. Sharp-edged anterior side.

anterior face. Same as anterior side.

anterior inner bar. Same as anterior inner lateral bar.

anterior inner lateral bar. Anteriormost of two or more lateral bars on inner side of asymmetrical compound conodont.

anterior limb. Same as anterior bar or anterior blade; also has been used to refer to anticusp.

anterior margin. Trace of anterior side of unit in lateral view.

anterior outer bar. Same as anterior outer lateral bar.

anterior outer lateral bar. Anteriormost of two or more lateral bars on outer side of asymmetrical compound conodont.

anterior process. Same as anterior bar, blade, or limb; also has been used to refer to anticusp.

anterior projection. Same as anterior bar, blade, or limb.

anterior side *(Vorderenkiel).* Front end of conodont; *(a)* in distacodontid conodonts, convex side of cusp, or side facing in direction opposite that toward which tip of cusp points; *(b)* in compound conodonts, convex side of cusp and denticles; in specimens with denticles not curved, end nearest pulp cavity; *(c)* in platelike conodonts, distal end of carina.

anterior wing. Enlarged anterior side of distacodontid; may be denticulated.

antero-inferior process. Same as anticusp.

antero-inner-lateral flange. Lobe just anterior to main cusp on inner platform of *Icriodella.*

anticusp. Downward projection of main cusp; commonly bears denticles.

apex. Tip of pulp cavity; also has been used re-

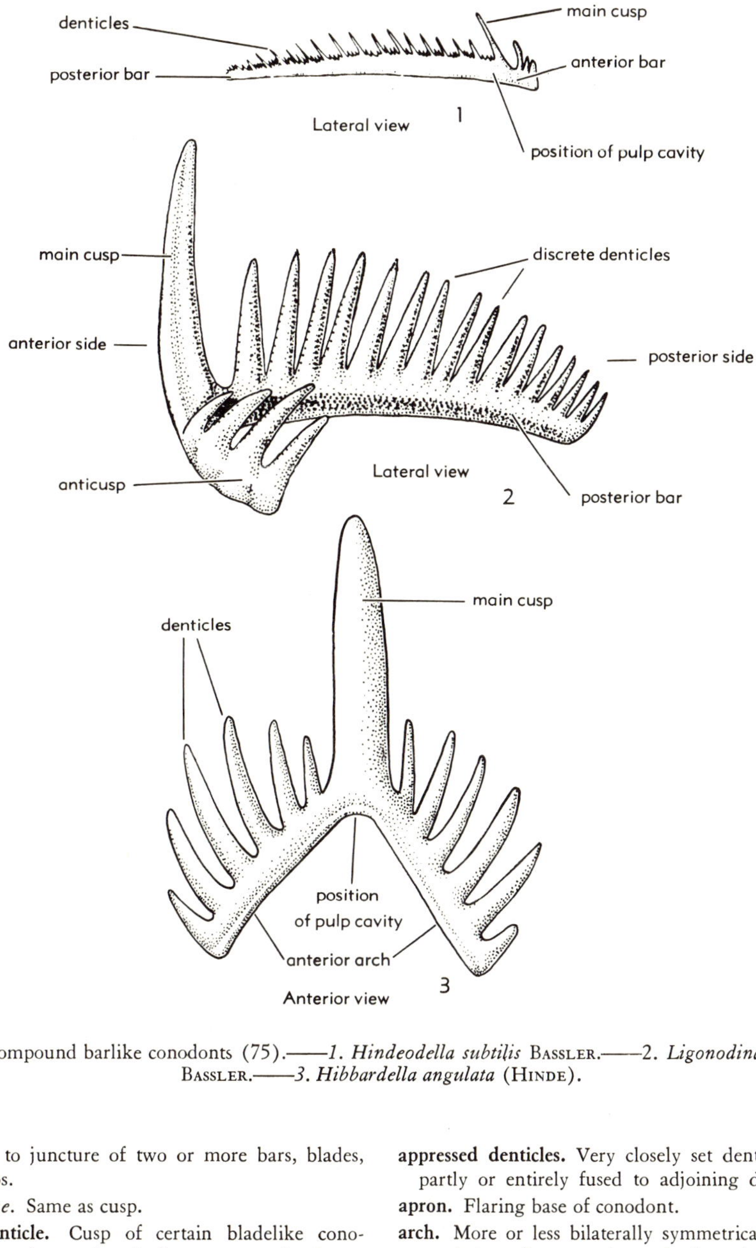

FIG. 2. Compound barlike conodonts (75).——*1. Hindeodella subtilis* BASSLER.——*2. Ligonodina pectinata* BASSLER.——*3. Hibbardella angulata* (HINDE).

ferring to juncture of two or more bars, blades, or limbs.

apical cone. Same as cusp.

apical denticle. Cusp of certain bladelike conodonts, such as *Ozarkodina;* commonly larger than adjacent denticles.

apical lamella. Small expansion or lip on anterior side of base of cusp of *Apatognathus* (*see* FAY, 27, p. 9).

apical lip *(Apicallippe).* Faint lateral ridge separating aboral side from lateral side of compound conodont.

apical pit. Same as pulp cavity.

appressed denticles. Very closely set denticles, each partly or entirely fused to adjoining denticles.

apron. Flaring base of conodont.

arch. More or less bilaterally symmetrical structure consisting of cusp and two backward- or downward-trending blades, bars, or limbs, each of which is joined to base of cusp and commonly bears denticles.

assemblage. Association of several kinds of discrete conodonts presumed to be structural parts of one animal.

attachment scar. Same as pulp cavity, especially an expanded pulp cavity or one which is larger

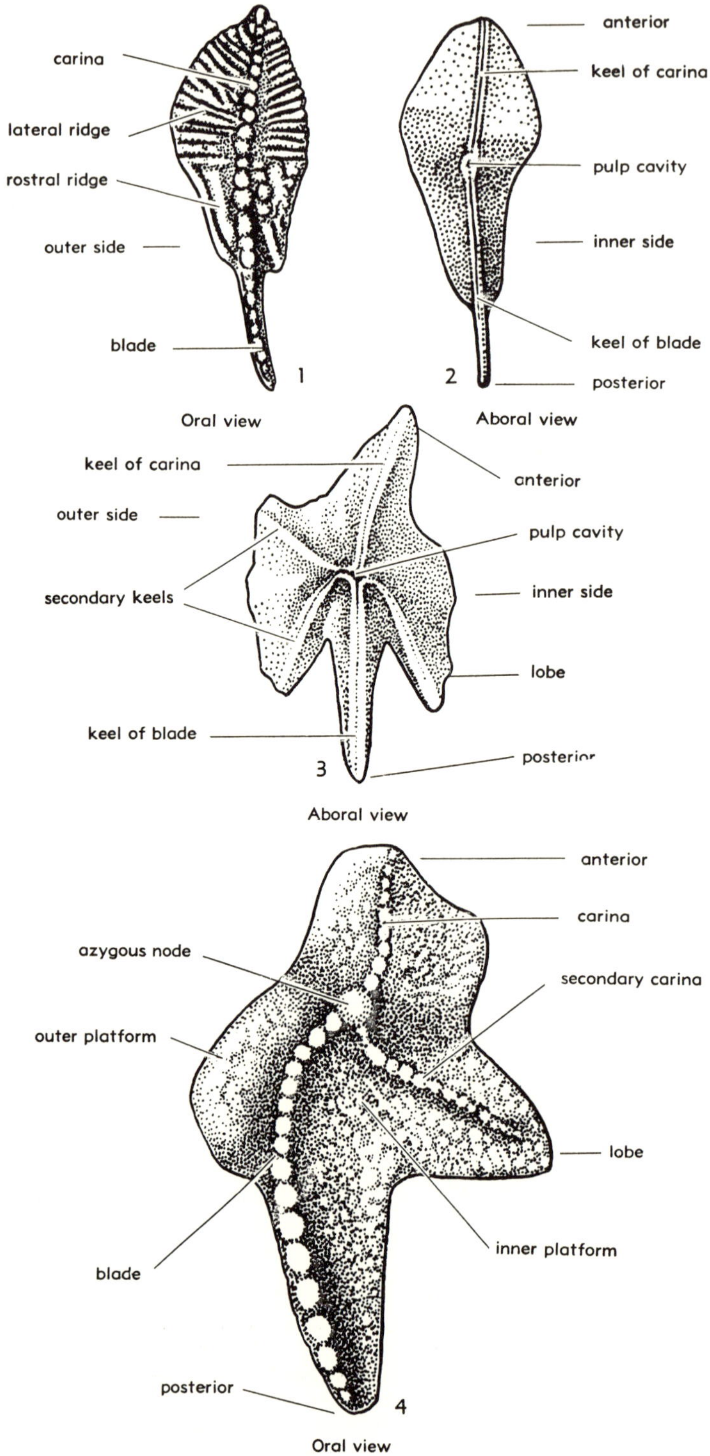

FIG. 3. Platelike conodonts with lateral platforms.——*1,2. Siphonodella duplicata* (BRANSON & MEHL) (10). *3. Ancyrodella* sp. (Hass, n).——*4. Palmatolepis perlobata* ULRICH & BASSLER (Hass, n).

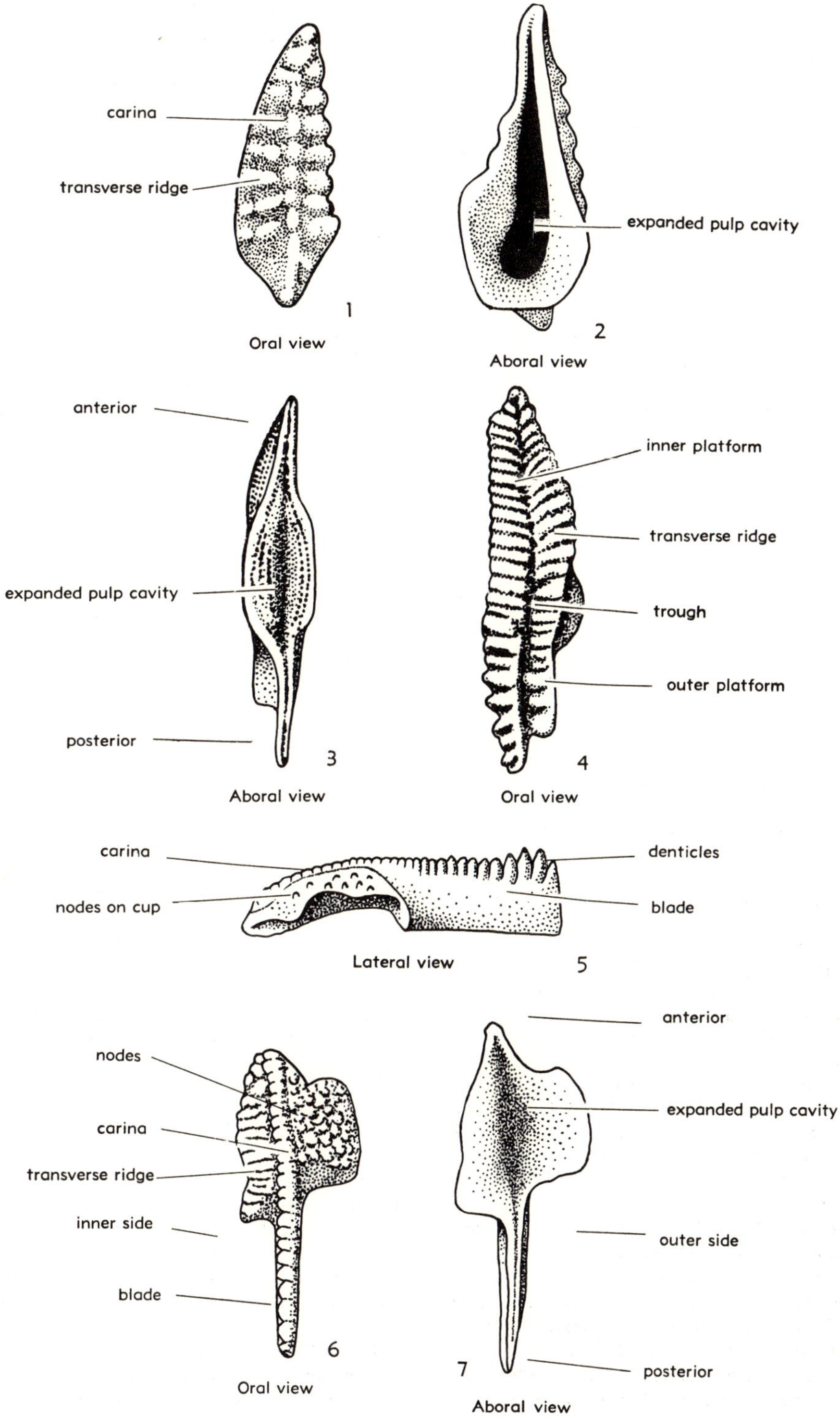

FIG. 4. Platelike conodonts with expanded pulp cavities.——*1,2. Icriodus expansus* BRANSON & MEHL (82). ——*3,4. Cavusgnathus cristata* BRANSON & MEHL (11).——*5-7. Gnathodus pustulosus* BRANSON & MEHL (11).

than small-sized pit; also has been used referring to portion of aboral side to which basal plate was attached.

axis. Structures located along anteroposterior midline of platelike conodont.

azygous node *(Zentralknötchen, Zentralknoten).* Node located directly above pulp cavity of *Palmatolepis* and *Panderodella;* special kind of cusp.

b-side. Same as posterior side, also has been used to refer to anterior side of platelike conodont.

bar. Shaft of compound conodont, commonly bearing denticles (*see* anterior bar, lateral bar, posterior bar).

bar teeth. Same as denticles.

basal attachment scar. Same as pulp cavity, especially an expanded pulp cavity or one which is larger than small-sized pit; also has been used referring to portion of aboral side to which basal plate was attached.

basal canalules (parasiten Gänge). Same as cells that are located in basal portion of distacodontid conodont.

basal cavity. Same as pulp cavity.

basal cavity inverted. Area about pulp cavity having striae on lateral sides of conodont, indicating that free edge of any lamella recedes orally from the free edges of previously accreted lamellae.

basal cone *(Basistrichter).* An excavated, conelike basal plate whose tip fits into the pulp cavity; also has been used to refer to the base of a conodont.

basal excavation. Same as pulp cavity.

basal expansion. Same as pulp cavity; also has been used to refer to base of conodont.

basal groove. Same as aboral groove.

basal margin. Same as aboral margin; also has been used to refer to aboral side.

basal plate. Laminated organic substance attached to aboral side of conodont (*see* basal cone and cone filling).

base. Area adjacent to aboral side; also has been used referring to aboral side.

basis. Same as base; also portion of conodont structure surrounding pulp cavity.

blade *(Blatt, Klinge).* Laterally compressed structure; *(a)* in compound conodonts divisible into posterior blade and anterior blade on basis of position with reference to pulp cavity, both blades commonly bearing denticles; *(b)* in platelike conodonts part of axis located posterior to pulp cavity, generally compressed and bearing denticles.

blade parapet. High narrow platform to which blade is joined, as in *Cavusgnathus.*

buttress. Same as linguiform process.

c-side. Same as inner side; also has been used to refer to outer side of some species of *Ancyrognathus, Ancyrodella,* and *Polygnathus.*

cancellated structure. Concentration of cells whereby lamellar structure of conodont is obscured.

carina. Row of nodes or low denticles on oral side of platelike conodont (*see* main carina, secondary carina); also has been used referring to portion of axis of platelike conodont flanked by platforms; also, for that portion of bar, blade, or limb of compound conodont flanked by flange; also, for ridge or costa.

cavity. Same as pulp cavity.

cells. Minute spherical or tubular voids within a conodont; in some specimens, cells so concentrated as to obscure lamellar structure, thereby forming cancellated structure.

central carina. Same as main carina.

central cusp. Same as cusp.

central pit. Same as pulp cavity.

central tooth. Same as cusp.

compound conodont. Bladelike or barlike unit, commonly bearing denticles (Coleodontidae, Prioniodinidae, and Prioniodontidae).

cone. Same as denticle or cusp.

cone axis. Same as growth axis if referring only to separation of lamellae along a line.

cone cavity *(Trichtergrube).* Excavation of basal cone, open aborally.

cone filling *(Trichterfüllung).* Portion of basal plate of some conodonts which is red-brown to dark brown, opaque to translucent, coarsely laminated and occupies cone cavity.

conical node. Same as azygous node.

costa. Long narrow raised area or ridge.

crest. Same as carina; originally proposed for high prominent carina.

crimp. Marginal band on aboral side of plate representing area covered by last lamella accreted to conodont structure.

cristula. Same as rostral ridge.

cup. Greatly expanded pulp cavity beneath anterior half of some conodonts (e.g., *Gnathodus*).

cusp *(grosser Zahn, Hauptzahn, Zahnspitze).* Spinelike, fanglike, or cone-shaped structure located above pulp cavity; *(a)* in distacodontid conodonts, this structure comprises entire specimen; *(b)* in most compound and some platelike conodonts, this structure commonly called main cusp, the apical denticle and azygous node being special kinds of cusps.

d-side. Same as outer side; also has been used to refer to inner side of some species of *Ancyrognathus, Ancyrodella,* and *Polygnathus.*

dental plate. Complete specimen; same as unit.

dental unit. Complete specimen; same as unit.

denticle *(Dentikel, kleiner Zahn, Zacken, Zahnchen).* Spinelike, needle-like, or sawtooth-like structure, similar to cusp but commonly smaller.

discrete denticles. Denticles of same conodont that are not closely set, each one being separated from adjacent denticle by open space.

distacodid. Incorrect spelling for distacodontid.

distacodontid. Type of conodont consisting of fanglike cusp (Distacodontidae, Belodontidae).

double keel. Keel split along its mid-line by pronounced groove.

erect. Trending upward, normal to oral edge of conodont.

escutcheon. Same as pulp cavity.

expanded pulp cavity. Greatly enlarged pulp cavity, aboral side of conodont being partly or entirely opened up into concavity.

fang. Same as cusp.

fibrous conodont. Specimen whose original lamellar structure has been obscured or replaced by fibrous structure through alteration.

first-order denticles *(grosse Zähnchen).* Larger-sized set of denticles on same blade, bar, or limb.

flange. Shelflike structure, broader than a lateral ridge, trending in anteroposterior direction along inner or outer side of blade, bar, or limb.

free blade *(freies Blatt).* Portions of blade not flanked by platforms.

furrow. Groove along anteroposterior oral mid-line of conodont; also, any narrow trench or long depression.

fused denticles. Same as appressed denticles.

germ denticles. Same as suppressed denticles.

growth axis. Direction of active growth indicated by separation of lamellae along a line or plane.

growth center. Point about which the conodont developed, that is, apex of pulp cavity.

growth lamella. Same as lamella.

growth lines. Traces of lamellae in section; also has been used referring to striae.

heel. Posteriorly extended base of *Belodus.*

height. Measurement in oral-aboral direction.

horizontal basis. Same as posterior bar.

horizontal section. Section parallel to oral side.

inferior side. Same as aboral side.

infero-anterior denticle. Same as cusp.

inner basal ridge. Same as lateral ridge; also has been used to refer to platform.

inner face. Lateral face of denticle or cusp on inner side of conodont.

inner lateral bar. Lateral bar on inner side of asymmetrical conodont.

inner lateral face. Same as inner face.

inner lateral lamina. Basal portion of inner side of cusp between anticusp and posterior bar of *Ligonodina.*

inner lateral process. Lateral process on inner side of asymmetrical conodont.

inner parapet. Parapet on inner side of axis.

inner platform. Shelf on concave side of axis of platelike conodont; commonly flanking carina but may flank part or all of blade as well.

inner side *(Innenseite).* Portion of conodont on concave side of anteroposterior mid-line (not applicable to bilaterally symmetrical conodont).

interior limb. Same as lateral bar.

jaw. Same as blade, bar, or limb.

keel *(Kiel).* Costa or rib on aboral side of platelike conodont (*see* main keel, secondary keel); also has been used referring to costa or ridge of distacodontid conodont.

keel angle. Angle opening posteriorly, between main keel and a secondary keel.

lamella *(Wachstumslamelle).* One of numerous thin layers or sheaths which constitute a conodont, each consisting of minute crystals of an apatite mineral that belongs to the dahlite-francolite isomorphous series; it is open toward aboral side of fossil and was accreted about the pulp cavity.

lamellar conodont. Specimen whose original lamellar structure is not obscured through alteration.

lateral bar. Bar whose proximal end joins anteroposterior mid-line; in asymmetrical unit, bar on inner side designated inner lateral bar, and bar on outer side designated outer lateral bar (*see* lateral process for exception).

lateral blade. Same as lateral bar.

lateral branch. Same as limb of anterior arch; also has been used to refer to secondary carina.

lateral carina. Ridge or costa extending from near base to near tip on lateral face of cusp.

lateral costae. Same as transverse ridges.

lateral edge. Sharp-edged lateral side.

lateral expansion. Large process flanking pulp cavity.

lateral face. Portion of denticle or cusp on lateral side of conodont (*see* inner face, outer face).

lateral keel. Same as ridge or costa.

lateral limb. Same as lateral bar.

lateral process *(Seitenast, Seitenfortsatz, Sporn).* Lateral bar, blade, or limb whose proximal end is joined to anteroposterior mid-line adjacent to pulp cavity; in asymmetrical conodont, structure on inner side of pulp cavity designated inner lateral process; structure on outer side of pulp cavity designated outer lateral process (*see* arch).

lateral ridge *(lateral Kante).* Ridge trending in anteroposterior direction along inner or outer side of blade, bar, or limb.

lateral side. Portion of conodont between anterior and posterior sides (*see* inner side, outer side).

lateral wing. Enlarged or expanded lateral side of distacodontid conodont; commonly denticulated.

length. Measurement in anteroposterior direction.

lesser denticles. Same as second-order denticles.

limb. Term used interchangeably with bar and blade of a compound conodont; used especially for unit with pulp cavity more or less equidistant from anterior and posterior ends (*see* anterior limb, posterior limb).

linguiform process. Tongue-shaped structure.

lip *(Lippe).* Small lateral expansion flanking pulp cavity.

lobe *(Lappen; inner, Innenlappen; outer, Aussenlappen).* Shelflike process; *(a)* in compound conodonts, it trends outward from mid-line, is more massive than a bar, and may support denticles or nodes; *(b)* in platelike conodonts, it is generally built up about a secondary carina as in *Palmatolepis,* and may be bifurcate, as in *Amorphognathus* and *Balognathus.*

longitudinal plication. Same as longitudinal ridge.

longitudinal ridge. Ridge on oral surface of platform trending parallel to axis of conodont.

longitudinal section. Section normal to oral side and parallel to anteroposterior direction.

lower anterior denticle. Same as cusp.

lower side. Same as aboral side.

lower surface. Same as aboral side.

main carina *(Kamm).* Portion of axis anterior to pulp cavity.

main crest. Same as main carina.

main cusp. Same as cusp.

main denticle. Same as cusp.

main keel. Keel along anteroposterior mid-line of platelike conodont.

main middle cusp. Same as cusp.

main series of denticles. Same as first-order denticles.

main trough. Trough along anteroposterior midline of platelike conodont.

major denticles. Same as first-order denticles.

median branch. Same as posterior bar.

median carina. Same as main carina.

median cone. Same as cusp.

median longitudinal section. Longitudinal section directly along anteroposterior mid-line of conodont.

median ridge. Lateral ridge approximately equidistant from oral and aboral sides; also has been used to refer to portion of axis of platelike conodont flanked by platforms.

middle cusp. Same as cusp.

navel. Same as pulp cavity.

node *(Knötchen, Tuberkel).* A protuberance, knob, or bump; some denticles, especially those of carina, are nodelike.

nodose denticles. Same as nodes.

oral *(oben).* Toward the upper side of conodont.

oral bar. Same as posterior bar; also has been used to refer to denticles.

oral denticles. Same as denticles.

oral edge. Basal stretch of posterior side of distacodontid conodont; also oral side of posterior bar, blade, or limb of compound conodont.

oral margin. Trace or outline of oral side of unit in lateral view; same as summit line; also has been used referring to oral side.

oral side *(Oralrand).* Upper surface or side opposite that onto which pulp cavity opens; in compound and platelike conodonts this side commonly supports such structures as denticles, nodes, and ridges.

oral surface. Same as oral side.

oral trough. Same as trough.

outer anterior spur. Portion of anterior bar, blade, or limb beyond its juncture with lateral bar, as in *Centrognathodus.*

outer basal ridge. Same as lateral ridge.

outer face. Lateral side of denticle or cusp on outer side of conodont.

outer lateral face. Same as outer face.

outer lateral lamina. Basal portion of outer side of cusp of *Ligonodina.*

outer lateral process. Lateral process on outer side of asymmetrical conodont.

outer parapet. Parapet on outer side of axis.

outer platform. Shelf on convex side of axis of platelike conodont; commonly flanking carina but may flank part or all of blade as well.

outer side *(Aussenseite).* Portion of conodont on convex side of anteroposterior mid-line (not applicable to bilaterally symmetrical conodont).

outward side. Same as anterior side.

parapet. Wall-like structure on platform of platelike conodont or on flange of compound conodont; also has been used referring to narrow platform separated from adjacent platform by deep trough (*see* blade parapet).

pinnate. Transverse rows of nodes or ridges which, together with axis, give featherlike appearance to conodont in oral view, as in *Siphonodella duplicata duplicata.*

pit. Small-sized pulp cavity.

plate *(Tafel).* Structure consisting of inner and outer platforms and adjoining portion of axis of platelike conodont; incorrectly used referring to platform.

platelike conodont. Unit having platforms or greatly expanded pulp cavity (cup); some units have both characteristics (Polygnathidae, Idiognathodontidae).

platform *(Plattform, Tafel).* Laterally broadened structure or shelf (*see* inner platform, outer platform); incorrectly used referring to plate.

posterior *(hinten).* Toward rear end of conodont.

posterior bar *(hinterast, hinter Hälfte).* Bar located along anteroposterior mid-line and posterior to pulp cavity.

posterior blade *(hinterast, hinter Hälfte).* Blade located along anteroposterior mid-line and posterior to pulp cavity.

posterior deflection *(Abbiegung, hintere Abwärtskrümmung).* Down-turned distal end of posterior blade, bar, or limb.

posterior denticles. Denticles of posterior blade, bar, or limb.

posterior downward deflection. Same as posterior deflection.

posterior edge *(hinterer Kiel).* Sharp-edged posterior side.

posterior inner bar. Same as posterior inner lateral bar.

posterior inner lateral bar. Posteriormost of two or more lateral bars on inner side of asymmetrical compound conodont.

posterior keel. Same as posterior side.

posterior limb *(hinterast, hinter Hälfte).* Bar or blade located along anteroposterior mid-line and posterior to pulp cavity.

posterior margin. Trace or outline of posterior side of unit in lateral view; incorrectly used for aboral margin.

posterior oral bar. Same as posterior bar.

posterior outer bar. Same as posterior outer lateral bar.

posterior outer lateral bar. Posteriormost of two or more lateral bars on outer side of asymmetrical compound conodont.

posterior outer lateral process. Same as posterior outer lateral bar.

posterior platform. Same as plate.

posterior process. Same as posterior bar, blade or limb.

posterior side *(hinterer Kiel).* Back or rear end of conodont; *(a)* in distacodontid conodonts, concave side of cusp, or side facing in direction toward which tip of cusp points; *(b)* in compound conodonts, concave side of cusp and denticles; in specimens with denticles not curved, end farthest from pulp cavity; *(c)* in platelike conodonts, distal end of blade.

posterior wing. Enlarged posterior side of distacodontid conodont, may be denticulated.

postero-outer-lateral flange. Lobe just posterior to cusp on outer platform of *Icriodella.*

principal denticle. Same as cusp.

proclined. Cusp trending upward and anteriorly with oral edge of unit oriented horizontally.

pulpa. Same as pulp cavity.

pulp cavity *(Basalgrube, Basisgrube, cavitas pulpae, Nabel, Schild).* Pit or concavity about which conodont was built through accretion of lamellae; this pit opens onto aboral side and is present on all true conodonts.

pustule. Minute, circumscribed elevation.

recline. Cusp trending upward and posteriorly to marked degree with oral edge of unit oriented horizontally.

recurved. Cusp trending upward and directed posteriorly to slight degree with oral edge of unit oriented horizontally.

restoration or regeneration of parts. Process whereby lost parts of damaged conodont were rebuilt by a localized separation of lamellae along one or more growth axes.

ridge. Long, narrow, raised zone or costa.

rim. Free edge or margin of platform of platelike conodont.

rostral ridge *(Diagonalleist).* Anteroposterior trending ridge adjacent to pulp cavity on oral side of inner or outer platform (as in *Siphonodella*).

rugae. Strong transverse ridges.

secondary carina *(Nebenkamm).* Noded or denticulated structure on oral side of platelike conodont, trending from axis to free margin of platform.

secondary keel *(Nebenkiel).* Keel trending from axis to free margin of platform, on aboral side of platelike conodont.

secondary keel angle. Angle, opening posteriorly, between secondary keels.

second-order denticles *(kleine Zähnchen).* Smaller-sized of two sets of denticles on blade, bar, or limb.

sheath. Expanded basal portion of cusp and adjacent part of blade, bar, or limb; also has been used to refer to translucent portion of blade, bar, or limb surrounding cancellated basal portion of denticle; also, same as lamella.

sheath lamella. Same as lamella.

sinus. Indentation in margin of plate.

slant. Pitch of posterior side of blade of compound or platelike conodont.

spur. Short blade, bar, or limb (*see* outer anterior spur).

striae *(Anwachsstreifen).* Free edges of lamellae, evident on aboral or lateral sides of conodont as faint parallel lines, arranged about pulp cavity as common point.

subapical aboral cavity. Same as pulp cavity.

subapical cavity. Same as pulp cavity.

subapical excavation. Same as pulp cavity.

subapical navel. Same as pulp cavity.

subapical pit. Same as pulp cavity.

sub-basal projection. Same as anticusp.

subsidiary denticles. Same as second-order denticles.

subterminal fang. Same as cusp.

sulcus *(Diagonalgrube).* Trough located immediately adjacent to carina or portion of blade flanked by platform.

summit line. Trace or outline of oral side of blade, bar, limb, or axis in lateral view; also, same as oral margin.

superior cusp. Same as cusp.

superior fang. Same as cusp.

superior side. Same as oral side.

supero-anterior denticle. Large denticle near anterior end of posterior bar of *Phragmodus.*

suppressed denticles *(Keimzähnchen).* Aborted denticles that could not develop into mature structures owing to crowded condition along growing edge of conodont; commonly called germ denticles.

suppression of parts. Process whereby some growth axes, chiefly those of small denticles, were incorporated by adjoining more favorably situated growth axes so that growth of the smaller denticles ceased; this condition, which resulted through lack of room along growing edge, is a characteristic of many compound and some platelike conodonts.

suture. Free edge of last lamella accreted to conodont structure.

terminal cusp. Same as cusp.

terminal denticle. Same as cusp.

terminal fang. Same as cusp.

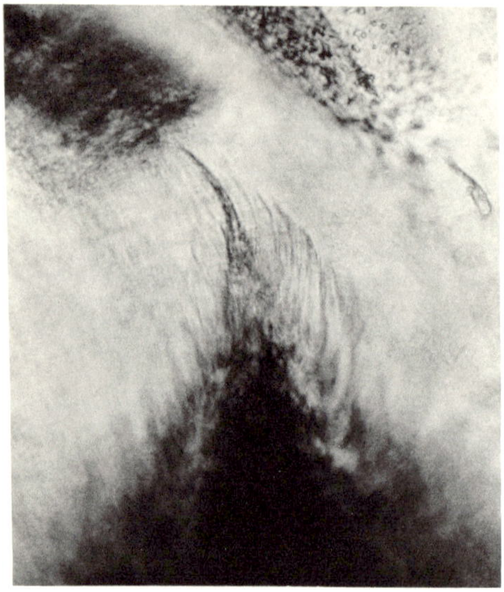

Fig. 5. Laminations about apex of pulp cavity of *Bryantodus* sp., lateral view, transmitted light, ×420 (34).

tip. Distal end of cusp or denticle.

tooth. Same as cusp or denticle; also used to refer to complete conodont specimen.

transverse ridge *(Querrippe).* Ridge on oral surface of platform of platelike conodont that trends approximately normal to axis.

transverse section. Section normal to both oral side and anteroposterior direction; also has been used to refer to any section normal to direction of active growth.

trough. Furrow on oral side of platelike conodont (*see* main trough, sulcus).

tubercle. Same as node.

unit. Complete specimen.

upper anterior denticle. Same as supero-anterior denticle.

width. Measurement at right angles to height and length of specimen.

STRUCTURAL FEATURES

The true conodonts of Pander are laminated structures, each one of them having been built up through the accretion of lamellae about the apex of the pulp cavity. In any conodont, these lamellae are open toward the aboral side of the fossil and are separated from each other along one or more growth axes. The very earliest growth stages of all conodonts were similar, for they are now represented by a series of cone-in-cone lamellae that cap the apex of the pulp cavity (Fig. 5). Later growth stages, however, were not the same in all specimens, for the lamellae recording them have many different shapes—as witness the large number of conodont species now recognized.

Based on form, conodonts have generally been divided into 3 large groups; the distacodontids, the compound bladelike and barlike conodonts, and the platelike conodonts. The distacodontids are fanglike structures, for they developed chiefly through the separation of lamellae along a single axis (Fig. 6,*1*). The compound and platelike conodonts are more variform than the distacodontids, for in their development, growth took place simultaneously in several directions and along numerous growth axes and resulted in the formation of the denticulated blades, bars, and limbs of the compound conodonts, and the platforms, expanded pulp cavities, and denticulated blades of the platelike conodonts.

Growth in the compound and platelike conodonts is herein illustrated by the longitudinal section of the blade of *Gnathodus texanus* Roundy (Fig. 7,*2*) and the horizontal section of the plate of *Siphonodella* sp. (Fig. 7,*1*). In the section of the blade of *Gnathodus texanus,* the trace of each lamella appears as a line that trends upward from the aboral side of the fossil into the growth axis of a denticle, where it is angular or chevron-shaped; any of these lines, moreover, can be traced from the basal part of a denticle into the denticle next closer to the proximal or anterior end. Unless the definition is very good, the traces of the lamellae of a conodont cannot be followed throughout their entire extent; but the observed portions of these traces invariably have a concentric or enclosing relationship (Figs. 6,*1*; 7,*1*; 8; 9).

The characteristics of the interlamellar areas of conodonts are herein illustrated by the longitudinal section of the blade of *Gnathodus texanus* Roundy (Fig. 7,*2*) and by the horizontal section of the plate of *Siphonodella* sp. (Fig. 7,*1*). These sections indicate that the interlamellar areas, in the directions of most active growth, are wide and either hollow or but slightly filled with structural material; whereas in the directions of less active growth, as well as in the later stages of growth along any growth axis, the interlamellar areas are either very narrow or, for all intents and purposes, entirely absent. The interlamellar areas are more or less tubelike in the blades, bars, and limbs of the compound conodonts and in the blades and platforms of the platelike conodonts; also, in the compound conodonts, these tubelike areas are open to the exterior along the aboral mid-line of the fossil, as well as along the aboral mid-line of any lateral blade or bar that may be

FIG. 6. Morphological features of conodonts.——*1. Oistodus lanceolatus* PANDER, longitudinal section showing lamellar mode of growth in a distacodontid conodont, ×135 (52).——*2. Subbryantodus* sp., lateral view of compound conodont showing suppression of parts, transmitted light, ×85 (34).——*3,4. Neoprioniodus* sp., lateral views of compound conodont showing effects produced by reflected *(3)*, and transmitted *(4)* light, ×85 (34).

FIG. 7. Morphological features of conodonts.——*1. Siphonodella* sp., horizontal section of plate showing lamellar structure, ×80 (34).——*2. Gnathodus texanus* ROUNDY, longitudinal section along blade showing lamellar structure, ×125 (34).

FIG. 8. *Lonchodus* sp., transverse section showing concentric relationship of lamellae, ×165 (34).

present. In the platelike conodonts, the interlamellar areas are also open to the exterior along the aboral mid-line of the fossil, as well as along the aboral mid-line of any secondary keel that may be present. In the distacodontids, as well as in the denticles and cusps of the compound and platelike conodonts, the hollow interlamellar areas are somewhat cone-shaped. Nodes, pustules, ridges, and most other features of conodonts were also formed through a localized separation of adjacent lamellae.

The lamellae of a conodont terminate along the aboral side of the fossil where their free edges appear as faint parallel lines. The position that each free edge has on the fossil with respect to all other free

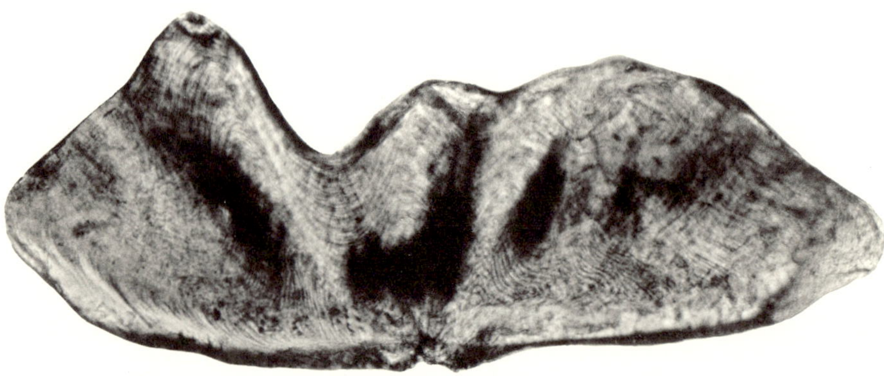

FIG. 9. *Siphonodella* sp., transverse section through rostral area showing concentric relationship of lamellae, ×150 (34).

edges determines the configuration of the aboral side of the fossil. In general terms, the aboral side can be concave, convex, or even. If the aboral side is concave or grooved, it is an indication that the free edge of any lamella extends aborally past the free edges of all previously accreted lamellae; for example, the expanded pulp cavity or cup of the Idiognathodontidae was formed in this manner. If the aboral side is convex or sharp-edged, it is an indication that the free edge of any lamella recedes orally from the free edges of all previously accreted lamellae; e.g., the inverted basal cavity described by LINDSTRÖM (44) was formed in this manner; and if the aboral side is more or less even, it is an indication that the free edge of any lamella neither extends past nor recedes from the free edges of all other lamellae (Fig. 8). Because of its mode of growth a conodont could have been attached to another structural hard part of the conodont-bearing animal only along its aboral side. This view is supported by the fact that many conodont specimens are found that still have a basal plate attached to the aboral side. This plate is variform, but the shape appears to be constant for any one species; for example, in *Palmatolepis perlobata* (Fig. 10) it is thin and solid, whereas in the distacodontid of Fig. 11, *1,2*, it is not only fanglike and hollow but also slit along the anterior side. Published views on the nature of the basal plate are given below:

STEWART & SWEET (71) who worked on some Middle Devonian conodonts from Ohio stated that in their samples the physical nature of the basal plate

> . . . is strikingly different from that of the conodont itself. In general, this substance is much softer, opalescent to waxy in luster, and apparently rather porous, for it has been conspicuously stained throughout by secondary iron oxide in most of our specimens.
>
> Professor Duncan McConnell, of the Department of Mineralogy of the Ohio State University, very kindly made petrographic and x-ray diffraction studies of this basal material for us. He reports (letter dated March 11, 1952) that the x-ray powder pattern is ". . . qualitatively identical with the pattern produced by collophane regardless of its source of origin." His petrographic examination further indicated that the material has a "lamellar structural arrangement and appears to be essentially isotropic, which is not characteristic of the conodont material itself." Both of these statements suggest to us that the animal, of which the conodonts were originally a part, had a skeletal (or exoskeletal) framework of lamellar collophane. However, the possibility remains that the basal substance of the conodonts was originally some type of tissue (perhaps cartilagenous) which has been converted to collophane during the process of fossilization (p. 262).

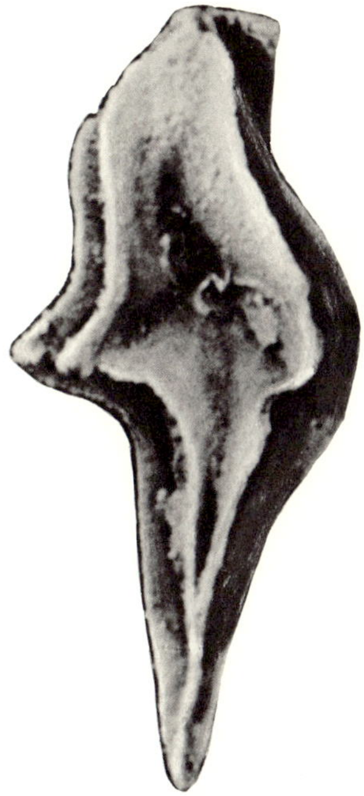

FIG. 10. Basal plate of *Palmatolepis perlobata* ULRICH & BASSLER, aboral side showing basal plate attached to underside of specimen, ×30 (Hass, n).

LINDSTRÖM (44, p. 537; 45), who studied some Ordovician conodonts from Sweden, found two kinds of basal plates in his material. One kind seemed to be composed of the same mineral matter as the conodont to which it is attached. He found it to be dense, homogeneous, soluble in hydrochloric acid, and definitely not bone. The other kind of basal plate seemed to be chitinous. This plate is not soluble in hydrochloric acid; it may be in the form of a cone with the conodont attached to its tip.

RHODES (59, p. 430) stated that the chemical composition of the basal plate "is essentially similar to that of the conodont to

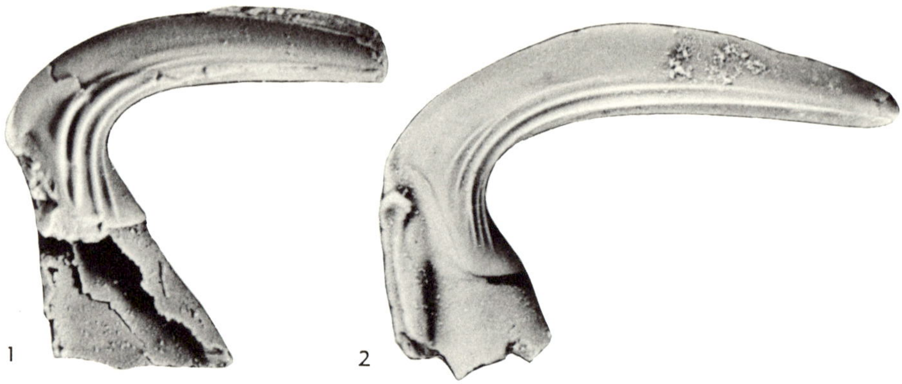

Fig. 11. Characteristics of basal plate of distacodontid conodont, *Scolopodus* sp., ×20 (Hass, n).——*1*. Lateral view with basal plate partly broken away.——*2*. Lateral view showing slit along anterior side of basal plate.

which it is attached." His conclusion was based on the X-ray diffraction work of Roy Phillips, who investigated some Silurian conodonts to each one of which the basal plate was still attached.

In 1957 Gross (33) reported on the basal plates of some Upper Silurian and Upper Devonian conodonts. Like McConnell (*in* Stewart & Sweet, 71, p. 262) Gross (Fig. 12) noted that the basal plate is laminated. He also observed that in some of his Upper Silurian conodonts, the basal plate is composed of two parts: a basal cone *(Basistrichter)* and a cone filling *(Trichterfüllung)*. The basal cone is somewhat transparent and consists of a series of thin conelike lamellae, set one within another. In many specimens, this basal cone extends far below the conodont to which it is attached; that is, only the tip of the basal cone is fitted into the pulp cavity. Also, this basal cone is open toward the underside of the fossil and may be deeply excavated (Fig. 11,*1*,*2*). The opening or excavation so formed is called the cone cavity *(Trichtergrube)* in order to distinguish it from the pulp cavity. An opaque to translucent, red-brown to dark-brown, coarsely laminated material may occupy the cone cavity. This material—the cone filling—is easily freed from the inner surface of the basal cone. In the case of his Upper Devonian specimens, Gross noted that the basal plate is indistinctly laminated, opaque, and dark brown. Moreover, it is homogeneous and could not be differentiated into a basal cone and a cone filling.

It appears that the basal plate increased in size through the accretion of lamellae to its undersurface. This surface is somewhat uneven and formerly may have merged into a softer non-preservable tissue of the conodont-bearing animal. Presumably the growth of a basal plate took place simultaneously with the development of the conodont to which it was attached.

The conodont-bearing animal was able to restore any part of a conodont structure that had been broken away and lost. The restoration of lost parts was accomplished through a localized separation of subsequently accreted lamellae along one or more growth axes and generally resulted in an atypical specimen, as the growth axes in the restored parts are commonly out of alignment with the stumps of the original growth axes (Fig. 13). Also, the conodont-bearing animal may have been able to re-fuse or knit the fractured parts of a conodont structure (Rhodes, 59, p. 431). The lost parts of a conodont structure were commonly restored and, therefore, it is the writer's opinion that they could not have functioned as teeth or other ingestive aids. This view is held because the lost parts could have been restored only as long as the conodont structure was covered by the tissues that secreted the lamellae; that is, at a time before the structure could have commenced to function as an ingestive aid. If conodonts actually were ingestive aids, the only real change they could have undergone after erupting from the jaw would have been to wear away through use, and the writer has found no conodont whose present condition can be so interpreted. Moreover, although the surface features of some large specimens may be weak, there is no criterion whereby a mature conodont can be distinguished from a large immature one, and it is therefore impossible to determine which conodonts could have functioned as teeth and which could not, at the time death overtook the animal that bore them.

RHODES (59, p. 440, 441), who is strongly of the opinion that conodonts are the ingestive aids of annelids, states that some specimens appear to show evidences of attrition (ELLISON, 24, pl. 22, fig. 17; BRANSON & MEHL, 7, p. 5; RHODES, 58, pl. 23, fig. 259) and that most conodonts exhibit some evidences of fracturing. He admits that this apparent wear could have resulted subsequent to the death of the conodont-bearing animal through "post-depositional physical processes"; but claims that, in any case, wear is probably not a deciding factor in determining the function of conodonts. He points out that conodonts could have been graspers of food instead of masticators. Thus their function would have been similar to that of scolecodonts, which seldom show evidences of attrition, despite the fact that they are composed of chitin—a material much softer than the apatite of conodonts.

As for the regeneration of the lost parts

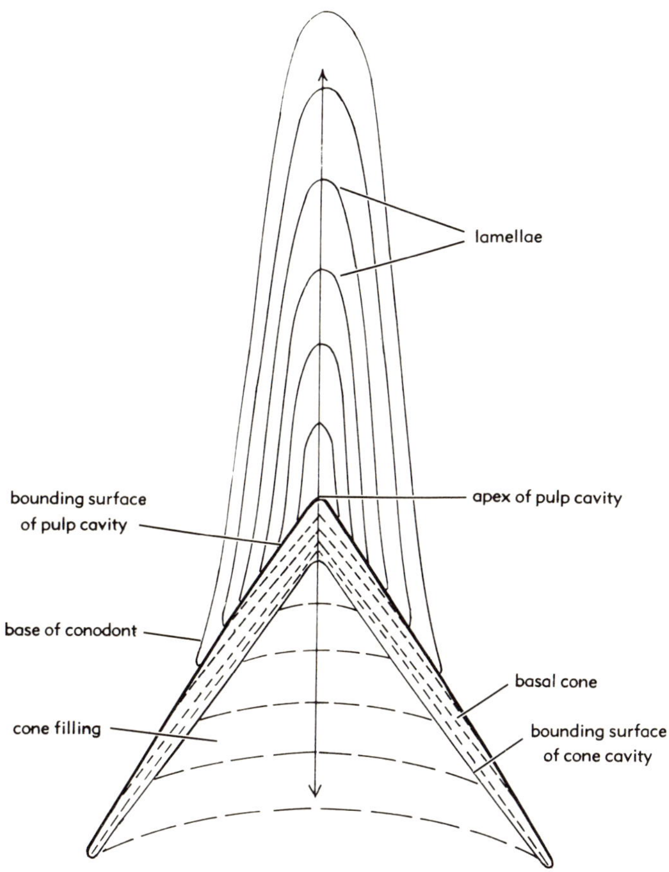

FIG. 12. Diagrammatic section showing morphological features and direction of growth of conodont and its basal plate (arrows) (33).

of a conodont, which the present writer (HASS, 34) regards as evidence for indicating that conodonts were supports for a covering tissue, RHODES (56, pl. 26, fig. 5) states that, with one possible exception, all repairs to a conodont structure appear to have involved the regeneration of missing parts rather than the refusion or knitting of fractured parts. In his opinion, regeneration would most likely have occurred after the structure had started to function as an ingestive aid, for under that condition, the fractured parts could easily have been permanently separated from each other. He suggests "that the secondary canals recorded by BECKMANN (1), which extend to the surface of the conodont, [might have had a] function in the regeneration of broken parts." BECKMANN'S (1) ideas are given on a later page of this paper.

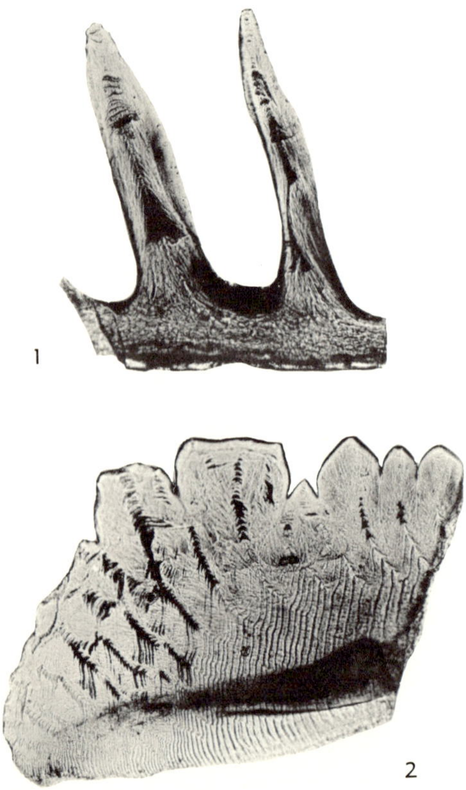

FIG. 13. Morphological features of conodonts.——*1. Lonchodus* sp., lateral view showing restoration of parts in compound conodont, transmitted light, ×70 (34).——*2. Elictognathus* sp., lateral view showing aberrant effects caused by suppression and restoration of parts in a fragment of a compound conodont, transmitted light, ×150 (34).

RHODES (59, p. 242) wrote: "The nature of the basal cavity in conodonts suggests that they were attached by the 'aboral' surface, rather than being surrounded by tissue (as an internal support would usually be). It seems equally unlikely that they functioned as supports on the body of some animal, since the plane of basal attachment is at right angles to the plane of maximum height of the conodont."

Not all growth axes of a compound or platelike conodont persisted throughout ontogeny if room was lacking along the growing edge. Under those conditions, some of the growth axes of a specimen were suppressed in favor of others. That is, during normal growth, the larger-sized denticles incorporated some of the adjoining smaller-sized denticles into their own structure. Evidence of suppression is commonly seen in the vicinity of the pulp cavity where the main cusp or apical denticle is located (see *Subbryantodus* sp., Fig. 6,*2*), but other parts of a conodont structure are similarly affected (see *Elictognathus* sp., Fig. 13,*2*). It thus appears that through a restoration of parts on the one hand and a suppression of parts on the other, the conodont-bearing animal attempted to maintain a maximum number of strong effective growth axes along the growing edge of the conodont structure.

As a general rule each lamella of a conodont is more or less transparent and has a slight brownish or grayish tint. Hence, most specimens range either between light tan and dark brown or between dark gray and grayish-black. But the appearance of some specimens is due to other causes. For example, some of the hollow interlamellar areas of a compound or platelike conodont are open along the aboral side of the fossil, making it possible for foreign substances to stain or react chemically with the interior as well as the exterior of a specimen. Specimens so affected commonly tend to be friable and light gray. In some conodonts, the lamellar structure has been locally obscured by numerous small spherical or tubular voids, which are called cells. Because of reflection from the surfaces of the cells, light rays cannot pass through a cellular area, and, as a result, the same specimen appears differently in transmitted light and reflected light (see *Neoprioniodus* sp., Fig. 6,*3,4*; Fig. 14). In transmitted light, a cellular area is brownish-gray or dark gray, whereas in reflected light the same area appears very light gray or yellowish-gray. In reflected light, denticles with a cellular structure appear peglike, and this feature has been interpreted by some investigators as proving that the denticles of a conodont are inserted into a blade or bar in a manner similar to that in which authentic teeth are set in a jaw. Actually the peglike appearance of a denticle is nothing more than an effect produced by reflected light on the cellular structure of a conodont.

The lamellar structure of some conodonts may be obscured by a series of alternate light and dark bands that trend outward from the growth axis of a cusp or denticle, or from the mid-line of a specimen (*see* *Cavusgnathus* sp., Fig. 15,*1,2*). These light and dark bands, however, are secondary features with respect to the lamellae, for at high magnification the lamellae can be seen passing through the bands. PANDER observed these light and dark bands in some of the specimens he studied and called such specimens "obliquely layered" conodonts. It was his opinion that the dark bands are composed of cells, whereas the light bands are composed of a homogeneous transparent substance. BECKMANN (1) believed that the dark bands are dentine tubules (*see* Fig. 16,*1*).

PANDER (52) published a rather detailed account of the morphology of conodonts. The whereabouts of PANDER's type material is not known (FAY, 27, p. 36), though it appears that at one time it was at Yale University (GRINNELL, 31, p. 229). The following is a translation of pertinent parts of PANDER's monograph (pages 5-8, 18, 19). This translation was made for the writer (34) by Mr. AYVAZOGLOU, formerly of the United States Geological Survey:

Conodonts are "minute fossils that closely resemble fish teeth in external form, [being] lustrous, elongated, sharply pointed upward or toward one of the extremities, gradually or rapidly expanded downward, more or less bent, and usually provided with sharp edges, one anteriorly, the other posteriorly. The lateral faces are very differently shaped, being symmetrical or asymmetrical, plain or lined along their length, and often having a carina projecting laterally from their smooth surface.

As in all teeth, a point and a base can be distinguished. The point is solid and the base hollow, forming the *cavitas pulpae*. The latter is differently shaped in the various types of conodonts. Usually the *cavitas pulpae* is rounded off in its upper part, often it becomes narrower, and sometimes it terminates in a blind point . . . . In most cases the upper solid end of the tooth is merely the gradual sharpening of the hollow base, but frequently the base is separated externally from the point by a constriction on the outer and inner borders or on the inner border only. The base, in some teeth, is extended at the lower borders of the tooth points in the same direction that the tooth is inclined, without causing much change in the outline of the latter [PANDER cites *Oistodus* as an example]; . . . in other cases this extension increased along a horizontal line [PANDER cites *Prioniodus*]; . . . or in an inclined or vertical line [PANDER cites *Cordy-*

FIG. 14. Morphological features of conodonts, *Neoprioniodus* sp., lateral view showing cellular structure of denticles, transmitted light, ×220 (34).

lodus] . . . . From this lateral elongation of the hollow base short and successive continuations are formed, which rise vertically, . . . obliquely, . . . or horizontally . . . . The many short additional denticles originated from these extensions. The compound teeth were formed in this manner. There can be no doubt that the hollow base was occupied by a pulp, and that a simple pulp formed a simple tooth and a compound pulp, a compound tooth.

The seeming slight change that conodonts have undergone is remarkable, for apparently their luster, color, and probably also their chemical composition are original, so that one might be tempted to ascribe them to the still living fishes. This complete preservation is also surprising, because these teeth can be traced from the oldest formations, the black slates, through all of the Lower Silurian [Ordovician] formations up into the Devonian limestones, that is, conodonts are found in beds that have entirely different chemical compositions and that certainly passed through manifold chemical changes. It is all the more striking that this substance appears to consist almost entirely of calcium carbonate, for, upon solution in acids, carbon dioxide is released, and the oxalates produce a very considerable precipitate.

The different substances of conodonts can be divided into three classes. These classes are based upon the external appearance and the more or less conformable internal structure: (1) snow-white, opaque, with translucent borders; . . . (2) yellow, entirely translucent, and appearing hornlike; . . . and finally (3) white-reddish, compact and entirely opaque . . . . We must admit that the white ones mentioned by us under number (1) were, during the early stages of their ontogeny, yellowish and transparent and became snow-white and opaque only at maturity. Therefore, taking the substance into consideration, we might conclude that those placed in the

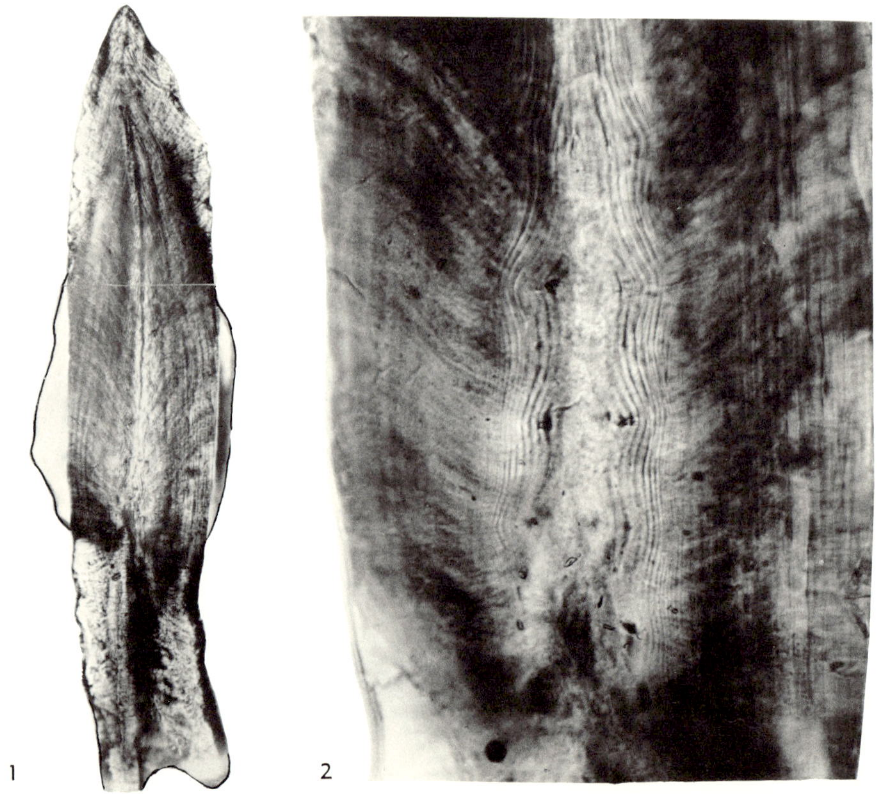

Fig. 15. Morphological feature of conodonts, *Cavusgnathus* sp.——*1*. Horizontal section of platelike conodont with expanded pulp cavity showing lamellar structure and alternating light and dark transverse bands, transmitted light, ×80 (34).——*2*. Enlargement of part of same specimen, transmitted light, ×300 (Hass, n).

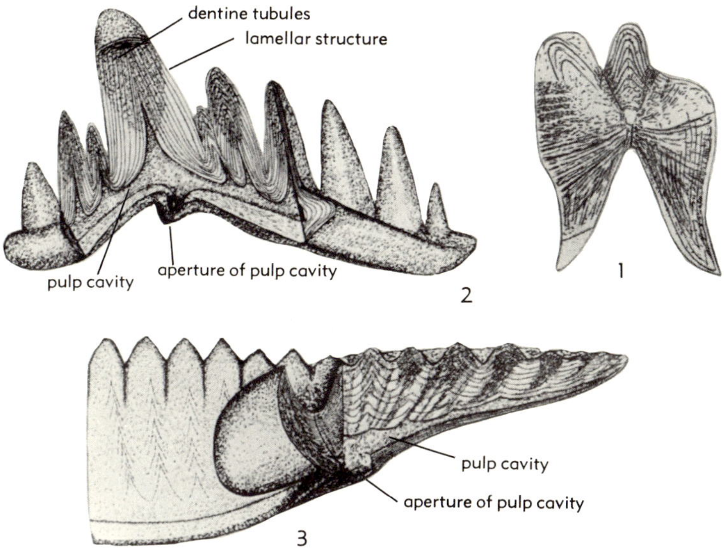

FIG. 16. Morphological features of conodonts according to BECKMANN (1). *1. Icriodus symmetricus* BRANSON & MEHL, transverse section showing lamellar mode of growth and darker transverse bands interpreted as dentine tubules, ×120.——*2. Bryantodus delicatus* BRANSON & MEHL, reconstruction with part of specimen cut away, ×90.——*3. Polygnathus pennata* HINDE, reconstruction with part of specimen cut away, ×90.

second category are the young of the first. This viewpoint, however, must be discarded, because not only are the mature forms of the second category the most numerous of all of the teeth found, but also they constantly differ from those in the first category by having a long, hollow base. The white-reddish, completely dull opaque forms are, as we shall see later, different in every respect from those in the first two categories.

We shall turn now to the structure of these teeth, which structure departs from any that up till now had been considered indicative of ichthyological character, and which principally supplies the reason for not considering these remains to be fish teeth.

We have seen that so far as the base is concerned, it is hollow and was occupied by a pulp. From the surface of this pulp, which persisted for a long time, the substance of the wall of the pulp and of the tooth point was formed in such a way that one lamella after another was deposited. A lamella formed subsequently was placed against the inner wall of that which was formed previously. From these cones, which are differently shaped, which lie one over another, and from which the name conodonts originated, one can obtain the clearest picture of the surface of the pulp during the formation of every single lamella. If we compare the shape of the pulp cavity in a conodont with that in a mature tooth, we see great differences insofar that the pulp cavity of a conodont does not extend very far up into the point. Thus it does not resemble the external form of the conodont as do the cavities in the teeth of most living fishes.

The successive formation of the lamellae can be clearly seen in the yellowish, transparent, flexible, hornlike teeth, especially when both surfaces of these teeth are slightly polished. It is much more difficult to observe the lamellar formation in the white, opaque teeth, which break easily owing to their brittleness. Since they are opaque, a much higher polish is required, and therefore they must be handled very carefully. A magnification of 100 diameters is sufficient to distinguish the concentric lamellae in both types; . . . and it seems that even at 300 diameters the yellowish types do not show anything more than homogeneous cones laid one upon another. . . . In the white types, the lamellae, or, more exactly, the spaces that lie between the cones, which alone we really see, are, instead of forming continuous interspaces, dissolved into small cells or bubbles, which are arranged regularly side by side along the length of the tooth.

In addition to the small cells discussed above, we see in the white teeth some other corresponding cells or cavities which are oval, larger, and distributed without any regular order. In general, they have their long axes parallel with the long axis of the tooth.

In the compound teeth . . . we could observe the cellular formation of the lamellae only in the large cusp and in the apical denticle, whereas in the denticles we observed only cells which, apparently, are distributed close to one another or above one another. . . . In the structure of the white-reddish, compact, opaque, mostly compound teeth, we have found a structure that was not clear to us, and therefore we have described it as we have seen it. It seems that their genesis from the surfaces of the pulp follows some other laws, which are still of a puzzling nature to us, namely: If we rub off both side surfaces, we find in the central plane, . . . at low magnifications, only alternating light and dark cross-striped areas, which are differently colored and pass from one rim of the tooth to the other. If a highly polished surface is studied at 300 magnifications, the dark stripes appear to be composed of small cells or cavities, whereas the light stripes represent the homogeneous, transparent basic substance.

Although the base is hollow and in most cases has smooth walls, at times one finds in this area little dark grains or cavities of various sizes, with a dark border . . . .

Among the many thousands of teeth that we studied with the aid of the magnifying glass and microscope we know of only three in which a disarrangement evidently occurred during their growth . . . . In all three the original lamellar formation began regularly at the point but was interrupted sooner or later . . . . This first stage, which did not advance farther, remained clear and transparent, while the structure of the lower part of the tooth which was formed subsequently progressed together with the base and became entirely cellular.

On pages 18 and 19 of his paper, PANDER stated that on the basis of internal structure he was able to

. . . establish two main classes of conodonts. The first class, the *lamellar teeth,* includes those teeth that consist of cones arranged in layers, one above the other, and nearly parallel to the external periphery of the tooth. This class has many representatives in the older periods. The second class, the *obliquely layered* teeth, includes those *teeth* in which the lamellar structure cannot be seen. These teeth are denser and apparently consist of alternating cellular and noncellular layers, which form the structure of the tooth by lying obliquely one over the other.

Simple and compound teeth are found in both classes. The simple teeth can hardly be separated from the class of lamellar teeth, for relatively few of them are obliquely layered . . . .

There are great numbers of simple and compound teeth in the first class. So far, the simple teeth have been found only in the lowest Silurian [Ordovician] formations; the compound teeth, on the other hand, pass from these beds through the Upper Silurian [Silurian] and Devonian formations into the Mountain [Lower Carboniferous] limestone.

There is not sufficient reason for us to erect a classification of these teeth that is based upon external form, that is, whether the teeth are straight or bent, crooked or inclined. Perhaps this feature can be used as an aid in determining species, but even here we had to be very cautious.

We found it more important to take into consideration the outlines we obtained by sectioning teeth through their middle part; for it can hardly be expected that smooth, ribbed, keeled, and truncated teeth could all exist in the mouth of the same animal.

BECKMANN (1) studied some well-preserved compound and platelike conodonts from the Upper Devonian of Germany. According to him, the first lamella of a conodont was secreted by a pulp that occupied the pulp cavity. This cavity in the compound and platelike conodonts was considered to be slitlike and present along the entire mid-line of the unit (*see* Fig. 16,*2,3*). The primary or first deposited lamella, as well as all others, was believed to have been broken through by pores. BECKMANN also believed that until it was fully formed, the conodont structure was covered by a meshlike tissue. This tissue was joined to the pulp through a system of canals and provided the medium whereby secretions were brought from the pulp to the outside surface of the growing conodont structure, where the fluids solidified to form the lamellae. In that way a second lamella was secreted on the outer surface of the primary lamella, a third lamella on the outer surface of the second lamella, and so forth. BECKMANN stated that the lamellae are thickest where the canals are most abundant, and that during ontogeny, the pulp cavity was gradually closed off from the aboral side of the conodont, so that in a mature compound or platelike unit, its opening is restricted to a small aperture. Because of the nature of the pulp cavity, the canal system, and the lamellae, BECKMANN assumed that the last-deposited lamella of a conodont is not much younger than the primary or first-deposited lamella. To him, these features proved that conodonts have a dentine structure.

BRANSON & MEHL (7) described the fibrous structure of some toothlike fossils from the Harding Sandstone of Colorado. According to them, these fossils, which are commonly found crushed and frayed instead of cleanly broken, are conodonts whose structure is composed of bundles of fibers instead of lamellae. "Fibrous" conodonts, however, appear to be specimens whose original lamellar structure has been obscured through alteration, though RHODES & WINGARD (60) suggested that such specimens, whose chemical composition approximates calcium metaphosphate, $Ca(PO_3)_2$, represent a group of primitive vertebrates distinct from the lamellar conodonts. In December, 1949, the writer examined BRANSON & MEHL's type specimens from the Harding Sandstone and found remnants of lamellae in some of the "fibrous" specimens illustrated by them. These specimens, with citation of their published figures, are listed below:

Supposed Fibrous Conodonts Figured by Branson & Mehl

Stereoconus robustus BRANSON & MEHL (pl. 1, fig. 28, 29), cone-in-cone laminations at tip of cusp.
Neocoleodus spicatus BRANSON & MEHL (pl. 1, fig. 37), free edges of lamellae of this specimen evident along aboral part of bar.
Chirognathus varians BRANSON & MEHL (pl. 2, fig. 6), cone-in-cone laminations in smallest denticle.
Chirognathus varians BRANSON & MEHL (pl. 2, fig. 7), cone-in-cone laminations in distal part of main cusp.
Chirognathus reversa BRANSON & MEHL (pl. 2, fig. 25), cone-in-cone laminations in largest denticle.
Chirognathus tridens BRANSON & MEHL (pl. 2, fig. 27), cone-in-cone laminations in distal part of main cusp.

Specimens other than those listed above have been identified with BRANSON & MEHL's "fibrous" conodonts. They belong to several genera and species, and the present writer believes that formerly they also possessed a lamellar structure; if, however, they are not laminated, they cannot be identified with the true conodonts of PANDER. It is the writer's opinion, therefore, that BRANSON & MEHL's (13) suborder Neurodontiformes (conodonts with a fibrous rather than a lamellar structure) has no place in conodont taxonomy, and that BRANSON & MEHL's (13) suborder Conodontiformes (conodonts with a lamellar structure) is unnecessary, as it, like the order Conodontophorida, includes all true conodonts.

PROPERTIES OF CONODONTS

Some investigators are of the opinion that the chemical properties of conodonts tend to align these fossils with the vertebrates despite the fact that some invertebrates possess a somewhat similar chemical composition. Many investigators have mentioned that conodonts consist chiefly of calcium phosphate, and some have reported that little or no organic matter is present. STAUFFER & PLUMMER (70, p. 21) have stated that if the present composition of conodonts is "an indication of the original composition [then] . . . they are far removed from the chitinous or horny teeth of the Arthropoda, the Chaetopoda, or the Mollusca." SCOTT (64, p. 450), on the other hand, held that the difference in the chemical composition of conodonts and scolecodonts had no greater taxonomic value than that of aiding "in placing various forms in different orders or families within a phylum."

ELLISON's paper *The composition of conodonts* (25) is a comprehensive treatment of that subject. His paper gives information obtained through chemical, mineralogical, petrographic, X-ray, and spectrographic means. However, it should be pointed out that according to ROY PHILLIPS (*in* RHODES, 59, p. 429) some of ELLISON's X-ray data are incorrectly given. Some of the data ELLISON recorded in his paper are listed below:

TABLE 1. *Properties of Conodonts*

PHYSICAL

Color: dark brown, light tan, clear amber.
Hardness:3 to 5 on Mohs scale.
Specific gravity: 2.84 to 3:10.
Fusibility: fuse with difficulty.
Indices of refraction: 1.595 to 1.612.
Birefringence: nil to weak, 0.000 to 0.003.
Crystallinity: composed of minute crystals.

CHEMICAL, QUANTITATIVE (average of two samples)

CaO	48.05 percent
P_2O_5	34.96
Insoluble	3.96
Remainder[1]	13.03
Total	100.00

[1] Probably CO_2, H_2O, F_2, Fe_2O_3, and organic and other matter.

CHEMICAL, QUALITATIVE

Water: droplets in heated closed tube.

Organic: Becomes dark gray in heated closed tube.

Soluble in: hydrochloric, sulphuric, and nitric acids.

Insoluble in: acetic and citric acids.

Positive test for: iron and fluorine.

Negative tests for: sulphur, chlorine, and manganese.

SPECTROGRAPHIC

Conodonts consist chiefly of calcium phosphate with iron, magnesium, sodium, and fluorine present as traces.

X-RAY

According to ELLISON (25, p. 138), the "diffraction pattern data on conodonts are very close to those . . . . for fluorapatite, chlorapatite, dahllite, and bone." DUNCAN MCCONNELL, of The Ohio State University, who, while at the University of Minnesota, did X-ray work on conodonts for STAUFFER (69), stated in a letter to ELLISON that "some conodonts are probably dahllite, others francolite and some probably lewistonite or dehrnite."

ELLISON (25, p. 139) concluded that conodonts are composed of the same mineral matter as that present in fossil and recent bones and teeth: "this mineral matter is similar to the minerals of the apatite group."

HASS & LINDBERG (39) presented corroborative evidence on the composition of conodonts and stated that conodonts are composed of a mineral of the apatite group that belongs to the dahllite-francolite isomorphous series. Inasmuch as the mineral matter of conodonts contains about 1 percent fluorine, it was identified as fluorine-bearing dahllite. HASS & LINDBERG also stated that each lamella of a conodont consists of innumerable dahllite crystals. Commenting on the orientation of these crystal units, they (p. 503, 504) stated that

> in dahllite, a uniaxial mineral, the optic and the crystallographic axes coincide; and as the dahllite crystals in a conodont are in extinction only when the growth axis of which they are a part is aligned with the vibration plane of the analyzer or the polarizer, it follows that the crystal units in each lamella of a conodont are oriented in conformity with the direction in which the conodont grew. The wave of extinction that moves through a conodont as the stage of the microscope is rotated is suggested by the four figures of the platforms of *Siphonodella duplicata* (BRANSON & MEHL). Figures 1 and 3 [*see* Fig. 17,*1*,*3*], are similar, for they record the two extinction positions of the same group of crystals; figures 2 and 4 [*see* Fig. 17,*2*,*4*] resemble each other for the same reason. In all four figures the darkened area along the carina is caused partly by extinction and partly by an excessive absorption of light.
>
> The retardation of light by a conodont specimen is slight. Most specimens appear gray or yellow between crossed nicols and only a few show a spot of first-order red. The retarding effect of the crystal units of a conodont on the gypsum plate is such that the predominant color resulting from subtraction is first-order yellow and the colors resulting from addition are first-order purple and second-order blue and green. Optically, dahllite is negative, and in conodonts the feature of subtraction results only if a direction of ontogenetic growth is aligned with the slow vibration plane of the gypsum plate. Hence, it follows that the *c*-axis of each dahllite crystal is invariably oriented in the direction in which the main ontogenetic growth occurred at the place in the lamella where the crystal is located. The color seen at any spot on a conodont is produced by the mass effect of a number of superimposed crystals. These crystals are not in exact alignment, but, as their birefringence is low, the resultant color approximates that which would be seen if the crystal units were actually parallel.

In 1954, ROY PHILLIPS (*in* RHODES, 59, p. 428-430) reported on chemical analyses and X-ray studies of conodonts. He believed that conodonts can be expected to show a variable composition within the hydroxy-carbonate—fluor apatite range of minerals, that, other things being equal, the fluorine content of conodonts should increase throughout geologic time, and that one is not justified "to apply the names of individual apatite species, such as dahlite and francolite, to the mineral content of conodonts," because the mineralogical nomenclature of the apatite group is in need of drastic revision.

Although conodonts have world-wide distribution and have been known to paleontologists for the last hundred years, they were not studied extensively until ULRICH & BASSLER'S (75) classification was pub-

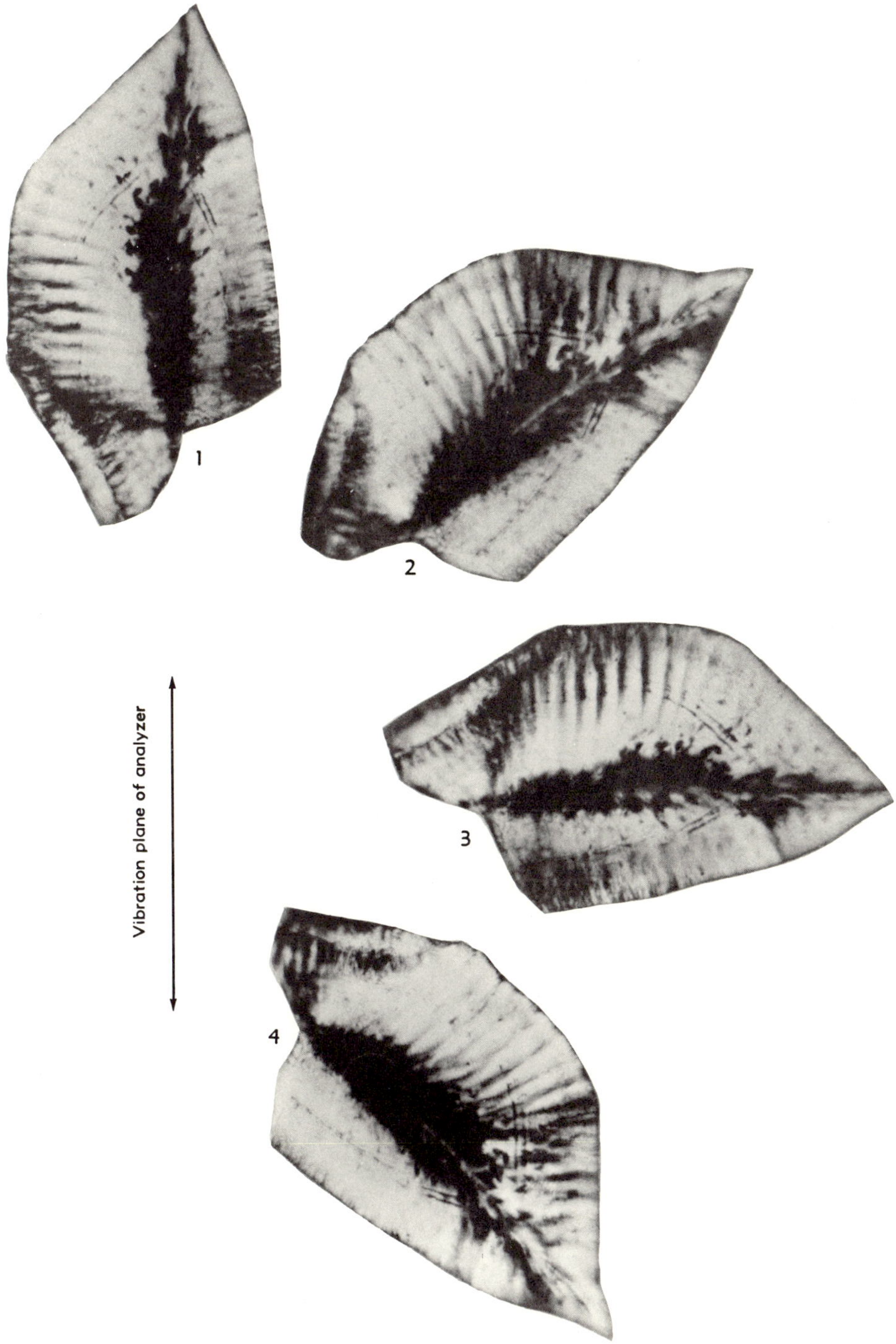

FIG. 17. Extinction of crystal units in platforms of *Siphonodella duplicata* (BRANSON & MEHL).——*1-4*. Extinction at 45° intervals, ×55 (39).

lished in 1926. Important advances have been made since that date but before this group of fossils can be of the utmost usefulness to science, problems concerned with the affinity of conodonts, the nomenclature of conodonts, and mixed conodont faunas must be resolved; also, much more information must be obtained pertaining to the stratigraphic ranges of discrete genera and species.

PROBLEMS IN CONODONT STUDIES

AFFINITY OF CONODONTS

Conodonts have few significant characteristics and presumably are but one of many anatomic parts that comprised the conodont-bearing animal. It is not surprising, therefore, that many conflicting views have been held concerning the zoological position of this animal and the function conodonts performed. Many published opinions on the affinity of conodonts are briefly stated and unsupported by data; only a few are based on detailed morphological studies. The many views that have been held on this subject are summarized below. Other summaries are given by STAUFFER & PLUMMER (70), SCOTT (64), and RHODES (59).

POSTULATE THAT CONODONTS BELONG TO MOLLUSCA

It has been suggested that conodonts are the spines, teeth, or hooklets of Mollusca, such as the Gastropoda and the Cephalopoda. LOOMIS (46) is a recent proponent of this view. To him (p. 663) it seemed "impossible that teeth so close in size, shape, and composition as those of the conodonts and the gastropods can belong to anything but the same group of organisms." He pointed out that gastropod denticles range from a quarter of a millimeter to a millimeter in length, are composed of horny or chitinous material, and are firmly inserted in a ribbon of like composition. He also stated that hundreds to several tens of thousands of denticles consisting of several structural types may be present on the median and lateral rows of the radula of a modern gastropod. Although the conodonts LOOMIS illustrated (*see* Fig. 18) are similar in size and shape to the gastropod denticles he figured, many other conodonts are quite dissimilar, both in size and shape. Moreover, the true conodonts of PANDER are not horny or chitinous as LOOMIS claimed but rather, are composed chiefly of calcium phosphate. According to PILSBRY (53), conodonts most closely resemble rachiglossate teeth, and, so far as he knew, gastropods having such teeth are post-Paleozoic. PILSBRY stated that "some conodonts resemble certain cephalopod teeth" but gave no supporting data.

POSTULATE THAT CONODONTS BELONG TO ANNELIDA

Some of the earlier investigators considered conodonts to be the hooklets or denticles of worms. ZITTEL & ROHON (80), for example, concluded that conodonts are the teeth of annelids or *Gephyrea*, but, as stated by BECKMANN (1), their investigation was limited in its scope and cannot be accepted as correctly interpreting the nature of conodonts.

Recent proponents of an annelid affinity are SCOTT (64, 65), DUBOIS (22), and RHODES (56-59). They stated that natural conodont assemblages have been found which seem to be most closely related to the jaw apparatus of the annelids. According to them, the number (14 to 22) and kind of paired components in an assemblage are constant and can be easily distinguished from random groupings, including those present in excreta. Presumably, because conodonts are composed chiefly of calcium phosphate, whereas scolecodonts are chitinous, SCOTT (64, p. 455) wrote that probably "one family of Paleozoic annelids possessed a jaw apparatus composed of teeth which we call conodonts; whereas, a second family possessed teeth known as scolecodonts." However, he (65, p. 298) stated later that "insofar as maneuverability is concerned, it [a conodont assemblage] could operate with equal ease either as the jaw apparatus of an annelid or as gill rakers of a fish." SCOTT's specimens came from some black shales in Montana. He (65, p. 295) designated the component parts of an assemblage "by com-

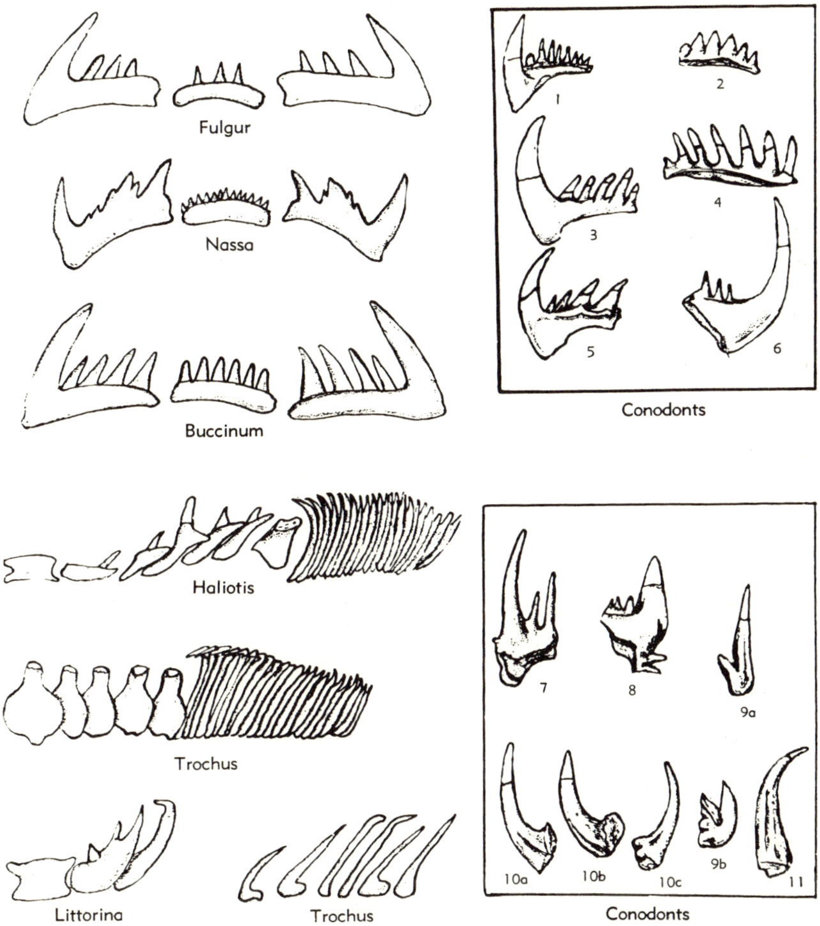

FIG. 18. Comparison of teeth on gastropod radulae with conodonts. Teeth on left belong to indicated genera of modern gastropods. Those on right are conodonts of the following genera: *1, Cyrtoniodus; 2, Neocoleodus; 3, Subcordylodus; 4, Lonchodus?; 5, Subcordylodus?; 6, Neoprioniodus; 7, Polycaulodus; 8, Pteroconus; 9, 10, Oistodus; 11, Paltodus* (46).

mon nouns derived from the names of the form genus to which similar parts have heretofore been referred," and described two new genera: *Lochriea,* which consists of hindeodells, prioniodells, neoprioniods, and spathognaths; and *Lewistownella,* which consists of cavusgnaths, hindeodells, neoprioniods, and subbryantods.

SCOTT (65, p. 297, 298) published a schematic representation of "the probable arrangement and relative position of conodonts in the genus *Lochriea.*" His drawing is reproduced as Figure 42,*2*; SCOTT had this to say about it: "There is not much question concerning the disposition of the hindeodells, but the position of the prioniods [neoprioniods], prioniodells, and spathognaths is a conjectural interpretation, though it is believed that their orientation is approximately correct. They probably operated as rights and lefts, or possibly they were placed in a circular position around an esophageal tract. All of the denticles were set on the soft parts of the animal . . . . but could be moved with considerable ease."

"Such an apparatus would not only form an excellent screen to prevent undesirable objects from entering, but would also present a formidable barrier for the escape of desirable food once it had passed beyond the battery of teeth."

DuBois (22) studied some conodonts from a Pennsylvanian black shale of Illinois. It was his opinion that most conodonts appear to be pharyngeal and buccal structures, and he suggested that they probably are parts of annelids. His conodonts were associated with numerous fossils assumed to be of probable annelid origin—such as trails, segmented impressions, and "problematic 'parapodia'." Most of his conodonts were believed to be parts of an assemblage that consisted of a pair of polygnathids (identified as belonging to *Streptognathodus* and considered to be the anteriormost unit of the assemblage), a pair of bryantodids (identified as belonging to *Ozarkodina*), and several pairs of hindeodellids (identified as belonging to *Hindeodella* and considered to be the posteriormost unit of the assemblage). He stated (p. 158) that

> If it is assumed that conodonts are associated with both the problematic parapodia and the worm trails [mentioned above], it is possible to erect a picture which may represent the appearance in life of the animal which bore the teeth. The adult was an elongate worm, seldom more than three millimeters in width, with a length of at least three centimeters, and probably five or more. It probably possessed a ventral nerve cord and resembled modern annelids in many other internal structures. Metamerism may have been indicated by the serial development of the jaws, in which each type of tooth was restricted to a separate metamere, and by the presence of regularly arranged parapodia.
>
> The anterior part of the digestive tract was divided into buccal and pharyngeal regions. The buccal cavity had a single (but perhaps more in some cases) polygnathid on either side, with the blade directed anteriorly. These jaws were probably covered with hypodermis and cuticle so that only the actual cusps were visible. Protractor and retractor muscles supported and moved the teeth. Anterior to the polygnathids there may have been one or two teeth of the symmetrical type illustrated by Scott's figure 3c (1935). [Scott's paper (64) appears to have been published in December 1934. The conodont referred to above was identified as *Prioniodella(?)*.] The pharyngeal region supported the hindeodellids which probably functioned in the final straining or comminution of the food.

Denham (20) wrote that conodonts appear to be grasping or holding organs and asked if conodonts might not be the copulatory structures of one or more groups of extinct worms. He observed that some "living worms, including the Nematoda and several groups of the Turbellaria," have chitinous structures associated with their reproductive organs. According to Denham, these structures are paired and range from single-spined spicules in the Nematoda to quite complex objects in some of the flatworms (Fig. 19). Some worms have a single pair of these structures, whereas others have a group of them. These structures are kept within the body except during copulation, when they are extruded and assist in the process of fertilization. Denham suggested that conodonts might have performed a similar function, that the accretionary mode of growth in a conodont could have been accomplished while the conodont was held within the body of the worm, and that if, during copulation, a part of the conodont structure had been broken away and lost, it could have been restored later within the body of the worm.

The so-called micro-conodonts of Wetzel are not related to the true conodonts of Pander. Wetzel's fossils are extremely minute chitinous objects that were first found in some Cretaceous rocks of the Baltic region of Europe. In a recent paper he (76, p. 803) stated that these

> comb-shaped, bristly, and obviously organic (chitinous) fragments which might be identified with masticatory organs of annelids are found occasionally in flints and chalks. These chaeta-combs, as well as single and double chaetae of pincer-like form . . . . have been classed as micro-conodonts . . . . in contrast to Paleozoic macro-conodonts already known for a long time. Recently, Cretaceous micro-forms have been . . . [classified] by American specialists as . . . scolecodonts.

POSTULATE THAT CONODONTS BELONG TO ARTHROPODA

A few investigators have related conodonts to the arthropoda. It has been suggested that conodonts are the tips of segments of the exoskeleton of trilobites; that some could be the claws of crustacea; that they are the internal jaws of crustacea; and that they are spines attached to the carapace of an arthropod. Nothing similar has been published since 1889.

POSTULATE THAT CONODONTS BELONG TO CHORDATA

The known range of conodonts (Lower Ordovician-Upper Triassic) does not coincide with that of any class of chordates with which conodonts have been identified. Hence, some investigators have been rather noncommittal on the subject of that affinity. Ellison (25) suggested that conodonts are hard parts of fish or lower vertebrates; Stauffer (69), that their composition tends to relate conodonts with the vertebrates; Youngquist (77), that conodonts may be the internal supporting structures of fish; and Stauffer & Plummer (70), that conodonts are the teeth, spines, and plates of an extinct group of primitive fishlike animals.

Pander (52) regarded conodonts as fish teeth. He admitted, however, that their systematic position was open to question because he had no information about the animal that bore conodonts and because he knew of no similar teeth in any possible descendants or living animals. He was undecided as to whether conodonts were situated on the jawbone, the palate, or the tongue. Moreover, he could not decide whether each kind of conodont represented a distinct biologic species or whether several different kinds of disjunct conodonts were present in the same animal.

Conodonts have been found associated with some fish plates in the Harding Sandstone of Colorado. These plates, according to Kirk (43), are generally referred to as ostracoderm remains. Because the composition of these plates is identical with that of the basal plates of the associated conodonts, Kirk (p. 495) stated that

> If the identification of the Harding sandstone plates with the ostracoderms be accepted, this discovery would seem to provide a new and important clue to the real nature of these minute, toothlike bodies. The suggestion contained in these specimens that some conodonts, at least, may be mouth parts of ostracoderm fishes is in general agreement with a view that has long been held by many authorities . . . that conodonts are the teeth of primitive fishes.

Kirk, however, was of the opinion that the information he presented was insufficient to permit a generalization being made on the nature of all conodonts.

Some workers have considered conodonts to be the teeth of the Cyclostomata (lampreys and hagfish) despite the fact that the living representatives of this class have horny teeth. Ulrich & Bassler (75) believed that conodonts are the teeth of several groups of primitive fishes and classified the fanglike conodonts (i.e., Distacodontidae) as probable relatives of the myxines. Huddle (42, p. 33) tentatively placed conodonts in "the Cyclostomata, because this class includes the most primitive vertebrates with similar tooth structures."

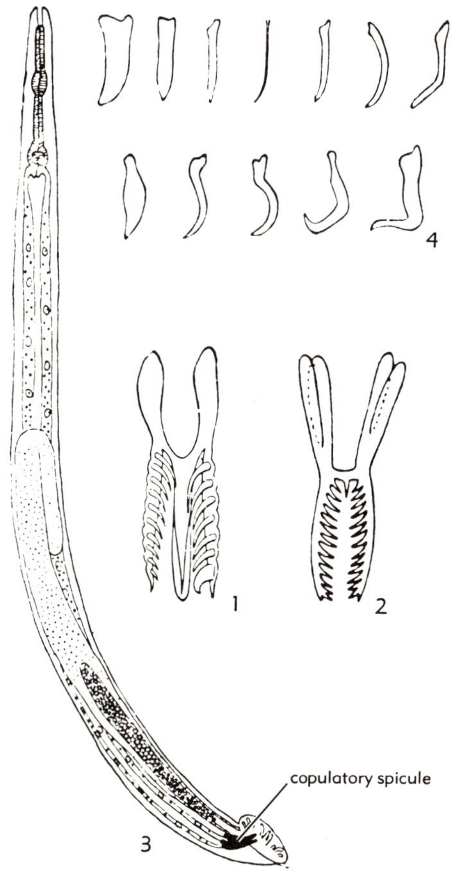

Fig. 19. Copulatory structures of some worms (20).——*1*. Chitinous copulatory structure of the turbellarian, *Dalyellia rossi,* ×285.——*2*. Similar structure of *Dalyellia viridis,* ×285.——*3*. Structure of adult male nematode, *Rhabditis* sp., ×200.——*4*. Various nematode spicules.

Schmidt (62, 63) studied some conodonts from a Carboniferous shale of Germany. He identified them as placoderm remains and believed that an assemblage of conodont

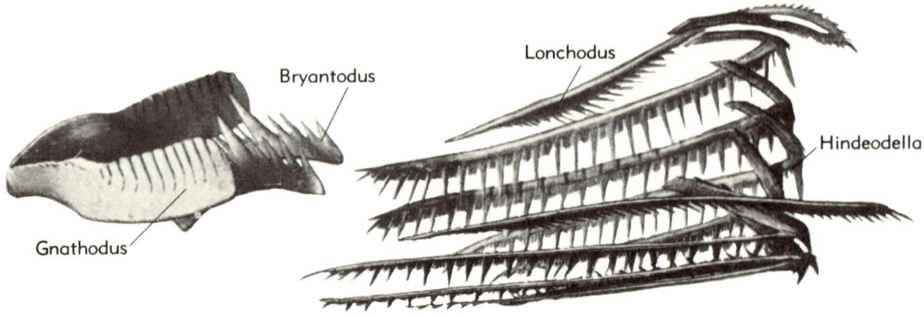

FIG. 20. SCHMIDT's reconstruction of conodont assemblage presumed to represent mouth and gill arch structures of a placoderm, *Westfalicus integer* (SCHMIDT): *Gnathodus,* mandibles; *Bryantodus,* teeth on hyoid arch; *Lonchodus* and *Hindeodella,* ceratobranchial and epibranchial gill arch structures, ×30 (62).

structures was present in each gnathostome, some *(Gnathodus)* being mandibles, others *(Bryantodus, Neoprioniodus)* teeth on the hyoid arch and still others *(Hindeodella, Lonchodus)* parts of the gill arches. His assemblage is now called *Westfalicus integer* (*see* Fig. 20).

The conodonts considered to have functioned as mandibles and identified as *Gnathodus* (discrete conodonts in the sense of PANDER) consist of a pair of platelike units, each of which has an expanded pulp cavity; in this respect they do resemble a gnathodid conodont, but SCHMIDT's specimens are too poorly illustrated to enable anyone to verify his identification. SCHMIDT regarded his *Gnathodus* element to be the anteriormost part of the assemblage and to be so oriented that the distal end of the cups of the two specimens formed the symphysis of the jaw. The conodonts described in his first paper and consisting of a pair of bladelike forms identified as the *Bryantodus element,* were considered to have functioned as teeth on the hyoid arch. In his second paper (63) he described an additional pair of neoprioniodids present in the assemblage. The *Neoprioniodus* element was presumed to have been located dorsal to the *Bryantodus* element. SCHMIDT believed that the above-mentioned conodonts are parts of the hyoid arch because, in his assemblages, they are located between conodonts identified as mandibles and others identified as parts of gill arches. The posteriormost part of his assemblage consists of five pairs of barlike conodonts that were believed to be parts of of five gill arches. These conodonts were referred to as the *Lonchodus (Hindeodella)* elements. Each one of the pair of conodonts assigned to the first gill arch is twisted and bears closely set denticles of one size; the main bar of each specimen was called the ceratobranchial and a shorter underslung bar, the epibranchial. Conodonts assigned to gill arches 2-5 belong to the genus *Hindeodella.* Here again SCHMIDT (63) considered the longer denticulated bar of specimen to be a ceratobranchial and the shorter denticulated bar to be an epibranchial. He believed that the Conodontophorida should probably be classified under the Aphetohyoidea, a class that includes the placoderms.

BECKMANN (1) accepted SCHMIDT's conclusions on the zoologic affinity and the function of conodonts. Specimens belonging to *Polygnathus, Ancyrodella,* and *Icriodus* were believed to have functioned as mandibles; those belonging to *Bryantodus* to have functioned as teeth on the hyoid arch; and those belonging to *Ligonodina* and *Neoprioniodus* to have functioned as ceratobranchials.

Conodonts have been identified as the teeth of Chondrichthyes. These fish have a cartilaginous skeleton and their teeth consist of dentine, a pulp cavity, and an enamel cap. ULRICH & BASSLER (75) considered those kinds of conodonts which in the present paper are assigned to the families Belodontidae, Coleodontidae, Prioniodontidae, and Prioniodinidae, as being the teeth of fish and as probably being related to the selachians.

They concluded that each kind of conodont was characteristic of a distinct animal. In their opinion, specimens classified as Polygnathidae resemble the dermal denticles of recent sharks and perhaps, therefore, should not be identified as true conodonts. EICHENBERG (23) favored the idea of conodont assemblages; he studied some recent fish remains and became convinced that conodonts functioned as the teeth, scales, and gill rakers of primitive elasmobranchs and teleosts. DEMANET (18) agreed with SCHMIDT as to the function of conodonts but preferred to classify them as elasmobranch remains instead of placoderm, as SCHMIDT had done. DEMANET (19) also found an object believed to be a conodont on the branchial arch of *Coelacanthus lepturus* AGASSIZ [=*Rhabdoderma elegans* (NEWBERRY)]; but RHODES (58) who examined the same specimen stated that the object in question is probably not a conodont.

GROSS (32) made a detailed histologic examination of *Spathognathodus murchisoni* (PANDER), a bladelike species from the Island of Oesel. He compared the structures present in his specimens with those of authentic teeth and the bones of some Paleozoic Agnatha and fishes. Although he was not able to solve the problem of zoological relationship, he was able, in his opinion, to eliminate some groups as possible close relatives of the conodont-bearing animals. His conclusions (GROSS, 32, p. 79) as given by MÜLLER (50, p. 1325) are:

1) Conodonts are not formed by a cuticula, as is the case in skeletons of arthropods and jaws of annelids. Those organs are secreted layer by layer from the epidermis-cells, and therefore become thicker toward the base.

2) Conodonts are neither mouth-teeth nor skin-scales of vertebrates. They are not composed of dentine, have no pulpa nor dentine channels, grow by outer instead of inner deposition, and are able to regenerate lost denticles as well as suppress others by the formation of germ denticles.

3) They are not a part of the endoskeleton of vertebrates. If so, in the case of Paleozoic Agnatha or fishes, they would form an ossification around a cartilaginous core, and therefore, as a fossil, would surround a cavity, filled with sediment. Otherwise, they would have the spongy texture of cartilagenous tissue; such is not the case. Also, the shape is not as would be expected in parts of an endoskeleton.

GROSS stated that conodonts probably were the sole preservable part of the animal that bore them and suggested that this animal belonged to a distinct stem of the chordates or jawless vertebrates.

POSTULATE THAT CONODONTS BELONG TO UNCLASSIFIED ANIMALS

Some investigators who consider conodonts to be ingestive aids have stated that the zoological position of the conodont-bearing animal is uncertain. Others have suggested that some conodonts could have been parts of the armor of an unknown animal. The present writer (34, p. 71) has published judgment that "conodonts functioned as internal supports for tissues that were located at a place exposed to stresses upon the exterior of or within the bodies of some genetically related group of marine animals."

Presumably this conodont-bearing animal was soft-bodied, bilaterally symmetrical, marine and pelagic. These opinions are held for the following reasons:

(1) The thesis that the conodont-bearing animal was soft-bodied—that is, that the organism consisted chiefly of structures and tissues incapable of fossilization under ordinary circumstances—is based on the fact that conodonts and the basal plate to which some conodonts are still attached are the only recognizable hard parts found after more than 100 years of research. The dark-brown to black carbonaceous substance associated with some conodonts in black shales may represent another part of the conodont-bearing animal but, as yet, the nature of this material has not been determined. Additional, and perhaps decisive, information on this subject could probably be obtained through a thorough examination of fine-grained rocks derived from sediments deposited in quiet marine and lagoonal environments; such rocks probably contain recognizable impressions and films of the softer parts. Concretions, especially those that developed in a reducing environment, might also contain impressions, films, and even mineralized replacements of the animal.

(2) The conodont-bearing animal is believed to have been bilaterally symmetrical because many species of disjunct conodonts contain both right-handed and left-handed

specimens. The arrangement of paired specimens in an assemblage, as in *Westfalicus integer* (SCHMIDT) and *Scottognathus typica* (RHODES), also indicates that the animal had this type of symmetry. Assemblages are quite scarce; they are also difficult to interpret. All those presently considered authentic are from the Carboniferous and contain between 7 and 11 pairs of components, assignable to three to five genera of disjunct conodonts; in addition, MÜLLER (50, p. 1326) suggests, some assemblages may have had an unpaired bilaterally symmetrical element, such as a roundyid or a hibbardellid. Nothing is known of the composition of pre-Mississippian, Permian, or Triassic assemblages, though it is generally assumed that the arrangement of their components is similar to those of the Carboniferous. However, it is also evident that the kinds of components in assemblages changed greatly throughout the phylogeny of the conodont-bearing clan.

Because some prepared collections consist chiefly, or even entirely, of one kind of conodont specimen, it has been suggested that some assemblages had only one or possibly two kinds of structures. This could be the case, though it is also possible that such singular associations resulted through winnowing. On the other hand, selective sorting may have played only a minor role in the concentration of specimens which exhibit no evidence of excessive fracturing.

It also has been suggested that the composition of a conodont assemblage might be worked out through a statistical study based on the relative abundance of the different kinds of disjunct conodonts in a large number of collections from the same bed, especially if the fauna of that bed consists of only a few different kinds of structures. The merit of this suggestion cannot be evaluated at present.

(3) The conodont-bearing animal is believed to have been both marine and pelagic. This view is held because conodonts have a world-wide distribution and are found associated with marine fossils in all of the ordinary kinds of marine sedimentary rocks. Conodonts, therefore, cannot be classified as facies fossils, and the animal that bore them must have been pelagic. The fact that conodonts are commonly found in black shales—which were derived from sediments deposited in an oxygen-deficient environment—gives support to this thesis, for it indicates that the animal lived in the oxygen-rich surface waters and only after death sank into the foul bottom waters.

Although conodonts are not facies fossils, they are more abundant in some kinds of marine rocks than in others. Argillaceous and arenaceous limestones are more likely to have an abundance of well-preserved specimens than are the purer denser kinds of calcareous rocks. Mudstones, sandstones, and conglomerates also contain a fair number of specimens, especially if the deposits are thin and lie directly on top of an eroded surface. Black shales commonly appear to contain a large number of conodonts due, at least in part, to the fact that most other kinds of fossils are either extremely scarce or entirely absent. Conodonts also tend to be more abundant in some beds of a formation than in others, even though all parts of the sequence have similar gross physical characteristics. This uneven distribution could have resulted through the introduction of reworked specimens into the natural fauna of a formation, through variations in the rate of accumulation of sediments—the slower the rate, the greater the concentration of conodonts—or through occasional explosive increases in, or wholesale deaths to, the conodont-bearing animal population of an area.

NOMENCLATURE

Most descriptive papers on conodonts are concerned with discrete specimens, but a few, including those by SCOTT (65), SCHMIDT (62), and RHODES (56), treat of assemblages; an assemblage consists of several different kinds of discrete conodonts that are presumed to represent parts of one animal (Figs. 20, 42). Most investigators are of the opinion that a system of dual nomenclature is needed to designate conodont material adequately. They hold that one set of names should be used in a utilitarian classification based on discrete specimens and that a second set of entirely different names should be used in a biologic classification based on assemblages; this view is held despite the fact that such a system is contrary to the Rules of Nomenclature as presently conceived.

Between 1856 and 1934 the binomina of disjunct conodonts were commonly treated as the names of whole animals, and were conceived as being subject to all of the procedures and rules of zoological nomenclature as laid down by the International Commission. But since 1934 many investigators have regarded the binomina of assemblages as referring to whole-animal taxa, and the binomina of disjunct conodonts as referring to form-taxa, that is, to form-genera and form-species, which are equivalent to the partial-genera and partial-species of MÜLLER (50). During the past 100 years approximately 160 generic names and over 2,500 specific names have been proposed for discrete conodonts, and although some of these names are obviously synonyms, they greatly outnumber the eight generic and nine specific names that have been given to approximately 250 observed conodont assemblages. Some of these associations of discrete conodonts, each originally described as representing an assemblage, are presently regarded as being accumulations that could not have been derived from one individual.

RHODES (56, 57) has stressed the need for devising a system of dual nomenclature. He intentionally proposed new generic and specific names for some conodont assemblages despite the fact that his synonymies include names of discrete conodonts which had been proposed previously in compliance with the Rules. RHODES objected to identifying any one of his conodont assemblages with the earliest validly proposed name of one of the components of that assemblage because had he done so he would have completely wrecked a well-established system of nomenclature which is extremely useful to the stratigraphic paleontologist. Moreover, he pointed out that representatives of the same genus or species of discrete conodonts could be present in several otherwise distinct assemblages, and that if such specimens were the first-named of the several components, the Rules would require placing unlike assemblages in the same generic or specific category.

SINCLAIR (66), however, has pointed out that the acceptance of a system of dual nomenclature contravenes an important principle, inasmuch as it would permit the same animal to have more than one valid name. He favored strict adherence to the International Rules of Zoological Nomenclature—that is, the Law of Priority must prevail—and suggested that each component of an assemblage be designated not by a generic and specific name but by a common noun derived from the name of the genus to which the component belongs. For example, he would use the name hindeodell element instead of *Hindeodella;* prioniod element instead of *Prioniodus;* and polygnath element instead of *Polygnathus.* SINCLAIR stated that a name does not belong to the material described, no matter how complete or incomplete that material may be, but to the animal possessing that material, and, also, that the name of a conodont assemblage cannot be placed in a higher nomenclatorial category than the names of its components. Moreover, he pointed out that all fossils are but parts of animals, even though some fossils may represent more of the whole of an animal than others do; also, zoological nomenclature would become quite transitory if the name of an animal were to be continually changed as more and more complete anatomical material is discovered and made known.

According to SYLVESTER-BRADLEY (73, p. 333), however, "There is no *legal* objection to the concurrent use of the two alternative systems of nomenclature" as long as the specific name of a conodont assemblage is not a junior objective synonym of the name of one of the components of the assemblage, and as long as the generic name of a conodont assemblage is not a junior objective synonym of the name of a genus based on discrete conodonts. All other names "are subjective synonyms and can always be validly used by a taxonomist who disagrees with the synonymy" as presented by another investigator. However, SYLVESTER-BRADLEY was well aware of the fact that any system of dual nomenclature would invite confusion unless regulatory provisions were written into the Rules.

Similar nomenclatorial problems confront specialists working with the discrete parts of some other groups of animals, such as annelid jaws (scolecodonts); radular elements and opercula of gastropoda and cephalopods (aptychi); ossicles of crinoids, cystoids, blastoids, echinoids, and asterozoans; spicules of sponges, octocorals, and holothurians; and isolated coccoliths. An adequate solution to the problem, therefore, is of concern to many zoologists and paleontologists. Some investigators, including FRIZZELL and EXLINE (29), are strongly in favor of a system of dual nomenclature that would function within the framework of the Rules of Zoological Nomenclature. Conversely, other investigators have recommended that the names of disjunct parts of animals be treated as technical terms rather than as zoological names. This recommendation, however, solves nothing, for should a student follow it, he must then employ a terminology—such as the military classification proposed by CRONEIS (17)—that falls outside the scope of accepted zoological nomenclature, thereby depriving himself and others of the protection, regulation, uniformity, and stability that the Rules give to students of whole animals. Obviously something must be done, for the existing situation leads to uncertainty in the application of the Law of Homonymy and thereby affects the nomenclature of all groups of animals.

In an attempt to resolve these nomenclatorial problems, MOORE & SYLVESTER-BRADLEY (47) submitted an application to the International Commission on Zoological Nomenclature in July, 1958. They requested that "a special category [be recognized] for the classification and nomenclature of discrete fragments or of life-stages of animals which [in the opinion of the Commission] are inadequate for identification of whole-animal taxa." MOORE & SYLVESTER-BRADLEY proposed the designation *parataxa* (associate taxa) for this new category. They stated that the "nomenclature applied to taxa and parataxa should be mutually exclusive and independent for the purposes of the Law of Priority, but coordinate for the purpose of the Law of Homonymy, names belonging to one category not being transferable to the other." If adopted, they believed, their proposal would provide a means of preventing "(a) the invalidation of names applied to terms of whole animals which are junior synonyms of parataxa; and (b) the invalidation of parataxa as synonyms by the discovery that more than one parataxon belongs to a single whole animal." MOORE & SYLVESTER-BRADLEY's application provided that "once the Commission has ruled that the classification of any group of animal fragments shall be in terms of parataxa, that ruling shall apply retroactively, as well as to future publication, irrespective of whether the author in question uses the term parataxa."

MOORE & SYLVESTER-BRADLEY (48) also submitted to the International Commission an "application for a ruling . . . directing that the classification and nomenclature of discrete conodonts are to be in terms of parataxa." This application gave a detailed account of the nomenclatural uncertainties that confront the conodont specialist under the existing Rules, uncertainties which leave a worker no alternatives other than the disrupting of conodont nomenclature or disregard of the Rules of Nomenclature. MOORE & SYLVESTER-BRADLEY's proposals on parataxa were rejected by the 15th International Congress on Zoology which met in London in July, 1958. However, the Congress did pass a resolution suggesting that the names of fragments (such as those of disjunct conodonts) should not be required to compete in synonymy with the names of genuine taxa, as would be the case under a strict interpretation of the Rules. Hence, in the present paper, a system of dual nomenclature, fashioned after the MOORE & SYLVESTER-BRADLEY proposals, is used in anticipation of its ultimate acceptance by the International Commission on Zoological Nomenclature.

In order to supply ready reference to the many changes which must be made before the conodont specialist is provided with a stable nomenclature, MOORE & SYLVESTER-BRADLEY's proposals are given below; these authors requested the Commission to direct that:

(1) the nomenclature of all categories based on types which, in the opinions of the original authors, are discrete conodonts, shall be in terms of parataxa and as such shall be unavailable as names of taxa based on conodont assemblages;

(2) the names of all categories based on types which, in the opinion of the original authors, are assemblages of conodonts derived from single animals, shall be unavailable for the designation of parataxa;

(3) notwithstanding (2) above, the generic name *Polygnathus* HINDE (1879:359) (gender: feminine) (type-species, by subsequent

designation by MILLER, 1889: 520, *Polygnathus dubius* HINDE, 1879) be placed on the *Official List of Generic Names in Zoology* as the name of a parataxon;

(4) the name *dubius* HINDE (1879: 362-365), published in the combination *Polygnathus dubius* HINDE, 1879 (type-species of *Polygnathus* HINDE, 1879) be placed on the *Official List of Specific Names in Zoology* as the name of a parataxon, this species to be interpreted by the specimen figured by HINDE as pl. 16, fig. 17, now preserved in the British Museum (Natural History) under Catalogue Number A.4211, which specimen is to rank as lectotype;

(5) the following generic names introduced for assemblages of conodonts believed by their authors to represent single animals, are not available as names of parataxa, and are to be entered in the *Official List of Generic Names in Zoology;*

Duboisella RHODES (1952: 895) (gender: feminine) (type-species, by original designation, *D. typica* RHODES, 1952);

Scottognathus RHODES (1953: 612) (gender: feminine) (type-species, by original designation, *Scottella typica* RHODES, 1952);

Illinella RHODES (1952: 898) (gender: feminine) (type-species, by original designation, *I. typica* RHODES, 1952);

Lochriea SCOTT (1942: 293) (gender: feminine) (type-species, by original designation, *L. montanaensis* SCOTT, 1942);

Lewistownella SCOTT (1942: 299) (gender: feminine) (type-species, by original designation, *L. agnewi* SCOTT, 1942);

Westfalicus SCHMIDT [? 1956] (gender: masculine) (type-species, by original designation, *Gnathodus integer* SCHMIDT, 1934).

(6) the following specific names, type-species of the genera listed in paragraph (5), being based on assemblages of conodonts presumed by their authors to represent single animals, are not available as names of parataxa, and are to be entered on the *Official List of Generic Names in Zoology:*

typica RHODES (1952: 895), as published in the combination *Duboisella typica* (type-species of *Duboisella*);

typica RHODES (1952: 891), as published in the combination *Scottella typica* (type-species of *Scottognathus* RHODES, 1953);

typica RHODES (1952: 899), as published in the combination *Illinella typica* (type-species of *Illinella*);

montanaensis SCOTT (1942: 295), as published in the combination *Lochriea montanaensis* (type-species of *Lochriea*);

agnewi SCOTT (1942: 300), as published in the combination *Lewistownella agnewi* (type-species of *Lewistownella*);

integer SCHMIDT (1934: 77), as published in the combination of *Gnathodus integer* (type-species of *Westfalicus*).

(7) the following generic names, being introduced as names of categories based on discrete conodonts, are to be entered as names of parataxa on the *Official List of Generic Names in Zoology:*

Prioniodus PANDER (1856: 29) (gender: masculine) (type-species by subsequent designation by MILLER, 1889: 520, *P. elegans* PANDER, 1856);

Gnathodus PANDER (1856: 33) (gender: masculine) (type-species, by monotypy, *G. mosquensis* PANDER, 1856);

(8) the following specific names, having as type specimens discrete conodonts, are to be entered as parataxa on the *Official List of Specific Names in Zoology:*

elegans PANDER (1856: 5), as published in the combination *Prioniodus elegans* (type-species of *Prioniodus* PANDER, 1856);

mosquensis PANDER (1856: 34) as published in the combination *Gnathodus mosquensis* (type-species of *Gnathodus* PANDER, 1856);

(9) the following family-group names, having as type-genera conodonts classed as parataxa, are to be entered as parataxa on the *Official List of Family-Group Names in Zoology:*

POLYGNATHIDAE BASSLER (1925: 219) (type-genus: *Polygnathus* HINDE, 1879);

PRIONIODONTIDAE (correction, first made herein, of PRIONIODIDAE) BASSLER (1925: 218) (type-genus: *Prioniodus* PANDER, 1856);

GNATHODONTIDAE CAMP, TAYLOR & WELLES (1942: 525) (type-genus: *Gnathodus* PANDER, 1856);

(10) the name *Scottella* RHODES, 1952, a junior homonym of *Scottella* ENDERLEIN, 1910, be entered on the *Official Index of Rejected and Invalid Generic Names in Zoology:*

(11) the names PRIONIODIDAE BASSLER, 1925 (an Invalid Original Spelling of PRIONIODONTIDAE) and GNATHODONTIDAE HUENE, 1929 (an unavailable name since not based on the stem of a type-genus) be entered on the *Official Index of Rejected and Invalid Family-Group Names in Zoology*.

MIXED FAUNAS

Two kinds of naturally-mixed conodont faunas are recognized: stratigraphic leaks and stratigraphic admixtures. A stratigraphic leak involving conodonts has been defined as "the introduction of conodonts of one age into association with beds of an earlier time" (BRANSON & MEHL, 11, p. 206). This kind of mixed fauna resulted through the filling of cavities in a formation with materials of a younger stratigraphic unit; the filling occurred either at the time of deposition of the younger unit or later, through the action of geologic agents. A stratigraphic admixture of conodonts has been defined as "the inclusion of an earlier assemblage of conodonts in the sediments and faunas of a later age" (BRANSON & MEHL, 11, p. 197). This kind of mixed fauna is common because conodonts are resistant to many kinds of chemical weathering. Acetic and similar acids do not affect conodonts, and, because these acids are stronger than those that usually have been active in the weathering of rocks throughout geologic time, conodonts generally have remained unchanged in the resulting residuum. Later, such conodonts were reworked into the sediments of a younger formation. Hence, it is possible for specimens from different sources and of different ages to be found together, especially in the basal beds of a formation.

Differences in the physical appearance (color, preservation, luster) of associated specimens are indicators of a mixed fauna; but the recognition of a mixed fauna is chiefly dependent upon one's knowledge of the true stratigraphic range of each kind of discrete conodont. This sort of knowledge can aid in correctly interpreting the geology of an area, for some of the reworked conodonts in a collection could have been derived from one or more unrecognized stratigraphic units. In all probability, when the presence of such a unit is established, the unit will be found either to be thin in comparison with adjacent stratigraphic units or to have a restricted areal distribution. It is also possible that the rocks of a formation could have been completely eroded from a given area. A formation missing from the stratigraphic succession of an area has been called a "phantom formation" by BRANSON & MEHL (11, p. 208, 209) if its former presence in the area is postulated on the basis of finding some conodonts that presumably could not have been derived from any of the known formations.

STRATIGRAPHIC RANGE OF DISCRETE CONODONT GENERA

At the present time there is a real need for papers that adequately describe and illustrate discrete conodonts. If possible, descriptive work should be based on suites of whole specimens collected from beds located in known intervals of measured sections. Moreover, each conodont collection should come from an undisturbed sequence of rocks, preferably one containing megafossils and resulting from the slow continuous deposition of sediments. Through the study of material in numerous collections meeting these specifications, it is possible not only to establish faunal zones in a formation, and to determine the stratigraphic ranges of discrete conodonts, but also to recognize exotic specimens that were introduced into a collection either by natural means or by man.

Conodonts definitely range from the Lower Ordovician into the Upper Triassic, and recent work indicates that they may range from the Upper Cambrian into the Upper Cretaceous (MÜLLER, 49; DIEBEL, 21). Conodont faunas are well diversified in the lowermost Ordovician and it is therefore reasonable to believe that they also occur in authentic Cambrian rocks. The writer, in 1954, examined some conodont-like objects from the Upper Cambrian

Conaspis Zone in the Conant Creek area of Freemont County, Wyoming, and from the *Conaspis* Zone in the Dugway Mountains of Utah. None of the specimens examined from these two localities has a laminated structure, but inasmuch as these specimens were prepared with formic acid it is possible that the objects in question are true conodonts with lamellar structure that has become obscured through alteration. It is the writer's opinion that the stratigraphic range of conodonts should not be recorded as definitely extending into the Cambrian until irrefutable, well-documented evidence has been published. Authentic conodonts, including *Gondolella mungoensis* (DIEBEL), have been reported from the upper Cretaceous of the Cameroons in West Africa (DIEBEL, 21). This reported occurrence greatly extends the known stratigraphic range of conodonts; hence, before it is accepted as being a valid record of the occurrence of conodonts, it should be substantiated through investigations of Cretaceous rocks in other parts of the world.

Figures 21, 21A, 23A, 32A, and 35A record the stratigraphic ranges of families and some genera of disjunct conodonts occurring in post-Cambrian and pre-Jurassic rocks. The indicated ranges are based on an evaluation of published information (other than faunal lists) and on an examination of specimens in numerous collections. As considerable new information has accumulated since ELLISON (26) published the first range chart of conodont genera, the indicated range of some of the genera listed differs from that given by ELLISON as well as from that shown in publications of several other authors.

Ordovician conodont faunas appear to be fairly well known, 57 genera presently being recognized in the rocks of the system. Many of these genera have been reported as occurring in the Ordovician of both North America and Europe, and a few have also been found in Australia (PANDER, 52; LINDSTRÖM, 44; RHODES, 58; BRANSON & MEHL, 7, 8, 9, 10; FURNISH, 30; and SWEET, 72). Also, a very large number of these genera are restricted to the Ordovician System, and some of them appear to range through only small parts of the system

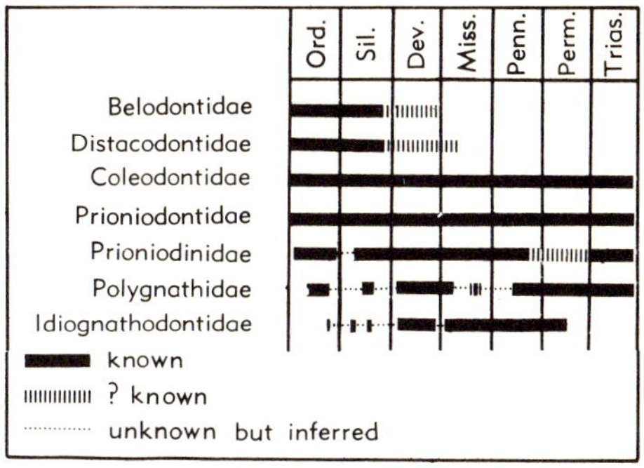

FIG. 21. Stratigraphic distribution of conodont families (Hass, n).

(e.g., *Loxodus, Leptochirognathus, Rhipidognathus, Scyphiodus, Balognathus, Icriodella*). Therefore, we find indication that eventually many conodont faunal zones will be recognized in the Ordovician, in addition to the four zones established by LINDSTRÖM (44), in the lowermost Ordovician strata of Sweden. Representatives of the Distacodontidae and Belodontidae are quite common, especially in the Lower Ordovician.

Silurian conodont faunas are not very well known but it appears that discrete conodonts are less abundant in the rocks of this system than in those of the Ordovician. The known Silurian faunas consist chiefly of bladelike and barlike conodonts, together with a lesser number of distacodontids. As would be expected, these faunas contain some genera in common with those from the Ordovician (e.g., *Acodus, Distacodus, Drepanodus, Paltodus*), as well as others that range into younger strata (e.g., *Ozarkodina, Ligonodina, Hindeodella, Spathognathodus*). To date, only three representatives of the platelike conodonts are known from the Silurian; one of these, *Icriodina,* appears to be restricted to a part of the Lower Silurian, and the other two, *Polygnathoides* and *Kockelella,* to higher beds. Except for a very few recorded occurrences, some or all of which may have resulted through reworking, the Distacodontidae and the Belodontidae are restricted to pre-Devonian rocks.

Conodonts appear to be fairly common in Lower and Middle Devonian, and are extremely abundant and diversified in the Upper Devonian. The Polygnathidae, or platelike conodonts with a restricted pulp cavity, are very common in most collections; and some genera assigned to this family, as well as their species, are considered to be very good index fossils for restricted intervals of the Devonian (e.g., *Icriodus, Ancyrodella, Ancyrognathus, Palmatolepis, Panderodella, Polylophodonta*). The stratigraphic importance of Devonian conodonts has been demonstrated by the investigations of HUDDLE (42) and HASS (36, 37) in the United States and by SANNEMANN (61), ZIEGLER (79), BISCHOFF (2, 3), BISCHOFF & ZIEGLER (6), BISCHOFF & SANNEMANN (4), and MÜLLER (51) in Europe.

HUDDLE (42) and HASS (36, 37) have reported on the succession of conodont faunas in the Devonian and Mississippian black-shale sequence. They have shown that this sequence contains many short-ranging, easily recognized species of disjunct conodonts; and this, in turn, indicates that through conodont studies, the long-standing controversial problems concerned with the age and correlation of the black shales can be solved. Recent descriptive works and stratigraphic studies based on the conodont faunas of the German Devonian succession are especially important, because in those rocks conodonts are associated with megafossils, including the cephalopods *Maenioceras* (Givetian); *Manticoceras* (Frasnian); and *Cheiloceras, Prolobites, Platyclymenia, Laevigites,* and *Wocklumeria* (Famennian). Ranges of conodonts in the German succession will thus influence biostratigraphic interpretations throughout the world. Much needs to be done, but it is already known that many genera and species of disjunct conodonts are common to the rocks of both Germany and the United States. These species include such easily recognized forms as *Panderodella truncata* and *P. gracilis, Palmatolepis perlobata* and *P. subperlobata; Ancyrognathus asymmetrica; Palmatodella delicatula; Spathognathodus jugosus; Branmehla inornata; Neoprioniodus alatus; Ancyrodella rotundiloba; Polygnathus ordinata, P. pennata,* and *P. linguiformis.* A few of these species have also been reported from other European countries and from Australia.

As in the Upper Devonian, conodonts are also extremely abundant and diversified in the Lower Mississippian; this is especially true of the Kinderhookian Series, from which BRANSON & MEHL (9), COOPER (16), HASS (38), and others have recorded several distinctive faunas. The Lower Mississippian faunas, moreover, are characterized by genera and species that easily distinguish them from those in the Upper Devonian. As shown in Table 1, many genera (including *Siphonodella, Elictognathus, Pseudopolygnathus, Pinacognathus, Scaliognathus, Dollymae,* and *Bactrognathus*) range throughout parts of the Lower Mississippian (Kinderhookian and Osagian) succession; and *Geniculatus* and *Kladognathus* range throughout parts of the Upper Mississippian (Meramecian and Chesteran) succession. *Gnathodus* and *Cavusgnathus* are representative of genera that range from the Mississippian into younger rocks. There are many distinctive short-ranging species in the Mississippian; these include such easily recognizable species as *Siphonodella duplicata, S. quadruplicata,* and *S. obsoleta; Pseudopolygnathus prima; Elictognathus lacerata; Dollymae sagittula; Geniculatus claviger; Staurognathus cruciformis; Gnathodus bilineatus, G. punctatus,* and *G. texanus.* Some of the species listed above, as well as other American species, have been reported from the Lower Carboniferous of Germany by BISCHOFF (3) and by BISCHOFF & ZIEGLER (5) and from the Lower Carboniferous of Austria by FLÜGEL & ZIEGLER (28). The Polygnathidae have not been reported from the Upper Mississippian, though *Mestognathus* is present in equivalent rocks of Germany, and, with the exception of *Gondolella,* none of the Polygnathidae has been found in Pennsylvanian or younger rocks. On the whole, conodonts are not especially abundant in the Upper Mississippian; several descriptive papers have been published, including those by REXROAD (54, 55) and HASS (35).

Conodonts are fairly abundant at some levels in the Pennsylvanian, though in many faunas the variety of genera and species is somewhat limited. Despite this, Pennsylvanian faunas commonly contain specimens of a distinctive character. Portions of the Lower Pennsylvanian, for example, are characterized by specimens of *Idiognathoides* (=*Polygnathodella*). Collections from the upper part of the Middle Pennsylvanian (Desmoinesian) commonly contain numerous specimens of *Idiognathodus,* a lesser number of specimens of *Streptognathodus,* and the first few representatives of *Gondolella.* In the lower part of the Upper Pennsylvanian, specimens of *Idiognathodus, Gondolella,* and *Streptognathodus* are the dominant components of most faunas. *Idiognathodus* is rarely found in the Upper Pennsylvanian and presumably does not range as high as the Permian (Ellison, 24).

Little is known concerning Permian conodont faunas. They have been reported from several formations, but only a few of them have been described and illustrated. Youngquist, Hawley, & Miller (78, p. 360) have stated that "in general, the Phosphoria conodonts show a reduction in size, diversity, ornamentation, and . . . abundance, when compared to Pennsylvanian faunas." *Gondolella* is present in the Phosphoria, and specimens of this genus in the writer's collections have a superficial resemblance to specimens of *Polygnathus;* however, the pulp cavity of these gondolellids is located near the anterior end of the unit and the keel is somewhat split along a portion of the mid-line, thereby easily differentiating these fossils from true specimens of *Polygnathus.* Ellison (24) has treated the conodont fauna of the Permian Big Blue group of Kansas. He did not record the presence of *Gondolella* and *Idiognathodus* in the Permian but did note the occurrence of *Streptognathodus.* However, *Gondolella* must be present in early Permian rocks, since representatives of this genus have been found in older as well as younger formations.

The presence of conodonts in the Triassic is now well established, as they have been found at several places in the United States, Europe, Egypt, and Asia (Youngquist, 77; Müller, 49; Tatge, 74; and Huckriede, 41). Most Triassic specimens have been identified with Paleozoic genera, though some appear to be sufficiently distinctive to permit their being placed in new categories. *Gondolella* is commonly represented in collections by numerous specimens; this genus is the only polygnathid so far recorded from the Triassic.

Huckriede (41) studied the conodonts of the Mediterranean Triassic. He found that conodonts are widely distributed throughout the Triassic, but that there are fewer species in these faunas than in those from the Paleozoic. The complete absence of conodonts in the Rhaetic, Jurassic, and Cretaceous beds of the Alps indicated to Huckriede that the conodont-bearing animal became extinct in the Late Triassic (Obernor).

CLASSIFICATION

Too little is known of the affinity of conodonts to warrant assigning them to any class of animals. Herein, they are placed in the order Conodontophorida, and a dual classification is used for categories below the rank of order, one being a utilitarian classification and the other a biologic one.

The utilitarian classification is based on the fact that each individual conodont was built up through the accretion of lamellae about the pulp cavity; the many genera and species of discrete conodonts now recognized resulted because the lamellae in any conodont specimen are separated from each other along one or more growth axes and in one or more directions. In this classification, the pulp cavity is of primary importance, as the location of all other parts of a conodont are referred to it; and species of discrete conodonts are broadly defined, because the characteristics of the individual specimen changed during ontogeny, as recorded by its lamellae, and because an atypical specimen generally resulted if parts of that specimen were either suppressed or restored. The general trend in conodont evolution seems to have been toward increasing the surface area of the individual specimen. That is, the distacodontid conodonts, which, with the possible exception of a few strays, are restricted to Ordovician and Silurian rocks, appear to have evolved into the compound bladelike and barlike conodonts and these, in turn, to have evolved into the platelike types, which are especially characteristic of Middle Devonian and younger rocks.

The biologic classification is concerned with conodont assemblages, each of which consists of discrete specimens that are presumed to represent parts of one animal. These assemblages are considered to be whole-animal taxa; and in this paper are listed alphabetically according to their generic name. Very little of a factual nature is known about assemblages and therefore they are classified as *incertae sedis.*

The major divisions of the classification proposed for this paper follow. For reasons previously given, Branson & Mehl's (13) suborders *Neurodontiformes* and *Conodontiformes* are not recognized. Figures in parentheses denote the number of genera presently known to belong in each division.

OUTLINE OF CLASSIFICATION

Conodontophorida (147)

Utilitarian (141)

Distacodontidae (11). Pulp cavity surmounted by a single, straight or curved, undenticulated, fanglike cusp. *L.Ord.-U.Sil., ?Dev., ?Miss.*

Belodontidae (6). Pulp cavity surmounted by a single, straight or curved, denticulated, fanglike cusp whose base may be greatly enlarged. *L.Ord.-U.Sil., ?Dev.*

Coleodontidae (47). Pulp cavity beneath main cusp at or near the anterior end of denticulated bladelike or barlike unit. *L.Ord.-U.Trias.*

Coleodontinae (7). Main cusp indistinct, not terminal; anterior bar or blade short. *M.Ord.-L.Miss.*

Hindeodellinae (5). Main cusp distinct, not terminal; anterior bar or blade short. *?L.Sil., M.Sil.-U.Trias.*

Neoprioniodontinae (5). Main cusp terminal; aboral side of posterior bar may be deeply grooved but is not expanded into a concavity; anticusp, if present, commonly undenticulated but may support nearly or completely fused denticles. *L.Ord.-M.Trias.*

Cyrtoniodontinae (10). Main cusp terminal; aboral side of posterior bar partly or wholly expanded into a concavity; anticusp, if present, may be denticulated. *Ord.*

Ligonodininae (5). Main cusp terminal; posterior blade or bar may be grooved but is not excavated; anticusp present, denticulated, well-formed. *M.Ord.-M.Trias.*

Hibbardellinae (6). Main cusp terminal, at apex of denticulated anterior arch; posterior bar present. *L.Ord.-U.Trias.*

Chirognathinae (4). Main cusp at apex of denticulated arch; unit tends to be palmate. *M.Ord.-M.Trias.*

Lonchodininae (5). Main cusp at apex of denticulated arch; unit is not palmate. *L.Ord.-U.Trias.*

Prioniodinidae (10). Pulp cavity beneath main cusp at or near the posterior end of denticulated bladelike or barlike unit. *L.Ord.-U. Trias.*

Prioniodontidae (22). Pulp cavity in middle third of bladelike or barlike unit. *L.Ord.-U.Trias.*

Prioniodontinae (9). Main cusp larger than denticles of blade or bar; denticulated lateral processes may be present; unit is not palmate. *L.Ord.-U.Trias.*

Spathognathodontinae (13). Main cusp inconspicuous or but slightly larger than denticles of blade or bar; unit is not palmate. *M.Ord.-M.Trias.*

Polygnathidae (21). Pulp cavity greatly restricted; platforms flank part or all of axis. *M.Ord.-U.Trias.*

Idiognathodontidae (12). Pulp cavity not greatly restricted, so that aboral side of unit is partly or entirely opened up into a large concavity; platforms may flank part or all of axis. *U. Ord.-Up.L.Perm.*

Idiognathodontinae (8). Blade present, denticulated, well-formed; expanded pulp cavity restricted, more or less, to the anterior end of the unit. *M.Sil.-Up.L.Perm.*

Balognathinae (2). Blade present; aboral side excavated. *U.Ord.*

Icriodontinae (2). Blade poorly developed or entirely absent; aboral side excavated, or nearly so. *L.Sil.-U.Dev.*

Incertae sedis, discrete forms (12).

Biologic (6)

Incertae sedis, natural assemblages (6). [Omits assemblages named *Polygnathus dubia* by Hinde, 1879 (40); and *Prioniodus hercynicus* by Eichenberg, 1930 (23).]

SYSTEMATIC DESCRIPTIONS

Order CONODONTOPHORIDA Eichenberg, 1930

Microscopic toothlike and platelike structures belonging to an unknown monophyletic group of extinct marine animals which probably were bilaterally symmetrical, soft-bodied, and pelagic. These structures, commonly called conodonts, are usually either grayish-black or some shade of brown, composed chiefly of calcium phosphate, and consist of lamellae, open aborally, that were accreted about an initial pit—the apex of the pulp cavity. Separation of the aforementioned lamellae from one another—along one or more growth axes and in one or more directions—resulted in the formation of fanglike structures without denticles, fanglike structures with denticles, denticulated blades and bars, and platelike structures with platforms and/or a greatly expanded pulp cavity. The function performed by conodonts is as yet undetermined. Generally accepted range, *L.Ord.-U.Trias.;* possible range, *U.Cam.-U.Cret.*

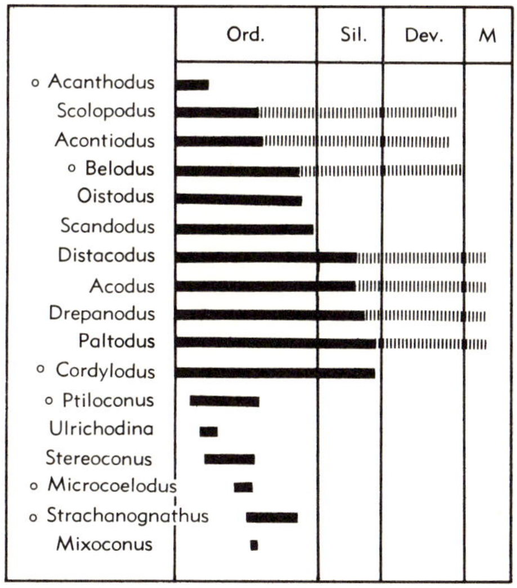

FIG. 21A. Stratigraphic distribution of conodonts (Hass, n). Classification of genera in families is indicated by presence or absence of symbol (Distacodontidae, o—Belodontidae).

UTILITARIAN CLASSIFICATION

Family DISTACODONTIDAE Bassler, 1925

[*nom. correct.* HASS, 1958 (*pro* Distacodidae BASSLER, 1925)]

Pulp cavity surmounted by a single, straight or curved, undenticulated, fanglike cusp. *L.Ord.-U.Sil., ?Dev., ?Miss.*

The stratigraphic distribution of genera included in the Distacodontidae and Belodontidae is shown graphically in Figure 21A.

Distacodus HINDE, 1879 (p. 357) [*pro Machairodus* PANDER, 1856 (*non* KAUP, 1833)] [**Machairodus incurvus* PANDER, 1856; SD MILLER, 1889 (p. 313)] [=*Machairodia* SMITH, 1907]. Bilaterally symmetrical; anterior and posterior sides sharp-edged; lateral sides with ridge along mid-line. *L.Ord.-M.Sil., ?U.Sil.-?L.Miss.,* N.Am.-Eu.——FIG. 22,*1a,b*. **D. incurvus* (PANDER); both lat., mag. unknown.——FIG. 22,*1c*. *D. ensiformis* (PANDER), L.Ord.(Glaukonitsand), Balt.; diagram. horiz. secs., mag. unknown (52).

Acodus PANDER, 1856 (p. 21) [**A. erectus;* SD ULRICH & BASSLER, 1926 (p. 7)]. Resembles *Distacodus* but asymmetrical, having ridge along mid-line of one lateral side only. *L.Ord.-M.Sil., ?U.Sil.-?L.Miss.,* N.Am.-Eu.——FIG. 22,*8*. **A. erectus,* L.Ord.(Glaukonitsand), Balt.; *8a,b,* inner, outer, mag. unknown (56).——FIG. 22,*8c*. *A. acutus* PANDER, L.Ord.(Glaukonitsand), Balt.; diagram. horiz. sec., mag. unknown (52).

Acontiodus PANDER, 1856 (p. 28) [**A. latus;* SD ULRICH & BASSLER, 1926 (p. 7)] [=*Acodina* STAUFFER, 1940]. Compressed anteroposteriorly; posterior side commonly with median ridge; in horizontal section outline of anterior side convex to obtuse; posterior side concave. *L.Ord.-M.Ord., ?U.Ord.-?U.Dev.,* N.Am.-Eu.——FIG. 22,*9a-c*. **A. latus,* L.Ord.(Glaukonitsand), Balt.; *9a-c,* post., ant., diagram. horiz. sec., mag. unknown (52).

Drepanodus PANDER, 1856 (p. 20) [*non* MENGE, 1869] [**D. arcuatus;* SD MILLER, 1889 (p. 313)] [=*Oneotodus* LINDSTRÖM, 1954]. Almost bilaterally symmetrical; outline biconvex to subcircular in horizontal section; anterior and posterior sides rounded or sharp-edged. *L.Ord.-U.Sil., ?L.Dev.-?L.Miss.,* N.Am.-S.Am.-Eu.-Austral.——FIG. 22, *10*. *D. subarcuatus* FURNISH, L.Ord.(Shakopee Dol.), USA(Wis.); *10a,b,* lat., lat. with diagram. horiz. sec., ×25 (30).

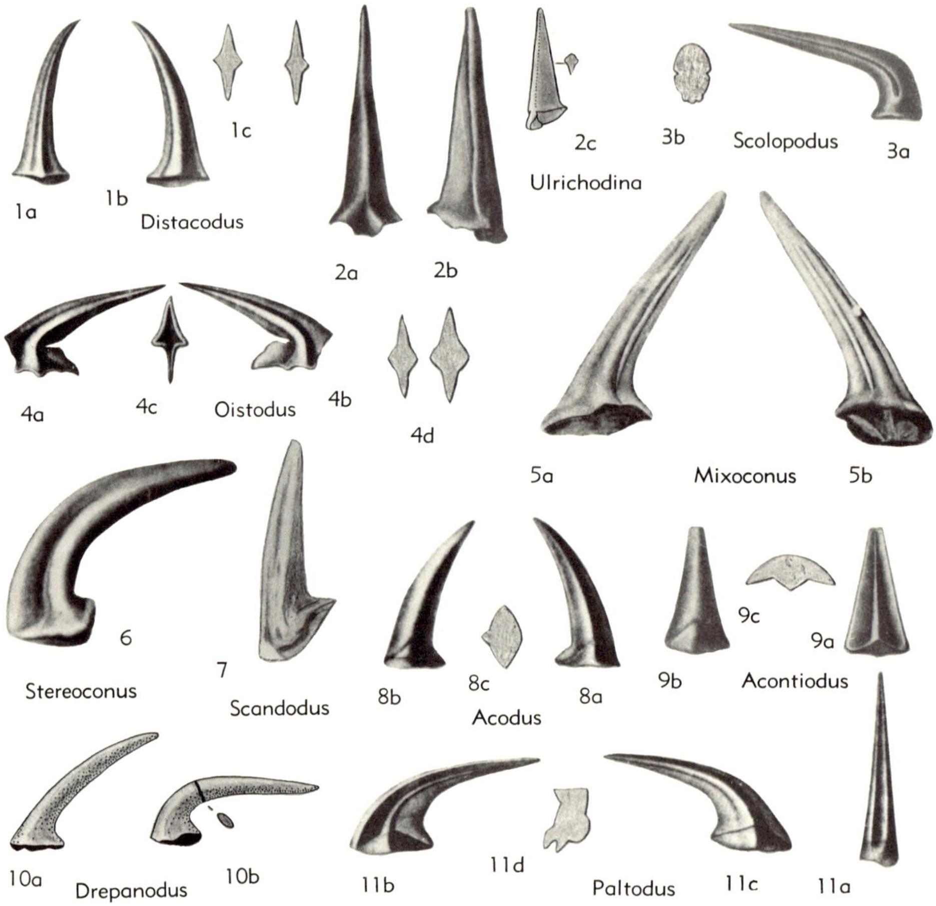

FIG. 22. Distacodontidae (p. W43-W45).

Mixoconus SWEET, 1955 (p. 244) [**M. primus*]. Resembles *Distacodus;* bilaterally symmetrical or nearly so; anterior and posterior sides rounded; lateral sides with broadly rounded, aborally widening ridge; pulp cavity shallow. *M.Ord.*, N.Am. ——FIG. 22,*5a,b*. **M. primus*, Harding Ss., USA (Colo.), both lat., ×27 (72).

Oistodus PANDER, 1856 (p. 27) [**O. lanceolatus;* SD ULRICH & BASSLER, 1926 (p. 7)]. Like *Distacodus* but with base greatly expanded posteriorly. *Ord.*, N.Am.-S.Am.-Eu.-Austral.——FIG. 22,*4a-c*. **O. lanceolatus*, L.Ord.(Glaukonitsand), Balt.; *4a-c*, lat., lat., aboral, mag. unknown (52).——FIG. 22,*4d*. *O. acuminatus* PANDER; diagram. horiz. sec. above base, mag. unknown (52).

Paltodus PANDER, 1856 (p. 24) [**P. subaequalis;* SD ULRICH & BASSLER, 1926 (p. 7)] [=*Panderodus* ETHINGTON, 1959]. Anterior and posterior sides truncated, rounded, grooved, or sharp-edged; lateral sides commonly costate. *L.Ord.-U.Sil., ?L.Dev.-?L.Miss.*, N.Am.-Eu.-Austral.——FIG. 22, *11a-c*. **P. subaequalis*, L.Ord.(Glaukonitsand), Balt.; *11a-c*, ant., lat., lat., mag. unknown (52). ——FIG. 22,*11d*. *P. truncatus* PANDER; diagram. horiz. sec., mag. unknown (52).

Scandodus LINDSTRÖM, 1954 (p. 592) [**S. furnishi*]. Asymmetrical with anterior and posterior sides sharp-edged. Base expanded on inner side. Carina may be present along mid-line of lateral sides. *Ord.*, N.Am.-Eu.-Austral.——FIG. 22,*7*. **S. furnishi*, L.Ord.(L. *Planilimbata* Z.), Swed.; inner lat., ×20 (44).

Scolopodus PANDER, 1856 (p. 25) [**S. sublaevis;* SD ULRICH & BASSLER, 1926 (p. 7)]. Anterior side rounded to sharp-edged; posterior and lateral sides costate, grooved. *L.Ord.-M.Ord., ?U.Ord.-*

?U.Dev., N.Am.-Eu.——FIG. 22,*3*. **S. sublaevis*, L.Ord.(Glaukonitsand), Balt.; *3a,b*, diagram. horiz. sec., mag. unknown (52).

Stereoconus BRANSON & MEHL, 1933 (p. 27) [**S. gracilis*]. Bilaterally symmetrical, broadest near rounded posterior side; aboral side with cordate outline; pulp cavity at notched posterior end. *L. Ord.-M.Ord.*, N.Am.-Eu.——FIG. 22,6. **S. gracilis*, M.Ord.(Harding Ss.), USA(Colo.); lat., ×25 (7).

Ulrichodina FURNISH, 1938 (p. 334) [**U. prima*]. Bilaterally symmetrical; broadest near rounded anterior side; posterior side sharp-edged; base indented anteriorly. *L.Ord.*, N.Am.——FIG. 22,2. **U. prima*, Shakopee Dol., USA(Wis.); *2a,b*, ant., lat., ×45; *2c*, diagram. lat. and horiz. sec. above base, ×25 (30).

Family BELODONTIDAE Huddle, 1934

[*nom. correct.* HASS, 1958 (*pro* Belodidae HUDDLE, 1934)]

Pulp cavity surmounted by single, straight or curved, denticulated, fanglike cusp whose base may be greatly enlarged. *L.Ord.-U.Sil., ?Dev.*

Belodus PANDER, 1856 (p. 30) [**B. gracilis*] [=*Multioistodus* CULLISON, 1938; *Belodina* ETHINGTON, 1959; *Belodella* ETHINGTON, 1959]. Bilaterally symmetrical or nearly so; one or more denticles along posterior side; lateral sides even, costate, or grooved. [*Belodina* ETHINGTON has one pulp cavity, not two. The "upper cavity" of ETHINGTON is located beneath the posteriorly extended base of the cusp and is an integral part of the pulp cavity ("lower cavity" of ETHINGTON). The posterior extension of the base is the "heel" of ETHINGTON's terminology.] *Ord., ?Sil., ?Dev.*, N.Am.-Eu.-Afr.——FIG. 23,*6a*. **B. gracilis*, L. Ord.(Glaukonitsand). Balt.; lat., mag. unknown (52).——FIG. 23,*6b*. *B.* sp., U.Ord.(Burnam Ls.), USA(Tex.); lat., ×20 (Hass, n).

Acanthodus FURNISH, 1938 (p. 336) [**A. uncinatus*]. Like *Belodus* but with serrations rather than well-formed denticles along posterior side. *L.Ord.*, N.Am.——FIG. 23,*3*. **A. uncinatus*, Stonehenge Ls., USA(Pa.); lat., ×20 (Hass, n).

Cordylodus PANDER, 1856 (p. 33) [**C. angulatus;* SD ULRICH & BASSLER, 1926 (p. 8)]. Resembles *Belodus* but has denticles on posteriorly extended base. *L.Ord.-U.Sil.*, N.Am.-Eu.——FIG. 23,*5*. **C. angulatus*, L.Ord.(Glaukonitsand), Balt.; lat., mag. unknown (52).

Micrococlodus BRANSON & MEHL, 1933 (p. 89) [**M. typus*]. Lateral sides of expanded base denticulated. *M.Ord.*, N.Am.——FIG. 23,*2*. **M. typus* Joachim Dol., USA(Mo.); post. lat., ×25 (8).

Ptiloconus SWEET, 1955 (p. 245) [*pro Pteroconus* BRANSON & MEHL, 1933 (*non* HINDE *in* FOX, 1900)] [**Pteroconus gracilis* BRANSON & MEHL,

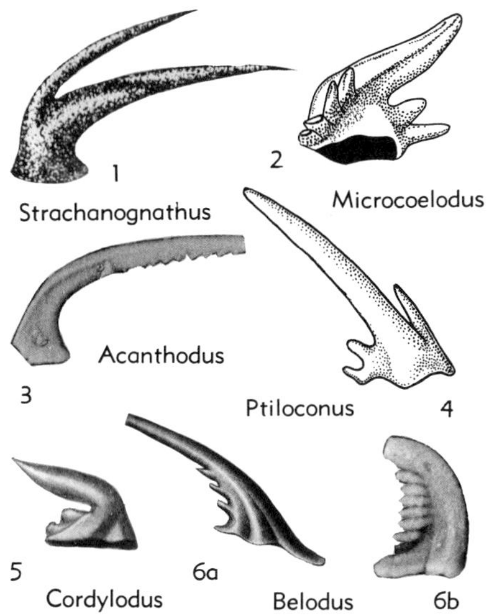

FIG. 23. Belodontidae (p. *W*45).

1933]. Anterior and posterior sides of base extended and denticulated; aboral side excavated. *L.Ord.-M.Ord.*, N.Am.——FIG. 23,*4*. **P. gracilis* (BRANSON & MEHL), M.Ord. (Plattin Ls.), USA (Mo.); outer, ×25 (8).

Strachanognathus RHODES, 1955 (p. 131) [**S. parvus*]. Cusp with one or more denticles along anterior side. *M.Ord.-U.Ord.*, N.Am.-Eu.——FIG. 23,*1*. *S.* sp., U.Ord.(Keisley), Eng.; inner lat., ×25 (95).

Family COLEODONTIDAE Branson & Mehl, 1944

[*nom. correct.* HASS, 1958 (*pro* Coleodidae BRANSON & MEHL, 1944)] [=Trucherognathidae BRANSON & MEHL, 1944]

Pulp cavity located beneath main cusp at or near the anterior end of denticulated bladelike or barlike unit. *L.Ord.-U.Trias.*

The stratigraphic distribution of genera included in the Coleodontidae is shown graphically in Figure 23A.

Subfamily COLEODONTINAE Branson & Mehl, 1944

[*nom. transl.* HASS, 1959 (*ex* Coleodontidae BRANSON & MEHL, 1944)] [=Trucherognathidae BRANSON & MEHL, 1944]

Main cusp indistinct, not terminal; anterior bar or blade short. *M.Ord.-L.Miss. (L. Osag.).*

Coleodus BRANSON & MEHL, 1953 (p. 24) [**C. simplex*]. Bladelike; denticles closely set; aboral side grooved posterior to pulp cavity and sharp-edged anterior to pulp cavity. *M.Ord.*, N.Am.——

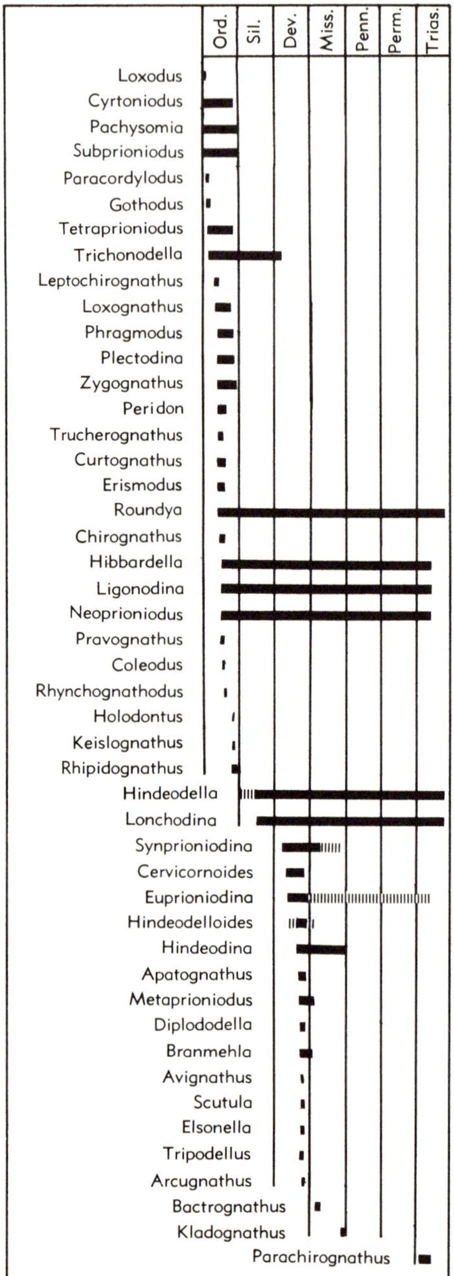

FIG. 23A. Stratigraphic distribution of conodonts (Hass, n). Classification of genera assigned to family Coleodontidae.

FIG. 24,*6*. **C. simplex,* Harding Ss., USA(Colo.); lat., ×15 (7).

Arcugnathus COOPER, 1943, *in* COOPER & SLOSS (p. 172) [**A. tenuis*]. COOPER's description; "Bar slender, regularly bowed upward; anterior end denticulated similar to *Hindeodella* with alternating upright teeth; posterior denticles strongly inclined backward; bar terminating in long slender denticle; no main cusp present." *U.Dev.,* N.Am. ——FIG. 24,*3*. **A. tenuis,* Can.(Alta.); lat., ×30 (84).

Bactrognathus BRANSON & MEHL, 1941 (p. 98) [**B. hamata*]. Posterior bar straight; anterior bar flexed inward. Denticles closely set. Lateral expansions of pulp cavity variform. *L.Miss.(uppermost Kinderhook.-lowermost Osag.),* N.Am.——FIG. 24,*7*. *B. penehamata* HASS, Chappel Ls., USA(Tex.); *7a-c,* oral, aboral, outer lat., ×20 (38).

Branmehla HASS, 1959 (p. 381) [**Spathodus inornatus* BRANSON & MEHL, 1934]. Bladelike; anterior end may be flexed inward slightly; denticles closely set; aboral side sharp-edged; lips of pulp cavity generally prominent. *U.Dev.-L.Miss.;* N. Am.-Eu.-Afr.——FIG. 24,*1*. **B. inornata* (BRANSON & MEHL), U.Dev.(Houy F.), USA(Tex.); inner lat., ×20 (38).

Hindeodina HASS, 1959 (p. 382) [**H. simplaria*]. Like *Hindeodella* but with aborted main cusp; aboral side sharp-edged; lips of pulp cavity extremely small or entirely absent. *U.Dev.-U.Miss.,* N.Am.-Eu.——FIG. 24,*5*. **H. simplaria,* L.Miss. (Chappel Ls.), USA(Tex.); inner lat., ×25 (38).

Pravognathus STAUFFER, 1936 (p. 79) [*pro Heterognathus* STAUFFER, 1935 (*non* GIRARD, 1854; *nec* SCHMARDA, 1859; *nec* KING, 1864; *nec* REY, 1888)] [**Heterognathus idonea* STAUFFER, 1935]. Bladelike, largest denticles above pulp cavity; aboral side grooved, broadly so at anterior end. *M.Ord.,* N.Am.——FIG. 24,*2*. **P. idonea* (STAUFFER), Decorah Sh., USA(Minn.); inner lat., ×35 (68).

Trucherognathus BRANSON & MEHL, 1933 (p. 84) [**T. distorta*]. Anterior and posterior bars aligned; denticles irregular; those of mature specimen may be located along inner side of massive platform. *M.Ord.,* N.Am.——FIG. 24,*4*. **T. distorta* Joachim Dol., USA(Mo.); outer lat., ×25 (8).

Subfamily HINDEODELLINAE Hass, 1959

Main cusp distinct, not terminal; anterior bar or blade short. *?L.Sil., M.Sil.-U.Trias.*

Hindeodella BASSLER, 1925 (p. 219) [**H. subtilis* ULRICH & BASSLER, 1926; SD (p. 38)]. Bar denticles closely set, commonly with group of smaller denticles alternating with larger ones; main cusp generally much larger than bar denticles; pulp cavity small. *?L.Sil., M.Sil.-U.Trias.;* N.Am.-Eu.-Afr.——FIG. 25,*2*. **H. subtilis,* U.Dev. (Chattanooga Sh.), USA(Tenn.); inner lat., ×20 (75).

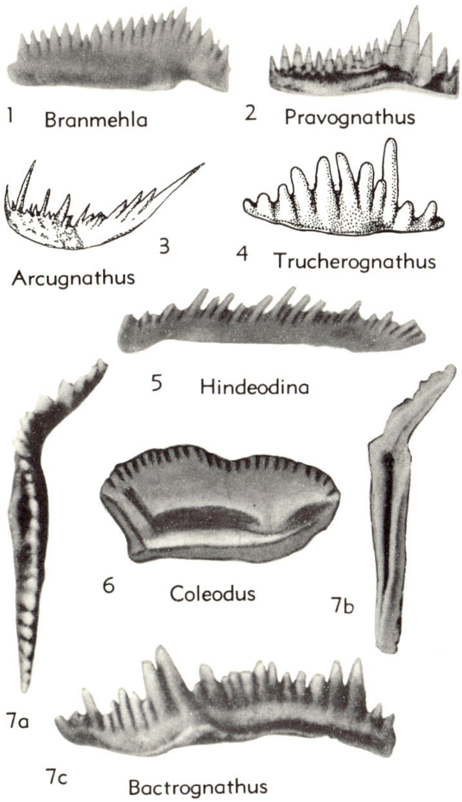

FIG. 24. Coleodontidae (Coleodontinae) (p. *W*45-*W*46).

Cervicornoides STAUFFER, 1938 (p. 424) [**C. alternatus*]. Resembles *Hindeodella* but is more sinuous; base of main cusp expanded on inner side; pulp cavity prominent. *M.Dev.-U.Dev.;* N. Am.-Eu.——FIG. 25,*5*. **C. alternatus,* U.Dev. (Olentangy Sh.), USA(Ohio); inner, ×50 (69).

Kladognathus REXROAD, 1958 (p. 19) [*pro Cladognathus* REXROAD, 1957 (*non* BURMEISTER, 1847)] [**Cladognathus prima* REXROAD, 1957]. Anterior bar aligned with posterior bar; inner lateral process directed downward and backward. *U.Miss. (Chester.);* N.Am.——FIG. 25,*3*. **K. prima, USA* (Ill.); *3a,b,* inner lat., aboral, ×40 (55).

Metaprioniodus HUDDLE, 1934 (p. 57) [**M. biangulatus*]. Resembles *Hindeodella* but has discrete denticles with largest denticles near posterior deflection. *U.Dev.-L.Miss.(Kinderhook.);* N.Am.——FIG. 25,*4*. **M. biangulatus,* L.Miss. (New Albany Sh.), USA(Ind.); *4a,b,* inner lat., outer lat., ×15 (42).

Tripodellus SANNEMANN, 1955 (p. 155) [**T. flexuosus*]. Posterior, anterior, and inner lateral bars compressed, denticulated; anterior bar much larger than posterior and inner lateral bars, directed downward, and oriented so as to face slightly toward outer side of unit; inner lateral bar joined to front basal portion of main cusp, curved backward and directed downward slightly; main cusp with sharp-edged anterior and posterior sides, curved inward and backward slightly; aboral side sharp-edged; pulp cavity very small. *U.Dev.,* Eu.——FIG. 25,*1*. **T. flexuosus,* L. *Cheiloceras* Z., Eu.; outer lat., ×40 (61).

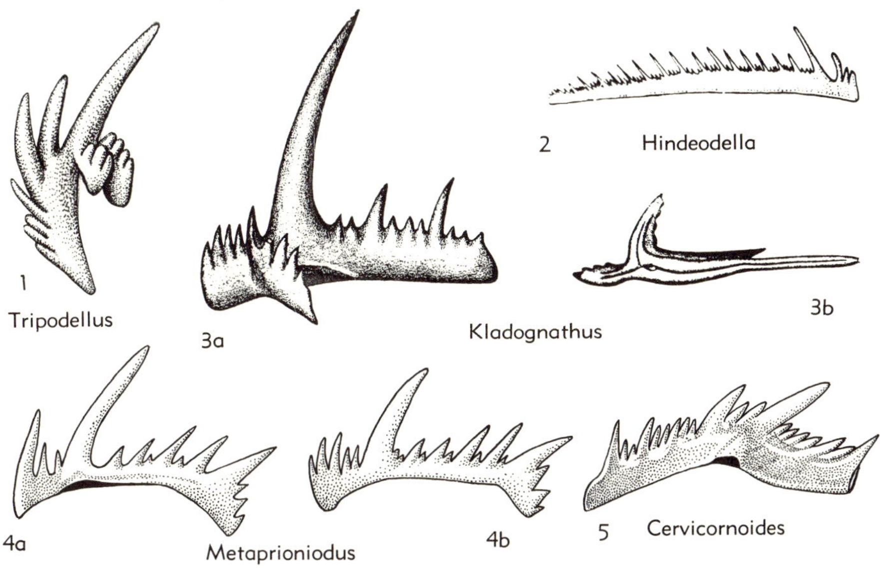

FIG. 25. Coleodontidae (Hindeodellinae) (p. *W*46-*W*47).

Subfamily NEOPRIONIODONTINAE Hass, 1959

Main cusp terminal; aboral side of posterior bar may be deeply grooved but is not expanded into a concavity; anticusp, if present, commonly undenticulated but may support nearly or completely fused denticles. *L.Ord.-M.Trias.*

Neoprioniodus Rhodes & Müller, 1956 (p. 698) [**Prioniodus conjunctus* Gunnell, 1931]. Somewhat pick-shaped; anticusp, if present, may support nearly or completely fused denticles along anterior side; posterior bar or blade either straight or slightly bowed inward; pulp cavity may have lip. *M.Ord.-M.Trias.*, N.Am.-Eu.-Afr.——Fig. 26,*1*. **N. conjunctus* (Gunnell), M.Penn.(Cherokee Sh.), USA(Mo.); lat., ×30 (24).——Fig. 26,*2*. *N. ligo* (Hass), U.Miss.(Barnett F.), USA(Tex.); lat., ×30 (35).

Leptochirognathus Branson & Mehl, 1943 (p. 377) [**L. quadrata*]. Young specimens palmate, mature ones bladelike; thickest adjacent to minute pulp cavity, which, in mature specimens, is located beneath anteriormost denticle; denticles tend to be broad, compressed, more or less fused. *M.Ord.;* N.Am.——Fig. 26,*3*. **L. quadrata*, McLish F., USA(Okla.); inner lat., ×25 (12).

Loxodus Furnish, 1938 (p. 338) [**L. bransoni*]. Bladelike; triangular outline in lateral view, highest anteriorly; denticles closely set. *L.Ord.;* N.Am.——Fig. 26,*5*. **L. bransoni*, Oneota Dol., USA(Iowa); *5a*, lat., ×50; *5b*, lat., ×50; *5c*, lat., diagram., ×25 (30).

Pachysomia Smith, 1907 (p. 246) [**P. wanlockensis*]. Smith's description: "Beam thick and curving, one extremity with a knoblike termination, the other end sharp." *Ord.*, Scot.——Fig. 26,*6*. **P. wanlockensis*, Arenig.-Llandeil., Scot.; inner lat., ×25 (96).

Subprioniodus Smith, 1907 (p. 247) [**S. paucidentatus;* SD Ulrich & Bassler, 1926 (p. 8)]. Smith's description: "This genus differs from *Prioniodus* Pander, in that the spike at the end of the beam is sharp-pointed above the beam, but not below it." [Today, Smith would probably have compared his genus with *Neoprioniodus*.] *Ord.;* Scot.——Fig. 26,*4*. **S. paucidentatus* Smith, Arenig.-Llandeil., Scot.; lat., ×25 (97).

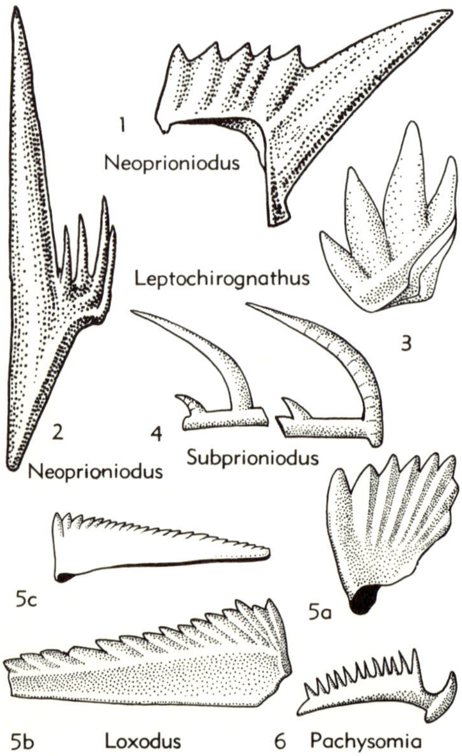

Fig. 26. Coleodontidae (Neoprioniodontinae) (p.W48).

Subfamily CYRTONIODONTINAE Hass, 1959

Main cusp terminal; aboral side of posterior bar partly or wholly expanded into a concavity; anticusp, if present, may be denticulated. *Ord.*

Cyrtoniodus Stauffer, 1935 (p. 140) [**C. complicatus*] [=*Barbarodina* Stauffer, 1935]. Main cusp may be flexed inward slightly, its base expanded on inner side; aboral side deeply excavated. *Ord.*, N.Am.-Eu.——Fig. 27,*5*. **C. complicatus*, M.Ord. (Glenwood Sh.), USA(Minn.); inner lat., ×50 (67).

Gothodus Lindström, 1954 (p. 569) [**G. costulatus*]. Similar to *Cyrtoniodus* but with anterior and outer sides of cusp costate, and with denticles along basal posterior side of cusp; aboral side excavated. *L.Ord.*, Eu.——Fig. 27,*2*. **G. costulatus*, U. *Planilimbata* Z., Swed.; *2a,b*, outer lat., inner lat., ×30 (44).

Holodontus Rhodes, 1953 (p. 303) [**H. superbus*]. Posterior bar appears to be undenticulated; inner lateral process with discrete denticles; denticles on anterior side of main cusp produced aborally; aboral side excavated. *U.Ord.*, Eu.——Fig. 27,*8*. **H. superbus*, Gelli-grin beds, Wales; outer post., ×90 (Hass, n).

Keislognathus Rhodes, 1955 (p. 130) [**K. gracilis*]. Like *Holodontus* but with denticulated posterior bar. *U.Ord.*, Eu.——Fig. 27,*4*. *K.* sp., Keisley Ls., Eng.; inner lat., ×25 (95).

Paracordylodus Lindström, 1954 (p. 584) [**P. gracilis*]. Unit compressed; denticles discrete, broad at base in anterior-posterior direction; anticusp well formed, undenticulated; pulp cavity minute. *L.Ord.*, Eu.——Fig. 27,*3*. **P. gracilis*, *L. Planilimbata* Z., Swed.; lat., ×30 (44).

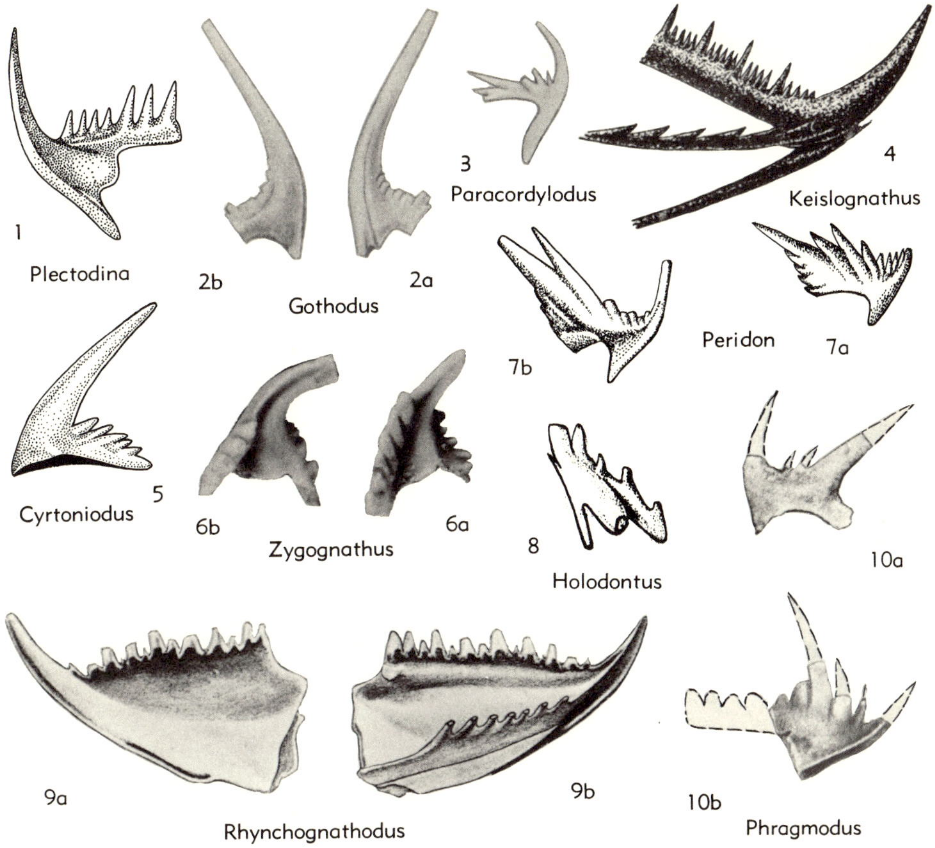

FIG. 27. Coleodontidae (Cyrtoniodontinae) (p. *W*48-*W*50).

Peridon HADDING, 1913 (p. 33) [**P. aculeatus*]. Like *Plectodina* but with anterior side very minutely denticulated; posterior bar or blade short, its denticles partly fused and largest near the distal end. *M.Ord.;* N.Am.-Eu.——FIG. 27,7. **P. aculeatus,* Swed.; *7a,b,* outer lat., inner lat., ×40 (90).

Phragmodus BRANSON & MEHL, 1933 (p. 98) [**P. primus*]. Posterior bar with large denticle near anterior end; anterior to this denticle, posterior bar is excavated and flexed downward; pulp cavity beneath main cusp. *M.Ord.-U.Ord.;* N.Am.-Eu.——FIG. 27,*10a.* **P. primus,* M.Ord.(Joachim Dol.), USA(Mo.); outer lat., ×40 (8).——FIG. 27,*10b. P. undatus* BRANSON & MEHL, M.Ord. (Plattin Ls.), USA(Mo.); inner lat., ×30 (12).

Plectodina STAUFFER, 1935 (p. 152) [**P. dilata*] [=*Subcordylodus* STAUFFER, 1935]. Main cusp flexed inward; its basal part lies beneath the anterior part of the posterior bar. *M.Ord.-U.Ord.,* N.Am.-Eu.——FIG. 27,*1.* **P. dilata,* M.Ord. (Glenwood Sh.), USA(Minn.); inner lat., ×25 (67).

Rhynchognathodus ETHINGTON, 1959 (p. 1128) [*pro Rhynchognathus* ETHINGTON, 1959 (*non* JAEKEL, 1929)] [**Rhynchognathus typica* ETHINGTON, 1959]. ETHINGTON'S description: "Asymmetrical dental units having a stout proclined curved cusp whose base is produced posteriorly as a shallow denticulate bar. The anterior margin of the cusp is continued posteriorly as a slender aboral process, which may be denticulate in the plane of the posterior bar. A lateral denticulate bar extends posteriorly, aborally, and laterally from the base of the cusp. The conical, peglike denticles of the posterior bar alternate in size and, in some specimens, a rudimentary hindeodellid arrangement may be developed. The three processes are mutually connected by sheath lamellae which enclose a hemipyramidal basal cavity extending anteriorly to a sharp point in the base of the cusp." *M.Ord.,* N.Am.——FIG. 27,9. **R. typica,* Galena F., USA; *9a,b,* inner lat., outer lat., ×100 (86).

Zygognathus BRANSON, MEHL, & C. C. BRANSON,

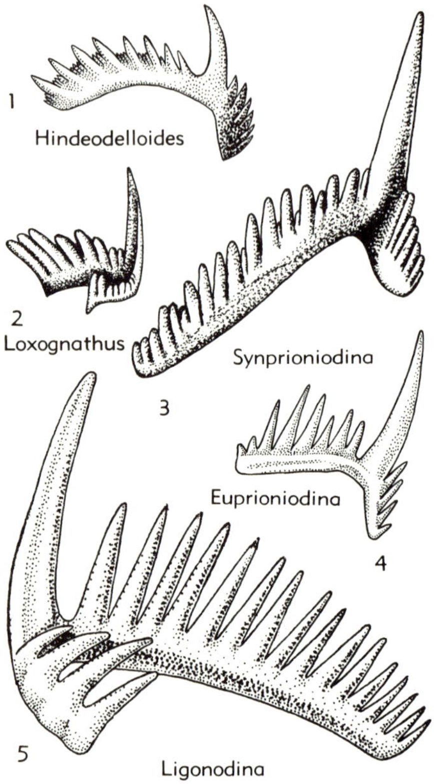

FIG. 28. Coleodontidae (Ligonodininae) (p. W50).

1951 (p. 11) [*Z. *pyramidalis*] [=*Eoligonodina* BRANSON, MEHL, & C. C. BRANSON, 1951]. Like *Plectodina* but with denticulated anticusp as in *Ligonodina*. *M.Ord.-U.Ord.*, N.Am.——FIG. 27, *6a*. *Z. *pyramidalis*, *U.Ord.* (Whitewater F.), USA(Ind.); inner lat., ×20 (14).——FIG. 27,*6b*. *Z. robusta* BRANSON, MEHL, & C. C. BRANSON, U. Ord.(Whitewater F.), USA(Ind.); inner lat., ×20 (14).

Subfamily LIGONODININAE Hass, 1959

Main cusp terminal; posterior bar or blade may be grooved but is not excavated; anticusp present, denticulated, well-formed. *M.Ord.-M.Trias.*

Ligonodina BASSLER, 1925 (p. 218) [*L. *pectinata* ULRICH & BASSLER, 1926; SD (p. 12)] [=*Hamulosodina* COOPER, 1931; *Idioprioniodus* GUNNELL, 1933; and *Neocordylodus* COOPER, 1939]. Like *Euprioniodina* but with anticusp so oriented that its denticulated oral side faces inward. *M.Ord.-M. Trias.;* N.Am.-Eu.——FIG. 28,*5*. *L. *pectinata*, U.Dev. (West Falls F.), USA(N.Y.); inner lat., ×25 (75).

Euprioniodina BASSLER, 1925 (p. 219) [*E. *deflecta* ULRICH & BASSLER, 1926; SD (p. 29)]. Pick-shaped unit with distinct, well-formed denticles along oral side of anticusp; these denticles are directed forward and not inward as in *Ligonodina*. *M.Dev.-U.Dev., ?L.Miss.-?L.Trias.;* N.Am.——FIG. 28,*4*. *E. *deflecta*, U.Dev.(West Falls F.), USA(N.Y.); lat., ×15 (75).

Hindeodelloides HUDDLE, 1934 (p. 48) [*H. *bicristatus*]. Differs from *Ligonodina* in being more compressed and in having closely set denticles that may alternate in size. *?M.Dev., U.Dev., ?L. Miss.*, N.Am.-Eu.-Afr.——FIG. 28,*1*. *H. *bicristatus*, U.Dev.(New Albany Sh.), USA (Ind.); outer lat., ×25 (42).

Loxognathus GRAVES & ELLISON, 1941 (p. 12) [*L. *flabellata*]. Base of main cusp extended on inner side into a posteriorly curved, denticulated, blade-like anticusp. *M.Ord.-U.Ord.*, N.Am.——FIG. 28, *2*. *L. *flabellata*, M.Ord.(Ft. Peña F.), USA(Tex.); inner lat., ×40 (87).

Synprioniodina BASSLER, 1925 (p. 219) [*S. *alternata* ULRICH & BASSLER, 1926; SD (p. 42)]. Like *Euprioniodina* but with closely set denticles. Main cusp points toward the anterior end. *L.Dev.-L.Miss., ?U.Miss.;* N.Am.-Eu.-Afr.——FIG. 28,*3*. *S. *alternata*, U.Dev.(Chattanooga Sh.), USA (Ala.); outer lat., ×30 (Hass, n).

Subfamily HIBBARDELLINAE Müller, 1956

[*nom. transl.* HASS, 1959 (*ex* Hibbardellidae MÜLLER, 1956)]

Main cusp terminal, located at apex of denticulated anterior arch. Posterior bar present. *L.Ord.-U.Trias.*

Hibbardella BASSLER, 1925 (p. 219) [**Prioniodus angulatus* HINDE, 1879] [=*Ellisonia* MÜLLER, 1956]. Denticles of anterior arch discrete; pulp cavity small. Denticulated posterior bar definitely present. *M.Ord.-M.Trias.;* N.Am.-Eu.-Afr.——FIG. 29,*9*. *H. *angulata* (HINDE), U.Dev.(West Falls F.), USA(N.Y.); ant., ×25 (75).——FIG. 29,*10*. *H. triassica* MÜLLER, L.Trias., USA(Nev.); oral, ×40 (49).

Avignathus LYS & SERRE, 1957 (p. 798) [*A. *beckmanni*]. Anterior arch and posterior blade with closely set, needle-like denticles; winglike denticulated lateral blades located near posterior end of unit; main cusp indistinct; pulp cavity extremely small. *U.Dev.-(Frasn.)*, Eu.——FIG. 29,*8*. *A. *beckmanni*, Fr.; *8a,b*, oral, lat., ×17 (91).

Diplododella BASSLER, 1925 (p. 219) [*D. *bilateralis* ULRICH & BASSLER, 1926; SD (p. 41)]. Anterior arch with 2 or more rows of closely set denticles that may alternate in size. *U.Dev.*, N.Am.——FIG. 29,*5*. *D. *bilateralis*, Chattanooga Sh., USA(Ala.); ant., ×30 (Hass, n).——FIG. 29,*3*. *D.* sp., Houy F., USA(Tex.); oral, ×30 (Hass, n).

——Fig. 29,*2*. *D.* sp., Houy F., USA(Tex.); lat., ×45 (Hass, n).

Elsonella Youngquist, 1945 (p. 358) [**E. prima*]. Nearly bilaterally symmetrical; resembles *Hibbardella* but with closely set denticles on anteroposteriorly compressed anterior arch; posterior bar present but characters not known. *U.Dev.;* N.Am.——Fig. 29,*4*. **E. prima*, USA(Iowa); ant., ×25 (101).

Roundya Hass, 1953 (p. 88) [**R. barnettana*]. Like *Hibbardella* but with very large pulp cavity. *M.Ord.-U.Trias.;* N.Am.-Eu.-Afr.-Asia.——Fig. 29,*1*. **R. barnettana*, U.Miss.(Barnett F.), USA (Tex.); *1a,b*, post., lat., ×25 (35).

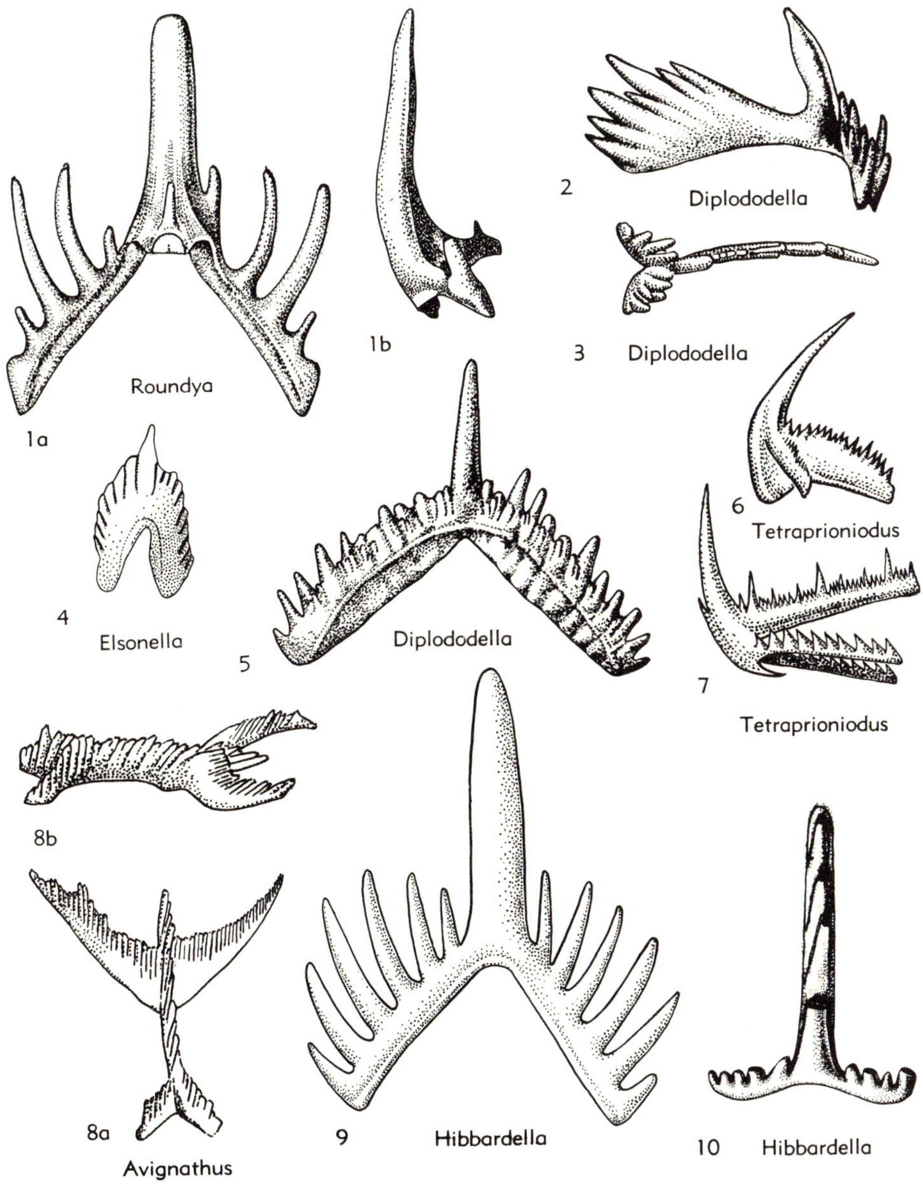

Fig. 29. Coleodontidae (Hibbardellinae) (p. *W*50-*W*52).

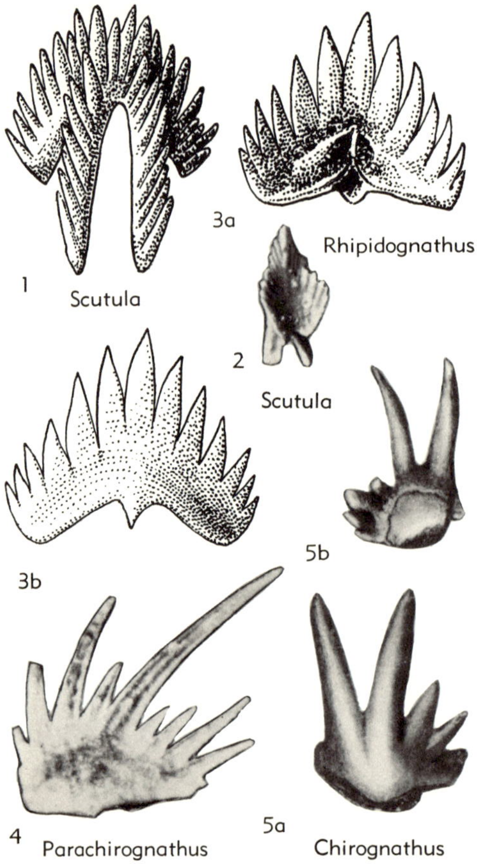

FIG. 30. Coleodontidae (Chirognathinae) (p. W52).

Tetraprioniodus LINDSTRÖM, 1954 (p. 596) [**T. robustus*] [=*Oepikodus* LINDSTRÖM, 1954; *Rosagnathus* RHODES, 1955]. Resembles *Hibbardella* but with closely spaced denticles on posterior bar, and with anticusp which may support denticle on its anterior side; lateral processes, if well developed, minutely denticulated. *Ord.;* N.Am.-Eu.——FIG. 29,*6*. **T. robustus,* L.Ord.(*L. Planilimbata* Z.), Swed.; lat., ×30 (44).——FIG. 29,*7*. *T.* sp., U.Ord.(Keisley Ls.), Eng.; inner lat., diagram. ×25 (95).

Subfamily CHIROGNATHINAE Branson & Mehl, 1944

[*nom. transl.* HASS, 1959 (*pro* Chirognathidae BRANSON & MEHL, 1944]

Main cusp at apex of denticulated arch; unit tends to be palmate. *M.Ord.-M.Trias.*

Chirognathus BRANSON & MEHL, 1933 (p. 28) [**C. duodactyla*]. Palmate, arched, broadest adjacent to main cusp; aboral side somewhat excavated. *M.Ord.;* N.Am.——FIG. 30,*5*. **C. duodactyla,* Harding Ss., USA(Colo.); *5a,b,* outer, inner, ×25 (7).

Parachirognathus CLARK, 1959 (p. 311) [**P. ethingtoni*]. Palmate unit resembling *Chirognathus,* but more compressed in the antero-posterior direction; pulp cavity minute; aboral side sharp-edged, straight; oral side of blade arched. *L.Trias.-M.Trias.;* N.Am.-Eu.——FIG. 30, *4*. **P. ethingtoni,* L.Trias. (*Meekoceras* Z.), USA (Nev.); ant., ×60 (83).

Rhipidognathus BRANSON, MEHL, & C. C. BRANSON, 1951 (p. 10) [**R. symmetrica*]. Unit tends to be palmate and concave toward posterior side, though some massive specimens may resemble *Bryantodus;* pulp cavity flanked by aborally trending lip; aboral side more or less excavated. *U.Ord.;* N.Am.——FIG. 30,*3*. **R. symmetrica,* Whitewater F., USA(Ind.); *3a,b,* post., ant., ×30, ×25 (14).

Scutula SANNEMANN, 1955 (p. 154) [**S. venusta*]. Anterior arch compressed, posteriorly bowed, den-

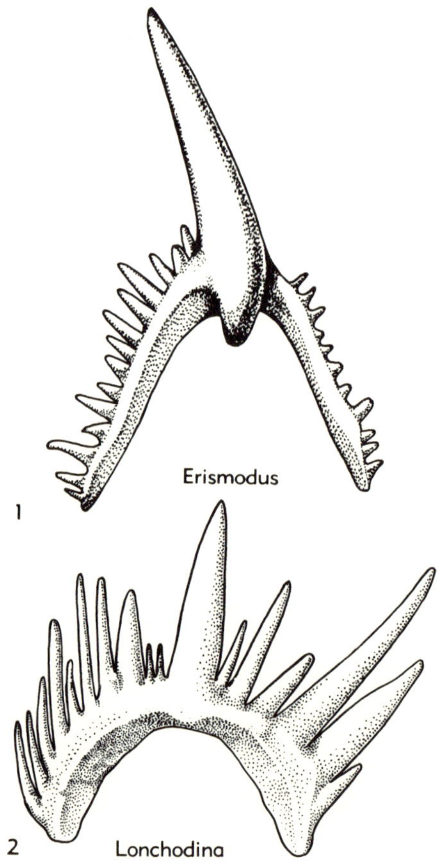

FIG. 31. Coleodontidae (Lonchodininae) (p. W53).

ticles closely set; main cusp may be indistinct, flanked by denticulated blades, bars, or limbs on anterior side of arch; pulp cavity minute. *U.Dev.;* N.Am.-Eu.——FIG. 30,*1*. *S.* sp., Houy F., USA Tex.); ant., ×40 (Hass, n).——FIG. 30,*2*. *S. bipennata* SANNEMANN, L. *Cheiloceras* Z., Eu.; post., ×40 (61).

Subfamily LONCHODININAE Hass, 1959

Main cusp at apex of denticulated arch; unit is not palmate. *L.Ord.-U.Trias.*

Lonchodina BASSLER, 1925 (p. 219) [**L. typicalis* ULRICH & BASSLER, 1926; SD (p. 31)]. Unit asymmetrical, barlike; bars of arch may be offset slightly with reference to each other; denticles discrete. *M.Sil.-U.Trias.;* N.Am.-Eu.-Asia-Afr.——FIG. 31,*2*. **L. typicalis,* U.Dev.(West Falls F.); USA(N.Y.); lat., ×25 (75).

Apatognathus BRANSON & MEHL, 1934 (p. 201) [**A. varians*]. Bars of anterior arch flexed so as to trend posteriorly. *U.Dev.,* N.Am.-Eu.-Afr.——FIG. 32,*1*. **A. varians,* USA(Mo.); *1a,b,* oral, aboral, ×37 (9).

Curtognathus BRANSON & MEHL, 1933 (p. 87) [**C. typa*]. Barlike; broadest along aboral side; denticles point outward. *M.Ord.,* N.Am.——FIG. 32, *3*. **C. typa,* Joachim Dol., USA(Mo.); ant., ×37 (8).

Erismodus BRANSON & MEHL, 1933 (p. 25) [**E. typus*]. Nearly bilaterally symmetrical; main cusp curved posteriorly, produced aborally on anterior side, forming linguiform process. *M.Ord.,* N.Am.——FIG. 32,*2*. **E. typus,* Harding Ss., USA (Colo.); ant., ×37 (7).——FIG. 31,*1*. *E. radicans* (HINDE), Can.(Que.); ant., ×25 (7).

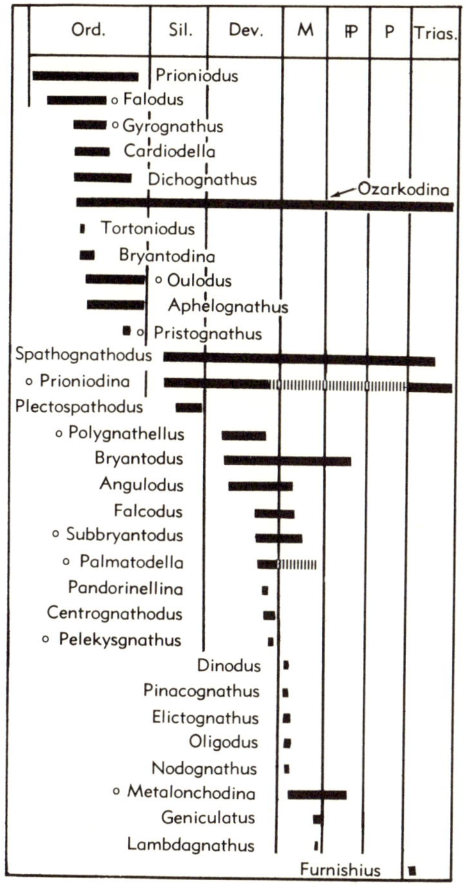

FIG. 32A. Stratigraphic distribution of conodonts (Hass, n). Classification of genera in families is indicated by presence or absence of symbol (o—Prionidinidae, Prioniodontidae).

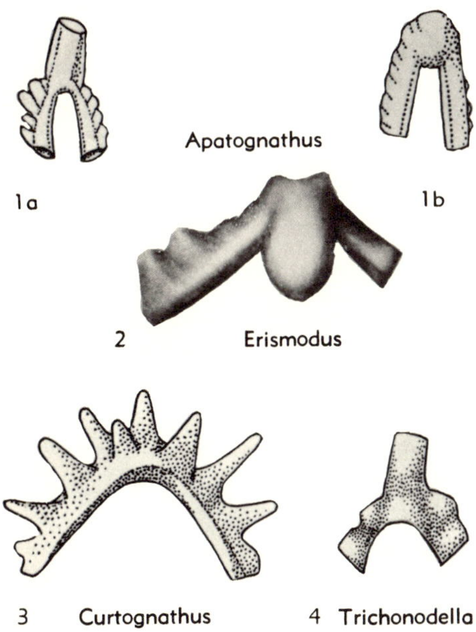

FIG. 32. Coleodontidae (Lonchodininae) (p. *W*53).

Trichonodella BRANSON & MEHL, 1948 (p. 527) [*pro Trichognathus* BRANSON & MEHL, 1933 (*non* BERTHOLD, 1827; *nec* GEMMINGER & HAROLD, 1868)] [**Trichognathus prima* BRANSON & MEHL, 1933]. Nearly symmetrical; main cusp curved posteriorly, its base produced posteriorly but undenticulated; pulp cavity large. *L.Ord.-L.Dev.,* N.Am.-Eu.——FIG. 32,*4*. **T. prima* BRANSON & MEHL, M.Ord.(Harding Sh.), USA(Colo.); post.; ×37 (7).

Family PRIONIODINIDAE Bassler, 1925

Pulp cavity beneath main cusp, at or near posterior end of denticulated bladelike or barlike unit. *L.Ord.-U.Trias.*

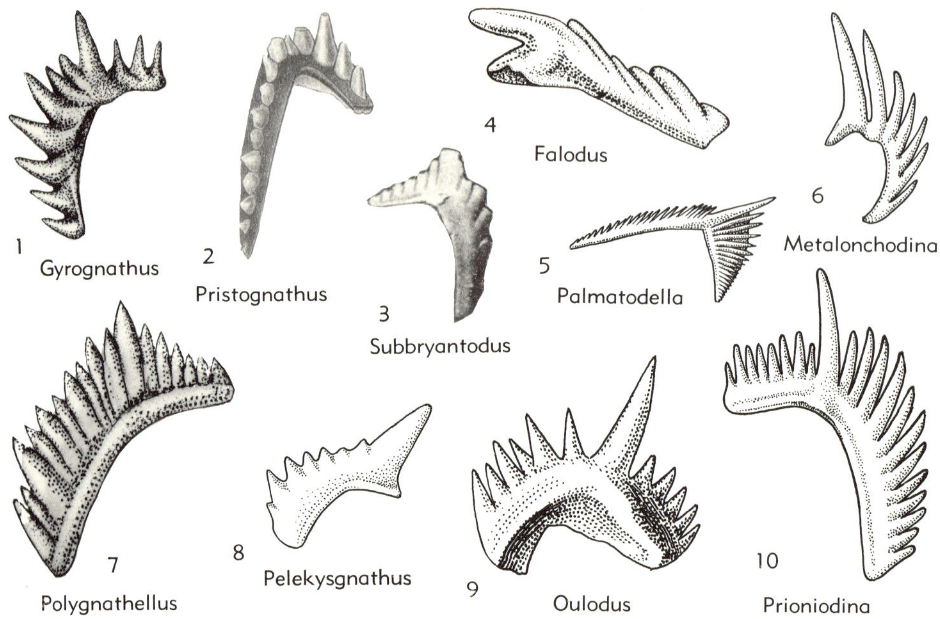

FIG. 33. Prioniodinidae (p. W54-W55).

Stratigraphic distribution of genera contained in the families Prioniodinidae and Prioniodontidae is illustrated graphically in Figure 32A.

Prioniodina BASSLER, 1925 (p. 219) [**P. subcurvata* ULRICH & BASSLER, 1926; SD (p. 18)]. [=*Prioniodella* BASSLER, 1925]. Denticles of posterior bar discrete, erect; anterior bar directed downward, its denticles discrete, curved upward; base of main cusp expanded. *L.Sil.-U.Dev., Trias.,* N.Am.-Eu.-Afr.——FIG. 33,*10*. **P. subcurvata,* U.Dev.(West Falls F.), USA(N.Y.); inner lat., ×25 (75).

Falodus LINDSTRÖM, 1954 (p. 569) [**Oistodus prodentatus* GRAVES & ELLISON, 1941]. Base of cusp produced posteriorly; its aboral side excavated; anterior bar or blade denticulated; unit lacks posterior bar or blade. *L.Ord.-M.Ord.,* N.Am.-Eu. ——FIG. 33,*4*. **F. prodentatus* (GRAVES & ELLISON), M.Ord.(Ft. Peña F.), USA(Tex.); lat., ×35 (87).

Gyrognathus STAUFFER, 1935 (p. 144) [**G. prima*]. Unit sinuous; anterior bar flexed inward and directed downward; posterior bar flexed outward; aboral side entirely excavated. *M.Ord.,* N.Am.-Eu.——FIG. 33,*1*. **G. prima,* Glenwood F., USA(Minn.); inner lat., ×30 (67).

Metalonchodina BRANSON & MEHL, 1941 (p. 105) [**Prioniodus bidentatus* GUNNELL, 1931]. Posterior bar supports large denticle; anterior bar trends downward, its denticles discrete; main cusp expanded on inner side at base. *L.Miss.-U.Penn.,* N.Am.-Eu.——FIG. 33,*6*. **M. bidentata* (GUNNELL), M.Penn.(Lexington Coal), USA(Mo.); lat., ×15 (24).

Oulodus BRANSON & MEHL, 1933 (p. 116) [**O. mediocris*]. Anterior blade or bar flexed outward and trending downward; posterior blade or bar short, straight; base of main cusp expanded on inner side; aboral side excavated. *M.Ord.-U.Ord.,* N.Am.——FIG. 33,*9*. **O. mediocris,* M.Ord. (Plattin Ls.), USA(Mo.); outer lat., ×40 (8).

Palmatodella BASSLER, 1925 (p. 219) [**P. delicatula* ULRICH & BASSLER, 1926; SD (p. 41)] [=*Telumodina* COOPER, 1931; *Ligonodinoides* STAUFFER, 1938]. Compressed anterior bar trending downward; bar denticles needle-like, directed upward. *U.Dev.-L.Miss.(Kinderhook.), ?L.Miss. (Osag.), ?U.Miss.,* N.Am.-Eu.-Afr.——FIG. 33,*5*. **P. delicatula,* U.Dev.(Houy F.), USA(Tex.); lat., ×15 (Hass, n).

Pelekysgnathus THOMAS, 1949 (p. 424) [**P. inclinata*]. Cusp with expanded base, aboral side of unit excavated. *U.Dev.,* N.Am.-Eu.——FIG. 33,*8*. **P. inclinata,* USA(Iowa); lat., ×25 (99).

Polygnathellus BASSLER, 1925 (p. 220) [**P. typicalis* ULRICH & BASSLER, 1926; SD (p. 53)]. Resembles *Bryantodus* but main cusp is inconspicuous or but slightly larger than denticles of the blades or bars, and pulp cavity is located in or very near posterior third of unit instead of approximately equidistant from anterior and pos-

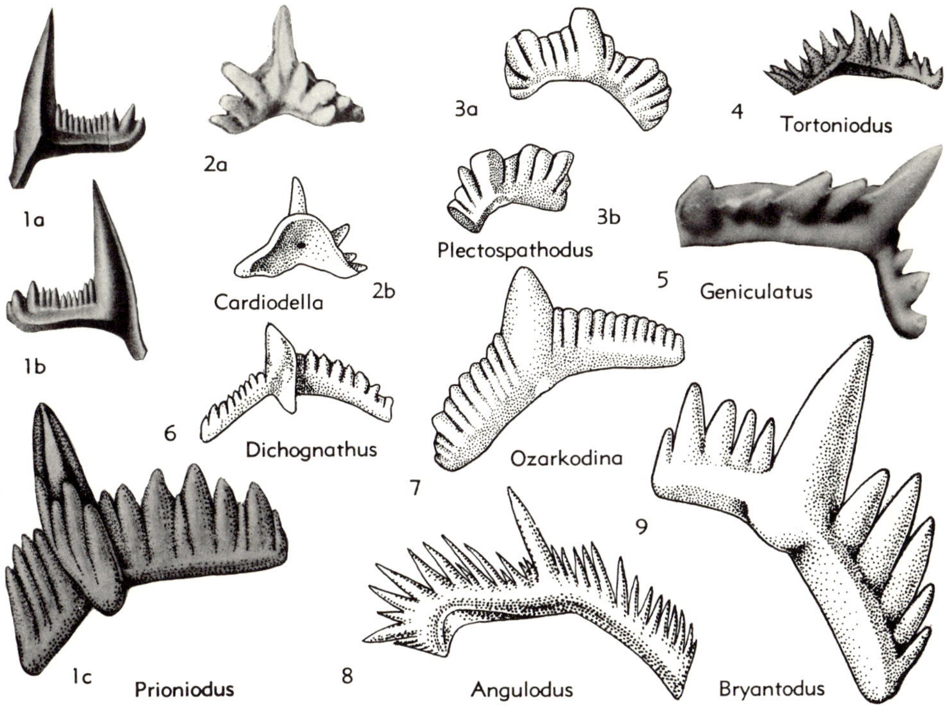

FIG. 34. Prioniodontidae (Prioniodontinae) (p. *W*55-*W*56).

terior ends of unit. *L.Dev.-U.Dev.*, N.Am.-Eu. ——FIG. 33,*7*. **P. typicalis*, U.Dev.(West Falls F.), USA(N.Y.); inner lat., ×30 (Hass, n).

Pristognathus STONE & FURNISH, 1959 (p. 226) [**P. bighornensis*]. Like *Gyrognathus* but lacking a distinct main cusp and having aboral side grooved along mid-line throughout most of unit instead of being decidedly excavated; unit not sinuous; also resembles *Polygnathellus* but its long anterior bar is flexed inward as well as downward. *U.Ord.*, N.Am.——FIG. 33,*2*. **P. bighornensis*, Bighorn Dol., USA(Wyo.); inner lat., ×30 (98).

Subbryantodus BRANSON & MEHL, 1934 (p. 285) [**S. arcuatus*]. Like *Prioniodina* but more compressed; denticles more or less fused; pulp cavity small. *U.Dev.-L.Miss.*, N.Am.-Eu.——FIG. 33,*3*. **S. arcuatus*, L.Miss.(Sulphur Springs F.), USA (Mo.); lat., ×25 (10).

Family PRIONIODONTIDAE Bassler, 1925

[*nom. correct.* MOORE & SYLVESTER-BRADLEY, 1957 (*pro* Prioniodidae BASSLER, 1925)]

Pulp cavity in middle third of bladelike or barlike unit. *L.Ord.-U.Trias.*

Subfamily PRIONIODONTINAE Bassler, 1925

[*nom. transl.* HASS, 1959; *nom. correct.* Prioniodontidae MOORE & SYLVESTER-BRADLEY, 1957 (*pro* Prioniodidae BASSLER, 1925)]

Main cusp larger than denticles of blade or bar; denticulated lateral process may be present; unit is not palmate. *L.Ord.-U. Trias.*

Prioniodus PANDER, 1856 (p. 29) [**P. elegans* PANDER, 1856; SD MILLER, 1889 (p. 315)]. Main cusp subcentral; anterior bar, posterior bar, and lateral process support closely set denticles; pulp cavity small. *L.Ord.-U.Ord.*, N.Am.-Eu.——FIG. 34,*1*. **P. elegans* PANDER, L.Ord.(Glaukonitsand), Balt.; *1a,b*, outer lat., inner lat., mag. unknown (52); *1c*, inner lat., ×100 (94).

Angulodus HUDDLE, 1934 (p. 76) [**A. demissus*]. Posterior bar with deflection at its distal end; anterior bar angled downward; denticles closely set; pulp cavity small. *M.Dev.-L.Miss.(Kinderhook.)*, N.Am.-Eu.——FIG. 34,*8*. **A. demissus*, U.Dev.(New Albany Sh.), USA(Ind.); lat., ×25 (42).

Bryantodus BASSLER, 1925 (p. 219) [**B. typicus* ULRICH & BASSLER, 1926; SD (p. 21)]. Bar triangular in transverse section with oral side broad and aboral side sharp-edged; pulp cavity may have lips. *L.Dev.-M.Penn.*, N.Am.-Eu.-Afr.-Austral.——FIG. 34,*9*. **B. typicus*, U.Dev.(West Falls F.), USA(N.Y.); inner lat., ×25 (75).

Cardiodella BRANSON & MEHL, 1944 (p. 237) [*pro Cardiodus* BRANSON & MEHL, 1933 (*non* TROUESSART, 1881)] [**Cardiodus tumidus*]. Young specimens barlike; mature ones somewhat platelike; main cusp more or less aligned with posterior bar; anterior bar joined to inner side of main cusp; aboral side concave. *M.Ord.*, N.Am.——FIG. 34,*2*. **C. tumida* (BRANSON & MEHL), Joachim Dol., USA(Mo.); *2a,b*, oral, aboral, ×25 (8).

Dichognathus BRANSON & MEHL, 1933 (p. 35) [**D. prima*]. Bladelike; denticles closely set; posterior bar merges into prominent main cusp which is flexed inward slightly; anterior bar joined to outer side of main cusp. *M.Ord.-U.Ord.*, N.Am.-Eu.——FIG. 34,*6*. *D. typica* BRANSON & MEHL, M.Ord. (Plattin F.), USA(Mo.); inner lat., ×25 (8).

Geniculatus HASS, 1953 (p. 77) [**Polygnathus? claviger* ROUNDY, 1926]. Geniculate, asymmetrical barlike unit, broadest at vertex where main cusp and pulp cavity are located; denticles variform. *U.Miss.*, N.Am.-Eu.——FIG. 34,*5*. **Geniculatus claviger* (ROUNDY), Barnett F., USA(Tex.); oral, ×20 (35).

Ozarkodina BRANSON & MEHL, 1933 (p. 51) [**O. typica*]. Resembles *Bryantodus* but compressed and bladelike; base of apical denticle slightly expanded. *M.Ord.-U.Trias.*, N.Am.-Eu.-Asia-Afr.——FIG. 34,*7*. **O. typica*, M.Sil.(Bainbridge Ls.), USA(Mo.); lat., ×25 (7).

Plectospathodus BRANSON & MEHL, 1933 (p. 47) [**P. flexuosus*]. Resembles *Ozarkodina* but bowed inward slightly, and expanded on inner side of main cusp; pulp cavity may have lip on inner side. *M.Sil.-U.Sil.*, N.Am.-Eu.——FIG. 34,*3*. **P. flexuosus*, M.Sil.(Bainbridge Ls.), USA(Mo.); *3a,b*, outer lat., inner lat., ×25 (7).

Tortoniodus STAUFFER, 1935 (p. 154) [**T. politus*]. Barlike, arched, twisted; pulp cavity prominent, flared. *M.Ord.*, N.Am.——FIG. 34,*4*. **T. politus*, Glenwood F., USA(Minn.); inner lat., ×30 (67).

Subfamily SPATHOGNATHODONTINAE Hass, 1959

Main cusp inconspicuous or but slightly larger than denticles of either the blade or bar; unit not palmate. *M.Ord.-M.Trias.*

Spathognathodus BRANSON & MEHL, 1941 (p. 98) [*pro Spathodus* BRANSON & MEHL, 1933 (*non* BOULENGER, 1900)] [**Spathodus primus* BRANSON & MEHL, 1933] [=*Ctenognathus* PANDER, 1856 (*non* FAIRMAIRE, 1843); *Mehlina* YOUNGQUIST, 1945; *Ctenognathodus* FAY, 1959]. Essentially straight in oral view; denticles along midline closely set and may be flanked laterally by one or more denticles; oral surface of lips or lateral expansions of pulp cavity may support denticles. *L.Sil.-M.Trias.*, N.Am.-Eu.-Afr.——FIG. 35,*9*. **S. primus* (BRANSON & MEHL), M.Sil.(Bainbridge Ls.), USA(Mo.); lat., ×25 (7).

Aphelognathus BRANSON, MEHL, & C. C. BRANSON, 1951 (p. 9)] [**A. grandis*]. Resembles *Spathognathodus*, but differs in having aboral side deeply excavated; unit somewhat expanded both about and anterior to pulp cavity; more massive and thicker than *Bryantodina*. *M.Ord.-U.Ord.*, N.Am.-Eu.——FIG. 35,*6*. **A. grandis*, U.Ord. (Richmond), USA(Ky.); *6a,b*, oral, inner lat., ×25 (14).

Bryantodina STAUFFER, 1935 (p. 131) [**B. typicalis*]. Like *Spathognathodus* but with aboral side more or less excavated and with denticles more discrete. *M.Ord.*, N.Am.——FIG. 35,*4*. **B. typicalis*, Glenwood F., USA(Minn.); inner lat., ×50 (67).

Centrognathodus BRANSON & MEHL, 1944 (p. 240) [*pro Centrognathus* BRANSON & MEHL, 1934 (*non* GUÉRIN-MENÉVILLE, 1840)] [**Centrognathus sinuosa* BRANSON & MEHL, 1934]. Sinuous in oral view; aboral side sharp-edged; pulp cavity small; one or more lateral bars may be present. *U.Dev.*, N.Am.——FIG. 35,*2*. **C. sinuosus* (BRANSON & MEHL), USA(Mo.), oral, ×22 (9).

Dinodus COOPER, 1939 (p. 386) [**D. leptus*]. Laterally compressed; anterior blade located beneath posterior blade; unit broadest adjacent to aboral side; denticles needle-like, closely set; main cusp indistinct; pulp cavity small. *L.Miss.*, N.Am.-Eu.——FIG. 35,*12*. *D. fragosus* (E. R. BRANSON), L. Miss.(Houy F.), USA(Tex.); lat., ×25 (Hass, n).

Elictognathus COOPER, 1939 (p. 386) [**Solenognathus bialata* BRANSON & MEHL, 1934] [=*Solenodella* E. R. BRANSON & MEHL, 1944 (*pro Solenognathus* BRANSON & MEHL, 1934; *non* AGASSIZ, 1846; *nec* BLEEKER, 1856-57; *nec* PICTET & HUMBERT, 1866)]. Compressed, arched slightly; basal part of posterior end may be flexed inward; inner side may have narrow platform and denticulated parapet; pulp cavity small. *L.Miss. (Kinderhook.)*, N.Am.-Eu.——FIG. 35,*1*. **E. bialata* (BRANSON & MEHL), L.Miss.(Chappel Ls.), USA(Tex.); inner lat., ×20 (38).

Falcodus HUDDLE, 1934 (p. 87) [**F. angulus*]. Laterally compressed; posterior bar straight to near posterior deflection where 1 or 2 large denticles commonly occur; anterior bar angled downward about 90 degrees; denticles closely set; pulp cavity small. *U.Dev.-L.Miss.(Kinderhook.)*; N. Am.-Eu.——FIG. 35,*7*. **F. angulus*, L.Miss.(U. New Albany Sh.), USA(Ind.); outer lat., ×25 (42).

Furnishius CLARK, 1959 (p. 310) [**F. triserratus*].

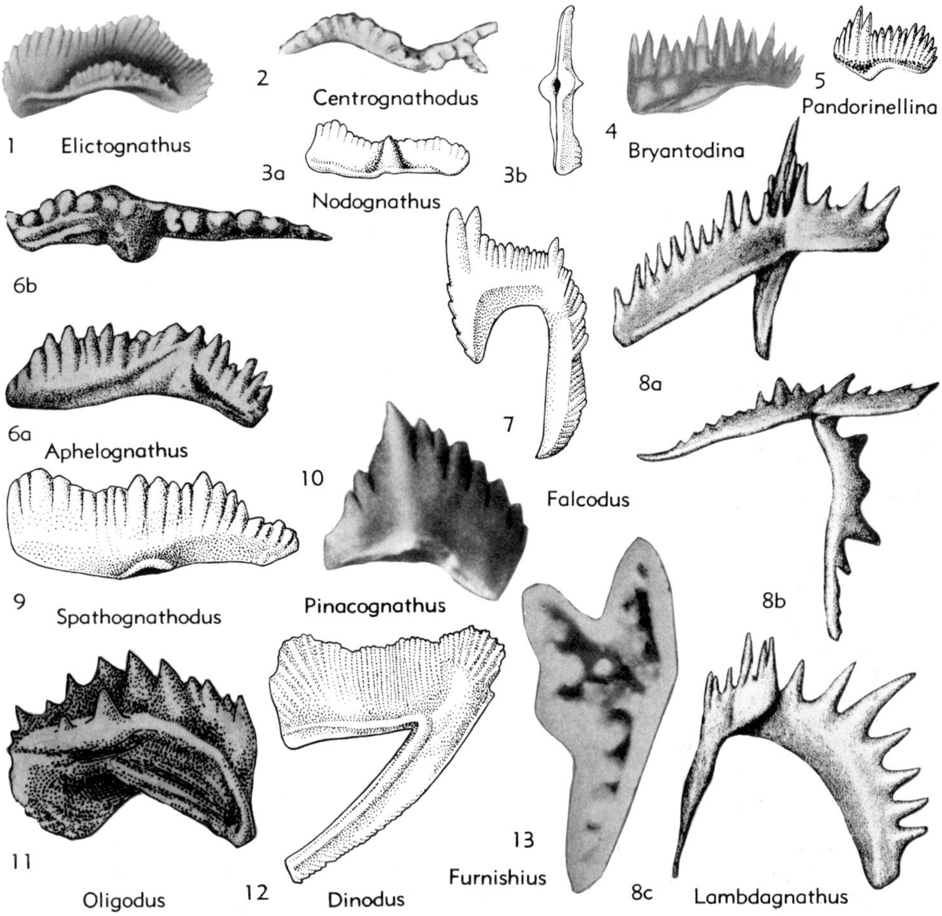

FIG. 35. Prioniodontidae (Spathognathodontinae) (p. *W*56-*W*58).

Unit "Y"-shaped, denticulated; posterior bar aligned with anterior bar; inner lateral process directed anteriorly; small plate with nodes and denticles developed in vicinity of pulp cavity; minute lobe may be present on inner side adjacent to pulp cavity; aboral side of plate not excavated; pulp cavity minute; aboral side of bars keeled. *L.Trias.*, N.Am.——FIG. 35,*13*. **F. triserratus, Meekoceras* Z., USA(Nev.); oral, ×88 (83).

Lambdagnathus REXROAD, 1958 (p. 19) [**L. fragilidens*]. Bladelike to barlike unit with large downward trending lateral process; denticles of unit discrete; aboral side of unit more or less sharp-edged; pulp cavity well formed, tends to be triangular in horizontal section. *U.Miss.(Chester.)*; N.Am.——FIG. 35,*8*. **L. fragilidens*, Glen Dean Ls., USA(Ill.,Ind.,Ky.); *8a-c*, outer lat., aboral, outer post., ×23 (55).

Nodognathus COOPER, 1939 (p. 397) [**N. spicata*]. Resembles *Spathognathodus;* blade thickened adjacent to prominent pulp cavity where each lateral expansion commonly supports long node or short transverse ridge, may grade into *Pseudopolygnathus*. *L.Miss.(Kinderhook.)*, N.Am.——FIG. 35,*3*. **N. spicata*, USA(Okla.); *3a,b*, lat., aboral, ×15 (38).

Oligodus COOPER, 1939 (p. 398) [**O. curtus*]. Like *Pinacognathus;* blade curved inward especially posterior to pulp cavity; flange on inner side prominent; aboral side wide, especially in posterior half of unit. *L.Miss.(Kinderhook.)*, N.Am.——FIG. 35,*11*. **O. curtus*, USA(Okla.); inner lat., ×40 (16).

Pandorinellina HASS, 1959 (p. 378) [*pro Pandorina* STAUFFER, 1940 (*non* BORY DE ST. VINCENT, 1827; *nec* SCACCHI, 1833] [**Pandorina insita* STAUFFER, 1940]. Compressed; denticles more or less fused; aboral side broad anterior to pulp cavity and sharp-edged posterior to pulp cavity. *U.Dev.*, N.Am.——Fig. 35,*5*. **P. insita* (Stauffer), Cedar Valley Ls., USA(Minn.); outer lat., ×25 (97).

Pinacognathus Branson & Mehl, 1944 (p. 244) [*pro Pinacodus* Branson & Mehl, 1934 (*non* Davis, 1883)] [**Pinacodus profundus* Branson & Mehl, 1934]. Short compressed; high anterior to small pulp cavity; denticles fused. *L.Miss. (Kinderhook.)*, N.Am.——Fig. 35,*10*. **P. profunda* (Branson & Mehl), L.Miss.(Sulphur Springs F.), USA(Mo.); lat., ×30 (38).

Family POLYGNATHIDAE Bassler, 1925

[=Polygnathinae Harris & Hollingsworth, 1933]

Pulp cavity greatly restricted; platforms flank part or all of axis. *M.Ord.-U.Trias.*

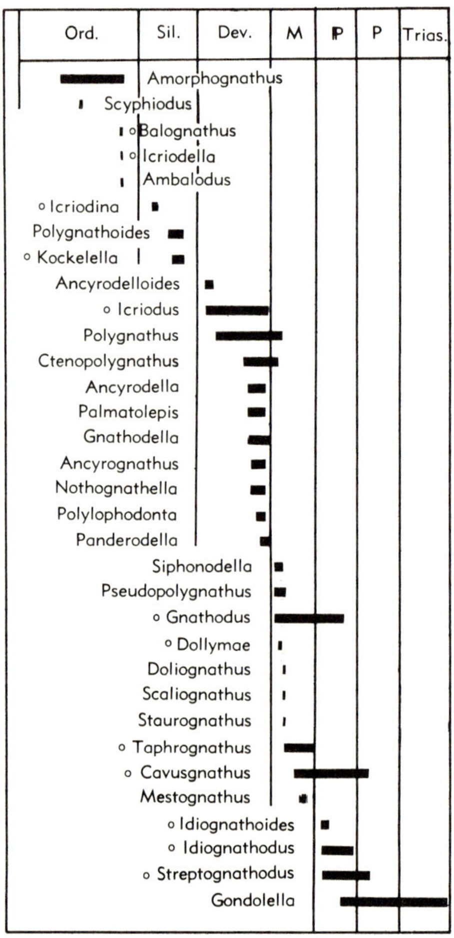

Fig. 35A. Stratigraphic distribution of conodonts (Hass, n). Classification of genera in families is indicated by presence or absence of symbol (Polygnathidae, o—Idiognathodontidae).

Stratigraphic distribution of genera contained in the families Polygnathidae and Idiognathodontidae is illustrated graphically in Figure 35A.

Polygnathus Hinde, 1879 (p. 359) [**P. dubia* Hinde, 1879, SD Miller, 1889 (p. 314). The lectotype of *P. dubia,* selected by Roundy, 1926, may be conspecific with specimens commonly identified as *Ancyrodella rotundiloba* Hinde. Should this observation prove to be correct, then *Ancyrodella* Ulrich & Bassler, 1926, is a junior subjective synonym of *Polygnathus* Hinde, and species presently assigned to *Polygnathus* would be transferred to *Macropolygnathus* Cooper, 1939]. [=*Macropolygnathus* Cooper, 1939]. Leaf-shaped to lanceolate; platforms variform, commonly flank all of carina and anterior part of blade, but in some specimens, platforms extend to posterior end of blade. *L.Dev.-L.Miss.*, N.Am.-Eu.-Afr.——Fig. 36,*5a*. **P. dubia* Hinde, U.Dev., USA(N.Y.); lat., ×25 (40).——Fig. 36,*5b*. *P. inornata* E. R. Branson, L.Miss.(U.Houy. F.), USA(Tex.); oral, ×25 (Hass, n).——Fig. 36,*5c*. *P. linguiformis* Hinde, Dev. (L.Houy F.), USA(Tex.); lat., ×25 (Hass, n).

Ambalodus Branson & Mehl, 1933 (p. 127) [**A. triangularis*]. Roughly triangular in oral view; arched about pulp cavity; narrow platforms flank axis; secondary carina may be present. *U.Ord.*, N.Am.-Eu.——Fig. 37,*2*. **A. triangularis,* Thebes Ss., USA(Mo.); oral, ×37 (8).

Amorphognathus Branson & Mehl, 1933 (p. 126) [**A. ordovicica*] [=*Polyplacognathus* Stauffer, 1935]. Depressed, irregular in outline; narrow platforms flank axis; secondary carinae radiate from pulp cavity. *M.Ord.-U.Ord.*, N.Am.-Eu.——Fig. 37,*5*. *A. ramosa* (Stauffer), M.Ord. Kimmswick Ls.), USA(Mo.); *5a,b,* aboral, oral, ×37 (8).

Ancyrodella Ulrich & Bassler, 1926 (p. 48) [**A. nodosa*] [=*Ancyropenta* Müller & Müller, 1957]. Sagittate to cordate, with variform platforms and secondary carinae trending from vicinity of apex of pulp cavity to margin of plate; oral surface of platforms with nodes and transverse ridges. *U.Dev.*, N.Am.-Eu.-Afr.-Austral.——Fig. 36,*1*. *A. lobata* Branson & Mehl, Houy F., USA(Tex.); *1a,b,* aboral, oral, ×25, ×20 (Hass, n).

Ancyrodelloides Bischoff & Sannemann, 1958 (p. 91) [**A. trigonica*]. Like *Ancyrodella*, but with oral surface of smooth platforms and aboral side of blade distinctly grooved along mid-line. *L.Dev.*, Eu.——Fig. 36,*6a-c*. **A. trigonica,* Siegenian, Ger.; *6a-c,* oral, aboral, lat., ×17 (4).

Ancyrognathus Branson & Mehl, 1934 (p. 240) [**A. symmetrica*] [=*Ancyroides* Miller & Youngquist, 1947]. Asymmetric, with platforms flanking main carina and part of blade; outer platform with lobe formed about anteriorly trending secondary carina. *U.Dev.*, N.Am.-Eu.-Afr.-Austral.——FIG. 37,*7*. *A. bifurcata* (ULRICH & BASSLER), Houy F., USA(Tex.); oral, ×30 (Hass, n).——FIG. 37,*8*. *A.* sp., Houy F., USA(Tex.); aboral, ×30 (Hass, n).

Ctenopolygnathus MÜLLER & MÜLLER, 1957 (p. 1084) [**Polygnathus angustidisca* YOUNGQUIST, 1945]. Characteristics indicate relationships with *Spathognathodus* and *Polygnathus;* platforms not extended to anterior end of unit. *M.Dev.-L.Miss.*, N.Am.-Eu.——FIG. 37,*4*. C. sp.; *4a-c,* inner, oral, aboral, diagram. (93).

Doliognathus BRANSON & MEHL, 1941 (p. 101) [**D. lata*]. Resembles *Ancyrognathus;* platforms extend to posterior end of blade; denticles of axis and secondary carina low and fused. *L.Miss.(uppermost Kinderhook.-lowermost Osag.)*, N.Am.-Eu.——FIG. 36,*3*. **D. lata*, Pierson Ls., USA (Mo.); *3a,b,* oral, aboral, ×15 (8).

Gnathodella MATERN, 1933 (p. 16) [**G. angulata*]. Translation of author's description: Compressed,

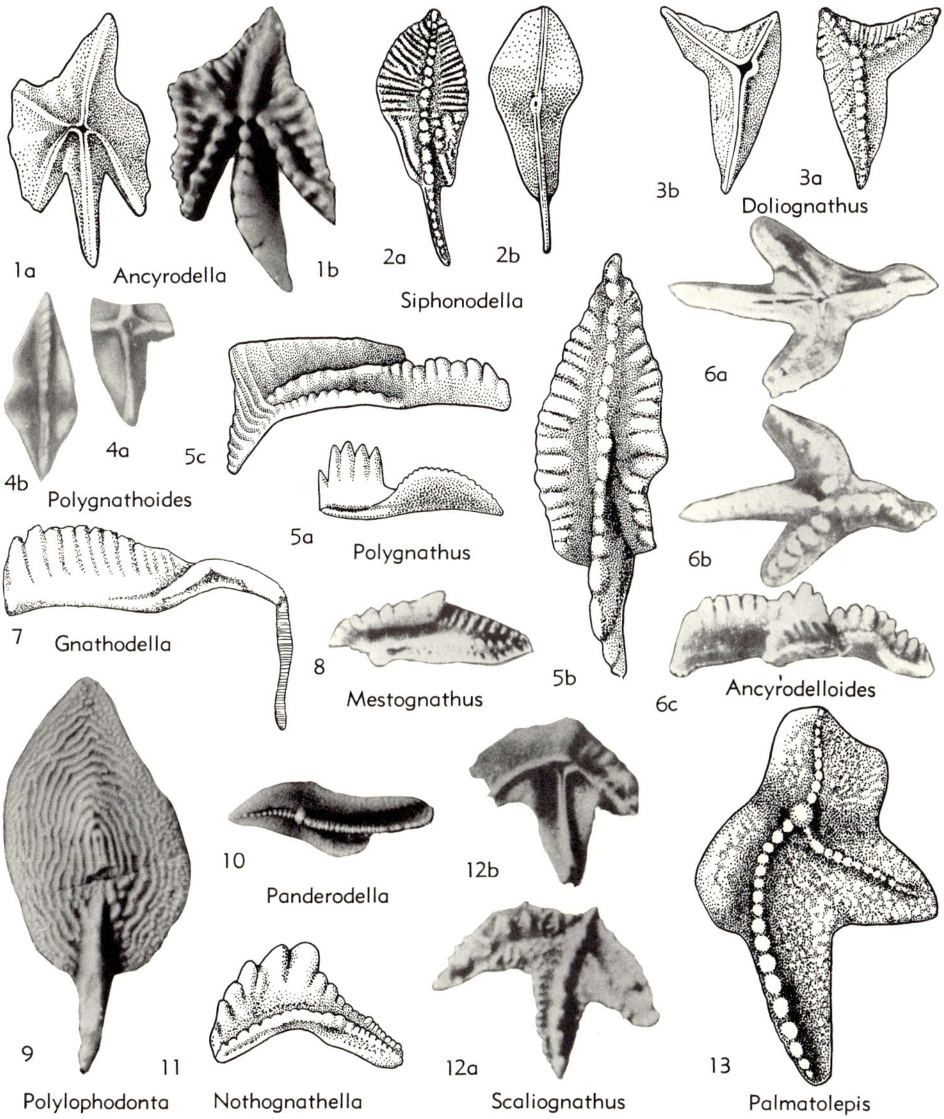

FIG. 36. Polygnathidae (p. *W*58-*W*61).

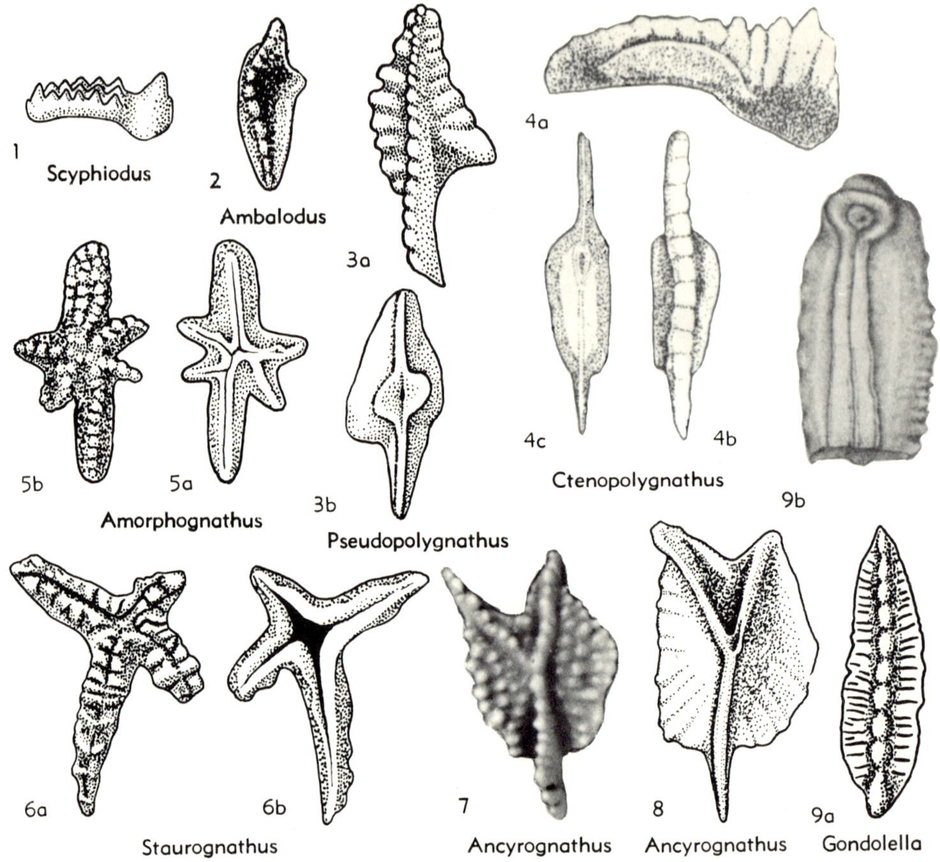

F. 3. 37. Polygnathidae (p. W58-W61).

blunt, high blade which merges int heavy symmetrical or truncated process. *U.D. .,* Eu.——FIG. 36,7. **G. angulata,* Ger.; lat., X.) (92).

Gondolella STAUFFER & PLUMMER, 19.. (p. 41) [**G. elegantula*]. Linguiform to somew at spatulate; platforms generally flank entire ax ; main cusp either terminal or located very near nterior end; keel and pulp cavity prominent. *M.Penn. (Desmoines.)-U.Trias.,* N.Am.-Eu.-Afr.-Asia.——FIG. 37,9. *G. curvata* STAUFFER & PLUMMER, M. Penn.(Labette Sh.), USA(Okla.); *9a,b,* oral, aboral, X37, X60 (70).

Mestognathus BISCHOFF, 1957 (p. 36) [**M. beckmanni*]. Like *Cavusgnathus* except aboral side not excavated; pulp cavity small. *L.Carb.,* Eu.——FIG. 36,8. **M. beckmanni,* Ger.; oral, X20 (3).

Nothognathella BRANSON & MEHL, 1934 (p. 226) [**N. typicalis*]. Resembles *Bryantodus* but has lateral platforms; denticles of axis tend to vary in size and shape. *U.Dev.,* N.Am.-Eu.-Afr.——FIG. 36,*11.* **N. typicalis,* USA(Mo.), inner lat., X25 (9).

Palmatolepis ULRICH & BASSLER, 1926 (p. 49) [**P. perlobata*] [=*Manticolepis* MÜLLER, 1956]. Asymmetric; axis generally sigmoid with carina concave toward outer platform and blade concave toward inner platform; azygous node located above minute pulp cavity; inner platform with lobe which may be built up about secondary carina and secondary keel. *U.Dev.,* N.Am.-Eu.-Afr.——FIG. 36,*13.* **P. perlobata,* Houy F., USA (Tex.); oral, X25 (Hass, n).

Panderodella BASSLER, 1925 (p. 220) [**P. truncata,* SD ULRICH & BASSLER, 1926 (p. 52) (=*Palmatolepis glabra* ULRICH & BASSLER, 1926)] [=*Deflectolepis* MÜLLER, 1956]. Like *Palmatolepis* but plate tends to be narrow and inner platform lacks lobe as well as secondary carina and secondary keel. *U.Dev.,* N.Am.-Eu.-Austral.——FIG. 36,*10.* **P. truncata,* Houy F., USA(Tex.); oral, X30 (Hass, n).

Polygnathoides BRANSON & MEHL, 1933 (p. 50) [**P. silurica*]. Axis straight to slightly angled inward anterior to pulp cavity; flanked by narrow platforms; denticles of axis nodelike; pulp cavity centrally located. *M.Sil.-U.Sil.*, N.Am.-Eu.——FIG. 36,*4*. **P. silurica,* M.Sil.(Bainbridge Ls.), USA (Mo.); *4a,b,* oral, aboral, ×25 (7).

Polylophodonta BRANSON & MEHL, 1934 (p. 242) [**Polygnathus gyratilineata* HOLMES, 1928 (=*Polygnathus confluens* ULRICH & BASSLER, 1926)]. Oral surface of plate with rows of nodes and ridges commonly arranged concentrically about apex of pulp cavity; carina generally suppressed; inner platform with short narrow trough adjacent to blade. *U.Dev.*, N.Am.——FIG. 36,*9*. **P. confluens* (ULRICH & BASSLER), Chattanooga Sh., USA(Ala.); oral, ×25 (89).

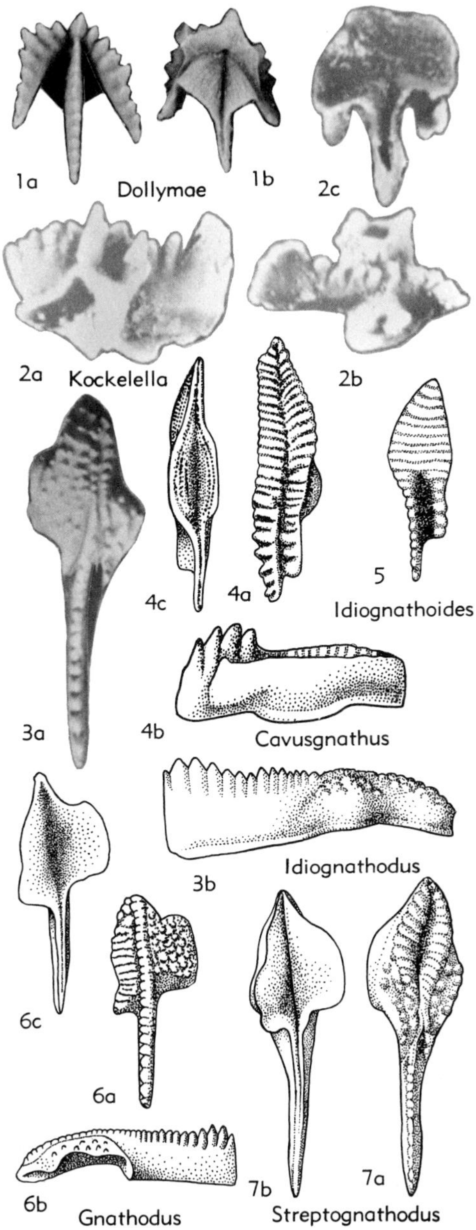

FIG. 38. Idiognathodontidae (Idiognathodontinae) (p. *W*62).

Pseudopolygnathus BRANSON & MEHL, 1934 (p. 297) [**P. prima*]. Oral surface of platforms with nodes and sturdy transverse ridges; pulp cavity prominent, its longer dimension, in aboral view, generally transverse to axis. *L.Miss.(Kinderhook.-lowermost Osag.),* N.Am.-Eu.——FIG. 37,*3*. **P. prima,* Sulphur Springs F., USA(Mo.); *3a,b,* oral, aboral, ×22 (10).

Scaliognathus BRANSON & MEHL, 1941 (p. 101) [**S. anchoralis*]. Anchor-shaped; pulp cavity near anterior end; axis and carinae flanked by narrow platforms. *L.Miss.(uppermost Kinderhook.-lowermost Osage.),* N.Am.-Eu.——FIG. 36,*12*. **S. anchoralis,* Pierson Ls., USA(Mo.); *12a,b,* oral, aboral, ×30 (38).

Scyphiodus STAUFFER, 1935 (p. 617) [**S. primus*]. Differs from *Icriodus* in having aboral side grooved instead of excavated, and in having main cusp distinctly set off from rest of fossil. *M.Ord.,* N.Am. ——FIG. 37,*1*. **S. primus,* Decorah Sh., USA (Minn.); lat., ×37 (68).

Siphonodella BRANSON & MEHL, 1944 (p. 245) [*pro Siphonognathus* BRANSON & MEHL, 1934 (*non* RICHARDSON, 1858)] [**Siphonognathus duplicata* BRANSON & MEHL, 1934]. Carina, with reference to blade, tends to be slightly angled downward and inward; posterior end of plate with rostral ridges; pulp cavity small. *L.Miss.(Kinderhook.),* N.Am.-Eu.——FIG. 36,*2a*. **S. duplicata* (BRANSON & MEHL), Sulphur Springs F., USA(Mo.); oral, ×15 (10).——FIG. 36,*2b*. *S. sexplicata* (BRANSON & MEHL), Sulphur Springs F., USA (Mo.); aboral, ×15 (10).

Staurognathus BRANSON & MEHL, 1941 (p. 102) [**S. cruciformis*]. Cruciform; aboral side grooved; pulp cavity fairly large; oral surface ornamented with low nodes and ridges. *L.Miss.(uppermost Kinderhook.-lowermost Osage.),* N.Am.-Eu.——FIG. 37,*6*. **S. cruciformis,* "Sycamore Ls.", USA (Okla.); *6a,b,* oral, aboral, ×22 (11).

Family IDIOGNATHODONTIDAE Harris & Hollingsworth, 1933

[*nom. transl.* HASS, 1959, *et correct.* HASS, 1958 (*pro* Idiognathinae HARRIS & HOLLINGSWORTH, 1933)] [=Gnathodontidae CAMP, TAYLOR & WELLES, 1942 (*non* Gnathodontidae VON HUENE, 1929, invalid designation of rhynchocephalian reptiles because not founded on a type genus); Gnathodontidae BRANSON & MEHL, 1944, *nom. correct.* herein (*pro* Gnathodidae BRANSON & MEHL, 1944)]

Pulp cavity not greatly restricted so that aboral side of unit is partly or entirely opened up into a large concavity; platforms may flank part or all of axis. *U.Ord.-Up.L. Perm.*

Subfamily IDIOGNATHODONTINAE Harris & Hollingsworth, 1933

[*nom. correct.* HASS, 1958 (*pro* Idiognathinae HARRIS & HOLLINGSWORTH, 1933)]

Blade present, denticulated, well-formed; expanded pulp cavity restricted, more or less, to anterior end of unit. *M.Sil.-Up.L. Perm.*

Idiognathodus GUNNELL, 1931 (p. 249) [**I. claviformis*]. Carina partly or completely suppressed; oral surface of cup, especially in anterior half, transversely ridged. *L.Penn.-U.Penn.*, N.Am.——FIG. 38,*3*. **I. claviformis*, M.Penn.(Ft.Scott Ls.), USA(Mo.); *3a,b*, oral, aboral, ×28, ×15 (24).

Cavusgnathus HARRIS & HOLLINGSWORTH, 1933 (p. 200) [**C. alta*]. Lanceolate, steep-sided, and troughlike in oral view; blade joined to outer platform; carina indistinct. *U.Miss.(Meramec.)-lowermost Perm.*, N.Am.-Eu.-Afr.——FIG. 38,*4*. *C. cristata* BRANSON & MEHL, U.Miss.(Caney Sh.), USA(Okla.); *4a-c;* oral, inner lat., aboral, ×25 (11).

Dollymae HASS, 1959 (p. 394) [**D. sagittula*]. Sagittate unit consisting of blade, terminal main cusp, and both inner and outer secondary carinae; each secondary carina joined to main cusp and flanks a lateral side of the blade; pulp cavity very large, its apex located very near anterior end of unit. *L.Miss.(Kinderhook.)*, N.Am.——FIG. 38, *1*. **D. sagittula*, Chappel Ls., USA(Tex.); *1a,b*, oral, aboral, ×20 (38).

Gnathodus PANDER, 1856 (p. 33) [*non* FIEBER, 1866] [**G. mosquensis*] [=*Dryphenotus* COOPER, 1939]. Carina evident; cup variform; its oral surface plain or ornamented with nodes and/or ridges. *L.Miss.(Kinderhook.)-U.Penn.*, N.Am.-Eu.-Afr.——FIG. 38,*6*. *G. pustulosus* BRANSON & MEHL, U.Miss.(Caney Sh.), USA(Okla.); *6a-c*, oral, lat., aboral, ×25 (11).

Idiognathoides HARRIS & HOLLINGSWORTH, 1933 (p. 201) [**I. sinuata*] [=*Polygnathodella* HARLTON, 1933]. Blade joined to outer platform; oral surface transversely ridged; trough of mature specimen restricted to posterior part of cup. *L.Penn.*, N.Am.——FIG. 38,*5*. *I. corrugata* (HARRIS & HOLLINGSWORTH), L.Penn.(Wapanucka Ls.), USA (Okla.); oral, ×25 (88).

Kockelella WALLISER, 1957 (p. 34) [**K. variabilis*]. Like *Gnathodus* but with aboral side of blade excavated. *M.Sil.-U.Sil.*, Eu.——FIG. 38,*2*. **K. variabilis*, M.Sil.(basal Orthocerenkalk), Ger.; *2a-c*, outer lat., oral, aboral, ×35 (100).

Streptognathodus STAUFFER & PLUMMER, 1932 (p. 47) [**S. excellsus*]. Differs from *Idiognathodus* in having trough along oral mid-line of cup. *L.Penn.-Up.L.Perm.*, N.Am.——FIG. 38,*7*. **S. excellsus*, U.Penn.(Graford F.), USA(Tex.); *7a,b*, oral, aboral, ×25 (33).

Taphrognathus BRANSON & MEHL, 1941 (p. 181) [*non* WELLES, 1947] [**T. varians*]. In oral view, plate lanceolate, steep-sided; trough split at posterior end by blade. *Miss.(Keokuk-Kinkaid)*, N. Am.——FIG. 39,*1*. **T. varians*, L.Miss.(Keokuk Ls.), USA(Mo.); *1a-c*, oral, aboral, lat., ×37 (10a).

Subfamily ICRIODONTINAE Müller & Müller, 1957

[*nom. transl. et correct.* HASS, 1959 (*pro* Icriodidae MÜLLER & MÜLLER, 1957)]

Blade poorly developed or entirely absent; aboral side excavated or nearly so. *L.Sil.-U.Dev.*

Icriodina BRANSON & C. C. BRANSON, 1947 (p. 550) [**I. irregularis*]. Differs from *Icriodus* in having irregularly arranged nodes on oral side; short, poorly developed blade, and a less expanded aboral side. *L.Sil.*, N.Am.——FIG. 39,*2*. **I. irregularis*, Brassfield Ls., USA(Ky.); *2a,b*, oral, aboral, ×37 (81).

Icriodus BRANSON & MEHL, 1938 (p. 159) [**I. expansus* (*non I. alternatus* BRANSON & MEHL)] [In 1934, BRANSON & MEHL erected the genus *Icriodus* and designated *Icriodus expansus*, a *nomen nudum*, as the type species. BRANSON & MEHL validated the generic name in 1938 when they published the characteristics of *Icriodus*, designated *I. expansus* as the type species, and described the characteristics of that species. In 1944, BRANSON & MEHL incorrectly cited *I. alternatus* as the type species of *Icriodus*]. Lanceolate, steep-sided in oral view; carina flanked on each side by a row of nodes or short transverse ridges; lateral process may trend outward from pulp cavity which is located at the expanded posterior end of the unit. *L.Dev.-U.Dev.*, N.Am.-Eu.-Afr.——FIG. 39,*3a*. **I. expansus*, U.Dev.(Snyder Creek Sh.), USA(Mo.); oral, ×37 (82).——FIG. 39,*3b*. **I. expansus*, M. Dev., USA(Mo.); aboral, ×37 (82).

Subfamily BALOGNATHINAE Hass, 1959

Blade present; aboral side excavated. *U. Ord.*

Balognathus RHODES, 1953 (p. 284) [**B. expansa*]. Resembles *Amorphognathus* but aboral side completely excavated; blade may rise above remainder of oral surface. *U.Ord.*, N.Am.-Eu.——FIG. 40,*2*. **B. expansa*, Gelli-grin Ls., Wales; *2a-c*, oral, oral, aboral, ×30 (58).

Icriodella RHODES, 1953 (p. 285) [**I. superba*]. Unit elongate; main cusp stout; blade transversely ridged; apex of pulp cavity in middle third of unit. *U.Ord.*, Eu.——FIG. 40,*1a*. **I. superba*, Gelli-grin Ls., Wales; oral, ×30 (58).——FIG. 40,*1b*.

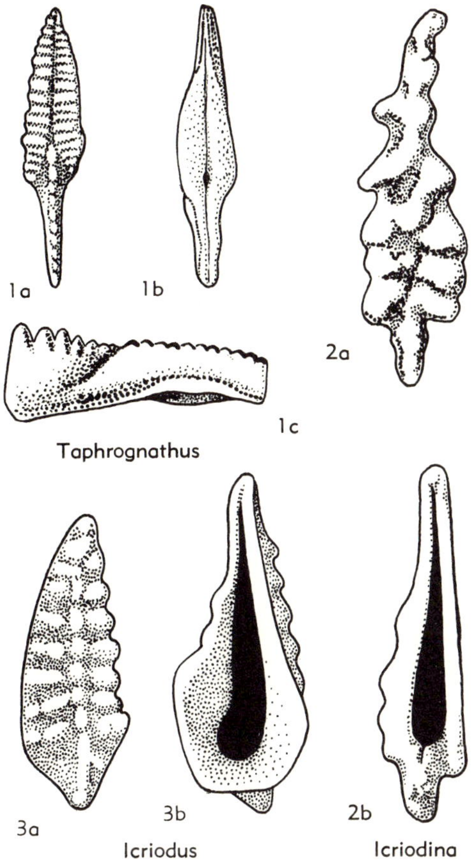

FIG. 39. Idiognathodontidae (Idiognathodontinae, Icriodontinae) (p. *W*62).

**I. superba* var. *acuta* RHODES, Gelli-grin Ls., Wales; lat., ×30 (58).

Family UNCERTAIN

Genera included in this division are not classified into families because their relationships are obscure, being based either on inadequate material or on eccentric specimens.

Coelocerodontus ETHINGTON, 1959 (p. 273) [**C. trigonius*]. ETHINGTON's description: "Simple hollow horn-shaped cones. Lateral walls are thin and enclose a central cavity which extends to the tip of the tooth. Edges of tooth are keeled." The above description suggests that this genus is based on the basal cones of conodonts (*basistrichter* of GROSS) rather than upon the conodont. *M.Ord.*, N. Am.——FIG. 41,*10*. **C. trigonius*, Galena F., USA(Iowa); lat., ×80 (86).

Cornuramia SMITH, 1907 (p. 246) [**C. monodonta;* SD ULRICH & BASSLER, 1926 (p. 42)]. SMITH's description: a "double-pointed, horn-like beam." *Ord.*, Scot.——FIG. 41,*2*. **C. monodonta,* Arenig.-Llandeil., Scot.; lat., ×25 (96).

Distomodus BRANSON & C. C. BRANSON, 1947 (p. 553) [**D. kentuckyensis*]. Original description: "Dental units are simple, curved or straight cones, with sharp or blunt anterior and posterior margins. One side nearly flat to gently convex in cross section, convex longitudinally; the other side gently convex in middle in cross section, gently concave longitudinally but curving out strongly near base. Outline of base triangular, one side of the cone turning in abruptly from the convexity to a plane to form one edge of the triangle. The front margin projects downward as a fragile prong but in most specimens this has been broken away. A depression, shaped like a hollow triangular pyramid, extends one-fourth to one-fifth the length of the cone from the base." *L.Sil.*, N.Am.——FIG. 41,*8*. **D. kentuckyensis,* Brassfield Ls., USA(Ky.); *8a,b,* inner lat., outer lat., ×25 (81).

Goniodontus ETHINGTON, 1959 (p. 278) [**G. superbus*]. Resembles the anterior end of *Phragmodus.* ETHINGTON's description: "Complex dental units having a stout cusp, an anterior outer denticulate basal process, and posterior and anterior inner undenticulate processes. Basal outline of the cusp is triangular. Cusp has plane anterior face, convex posterior face, and sharp anterolateral edges resulting in a subtriangular cross section. Steeply inclined anterior process bears stout erect denticles

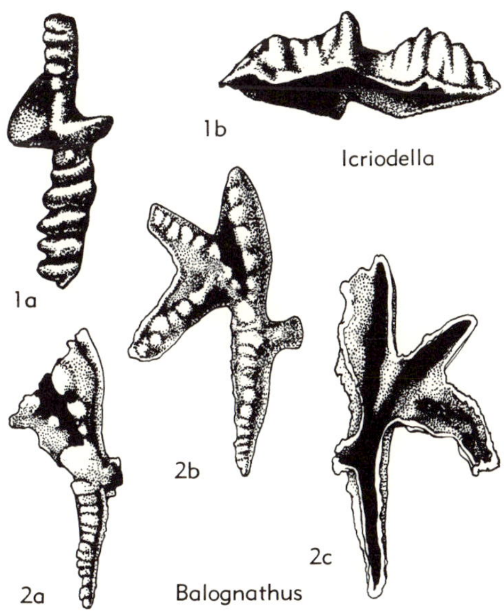

FIG. 40. Idiognathodontidae (Balognathinae) (p. *W*62).

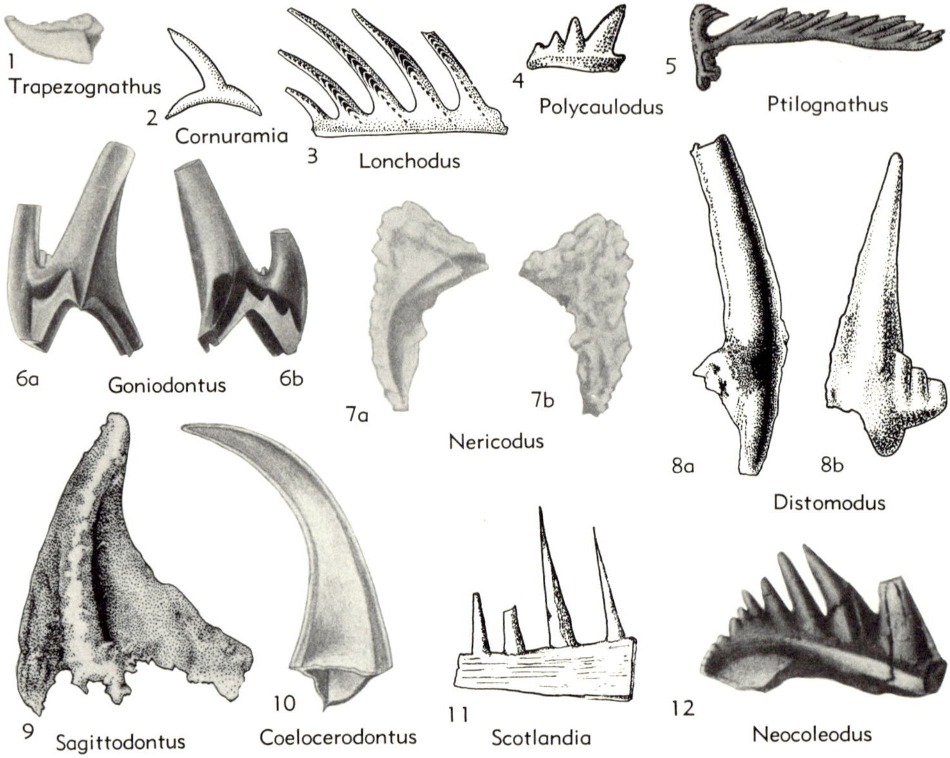

FIG. 41. Family Uncertain (p. *W63-W65*).

alternating with one or two germ denticles. Inner anterior process is short. Posterior process has sharp oral edge which is continued as a sharp carina up the posterior face of the cusp and merges with the outer lateral edge at mid-height. A large hemipyramidal basal cavity beneath the cusp is extended as a wide shallow groove beneath each of the three processes." *M.Ord.*, N.Am.——FIG. 41, *6.* **G. superbus*, Galena F., USA(Iowa); *6a,b*, outer anterolat., inner posterolat., ×80 (86).

Lonchodus PANDER, 1856 (p. 80) [*pro Centrodus* PANDER, 1856 (*non* GIEBEL, 1847; *nec* M'COY, 1848)] [**Centrodus simplex;* SD ULRICH & BASSLER, 1926 (p. 42)]. Straight or bowed bladelike and barlike fragments with discrete or closely set denticles. *L.Ord.-U.Trias.*, cosmop.——FIG. 41,*3.* **L. simplex* (PANDER), Carb.(Bergkalk), USSR; lat., mag. unknown (52).

Neocoleodus BRANSON & MEHL, 1933 (p. 24) [**N. spicatus*]. Barlike fragment curved inward at posterior end; aboral side deeply grooved; denticles discrete, decreasing in size posteriorly. *M.Ord.*, N. Am.——FIG. 41,*12.* **N. spicatus*, Harding Ss., USA(Colo.); inner lat., ×17 (7).

Nericodus LINDSTRÖM, 1954 (p. 570) [**N. capillamentum*]. LINDSTRÖM'S description: "Dome-shaped conodonts with numerous, mostly irregularly distributed little nodes that may fall into winding rows but are never developed as rows of denticles." Pulp cavity shallow. *L.Ord.*, Swed.——FIG. 41,*7.* **N. capillamentum; 7a,b*, inner lat., outer lat., ×30 (44).

Ptilognathus ELIAS, 1956 (p. 114) [**P. fayi*]. Posterior bar with closely set, broadly compressed, posteriorly directed denticles. "Transverse bar" (?=anterior arch) present. *U.Miss.*, N.Am.——FIG. 41,*5.* **P. fayi*, Goddard Sh., USA(Okla.); aboral, ×20 (85).

Polycaulodus BRANSON & MEHL, 1933 (p. 86) [**P. inclinatus*]. Barlike or platelike; denticles discrete, aligned; aboral side even. *M.Ord.*, N.Am.——FIG. 41,*4.* **P. inclinatus*, Joachim Dol., USA(Mo.); inner lat., ×25 (8).

Sagittodontus RHODES, 1953 (p. 310) [**S. robustus*]. RHODES' description: "General appearance barblike; single, large, stout denticle, triangular in cross-section with three more or less flattened faces and sharp dividing edges, the lower part of each face usually having a wide, shallow depression. Unit expanded at base into hemi-pyramidal form. Edges gently curved. Irregular aboral margin; aboral surface deeply excavated so that whole unit is hollow." *U.Ord.*, Wales.——FIG. 41,*9*. **S. robustus*, Gelli-grin Ls., inner lat., ×50 (58).

Scotlandia COSSMANN, 1909 (p. 68) [*pro Valentia* SMITH, 1907 (*non* STÅL, 1856; *nec* SMITH, 1901)] [**Valentia morrochensis* SMITH, 1907]. SMITH's description: "From a deep, thin plate a few long, slender teeth spring." *Ord.*, Scot.——FIG. 41,*11*. **S. morrochensis* (SMITH), Arenig.-Llandeil.; lat., ×40 (96).

Trapezognathus LINDSTRÖM, 1954 (p. 597) [**T. quadrangulum*]. Compound conodonts with a cusp and four denticulated edges or processes, two of which are anterior and two posterior. *L.Ord.*, Swed.——FIG. 41,*1*. **T. quadrangulum*, *Limbata* Z.; lat., ×30 (44).

BIOLOGIC CLASSIFICATION

Natural Assemblages—Family UNCERTAIN

Genera included in this division are not grouped taxonomically. These genera are considered by their authors to represent natural associations of several kinds of discrete conodonts. Each natural assemblage is said to consist of 14 to 22 discrete parts, assignable to 3 to 5 genera of disjunct conodonts.

Duboisella RHODES, 1952 (p. 895) [**D. typica*]. Considered to be a natural conodont assemblage composed of discrete specimens belonging to the genera *Ligonodina, Lonchodina, Hibbardella, Metalonchodina,* and *Neoprioniodus*. *U.Penn.*, N.Am. ——FIG. 42,*4*. **D. typica*, McLeansboro F., USA (Ill.); diagram., approx. ×15 (59).

Illinella RHODES, 1952 (p. 898) [**I. typica*]. Considered to be a natural conodont assemblage composed of discrete specimens belonging to the genera *Gondolella, Lonchodina,* and *Lonchodus*. *M. Penn.*, N.Am.——FIG. 42,*3*. **I. typica*, USA(Ill.); diagram., approx. ×15 (59).

Lewistownella SCOTT, 1942 (p. 299) [**L. agnewi*]. Considered to be a natural conodont assemblage composed of discrete specimens belonging to the genera *Hindeodella, Neoprioniodus, Subbryantodus,* and *Cavusgnathus*. U.Miss., N.Am.——FIG. 43,*1*. **L. agnewi*, Heath F., USA(Mont.), ×15 (59).

Lochriea SCOTT, 1942 (p. 298) [**L. montanaensis*]. Considered to be a natural conodont assemblage composed of discrete specimens belonging to the genera *Hindeodella, Spathognathodus, Neoprioniodus,* and *Prioniodella* [=*Prioniodina*]. *U.Miss.*, N. Am.——FIG. 42,*2*. **L. montanaensis*, Heath F., USA(Mont.), diagram., ×30 (65).——FIG. 43, *2*. **L. montanaensis;* alter. orient., ×15 (59).

Scottognathus RHODES, 1953 (p. 612) [*pro Scottella* RHODES, 1952 (*non* ENDERLEIN, 1910)] [**Scottella typica* RHODES, 1952]. Considered to be a natural conodont assemblage composed of discrete specimens belonging to the genera *Idiognathodus,* or *Streptognathodus, Ozarkodina, Synprioniodina,* and *Hindeodella*. *U.Penn.*, N.Am.——FIG. 42,*1*. **S. typica* (RHODES), McLeansboro F., USA(Ill.); diagram., approx. ×15 (56).

Westfalicus SCHMIDT *in* MOORE & SYLVESTER-BRADLEY, 1957 (p. 21) [*pro Gnathodus* SCHMIDT, 1934 (*non* PANDER, 1856)] [**Gnathodus integer* SCHMIDT, 1934]. Composed of discrete specimens belonging to *Gnathodus, Bryantodus, Hindeodella, Neoprioniodus,* and *"Lonchodus." U.Carb.(L. Namur.)* Ger.——FIGS. 20, 43. **W. integer* (SCHMIDT), Westphalia; diagram., ×30 (62), ×15 (59).

REJECTED GENERIC NAMES

The following names, published in articles on conodonts, do not refer to conodonts. [See FAY (27) for bibliographic information about names published prior to 1950.]

Archeognathus CULLISON, 1938.
Astacoderma HARLEY, 1861.
Bransonella HARLTON, 1933.
Clavohamulus FURNISH, 1938.
Dermatolithis EHRENBERG, 1854.
Fortscottella GUNNELL, 1931.
Holmesella GUNNELL, 1931.
Icthyodus HARRIS & HOLLINGSWORTH, 1933.
Lepodus E. B. BRANSON & MEHL, 1933.
Lepognathodus MEHL *in* FAY, 1959.
Multidentodus HARLTON, 1933.
Prionognathus PANDER, 1856 [*non* FERTÉ-SÉNÈCTERE, 1851 (=*Prionognathodus* FAY, 1959)].
Prionognathodus FAY, 1959.
Pygodus LAMONT & LINDSTRÖM, 1957.
Scolopodella STAUFFER & PLUMMER, 1932.
Stephanodella MATERN, 1933.

REFERENCES

Beckman, Heinz

(1) 1949, *Conodonten aus dem Iberger Kalk (Ober-Devon) des Bergischen Landes und ihr Feinbau:* Senckenbergiana, Bd. 30, Nr. 1-3, p. 153-168, pl. 1-4.

Bischoff, Günther

(2) 1956, *Oberdevonische Conodenten (to Iδ) aus dem rheinischen Schiefergebirge:* Hess. Landesamt Bodenf., Notizbl., Bd. 84, p. 115-137, pl. 8-10.

(3) 1957, *Die Conodenten-Stratigraphie des rheno-herzynischen Unterkarbons mit Berücksichtigung der Wocklumeria-Stufe und der Devon/Karbon-Grenze:* Hess. Landesamt Bodenf., Abhandl., H. 19, p. 1-64, pl. 1-6.

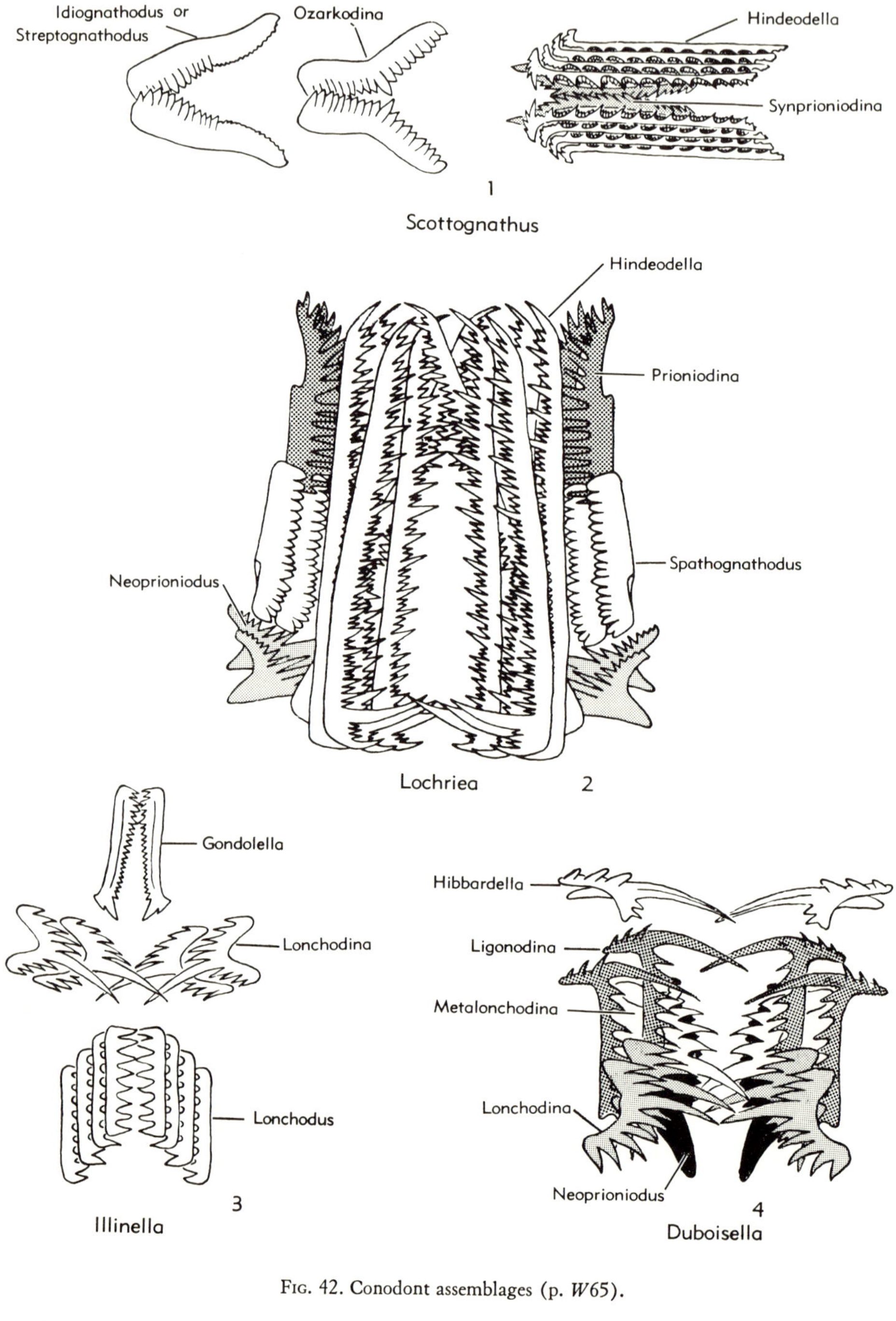

FIG. 42. Conodont assemblages (p. W65).

———, & Sannemann, Dietrich

(4) 1958, *Unterdevonische Conodonten aus dem Frankenwald:* Hess. Landesamt Bodenf., Notizbl., Bd. 86, p. 87-110, pl. 12-15.

———, & Ziegler, Willi

(5) 1956, *Das Alter der "Urfer Schichten" im Marburger Hinterland nach Conodonten:* Hess. Landesamt Bodenf., Notizbl., Bd. 84, p. 138-169, pl. 11-14.

(6) 1957, *Die Conodontenchronologie des Mitteldevons und des tiefsten Oberdevons:* Hess. Landesamt Bodenf., Abhandl., H. 22, p. 1-136, pl. 1-21.

Branson, E. B., & Mehl, M. G.

(7) 1933(Jan.), *Conodonts from the Harding sandstone of Colorado* (p. 19-38, pl. 1-2); *Conodonts from the Bainbridge (Silurian) of Missouri* (p. 39-52, pl. 3); *Conodonts from the Jefferson City (lower Ordovician) of Missouri* (p. 53-64, pl. 4): Univ. Missouri Studies, v. 8, no. 1, p. 1-72; pl. 1-4.

(8) 1933(April), *Conodonts from the Joachim (middle Ordovician) of Missouri* (p. 77-100, pl. 5-7); *Conodonts from the Plattin (middle Ordovician) of Missouri* (p. 101-119, pl. 8-10); *Conodonts from the Maquoketa-Thebes (upper Ordovician) of Missouri* (p. 121-131, pl. 10); *A study of Hinde's types of conodonts preserved in the British Museum* (p. 133-156, pl. 11-12): Univ. Missouri Studies, v. 8, no. 2, p. 77-167, pl. 5-12.

(9) 1933(July)[May, 1934], *Conodonts from the Grassy Creek shale of Missouri:* Univ. Missouri Studies, v. 8, no. 3, p. 171-259, pl. 13-21.

(10) 1933(Oct.)[1934], *Conodonts from the Bushberg and equivalent formations of Missouri:* Univ. Missouri Studies, v. 8, no. 4, p. 265-299, pl. 22-24.

(10a) 1940[1941], *Conodonts from the Keokuk formation:* Denison Univ., Bull., Jour. Sci. Lab., v. 35, p. 179-188.

(11) 1940[1941], *The recognition and interpretation of mixed conodont faunas:* Denison Univ., Bull., Jour. Sci. Lab., v. 35, p. 195-209.

(12) 1943, *Ordovician conodont faunas from Oklahoma:* Jour. Paleontology, v. 17, p. 374-387, pl. 63-64.

(13) 1944, *Conodonts:* in Shimer, H. W., & Shrock, R. R., Index Fossils of North America, p. 235-246, pl. 93-94, Wiley & Sons (New York).

———, ———, & Branson, C. C.

(14) 1951, *Richmond conodonts of Kentucky and Indiana:* Jour. Paleontology, v. 25, p. 1-17, pl. 1-4.

Branson, E. R.

(15) 1933(Oct.)[1934], *Conodonts from the Hannibal formation of Missouri:* Univ. Missouri Studies, v. 8, no. 4, p. 301-343, pl. 25-28.

Cooper, C. L.

(16) 1939, *Conodonts from a Bushberg-Hannibal horizon in Oklahoma:* Jour. Paleontology, v. 13, p. 379-422, pl. 39-47.

Croneis, Carey

(17) 1939, *A military classification for fossil fragments:* Science, v. 89, no. 2310, p. 314-315.

Demanet, Felix

(18) 1938, *La Faune des Couches de passage du Dinantien au Namurien dans le synclinorium de Dinant:* Mus. roy. Hist. nat. Belgique, Mém., t. 84, p. 1-201, pl. 1-14.

(19) 1939, *Filtering appendices on the branchial arches of Coelacanthus lepturus Agassiz:* Geol. Mag., v. 76, no. 899, p. 215-219, fig. 1-2.

Denham, R. L.

(20) 1944, *Conodonts:* Jour. Paleontology, v. 18, p. 216-218, 1 fig.

Diebel, Kurt

(21) 1956, *Conodonten in der Oberkreide von Kamerun:* Geologie, Bd. 5, Nr. 4/5, p. 424-450, pl. 1-6.

DuBois, E. P.

(22) 1943, *Evidence on the nature of conodonts:* Jour. Paleontology, v. 17, p. 155-159, pl. 25.

Eichenberg, W.

(23) 1930, *Conodonten aus dem Culm des Harzes:* Paläont. Zeitschr., Bd. 12, p. 177-182, fig. 1-30.

Ellison, S. P.

(24) 1941, *Revision of the Pennsylvanian conodonts:* Jour. Paleontology, v. 15, p. 107-143, pl. 20-23.

(25) 1944, *The composition of conodonts:* Jour. Paleontology, v. 18, p. 133-140.

(26) 1946, *Conodonts as Paleozoic guide fossils:* Am. Assoc. Petroleum Geologists, Bull., v. 30, p. 93-110, fig. 1-3.

Fay, R. O.

(27) 1952, *Catalogue of conodonts:* Univ. Kansas Paleont. Contr., Vertebrata, art. 3, 206 p., 109 fig.

(27a) 1958, *A key to conodont genera and subgenera:* Oklahoma Geol. Notes, v. 18, p. 103-120, 132 fig.

Flügel, Helmut, & Ziegler, Willi

(28) 1957, *Die gliederung des Oberdevon und Unterkarbons am Steinberg westlich von Graz mit Conodonten:* Mit. naturwiss. Ver. Steiermark, Bd. 87, p. 25-60, pl. 1-5.

Frizzell, D. L., & Exline, Harriet

(29) 1955, Monograph of fossil holothurian sclerites: Univ. Missouri School of Mines and Metallurgy, tech. ser., no. 89, 204 p., 11 pl.

Furnish, W. M.

(30) 1938, *Conodonts from the Prairie du Chien (Lower Ordovician) beds of the Upper Mississippi Valley:* Jour. Paleontology, v. 12, p. 318-340, pl. 41-42.

Grinnell, G. B.

(31) 1877, *Notice of a new genus of annelids from the Lower Silurian:* Am. Jour. Sci., ser. 3, v. 14, no. 81, p. 229-230.

Gross, Walter

(32) 1954, *Zur Conodonten-Frage:* Senckenbergiana, Bd. 35, Nr. 1/2, p. 73-85, pl. 1-5.

(33) 1957, *Über die Basis der Conodonten:* Paläont. Zeitschr., Bd. 31, p. 78-91, pl. 7-9.

Hass, W. H.

(34) 1941, *Morphology of conodonts:* Jour. Paleontology, v. 15, p. 71-81, pl. 12-16.

(35) 1953, *Conodonts of the Barnett formation of Texas:* U.S. Geol. Survey, Prof. Paper 243-F, p. 69-94, pl. 14-16.

(36) 1956, *Age and correlation of the Chattanooga shale and the Maury formation:* U.S. Geol. Survey, Prof. Paper 286, p. 1-47, pl. 1-5.

(37) 1958, *Upper Devonian conodonts of New York, Pennsylvania, and interior states:* Jour. Paleontology, v. 32, p. 765-769.

(38) 1959, *Conodonts from the Chappel limestone of Texas:* U.S. Geol. Survey, Prof. Paper 294-J, p. 365-399, pl. 46-50.

———, & Lindberg, M. L.

(39) 1946, *Orientation of the crystal units of conodonts:* Jour. Paleontology, v. 20, p. 501-504, fig. 1-4.

Hinde, G. J.

(40) 1879, *On conodonts from the Chazy and Cincinnati group of the Cambro-Silurian, and from the Hamilton and Genesee-shale divisions of the Devonian, in Canada and the United States:* Geol. Soc. London, Quart. Jour., v. 35, art. 29, p. 351-369, pl. 15-17.

Huckriede, Reinhold

(41) 1958, *Die Conodonten der mediterranen Trias. und ihr stratigraphischer Wert:* Paläont. Zeitschr., Bd. 32, Nr. 3/4, p. 141-175, pl. 10-14.

Huddle, J. W.

(42) 1934, *Conodonts from the New Albany shale of Indiana:* Bull. Am. Paleont., v. 21, no. 72, p. 1-136, pl. 1-12.

Kirk, S. R.

(43) 1929, *Conodonts associated with the Ordovician fish fauna of Colorado—a preliminary note:* Am. Jour. Sci., ser. 5, v. 18, no. 108, p. 493-496, fig. 1-14.

Lindström, Maurits

(44) 1954, *Conodonts from the lowermost Ordovician strata of south-central Sweden:* Geol. Fören. Stockholm, Förhandl., bd. 76, h. 4, no. 21, p. 517-603, pl. 1-10.

(45) 1955, *Om conodonter:* Svensk Faunistisk Revy, bd. 3, p. 1-4.

Loomis, F. B.

(46) 1936, *Are conodonts gastropods?:* Jour. Paleontology, v. 10, p. 663-664, 1 fig.

Moore, R. C., & Sylvester-Bradley, P. C.

(47) 1957a, *Proposed insertion in the "Règles" of provisions recognizing "Parataxa" as a special category for the classification and nomenclature of discrete fragments or of life-stages of animals which are inadequate for identification of whole-animal taxa, with proposals of procedure for the nomenclature of "Parataxa":* Bull. Zool. Nomenclature, v. 15, sec. A, pt. 1/4, Document 1/1, p. 5-13.

(48) 1957b, *Application for a ruling of the International Commission directing that the classification and nomenclature of discrete conodonts be in terms of "Parataxa":* Bull. Zool. Nomenclature, v. 15, sec. A, pt. 1/4, Document 1/1, p. 15-34.

Müller, K. J.

(49) 1956a, *Triassic conodonts from Nevada:* Jour. Paleontology, v. 30, p. 818-830, pl. 95-96.

(50) 1956b, *Taxonomy, nomenclature, orientation, and stratigraphic evaluation of conodonts:* Jour. Paleontology, v. 30, p. 1324-1340, pl. 145.

(51) 1956c, *Zur Kenntnis der Conodonten-Fauna des europäischen Devons. 1., Die Gattung Palmatolepis:* Senckenberg. Naturf. Gesell., Abhandl., Bd. 494, p. 1-70, pl. 1-11.

Pander, C. H.

(52) 1856, *Monographie der fossilen Fische des silurischen Systems der russisch-baltischen Gouvernements:* Akad. Wiss. St. Petersburg, p. 1-91, pl. 1-7, table.

Pilsbry, H. A.

(53) 1937, *Are the conodonts molluscan teeth?:* The Nautilus, v. 50, no. 3, p. 101.

Rexroad, C. B.

(54) 1957, *Conodonts from the Chester series in the type area of southwestern Illinois:* Illinois State Geol. Survey, Rept. Inv. 199, p. 1-43, pl. 1-4.

(55) 1958, *Conodonts from the Glen Dean formation (Chester) of the Illinois basin:* Illinois State Geol. Survey, Rept. Inv. 209, p. 1-27, pl. 1-6.

Rhodes, F. H. T.

(56) 1952, *A classification of Pennsylvanian conodont assemblages:* Jour. Paleontology, v. 26, p. 886-901, pl. 126-129.

(57) 1953a, *Nomenclature of conodont assemblages:* Jour. Paleontology, v. 27, p. 610-612.

(58) 1953b, *Some British Lower Palaeozoic conodont faunas:* Royal Soc. London, Philos. Trans., ser. B, v. 237, no. 647, p. 261-334, pl. 20-23.

(59) 1954, *The zoological affinities of the conodonts:* Cambridge Philos. Soc., Biol. Reviews, v. 29, p. 419-452.

———, & Wingard, Paul

(60) 1957, *The structure, chemical composition, and affinities of the Neurodontiformes:* Jour. Paleontology, v. 27, p. 448-454.

Sannemann, Dietrich

(61) 1955, *Oberdevonische conodonten* (to IIα): Senckenbergiana, Bd. 36, Nr. 1/2, p. 123-156, pl. 1-6.

Schmidt, Hermann

(62) 1934, *Conodonten-Funde in ursprünglichen Zusammenhang:* Paläont. Zeitschr., Bd. 16, Nr. 1/2, p. 76-85, pl. 6.

(63) 1950, *Nachträge zur Deutung der Conodonten:* Decheniana, Bd. 104, p. 11-19, fig. 1-7.

Scott, H. W.

(64) 1934, *The zoological relationships of the conodonts:* Jour. Paleontology, v. 8, p. 448-455, pl. 58-59.

(65) 1942, *Conodont assemblages from the Heath formation, Montana:* Jour. Paleontology, v. 16, p. 293-300, pl. 37-40.

Sinclair, G. W.

(66) 1953, *The naming of conodont assemblages:* Jour. Paleontology, v. 27, p. 489-490.

Stauffer, C. R.

(67) 1935a, *Conodonts of the Glenwood beds:* Geol. Soc. America, Bull., v. 46, p. 125-168, pl. 9-12.

(68) 1935b, *The conodont fauna of the Decorah shale (Ordovician):* Jour. Paleontology, v. 9, p. 596-620, pl. 71-75.

(69) 1938, *Conodonts from the Olentangy shale:* Jour. Paleontology, v. 12, p. 411-443, pl. 48-53.

———, & Plummer, H. J.

(70) 1932, *Texas Pennsylvanian conodonts and their stratigraphic relations:* Univ. Texas, Bull. 3201, p. 13-50, pl. 1-4.

Stewart, G. A., & Sweet, W. C.

(71) 1956, *Conodonts from the Middle Devonian bone beds of central and west-central Ohio:* Jour. Paleontology, v. 30, p. 261-273, pl. 33-34.

Sweet, W. C.

(72) 1955, *Conodonts from the Harding formation (Middle Ordovician) of Colorado:* Jour. Paleontology, v. 29, p. 226-262, pl. 27-29.

Sylvester-Bradley, P. C.

(73) 1954, *Form-genera in paleontology:* Jour. Paleontology, v. 28, p. 333-336.

Tatge, Ursula

(74) 1956, *Conodonten aus dem germanischen Muschelkalk:* Paläont. Zeitschr., Bd. 30, Nr. 1/2, T. 1, p. 108-127, pl. 5-6; Nr. 3/4, T. 2, p. 129-147.

Ulrich, E. O., & Bassler, R. S.

(75) 1926, *A classification of the toothlike fossils, conodonts, with description of American Devonian and Mississippian species:* U.S. Natl. Museum, Proc., v. 68, no. 2613, p. 1-63, pl. 1-11, fig. 1-5.

Wetzel, Otto

(76) 1953, *Résumé of microfossils from Upper Cretaceous flints and chalks of Europe:* Jour. Paleontology, v. 27, p. 800-804, pl. 83.

Youngquist, W. L.

(77) 1952, *Triassic conodonts from southeastern Idaho:* Jour. Paleontology, v. 26, p. 650-655.

———, Hawley, R. W., & Miller, A. K.

(78) 1951, *Phosphoria conodonts from southeastern Idaho:* Jour. Paleontology, v. 25, p. 356-364, pl. 54.

Ziegler, Willi

(79) 1956, *Unterdevonische Conodonten, insbesondere aus dem Schonauer und dem Zorgensis-Kalk:* Hess. Landesamt Bodenf., Notizbl., Bd. 84, p. 93-106, pl. 6-7.

Zittel, K. A. von, & Rohon, J. V.

(80) 1886(1887), *Ueber Conodonten:* Akad. Wiss., München, Math.-Phys. Kl., Sitzungsber., Bd. 16, p. 108-136, pl. 1-2.

SOURCES OF ILLUSTRATIONS

(81) **Branson, E. B., & Branson, C. C.,** 1947
(82) **———, & Mehl, M. G.,** 1938
(83) **Clark, D. L.,** 1959
(84) **Cooper, C. L., & Sloss, L. L.,** 1943
(85) **Elias, M. K.,** 1956
(86) **Ethington, R. L.,** 1959
(87) **Graves, R. W., & Ellison, S. P.,** 1941
(88) **Harris, R. W., & Hollingsworth, R. V.,** 1933
(89) **Holmes, G. B.,** 1928
(90) **Lindström, Maurits,** 1955
(91) **Lys, Maurice, & Serre, Bernadette,** 1957
(92) **Matern, Hans,** 1933
(93) **Müller, K. J., & Müller, E. M.,** 1957
(94) **Öpik, A. A.,** 1936
(95) **Rhodes, F. H. T.,** 1955
(96) **Smith, John,** 1907
(97) **Stauffer, C. R.,** 1940
(98) **Stone, G. L., & Furnish, W. M.,** 1959
(99) **Thomas, L. A.,** 1949
(100) **Walliser, O. H.,** 1957
(101) **Youngquist, W. L.,** 1945

RECOGNITION, INTERPRETATION, AND TAXONOMIC POSITION OF CONODONT ASSEMBLAGES

By F. H. T. Rhodes
[University College, Swansea]

CONTENTS

INTRODUCTION

Since conodonts were first described by Pander in 1856 almost 700 papers have been devoted to them. Because of the vagaries of fossilization and the techniques employed in the extraction of microfossils, the great majority of conodonts are known as single isolated specimens. They have proved to be a varied and stratigraphically useful group, and a binominal system of classification established upon these isolated specimens includes about 2,000 species. These taxa are distinguished in the present paper by the use of quotation marks ("genera" and "species").

Systematic treatment of conodont "genera" and "species" is given in the preceding chapter by Wilbert H. Hass. Most conodont workers have recognized that the isolated conodonts upon which such "taxa" have been established may be fragmentary fossils, but the nature and distribution of conodonts is such that this method of classification has proved to be both readily applicable and stratigraphically useful. Some "genera" and "species" have a short stratigraphic range accompanied by wide geographic distribution and hence have great value in stratigraphic paleontology. A number of "suprageneric taxa" have also been established.

Several workers (e.g., Schmidt, 17, 18; Scott, 19, 20; Eichenberg, 7; DuBois, 6; Rhodes, 12, 13, 15, 16) have described what they claim to be natural conodont assemblages, and have shown that a single assemblage, which they interpret as representing the remains of an individual animal, may contain discrete elements classified in as many as five conodont "genera." A number of different genera have been based upon the recognition of these natural assemblages as taxonomic units, and there has therefore grown up a second taxonomic framework. The consequent taxonomic problems are complex and are discussed later. Most conodont students have accepted the interpretation of such remains as natural assemblages but a few (e.g., Branson & Mehl, 2; Branson, 1; Fay, 8) have suggested that they may be coprolitic in origin.

The purpose of the present paper, written at the invitation of Dr. R. C. Moore as an addendum to Dr. Hass's main contribution, is to describe these conodont assemblages and to examine the problems of their interpretation and taxonomic treatment.

DESCRIPTIONS OF CONODONT ASSEMBLAGES

POLYGNATHUS

The first conodont assemblage recorded was one from the Devonian Genesee Shale at North Evans, New York, described by HINDE in 1879 (9, p. 361-364, pl. 16, fig. 6-18). HINDE proposed the genus *Polygnathus* for "an animal possessing numerous minute and variously formed Conodont teeth and similarly minute tuberculated plates grouped together, but of which the natural arrangement is not at present known" (9, p. 361). HINDE observed that the single specimen of the genus contained "about twenty-four entire and fragmentary teeth and six plates . . . crushed together in a small patch of about one fourth of an inch in diameter in black shale" (9, p. 361). He agreed that, although no indication could be seen of the natural position of the teeth and plates, it could "hardly be doubted that they all belonged to one individual, as it would be beyond all reasonable probability that so many diversely formed teeth, of such delicate structure, could have been thus brought together into so small a space by mechanical means, more particularly when it is a very rare circumstance to find, in the same rock, even two detached teeth at all close together" (9, p. 362).

HINDE did not figure the single assemblage on which he based his conclusions, and he noted that the single specimen "in which the teeth of this remarkable form are grouped together has been crushed to such an extent that individual teeth and plates can be only partially distinguished, but the various kinds are met with in a very perfect condition, as so many separate specimens scattered through the rock" (9, p. 362). Three main types of conodonts were described in the assemblage: (1) pectinate teeth (arched blades such as ozarkodinids and bryantodids, illustrated in his pl. 16, fig. 6-9, and bar types, some of which seem to be broken posterior bars of ligonodinids); (2) fimbriate teeth (hindeodellids, pl. 16, fig. 13, 14); and (3) crested teeth of two varieties (one probably a spathognathodid and the other a polygnathid, pl. 16, fig. 15-17). The small plates, six of which HINDE described in the assemblage (e.g., pl. 16, fig. 18) are apparently the broken posterior platforms of these polygnathids.

BRYANT (4, p. 22-23) suggested that HINDE's assemblage may have been coprolitic in origin, and SCHMIDT (17, p. 76) considered that the forms represented in HINDE's assemblage represented the remains of different individuals.

BRANSON (in BRANSON & MEHL, 2, p. 136-137, 140, 142-143, 146-147, 152-153) re-examined and redescribed HINDE's specimens. He illustrated the isolated specimens interpreted by HINDE as similar to the components of *Polygnathus dubius* and referred them to the "genera" *Lonchodina, Hindeodella, Polygnathus, Bryantodus,* and *Spathodus.*

BRYANT (4, p. 9-23) and ULRICH & BASSLER (22, p. 43) revised HINDE's generic description of *Polygnathus* and restricted it to bladed-platform type conodonts, in which the platform is ornamented by a straight median carina and lateral, transverse ridges.

DR. H. W. BALL of the Palaeontology Department of the British Museum (Natural History) has kindly allowed me to examine the "assemblage" specimen (BMNH no. A4305-6) described by HINDE (9, p. 361). It occurs in a dark shale matrix and consists of about 48 individual conodonts. These show no alignment or paired relationship to one another, and they are not confined to a single horizon in the shale. Numbers of them are broken and most are difficult to identify with any certainty. Some forms which, if the assemblage were natural, would be complementary, show very considerable variation in size. Only one specimen of *Hindeodella* is present. The remaining specimens represent the "genera" *Neoprioniodus, Hibbardella?, Lonchodina, Polygnathus* (as used for a discrete conodont), *Ozarkodina* or *Bryantodus* and possibly other "genera." I believe that the assemblage is fortuitous, and not "natural," in the present sense.

It therefore seems proper that the "generic" name *Polygnathus* HINDE, 1879, type-species *P. dubius (recte P. dubia)* HINDE, by subsequent designation of BASSLER, 1915, should be restricted to isolated conodonts as defined by ULRICH & BASSLER (22, p. 43).

PAIRED CONODONTS

Bryant (4, p. 24) was one of the earlier workers to recognize the presence of right and left forms of conodonts, and to appreciate that this implied that they must have been paired structures in the body of the conodont-bearing animal.

WESTFALICUS

Schmidt (17) described nine conodont assemblages from the lower Namurian, Upper Carboniferous, of Germany. He described them as containing one pair of *Gnathodus,* one pair of *Bryantodus,* and a number of pairs of *Lonchodus* (including *Hindeodella*). It is difficult to check Schmidt's determinations from his figures, but his *Bryantodus* seems to be *Ozarkodina,* and the *Lonchodus* blades include *Hindeodella* and *Synprioniodina.* The determination of the polygnathid component is impossible from the figures. Schmidt (18) later amplified his descriptions and offered detailed interpretations (see Hass, this volume).

LOCHRIEA and LEWISTOWNELLA

Scott (19, 20) described 180 conodont assemblages from the Heath Shale (Upper Mississippian or Lower Pennsylvanian) of Montana, which he interpreted as representing two distinct genera. These he named, and he described their component discrete conodonts by reducing their "generic" names to nouns (e.g., "*Hindeodella*" became "hindeodells"). *Lochriea* comprised pairs of conodonts representing the "genera" *Spathognathodus, Prioniodella, Prioniodus,* and *Hindeodella. Lewistownella* contained pairs of *Cavusgnathus, Subbryantodus, Prioniodus,* and *Hindeodella.* Scott recognized two species of *Lochriea,* which he based on minor variations in the form of the component conodonts.

SCOTTOGNATHUS, DUBOISELLA, and ILLINELLA

Rhodes (12, 13) described three genera of assemblages from the Pennsylvanian of Illinois. *Scottognathus* (initially published as *Scottella,* a junior homonymous name) contained paired discrete conodonts belonging to the "genera" *Idiognathodus* or *Steptrognathodus, Ozarkodina, Synprioniodina,* and probably four pairs of *Hindeodella. Duboisella* contained two pairs each of *Ligonodina* and *Lonchodina* and one pair each of *Hibbardella, Metalonchodina,* and *Neoprioniodus. Illinella* contained four pairs of *Lonchodus,* two pairs of *Lonchodina,* and one pair of *Gondolella.*

PRIONIODUS HERCYNICUS

Eichenberg (7) described a collection of conodonts from the Culm (Lower Carboniferous) of the Harz Mountains, Germany. The material was poorly preserved and his description suggests that it was collected from a number of horizons. It is difficult to identify all the specimens illustrated by Eichenberg, but representatives of the following discrete conodont "genera" are included: *Hindeodella, Neoprioniodus, Ozarkodina* or *Bryantodus, Falcodus?, Ancyrodella* and other platform types. Eichenberg's description does not suggest that he regarded these "genera" as representing a natural assemblage in the present sense of the phrase, although he presumably assumed them to have come from a single animal, to which he gave the name *Prioniodus hercynicus.* Existing knowledge of conodont assemblages is inadequate to determine with certainty whether or not this assumption is correct, but it seems unlikely, and Eichenberg does not record any intimate association of the various components.

CHARACTERISTICS OF CONODONT ASSEMBLAGES

A study of the various assemblages described above permits the following general observations.

1. Conodonts are paired, the right and left forms being mirror images of one another. They are alike in major morphological features, but show minor differences, which are of only "infraspecific" value. In addition to such difference as this, complementary differential curvature and reversed ornamentation or node development are often characteristic of the opposed pairs.

2. The number and arrangement of many conodonts in assemblages suggest lateral opposition (as left and right forms), rather than the duplication and opposition characteristic of elements of upper and lower "jaws."

3. Assemblages contain components representing a number of distinct "genera." Most have four such "genera," but *Illinella* has three and *Duboisella* five.

4. The same "genus" may be present in more than one natural assemblage. Thus the *Hindeodella* component is present in four genera of assemblages, and a form structurally analogous to it *(Lonchodus)* in a fifth. This clearly implies a functional similarity both within and between these assemblages.

5. Where the same "genus" is not present in two distinct genera of assemblages, it is sometimes found that it is represented by a similar "genus" of the same structural type. Thus in four assemblage genera, a closely similar platform "genus" of conodont is present: *Cavusgnathus* in *Lewistownella, Spathognathodus* in *Lochriea, Streptognathodus* or *Idiognathodus* in *Scottognathus,* and *Gnathodus* in *Westfalicus.* This close structural analogy of components in some assemblages is best illustrated by the tabular representation below.

I have revised some of the discrete conodont "generic" names in the table, to conform with existing nomenclature. For the sake of convenience I shall refer to these four similar (assemblage) genera as Class A.

6. *Illinella* shows some resemblance to the four related Class A genera described above. It has, for example, an arched-blade component *(Lonchodina)* analogous to those of Class A genera. It has paired platform-type components *(Gondollella),* but they are not closely analogous in structure to the platform blades of Class A. It has a battery of elongated blade components, but they are not the typical *Hindeodella* type of Class A. It apparently lacks the pick-shaped blades of Class A assemblages, though this may be the result of nonpreservation in the known specimens of the genus.

7. *Duboisella* appears to be quite distinct in general structure from both Class A assemblages and *Illinella.* As known at present, it lacks any obvious battery of elongated blades and platform-type components. This difference should be an important factor in any attempt to interpret the function of assemblages on the basis of analogy of form with structures in known organisms.

8. It is difficult to determine with certainty the numbers of individual component conodonts and the total number of the various component conodonts present in conodont assemblages. The following num-

Discrete Conodont "Genera" as Components of Assemblages

Component conodont type	*Lochriea* SCOTT	*Lewistownella* SCOTT	*Westfalicus* SCHMIDT	*Scottognathus* RHODES
Elongated blades	4 pairs *Hindeodella*	4 pairs *Hindeodella*	4 pairs *Hindeodella*	4 pairs *Hindeodella*
Arched blades	*Prioniodella*	*Ozarkodina*	*Ozarkodina*	*Ozarkodina*
Pick-shaped blades	*Neoprioniodus*	*Neoprioniodus*	*Synprioniodina*	*Synprioniodina*
Platform blades	*Spathognathodus*	*Cavusgnathus*	*Gnathodus*	*Streptognathodus* or *Idiognathodus*

bers are provisional. The total numbers range from a minimum of 12 to a maximum of 22. *Illinella* has 12; *Scottognathus, Lewistownella, Duboisella,* and probably *Westfalicus* have 14, and *Lochriea* 22.

9. The *Hindeodella* components are represented by a battery of four pairs of discrete conodonts. Other "genera" are represented by one or two pairs.

10. Data are inadequate with respect to the extent of variation in component conodonts both within and between species of natural assemblages. SCOTT (20, p. 297) has distinguished two species of the genus *Lochriea, L. bigsnowyensis* and *L. montanaensis,* in both of which the same "generic" components are present, but in which they are "specifically" distinct. RHODES (12) has given details of extensive "specific" variation in components of *Scottognathus, Illinella* and *Duboisella,* but has suggested that such variation may represent the extent of infraspecific variation within a single assemblage. Further collecting and study are needed to assess the true taxonomic significance of this "specific" variation of assemblage components.

11. The general alignment and arrangement of conodonts within assemblages tend to suggest an anteroposteriorly elongated arrangement within the animal.

12. The overall size of assemblages is small. The largest are about 9 mm. in length and 2 to 3 mm. in width.

13. No assemblages yet discovered contain conodonts attached to any basal bonelike substance. This basal material is common in neurodontiform conodonts and is rarely present in true conodonts (RHODES 14, p. 325). In view of the undisturbed condition of the assemblages, it seems unlikely that such material was present in their component conodonts.

INTERPRETATION OF CONODONT ASSEMBLAGES

That conodont assemblages occur is indisputable. They are not common, but more than 300 have been recorded from black shales of Carboniferous age in both North America and Europe (for details see RHODES, 12, p. 886-887). Their apparent restriction to black shales is probably the result both of the quiet conditions under which such strata frequently accumulated, and of the fact that fissility of the shales lends itself to study of their bedding planes under a binocular microscope. Other strata from which conodonts have been collected (such as limestones and sandstones) commonly accumulated under more disturbed conditions and are invariably subjected by paleontologists to such violent chemical and physical methods of disintegration, that there is little hope of recovering from them the conodonts which they may contain in anything but an isolated condition. The present lack of assemblages from strata other than those of Carboniferous age is probably more apparent than real, although there is perhaps a relatively higher proportion of black shales in the Carboniferous System than in other systems (Cambrian-Triassic) in which conodonts occur. Need exists, however, for careful study of such rich conodont-bearing black shales as those of the Upper Devonian of eastern and central North America.

Those who deny the validity of conodont assemblages do so, not because they deny their existence, but because they regard them as fortuitous rather than "natural" associations. The word "natural" could, in one sense, be used to describe any occurrence (whether random or not), but it is used here to describe an association which is the direct result of the original association of a variety of individual conodonts within the body of one conodont-bearing animal. I propose to examine the evidence which supports the recognition of natural conodont assemblages and then to consider the arguments of those who reject such an interpretation.

There are at least seven distinct aspects of the occurrence of conodont assemblages which support their interpretation as natural assemblages.

ASSOCIATION OF "GENERA" IN ASSEMBLAGES

The same genera (founded, that is, on "natural assemblages"), both from the same and from different localities and horizons, prove to contain the same component "genera" of isolated conodonts. Assemblages (at present undescribed) from upper Carboniferous Coal Measures of Britain, for example, are exactly similar to those described from the Pennsylvanian of Illinois and Kentucky (12). Thus, in both occurrences the assemblage *Scottognathus* contains component conodonts representing the same five "genera." This is not to imply that every assemblage studied contains all five components, for the degree of completeness is very variable. Sometimes, for example, only a single pair of components is found. The degree of resemblance is best illustrated by the uniformity of association rather than its completeness. In a detailed study of more than 200 assemblages, Rhodes (12) found that only two of them revealed the admixture of genera not commonly associated together in the same natural assemblage. It should also be noted that the overall numbers of components in these various assemblages are broadly consistent. The "specific" identities of component conodonts from British assemblages agree closely with comparable assemblages from the Pennsylvanian of North America (Rhodes, 12, p. 891-895). In these latter the individual components are variable in "specific" form, but it is not yet possible to assess the significance of this in the character of the assemblage variation and taxa.

It has been noted that the same "genus" may be present in more than one kind (genus) of natural assemblage. Thus, *Hindeodella* is present in *Scottognathus, Gnathodus* (Schmidt *non* Hinde), *Lewistownella,* and *Lochriea,* although it appears that the "species" of *Hindeodella* represented are different in each case. This complicates the evaluation, but in no way detracts from the importance of the regularity of association discussed above.

RATIOS OF ISOLATED COMPONENT "GENERA"

If component discrete "genera" occur in a fixed proportion within a natural conodont assemblage, it is probable that isolated "genera" would also be found in fixed proportions. There is a conspicuous lack of published data on this subject.

Scott (20, p. 295) studied 3,000 isolated conodonts from the Heath Shale and noted that "most of the different kinds of individual conodonts can be recognized in the assemblages; furthermore, the kinds found as individuals are proportional in numbers to those represented in the assemblages, i.e., hindeodells are found approximately three times as often as spathnognaths." DuBois (6, p. 157) studied 479 isolated conodonts from the fissile black shales below the La Salle Limestone (Middle Pennsylvanian) of Illinois. Of these he identified 108 polygnathids, 67 ozarkodinas, and 304 hindeodellas, or a ratio of roughly 1.6: 1: 4.5. This contrasts with the ratio 1: 1: 4 which Du Bois established by analysis of the conodont assemblages. He explained the apparent anomaly by the "differential ability of the teeth to withstand fragmentation."

The results of Du Bois' analysis are difficult to evaluate, but the deviation from the predicted ratio could be explained by the differential hazards of preservation. The more massive polygnathid components are undoubtedly more resistant to abrasion and probably less liable to transportation than the more delicate components. Indeed, the vagaries of fossilization are such that it may be doubted whether any consistent ratios should be expected. The ratio obtained from a limestone, for example, may be quite different from that obtained from a black shale. My own preliminary studies of the ratios between isolated components are not conclusive. There is a need for an extensive study of the ratios of isolated "genera."

PAIRED OCCURRENCE OF COMPONENTS AND THEIR ALIGNMENT

Assemblages are frequently readily recognizable by the paired arrangement of their components. These components are not only of the same size and general form but may sometimes be shown to be paired in such a way that one is the mirror image of the other. Sometimes other very minor morphological differences are observable between two such paired components, but these are no greater than those found, for example, between comparable complementary teeth in craniate skulls. No similarity of function is implied by this analogy, but it is useful in indicating the extent of this variation.

These paired components are frequently aligned with others in such a way as to form an elongated series. It would be difficult to account for such alignment, and virtually impossible to account for the paired relationship, except by the acceptance of these associations as natural assemblages.

STRUCTURAL SIMILARITY OF ASSEMBLAGES

Six genera of assemblages, all of Carboniferous age, are now known in sufficient detail to provide a comparison of their components. Of these assemblages, four are closely similar in their general make-up, another is broadly similar to them, and one is quite different. It is unnecessary here to discuss these resemblances in detail but they are illustrated diagrammatically in Fig. 43. Such similarity is very difficult to explain if the assemblages are interpreted as random associations.

OPINIONS OF INDEPENDENT STUDENTS

The first assemblages to be described were those from the Heath Formation of Montana (Scott, 19, 20) and the lower Namurian of Germany (Schmidt, 17), which were described in simultaneous but entirely independent publications. Scott and Schmidt differed in their interpretations of the zoological affinities of the conodonts, but both were in complete agreement that the assemblages which they described represented natural associations. Du Bois' (6) study of conodont assemblages from the McLeansboro Formation (Pennsylvanian) of Illinois convinced him that they were natural, rather than random. Rhodes was originally unwilling to accept the interpretation of natural conodont assemblages, but became convinced of its validity as a result of a study of Pennsylvanian assemblages from Illinois and Kentucky (12).

The independent conclusions of these workers who have studied conodont assemblages are thus in agreement in regarding them as natural associations.

GEOGRAPHICAL DISTRIBUTION OF ASSEMBLAGES

The assemblages described above come from the western and midwestern United States, from Germany, and (in the case of some still undescribed forms) from England and Wales. This widespread geographical distribution is another factor which supports their interpretation as natural associations. One occurrence of an assemblage might perhaps be accepted as fortuitous, but the occurrence of several hundred assemblages, composed of similar components, many of them paired, in broadly similar numbers, in similar alignment, described by half a dozen workers, from different parts of the column, in different continents, makes it difficult to maintain such a conclusion.

COPROLITIC ASSOCIATIONS

Coprolitic associations of conodonts are known, and indeed are described from strata which also yield natural assemblages. They are generally characterized by three features: (1) The very large number of isolated conodonts which they contain (up to 150 in comparison with a present maximum of 22 described from natural assemblages). (2) These conodonts show no alignment or obvious pairing, and may sometimes (but not always) represent more than one natural genus. (3) There may sometimes be a slight discoloration associated with the matrix around coprolitic associations. These criteria provide distinctive features by which two types of assemblage, one natural, the other coprolitic, may be differentiated.

It is now proper to consider the objections of those who do not accept the interpretation of natural conodont assemblages.

C. C. Branson (1, p. 169), in discussing the establishment of parataxa, writes: "Scott's assemblages are coprolite [coprolitic] associations. The validity of other assemblages is not demonstrated." He thus makes two distinct claims, the second of which may be assessed in the light of the detailed discussion above. The first statement—that Scott's assemblages are coprolitic associations—is curiously dogmatic in that it is unsupported by any evidence. The burden of data outlined above is very strongly in support of SCOTT's interpretation of the assemblages as natural, rather than coprolitic in origin. SCOTT (20, p. 296) concluded ". . . it would be strange indeed to find a group of animals with such a balanced diet that the excretal material would consist time after time of one pair of prioniodids, one pair of spathognaths, one pair of prioniodells, and approximately four pairs of hindeodells" (the components of the natural assemblage genus *Lochriea*). It would be of the greatest interest to know the evidence which persuades some other students to the contrary.

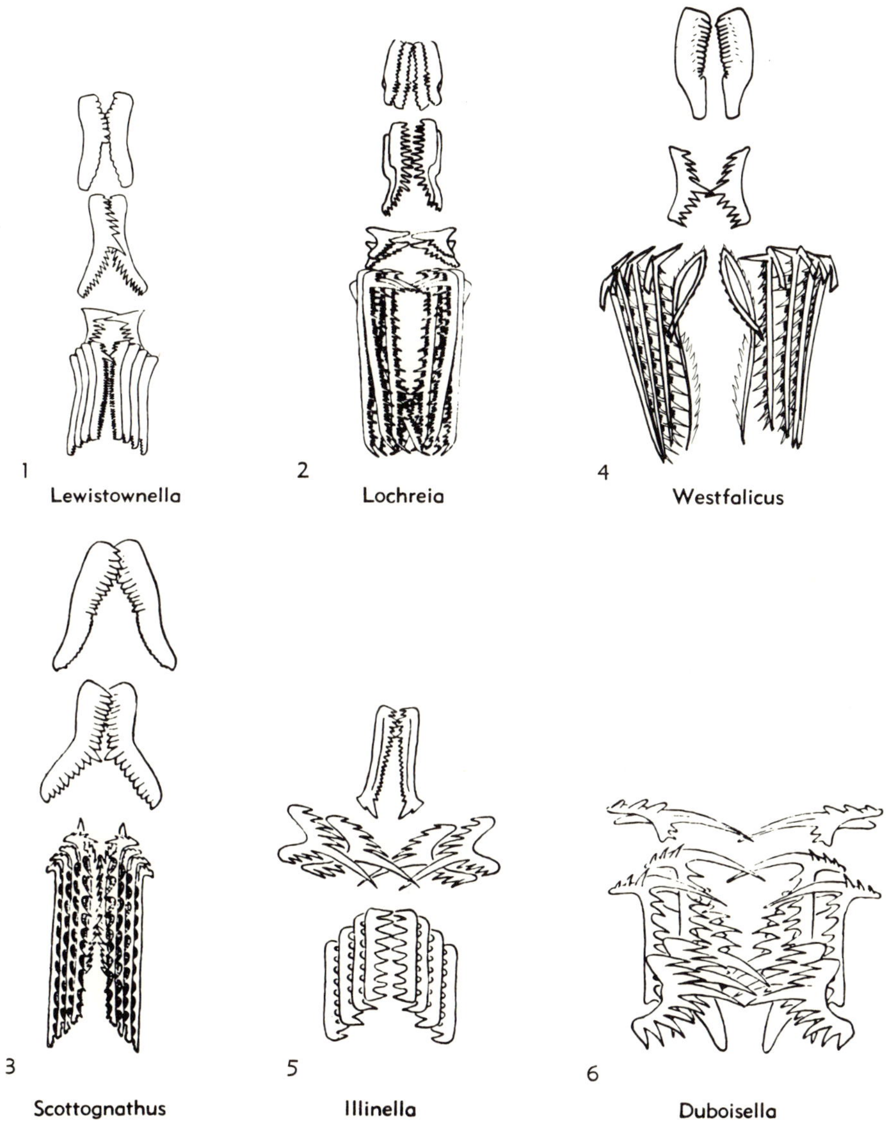

FIG. 43. Diagrammatic illustrations of natural conodont assemblages: *1, Lewistownella agnewi* SCOTT; *2, Lochriea montanaensis* SCOTT; *3, Scottognathus typica* (RHODES); *4, Westfalicus integer* (H. SCHMIDT); *5, Illinella typica* RHODES; *6, Duboisella typica* RHODES; all ×15 (approx.) (*1,2,4-6* after Rhodes, 1954; *3,* mod. from Schmidt, 1934).

Branson & Mehl (3) wrote: "An apparently insurmountable difficulty to the group assemblage is the fact that the involved genera are not co-extensive in their stratigraphic range." But, as other workers have already remarked, the addition or subtraction of certain kinds of "teeth" would be a normal consequence of the evolution of the group, and some component "genera" are present in more than one genus of natural assemblage. If the validity of the association of component bones in vertebrate skeletons were determined by the co-extensive ranges of the "genera of bones involved," our interpretation of vertebrate palaeontology would be in need of drastic revision. But most paleontologists accept the fact that parts of an organism may evolve at differential rates, and that such parts ("genera") as a pelvis or a scapula may be present in more than one species of organism.

Branson & Mehl (3, p. 233) further commented, "Doubt is cast on the assemblage finds as normal associations because the analysis of one, or many, samples from a productive zone . . . fails to show proportional numbers of kinds supposedly found in one animal." This is the only objection to the natural assemblage interpretation which has any weight. As shown above, however, the present evidence is not conclusive and to expect absolute constancy of ratios of isolated components is to underestimate the hazards, vagaries, and selective nature of the processes of both fossilization and micropaleontological extraction.

ZOOLOGICAL AFFINITIES

Workers who have studied conodont assemblages are divided in their interpretation of their zoological affinities. Scott (19, 20) and Du Bois (6) regarded them as characteristic of the annelids. Schmidt (17, 18) favored an association with fish. Hass has discussed these interpretations in the preceding chapter.

TAXONOMIC PROBLEMS

One of the most difficult problems raised by the acceptance of natural conodont assemblages is that of their nomenclature. I propose to consider this in some detail.

PRESENT TAXONOMIC POSITION

It has already been noted that an extensive "taxonomy" has been established upon isolated conodont specimens. This nomenclature, which includes about 2,000 "species," has been established by workers who have rigidly observed the code of zoological nomenclature. The suggestion of Croneis (5) for an independent "military classification" has not been generally followed.

The acceptance of natural conodont assemblages, containing up to five component "genera," has led some workers (Scott, 20; Schmidt, 17, 18; Eichenberg, 7; Rhodes, 12) to propose a new classification, based on the recognition of conodont assemblages as the remains of individual organisms, and consequently as natural taxonomic units. Therefore, two classifications exist and it is necessary to consider their implications.

The nomenclature of natural conodont assemblages has been established in three more or less distinct ways.

Method 1. Assemblages have been named after the earliest applicable name of any component which they contain (e.g., Eichenberg, 7; Schmidt, 17; Sinclair, 21).

Method 2. Assemblages have been given new binomina, and the component conodonts have been designated by descriptive technical terms. Scott (20) followed essentially this practice, identifying the "genera" (but not the "species") represented in two genera of natural assemblages, and describing the components by common nouns coined from the "generic" names. Thus specimens of *Hindeodella* were termed hindeodells, etc. Scott emphasized, however, that he considered it desirable that the earlier "form-classification" should be retained (20, p. 295), even though he found it inconvenient to employ it for assemblages.

Method 3. Assemblages have been given new binomina and the component conodonts have been designated by their previously established "generic" and "specific" names (if any) (e.g., RHODES, 12).

PROBLEMS OF DUAL CONODONT CLASSIFICATION

These problems have been discussed in detail by MOORE & SYLVESTER-BRADLEY (10, Supplemental Application "A"). It is useful to summarize them for the present discussion.

(1) The existence of two systems of nomenclature is illegal under the Rules and consequently confusing and unstable. Both are necessary and useful, but they can only exist because conodont workers are forced to ignore the Rules.

(2) In some cases the same generic and specific names have been applied both to assemblages and to some of their isolated components (e.g., *Gnathodus, Prioniodus,* and *Polygnathus dubius*). This is clearly undesirable. The application of Article 27 leaves one or the other taxon without a name. [The problems of applying new names are discussed subsequently.]

(3) In cases in which new generic names have been used for assemblages there are also serious problems if the Rules are applied. Thus, the genus *Duboisella* RHODES (type-species, *D. typica*) contains components representing five "genera" of previously described component conodonts. If Article 27 is applied, these five genera and species should be placed in synonomy with *D. typica,* which name should itself be replaced by that of the earliest described component. But generic identity between discrete conodonts and natural assemblages can only be recognized if the type species of the discrete genus is present in an assemblage of the genus. In the case of *Duboisella,* the type species *"Neoprioniodus conjunctus"* and *"Metalonchodina bidentata"* have been recognized in assemblages. According to the Rules, the name *"Neoprioniodus"* should therefore take priority over *Duboisella,* and *"Metalonchodina"* would also be regarded as a junior subjective synonym of *"Neoprioniodus."*[1] The specific name *"typica"* would also be replaced by its earliest synonym. New names would be needed to designate all the other discrete conodonts represented in the assemblage. The type "species" *"Metalonchodina bidentata"* is present in the assemblage named *Duboisella,* but other species of the "genus" have not been so identified. They may or may not be congeneric with *D. typica,* and, unless they are found intimately associated with an assemblage, they cannot be placed in an assemblage genus. The rarity of natural asemblages may mean that it will never be possible to identify the other named "species" of *"Metalonchodina"* with whole-animal taxa. What name is to be used for them?

NEED FOR REVISION OF EXISTING TAXONOMY

I have used the examples cited above to illustrate the illegal and unstable nature of the present position, the drastic revision and utter confusion that would result from application of the Rules, and the fact that in some cases their rigid application would lead to a nonsensical taxonomy.

Any solution to the problem must provide freedom of taxonomic expression to those who work with natural conodont assemblages and to those who work with discrete conodonts. Such a solution must satisfy five requirements: (1) It must provide a method for the recognition and classification of natural conodont assemblages. (2) It must provide a name to differentiate each of the diagnostic forms of individual, isolated conodonts, which are of value in stratigraphy. (3) Homonymy between these two systems of nomenclature must be avoided. (4) Both systems must exist within the legal framework of the International Code, and must derive the protection, stability, and uniformity which the Code provides. (5) Any changes in procedures should be such as to produce the minimum possible disturbance in the existing nomenclature.

---

[1] Since *Metalonchodina* BRANSON & MEHL was introduced in 1941, with *Prioniodus bidentatus* GUNNELL, 1931, as type species, and *Neoprioniodus* RHODES & MÜLLER was first published in 1956, with *Prioniodus conjunctus* GUNNELL, 1931, as type-species, *Metalonchodina bidentata* should replace *Duboisella typica* according to the Rules. *Metalonchodina* clearly has priority over *Neoprioniodus.*—EDITOR.

PROPOSALS FOR TAXONOMIC REFORM

The following remarks are intended as a generalized but critical review of each of the three possible methods of naming assemblages outlined above. This is based partly on comments published in Document 1/47 of the *Bulletin of Zoological Nomenclature* (RHODES, 16).

(1) Of the three ways of dealing with the present taxonomic problem previously indicated, Method 1, in which the assemblages are named after the earliest applicable name of any discrete conodont which they contain is the "legal" solution under the existing Rules (Article 27). However, it involves serious difficulties of two main types.

(a) What name is correctly applicable to an assemblage? Clearly the name that must be applied to an assemblage is that given to the *first-named* part of the animal. If this is done, the following considerations arise:

(i) Objective identification with a natural genus can therefore only be made if the type-species is present in the assemblage.[1]

(ii) One "genus" may be present in more than one type of natural assemblage (i.e., in more than one natural genus).

(iii) It might be suggested that this problem could be overcome by a modified application of the Law of Priority, according to which the name of a unique "genus" among the component discrete conodonts would be chosen to be the type of the natural assemblage selected. Conodont assemblages are rare, however, and it is quite impossible to predict whether or not any such component "genus" would prove to be peculiar to a single type of assemblage.

(iv) Conodont specialists find it convenient to distinguish the two discrete conodont "genera," *Streptognathodus* and *Idiognathodus,* on minor morphological features, in spite of the fact that these two "genera" are transitional. RHODES (12, 13) has shown, however, that *Scottognathus,* a genus represented by natural conodont assemblages, may contain either one or the other of these "genera," which are transitional within the assemblages. Similar cases may also exist, and it would be misleading if one of these "generic" names were applied to assemblages in which the "genus" itself was not present. It may be argued that the "genera," if transitional, must *ipso facto* be synonymous, but practicing paleontologists would reserve the right to dispute this principle. Chronological fossil sequences show all grades of transition, and in the most complete successions taxonomic units are more or less arbitrary subdivisions of more or less continuous fossil sequences.

(v) Similar problems to the three noted above arise in the choice of a specific name. Other aspects of the problem of the choice of a specific name have been discussed by MOORE & SYLVESTER-BRADLEY (10).

(b) What name is correctly applicable to an isolated conodont?

(i) If the earliest applicable name of a discrete conodont contained in an assemblage should be applied to the whole assemblage, all other differently named "genera" and "species" of discrete conodont identified within the assemblage would be junior synonyms of the name given to the assemblage. This would require drastic revision in the nomenclature of isolated conodonts. Some names must be available to designate the different kinds of isolated conodonts, which are of great stratigraphic importance.

[1] It is important to take note of the fact that the asserted presence of the type-species of a discrete conodont genus is a subjecive identification.—EDITOR.

(ii) Some writers (e.g., SINCLAIR, 21, p. 489) have argued that, if this method were adopted, it would be possible to designate individual conodonts as (for example) the "subbryantod element of *Streptognathodus elegans*" (where the binomen is that given to a natural assemblage). The term "subbryantod" is coined from the name of the discrete conodont "genus" *Subbryantodus*. This might appear satisfactory for conodont components which show little variation in natural assemblages. Some elements, however, are present in more than one genus, so that two or more names are applicable. This results in three possible "states of synonomy," which are at three categorical levels. In any given case all three might apply. Thus, two "species" of isolated conodonts, *Xognathus aa* and *Xognathus ba,* may be regarded as synonymous. Both might be shown to occur in the same assemblage, to the selected name of which one or possibly both would then be synonymous. But thirdly, they might also be found to be present in more than one type of natural assemblage, in which case the descriptions "xognathid element of *Alphagnathus beta*" and "xognathid element of *Gammagnathus delta*" would be synonymous at a third (quite different) taxonomic level. One need not elaborate the taxonomic confusion that would result from such a situation.

(iii) This would be only the beginning of confusion, for only such isolated conodonts as show little variation in natural assemblages have so far been considered. In many cases variation is considerable and a "specific" qualification would be necessary to designate any particular form (e.g., "the subbryantod type 23 element of *Streptognathodus elegantulus*"). This would not only involve a complete revision of conodont terminology and the substitution of a clumsy, very unsatisfactory system of nomenclature for that now used, but it would deprive the new system of nomenclature of uniformity and protection which the Rules are designed to afford.

(iv) Even if, in spite of this, the suggested solution were accepted, one insurmountable problem would remain. Only a very few "genera" and "species" (perhaps fewer than 5 percent of the "species") are at present known as components of natural assemblages. For the great majority of isolated conodonts, therefore, *no* names would be available.

(2) It may be suggested that in order to avoid confusion, all conodont "genera" and "species" not based on natural assemblages should be regarded as invalid. This would mean that zoological names should be applied only to assemblages, and suitable technical terms then would be employed to designate isolated component conodonts (Method 2, previously outlined). Such a solution would reduce problems of synonymy, but all the other major problems discussed above would remain.

(3) The third possible method would be to give new names to natural conodont assemblages and to retain the existing system of nomenclature for isolated conodonts (Method 3, previously suggested). In view of the problems discussed above, this is clearly the most desirable solution. Indeed, it is the only solution that will permit satisfactory continuation and development of conodont studies.

This is the method suggested and cogently supported by MOORE & SYLVESTER-BRADLEY (10) in an application to the International Commission on Zoological Nomenclature for establishment of parataxa. Their carefully reasoned document proposed that discrete conodont "taxa" should be designated as parataxa and should exist as categories within, subject to, and protected by the

Rules. They urged that parataxa should be regarded as a special taxonomic category, and that use of them should be restricted to discrete fragments or life stages of animals which are inadequate for identification of whole-animal taxa. They considered that such a system of nomenclature would be useful in the classification of coccoliths, sponge and octocoral spicules, holothurian sclerites, ossicles of crinoids, cystoids, blastoids, echinoids and asterozoans, scolecodonts, gastropod radular elements and opercula, and cephalopod aptychi. Detailed safeguards, such as the mutual nonavailability for parataxa of taxonomic names employed for whole animal taxa, and *vice versa,* were included in the proposals.

The proposals were presented at meetings of a Colloquium on Zoological Nomenclature held at the Fifteenth International Congress of Zoology in London in 1958. After very brief consideration the Colloquium rejected the proposals, yet offered no alternative solution. This action leaves conodont nomenclature in a confused and unstable position. Presumably students of discrete conodonts will ignore the decision and continue to use a binomial system of nomenclature for the isolated conodonts which they study. But the continuing study of assemblages will show that more and more discrete "species" and "genera" are parts of whole animals, and therefore are synonyms of whole-animal taxa. It is hoped that students of conodonts and other discrete fossil fragments will continue to press for the recognition of parataxa or whatever other means may be devised for practicable classificatory and nomenclatural procedure not in conflict with the International Rules.

REFERENCES

Branson, C. C.

(1) 1957, *Comment on the Moore/Sylvester-Bradley "Parataxa Plan":* Bull. Zool. Nomenclature, v. 15, Document 1/25, p. 169.

Branson, E. B., & Mehl, M. G.

(2) 1933, *A study of Hinde's types of conodonts preserved in the British Museum:* Univ. Missouri Studies, v. 8, p. 133-167, pl. 11-12.

(3) 1936, *Geological affinities and taxonomy of conodonts* (abs.): Pan-Am. Geologist, v. 65, p. 233.

Bryant, W. L.

(4) 1921, *The Genesee conodonts:* Buffalo Soc. Nat. Sci., Bull., v. 13, p. 1-59.

Croneis, Carey

(5) 1938, *Utilitarian classification for fragmentary fossils:* Jour. Geology, v. 46, p. 975-984.

Du Bois, E. P.

(6) 1943, *Evidence on the nature of conodonts:* Jour. Paleontology, v. 17, p. 155-159, pl. 25.

Eichenberg, W.

(7) 1930, *Conodonten aus dem Culm des Harzes:* Paläont. Zeitschr., Bd. 12, p. 177-182, fig. 1-30.

Fay, R. O.

(8) 1957, *Comment on the Moore/Sylvester-Bradley "Parataxa Plan":* Bull. Zool. Nomenclature, v. 15, Document 1/26, p. 170-171.

Hinde, G. J.

(9) 1879, *On conodonts from the Chazy and Cincinnati group of the Cambro-Silurian, and from the Hamilton and Genesee-shale divisions of the Devonian, in Canada and the United States:* Geol. Soc. London, Quart. Jour., v. 35, p. 351-369, pl. 15-17.

Moore, R. C., & Sylvester-Bradley, P. C.

(10) 1957, *Proposed insertion in the "Règles" of provisions recognizing "Parataxa" as a special category for the classification and nomenclature of discrete fragments or of life-stages of animals which are inadequate for identification of whole-animal taxa, with proposals of procedure for the nomenclature of "Parataxa":* Bull. Zool. Nomenclature, v. 15, sec. A, pt. 1/4, Document 1/1, p. 5-13.

Pander, C. H.

(11) 1856, *Monographie der fossilen Fische des silurischen Systems der russisch-baltischen Gouvernements:* Akad. Wiss. St. Petersburg, p. 1-91, pl. 1-7, table.

Rhodes, F. H. T.

(12) 1952, *A classification of Pennsylvanian conodont assemblages:* Jour. Paleontology, v. 26, p. 886-901, pl. 126-129.

(13) 1953, *Nomenclature of conodont assemblages:* Jour. Paleontology, v. 27, p. 610-612.

(14) 1953, *Some British Lower Palaeozoic conodont faunas:* Royal Soc. London, Philos. Trans., ser. B, v. 237, no. 647, p. 261-334, pl. 20-23.

(15) 1954, *The zoological affinities of the conodonts:* Cambridge Philos. Soc., Biol. Reviews, v. 29, p. 419-452, fig. 1-6.

(16) 1957, *Comment on the Moore/Sylvester-*

Bradley "Parataxa Plan": Bull. Zool. Nomenclature, v. 15, Document 1/47, p. 305-312.

Schmidt, Hermann

(17) 1934, *Conodonten-Funde in ursprünglichen Zusammenhang*: Paläont. Zeitschr., Bd. 16, p. 76-85, pl. 6.

(18) 1950, *Nachträge zur Deutung der Conodonten*: Decheniana, Bd. 104, p. 11-19, fig. 1-7.

Scott, H. W.

(19) 1934, *The zoological relationships of the conodonts*: Jour. Paleontology, v. 8, p. 448-455, pl. 58-59.

(20) 1942, *Conodont assemblages from the Heath Formation, Montana*: Jour. Paleontology, v. 16, p. 293-300, pl. 37-40.

Sinclair, G. W.

(21) 1953, *The naming of conodont assemblages*: Jour. Paleontology, v. 27, p. 489-490.

Ulrich, E. O., & Bassler, R. S.

(22) 1926, *A classification of the toothlike fossils, conodonts, with descriptions of American Devonian and Mississippian species*: U. S. Natl. Museum, Proc., no. 2613, v. 68, p. 1-63, pl. 1-11, fig. 1-5.

TAXONOMY, EVOLUTION, AND ECOLOGY OF CONODONTS

By Klaus J. Müller

[Technische Universität Berlin, West Berlin, Germany]

CONCEPTS IN TAXONOMY OF DISJUNCT CONODONTS

Species are to be defined empirically, by observation. Illustrations of the variation of conodonts within a single animal, as demonstrated by a conodont assemblage, have been given by Rhodes (11).

The range of variability within a population is quite wide, but differs considerably from one to another. The most advanced forms within a branch of evolution generally are more stable morphologically than the primitive ones.

In establishing species, ontogenetic transformation has to be considered. The growth lamellae of conodonts are not exactly parallel to each other, but are variable in their width. New denticles originate and in some specimens denticles are overgrown, thus forming germ denticles. The number of denticles on a conodont, therefore, is also dependent on its size, and for comparison not just the number of denticles has to be given, but also the measurements. Sculpture on the plate of platform types also changes during growth. In *Polygnathus* and *Palmatolepis* the smallest growth stages are unsculptured, whereas later stages more or less suddenly exhibit a distinct sculpture, which in the most mature stages is somewhat smoothened off again. This can be observed by comparison of the different growth stages in a population, as well as of growth lines seen in a thin section through a mature specimen.

In other species a direction of most prominent growth has been observed. The outline (e.g., *Ancyrognathus triangularis* Youngquist) changes considerably during growth, because in the anterior portion the growth lamellae are much wider apart than on the rest of the unit. For comparison, therefore, it is necessary to select specimens of the same size or at least to take account of transformation by observing the growth lines.

Occasionally populations contain a few atypical specimens, which do not fall in the range of species present. The differences are due to abnormal growth. Redefinition of species in order to include these specimens would blur the picture and would make it difficult to recognize the species elsewhere. Therefore, these specimens should be regarded as indeterminable.

Subspecies based on minor differences have been erected by some authors. However, at the present stage of description no agreement exists as to what may be regarded as minor differences suitable for subspecific distinction only. Therefore, some established species deserve subspecific rank only and units defined as subspecies by some authors are regarded as species by other students.

Genera have been abstracted to assist determination. For grouping a number of species into a form-genus, features have to be found which are present in all species thus grouped together. These features do not necessarily have to be the most obvious ones, for only their persistence within the group is important.

Genera of disjunct conodonts are artificial, even if they are based on natural relationship in an increasing number of cases. Monotypic genera in conodonts have little meaning, for they will be of no help in determination. Proposal of them only rarely is justified (e.g., if a rather common species, generally of short vertical range, is considerably different from established genera). However, it is of no advantage for development of a clear system if four of seven new genera erected in 1957 are respectively based on a single species only.

Families and subfamilies of conodonts recognized by Hass in this volume are defined mainly on a single feature—position of the basal cavity in relation to other parts of the conodont structure. This detail is easily recognizable in nearly every specimen, but dependence on it for systematic arrangement leads to assignment of closely related genera with intergrading species to different families (e.g., *Prioniodina* to Prioniodinidae, *Prioniodus* to Prioniodontidae, *Neoprioniodus* to Coleodontidae). Besides, it has little meaning as a help for determination.

For most platform and single-cone types of conodonts, I judge that a more natural system can be attained by grouping genera according to phylomorphogenesis. However, such a system hardly can be established for all conodonts, because many bars and blades of similar form occur in different assemblages. In an artificial system, such as that proposed by Ulrich & Bassler and accepted by Hass, the bar and blade types are grouped in families also. Then various parts of the same animal may belong to different "families" and "subfamilies."

The various conodont elements in an assemblage do not have equal meaning for taxonomy. Some are inconstant even in closely related assemblages, whereas others are seen to be quite stable in unrelated assemblages and thus are believed to have little significance for systematics (and as a consequence for stratigraphy). For approach to a natural system of conodonts it is necessary to take account of this observation. Taxonomy needs to be based mainly on features which exhibit change during evolution, stable elements having only minor systematic value.

At present, conodont assemblages are described only from the Pennsylvanian and beds of equivalent age. When more complete knowledge of the assemblage occurrences is attained, it should be possible to sort out the "good" and "inferior" form-genera of disjunct conodonts in the entire system. Taxonomy could be simplified considerably by suspending the generic names of stable elements (through action of the International Commission on Zoological Nomenclature) and recoining them as morphologic terms (e.g., hindeodellids). It is believed that such a system would not differ from a system of natural assemblages, which thus would be named like the distinctive disjunct parts.

EVOLUTION

Although conodonts comprise parts of an animal yet unknown, they are nevertheless well suited for tracing evolution. Ellison (2) stated: "Conodonts are among the best fossils for family-tree studies." Similar to solitary corals and ammonoids, the preserved hard parts give evidence of ontogeny. Their structure is composed of growth lamellae which center around a nucleus. By observation of these lamellae changes of morphological features such as outline, pattern of sculpture, and mode of insertion of denticles during ontogeny can be observed. Many specimens show that the earliest growth stages of related species are quite similar—indeed, only the more mature ones can be distinguished, a fact which can be observed quite frequently in etching residues.

Furthermore, many genera and species of conodonts are linked together by transitional form types, some of which are excellent markers for certain time intervals. An example is the link between *Polygnathus dubius* Hinde and *Palmatolepis transitans* Müller. Samples of several sequences in cephalopod-bearing limestones at the Middle-Upper Devonian boundary have yielded intermediate forms in abundance (*see* Fig. 47, bottom).

Some difficulty in establishing a natural system of disjunct-conodont classification arises from the fact that certain tendencies of development occur in different groups independently, thus producing homeomorph "genera" and "species." The most striking example is the similarity between the Devonian-Lower Carboniferous genus *Polygnathus* and some Upper Triassic species. There are no representatives similar to *Polygnathus* in the long time interval from early Pennsylvanian until Middle Triassic, and the branch which leads to *Polygnathus mungeonsis* DIEBEL evolves from *Gondolella,* as is demonstrated by species of this form-genus.

The genus *Taphrognathus* was erected by BRANSON & MEHL, 1941, who stated in the original description: "Little can be offered in the way of generic analysis to separate [it from *Streptognathodus*] satisfactorily. We interpret these two groups as parallel developments, originating at two entirely different times, probably from the same stock." As was pointed out by REXROAD (10), *Taphrognathus* gave rise to *Cavusgnathus,* and in uppermost Chesteran beds transitional forms between *Cavusgnathus* and *Streptognathodus* have been observed.

Another example of homeomorphy is the striking similarity in many features exhibited by *Icriodus pesavis* BISCHOFF & SANNEMANN, from the Lower Devonian, and *Staurognathus anchoraria* HASS, from the Mississippian (Fig. 44). These forms are widely apart systematically and it is not clear whether the convergence is due to some functional reason or merely to reiteration because of the limited possibilities in form variety within the group, as seen in some Ammonoidea. The latter seems to me more probable.

As is true also in evolution of the Ammonoidea, repeated features may have a

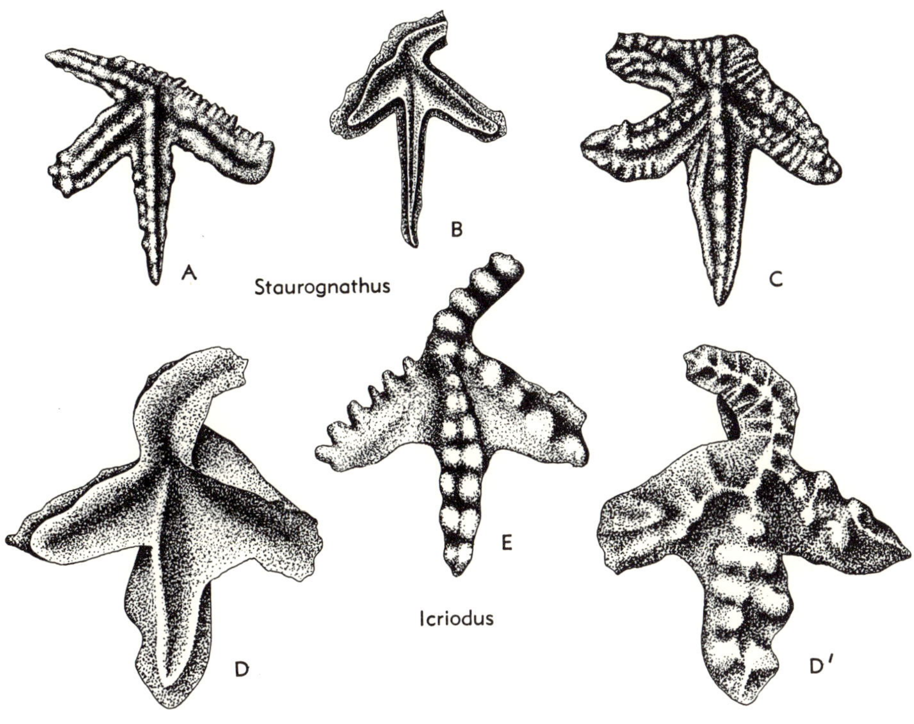

FIG. 44. Homeomorphy in conodonts.——*A-C. Staurognauthus anchoraria* HASS, Miss., USA(Tex.), three specimens demonstrating interspecific variability, ×35 (Hass, 1959).——*D-E. Icriodus pesavis* BISCHOFF & SANNEMAN, L. Dev., Ger., two specimens, ×27 (Bischoff and Sanneman, 1958). Both forms, phylogenetically of quite different origin, are considered to be index fossils for narrow zones. Similar forms are unknown in the long interval from the upper part of the Lower Devonian to lowermost Mississippian strata.

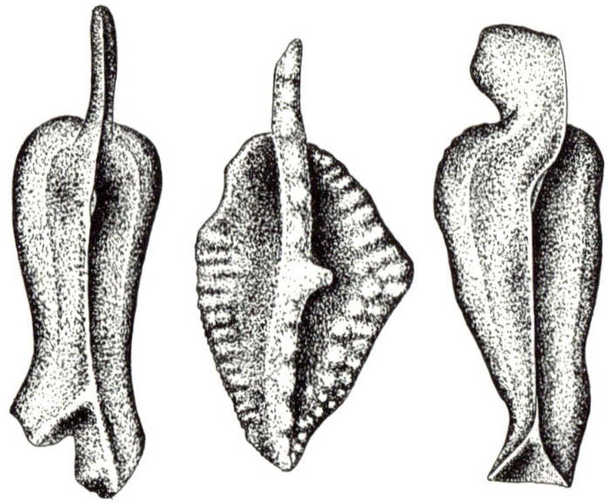

Fig. 45. Intraspecific variation of conodonts illustrated by three specimens of *Polygnathus unicornis* Müller & Müller, with secondary carinae or keels in different positions; only 5 percent of the available specimens have secondary keels or carinae, a feature which would put them into *Ancyrognathus* if it were stable. U.Dev., USA(Iowa), ×30 (9).

different taxonomic meaning in various phylogenetic lines. *Ancyrognathus* is distinguished from *Polygnathus* by the presence of a secondary branch on the "posterior" portion. The same feature occurs in the intraspecific variation of *Polygnathus unicornis* Müller & Müller (Fig. 45). Only 5 percent of the specimens have a secondary keel or carina, or both, a feature which would put them into *Ancyrognathus*. However, the position of the secondary elements is different in nearly every specimen, and therefore the feature has no bearing on taxonomy.

Close examination of homeomorphic forms generally reveals that only a part of their observed features are duplicated, while other more stable (though not always easily recognizable) features are distinct. Careful study of these features (e.g., crimp, mode of insertion of denticles) will help to attain a more natural classificatory arrangement of these fossils.

Simple cones (Distacodidae) have been little studied as yet in regard to their morphogenesis, but possibilities of finding connections between many form-genera seem to be good. General developments, which most probably have been attained in different branches within this group are (1) reduction in size of the basal cavity, which can be observed in specimens ranging from Cambrian to Devonian; and (2) differentiation of the sides by keels, carinae, and furrows, leading to the peak development of the Distacodidae in Ordovician time but decreasing somewhat in importance during the Silurian.

The simple cones gave rise to bar and blade types of conodonts by addition of denticles to the basic cone. Obviously, this happened in different branches of evolution independently at various times and in a different manner by (1) bowing flanges of the cone upward (e.g., *Westergaardodina*); (2) sudden reiteration of the denticle (e.g., *Loxodus*); (3) forming an extension on one side, later adorned by new and more or less widely spaced denticles (e.g., *Cordylodus*); (4) differentiation of a carina or keel to form a thin lamella which breaks up into rather small, somewhat irregular, closely spaced denticles (e.g., some Silurian form-species referred to "*Belodus*").

The bars generally seem to be quite stable and therefore their value as index fossils is much smaller than that of platform types. They have only a few features which are suited for establishing morphogenesis within the group. Convergence has been observed frequently. However, even here it may be possible to recognize gradational forms between bar and platform types. Ellison demonstrated perfect transition between the bar type, *Prioniodina,* and the platform type, *Gondolella.*

The blades gave rise to platform types in different lines of evolution. The latter are best suited for tracing the morphogenesis, since they possess many features which underwent gradual change during evolution. This is particularly true of form-genera derived directly or indirectly from *Spathognathodus.* Their probable relationship, as demonstrated by intergrading form-species, may be represented diagrammatically (Fig. 46). Within this group a natural system can be attained with the present state of knowledge.

Not only between form-genera but also, within some of them, transitional stages between species and subspecies can be established. Diagrams showing the morphogenetic development of *Palmatolepis* have been published by Müller (6), Scott & Collinson (12), and Helms (5), the last reproduced here (Fig. 47). This example demonstrates excellent suitability of a widely distributed, common form-genus for tracing evolutionary developments, as well as for subdivision of an epoch.

ECOLOGY

Conodonts have not yet been found in sediments which are thought to be nonmarine. In Upper Carboniferous coal measures of England, Western Germany, Kansas, and Illinois, their presence is regarded as proof that the containing beds were deposited in a paralic environment.

Quite commonly they are associated with cephalopods, tending to be particularly abundant in cephalopod-bearing limestones. Further, they are often associated with fish remains and ostracodes. However, in most Paleozoic sediments the distribution of conodonts is much wider than that of these other fossils. They can be secured also from black shales, in which few fossils are preserved.

Bioherms and biostromes composed largely of corals, stromatoporoids, sponges, and calcareous algae, contain very few conodonts. Fusulinid-bearing limestones almost nowhere yield conodonts, as demonstrated by many samples from North America, South America, and Europe.

That conodont-bearing animals had a pronounced bilateral symmetry, is concluded from the following observations. (1) The majority of conodonts occur in "right" and "left" specimens, which have mirror-image similarity. (2) The "right" and "left" specimens of most species occur in statis-

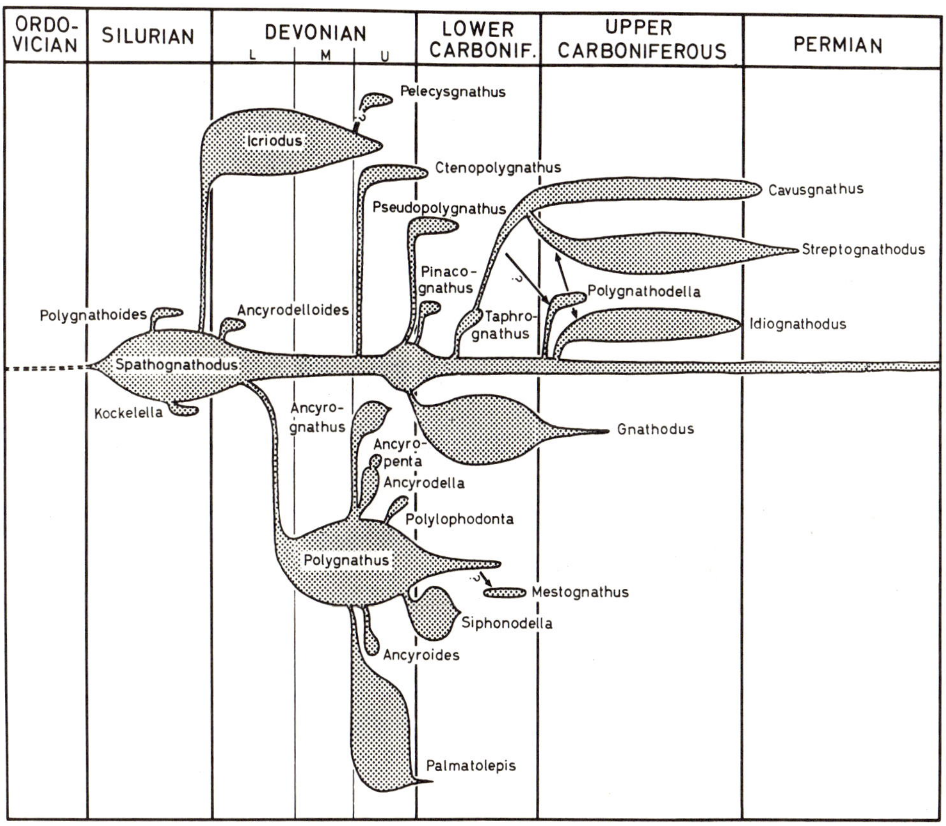

FIG. 46. Systematic relationship between genera derived from *Spathognathodus* as demonstrated by morphological intergrading. *Spathognathodus* includes the subgenera *Pandorinellina* and *Branmehla; Ancyroides* includes *Ancyrolepis.* (Data from Branson & Mehl, Bischoff & Sannemann, Müller, Rexroad, Thomas, Walliser, Ziegler, and others) [Müller, n].

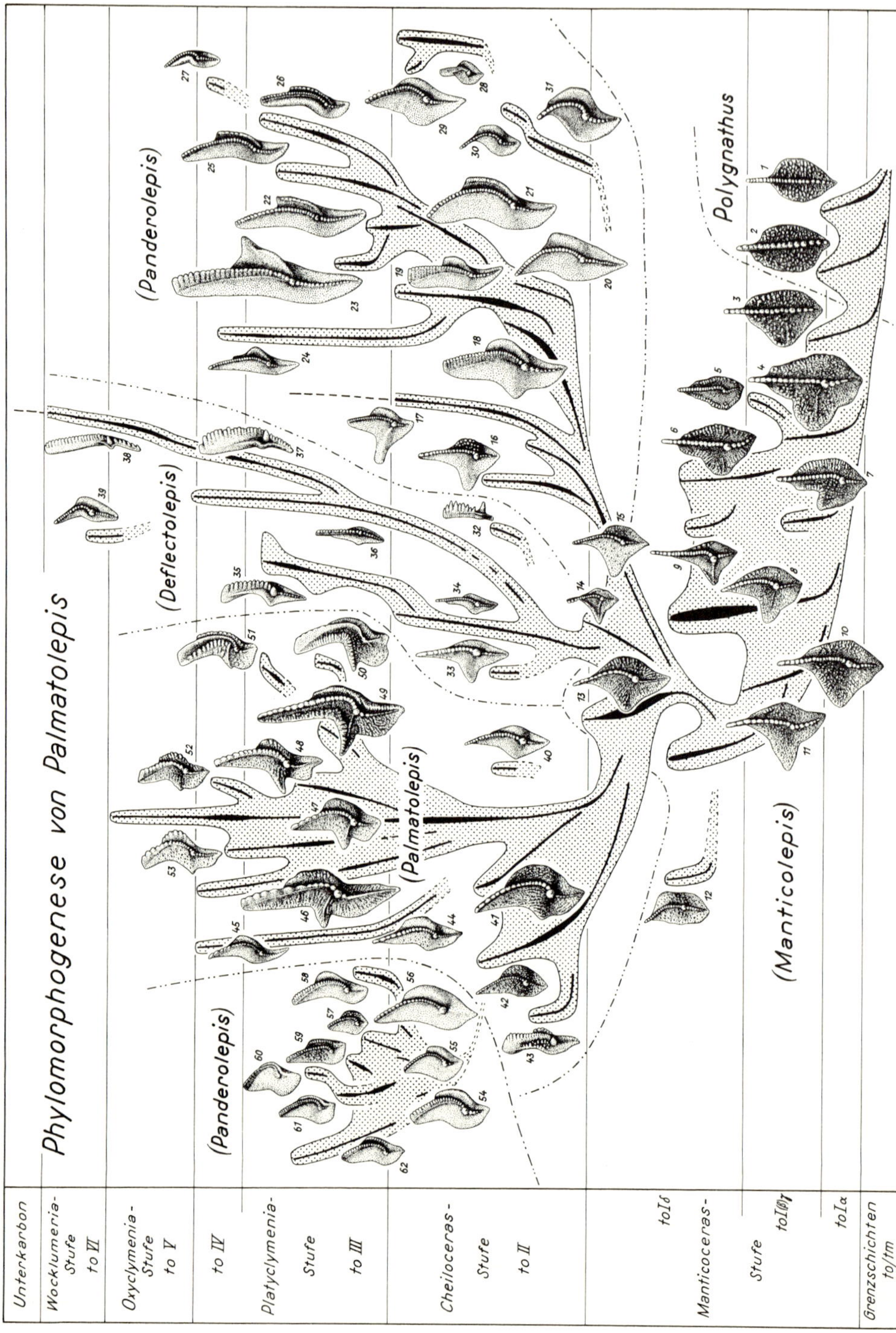
Phylomorphogenese von Palmatolepis
(Panderolepis)
(Deflectolepis)
(Palmatolepis)
(Panderolepis)
Polygnathus
(Manticolepis)
Unterkarbon
Wocklumeria-Stufe to VI
Oxyclymenia-Stufe to V
to IV
Platyclymenia-Stufe to III
Cheiloceras-Stufe to II
toIδ
Manticoceras-Stufe toI(β)γ
toIα
Grenzschichten to/tm

tically equal numbers in washed residues. Only in certain species of *Polygnathus* has deviation from the symmetry been described. In these cases the "right" and "left" specimens which most probably belonged to the same animal are different, which is regarded as secondary adaption. (3) A minority of conodonts do not occur in pairs but show more or less pronounced bilateral symmetry in themselves (e.g., *Hibbardella, Roundya*). These elements might have been arranged along the median line of the animal.

The pronounced bilateral symmetry suggests that the conodont animal was able to move about actively. This is suggested also by the shape of many platform types, which must have supported movable soft parts, particularly in the unsculptured grooves on one or both sides of the blade. In some "genera" (e.g., *Ancyrodella, Palmatolepis*) the "anterior" part of these grooves is fortified by much thickening, which is only understandable if this portion was subject to considerably more strain than the remainder of the conodont.

Conodonts commonly are not confined to sedimentary facies, since the same species is found to occur in different lithologies (e.g., limestone, shale, sandstone). This fact is of great advantage in using conodonts for correlation. The independence of lithofacies suggests that the conodont animals were free-swimming creatures. Because many species have a world-wide distribution, they may be interpreted as having been pelagic.

However, there are some exceptions which probably denote a change in habitat. Some form-genera are fairly abundant in a certain facies only (e.g., near-reef), and it may well be that these became adapted to a benthonic mode of life, or at least lived near the bottom (e.g., most species of *Icriodus, "Belodus"* from the Silurian). This may explain why a given species of *Icriodus* seems to have divergent ranges in different areas (e.g., *I. latericrescens* Branson & Mehl, which disappeared from middle Europe in late Early Devonian but is present in Middle Devonian and probably even earliest Late Devonian strata of North America).

CHARACTERS OF BASAL PLATE

Some discussion of characters of the basal plate of platelike conodonts, additional to that given by Hass, seems desirable. As has been proved by X-ray analysis of platelike conodonts and their basal elements or organs from the same specimens of various localities and ages, there is no difference in mineralogical composition between these parts. The obvious variation in hardness and texture between the conodont and basal plate most probably is the result of differences in infrastructure, which also may account for diverse receptability of coloring agencies.

The histology of basal plates has been studied from oriented thin sections, mainly of *Palmatolepis* and *Polygnathus,* by Gross (3). This author has introduced a term holoconodont for the fossil consisting of conodont proper and basal plate.

In first stages the growth lamellae of the conodont are concentric and surround a

---

Fig. 47. Phylomorphogenesis of *Palmatolepis*. This genus is among the best of all fossils for subdivision of Upper Devonian deposits. The dotted field symbolizes variability of form, width of black lines relative abundance (Helms, n). The figured species are as follows:

1, Polygnathus dubia dubia Hinde.——*2, P. dubia asymmetrica* Bischoff & Ziegler.——*3, Palmatolepis (Manticolepis) transitans* Müller.——*4, P. (M.) martenbergensis* Müller.——*5, P. (M.) foliacea* (Youngquist).——*6, P. (M.) unicornis* (Miller & Youngquist).——*7, P. (M.) proversa* (Ziegler).——*8, P. (M.) subrecta* (Miller & Youngquist).——*9, P. (M.) coronata* Müller.——*10, P. (M.) hassi* Müller & Müller.——*11. P. (M.) rhenana* (Bischoff).——*12. P. (M.) linguiformis* Müller.——*13, P. (M.) triangularis* (Sannemann).——*14, P. (M.) delicatula* (Branson & Mehl).——*15, 17, P. (M.) subperlobata* (Branson & Mehl) (2 subsp.).——*16, P. (M.) quadrantinodosalobata* (Sannemann).——*18, P. (Panderolepis) tenuipunctata* (Sannemann).——*19-23, P. (Pand.) serrata* (Hinde) [=*P. glabra* Ulrich & Bassler] (5 subsp.).——*24, P. (Pand.) elongata* (Holmes).——*25, P. (Pand.) serrata pectinata* (Ziegler).——*26,27, P. (Pand.) distorta* (Branson & Mehl) (2 subsp.).——*28,29, P. (Pand.) rhomboidea* (Sannemann) (2 subsp.).——*30,31, P. (Pand.) regularis* (Cooper) (2 subsp.).——*32, P. (Deflectolepis) subgracilis* (Bischoff).——*33,34,36, P. (D.) minuta* (Branson & Mehl) (3 subsp.).——*35, P. (D.) schleizia* Helms.——*37,38, P. (D.) gracilis deflectens* Müller.——*39, P. (D.) gonioclymeniae* Müller.——*40, P. (Palmatolepis)* sp. 4,——*41,48,52,53, P. (P.) perlobata* Ulrich & Bassler (4 subsp.), *42, P. (P.) crepida* Sannemann.——*43, P. (P.) termini* Sannemann.——*44,45, P. (P.) cymbula* Huddle (2 subsp.).——*46, P. (P.) perlobata maxima* Müller.——*47, P. (P.) perlobata perlobata* Ulrich & Bassler.——*49, P. (P.) humboldti* Helms.——*50, P. (P.) ampla* Müller.——*51, P. (P.) rugosa* Branson & Mehl.——*54,59,60,62, P. (Panderolepis) marginifera* (Ziegler) (4 subsp.).——*55,56, P. (Pand.) inflexa* (Müller) (2 subsp.).——*57, P. (Pand.) quadrantinodosa* (Branson & Mehl).——*58, P. (Pand.) elegans* Helms.——*61, P. (Pand.) marginifera marginifera* (Ziegler).

nucleus *("Erstanlage")* (Fig. 48). Later lamellae are also formed by concentric apposition on the upper side and laterally, but they are cut off on the lower side of the conodont. Fine ridges and furrows commonly have been observed on this surface. They are the result of periodical resorption, and their position is in accordance with the growth lines of the basal plate. Although obvious on *Palmatolepis* and *Polygnathus,* it is not yet clear that resorption takes place on all other conodonts also.

The basal plate is developed only beneath the area of resorption of the conodont and is loosely attached *("Basishaftfläche").* The basal plate is formed at a later stage of development than the conodont and seems to be related to resorption of the lower portion of the conodont, since it has been observed only in growth stages after the beginning of resorption. The growth nucleus of the basal plate lies beneath the growth nucleus of the conodont. The basal plate is thickest near the rim and somewhat thinner toward the middle. The underside of mature basal plates is marked by a furrow that runs beneath the growth nucleus (Fig. 48*C*).

Growth of the basal plate has taken place by outer apposition of lamellae, in a similar fashion as on the conodont. The basal plate is formed by an independent set of growth

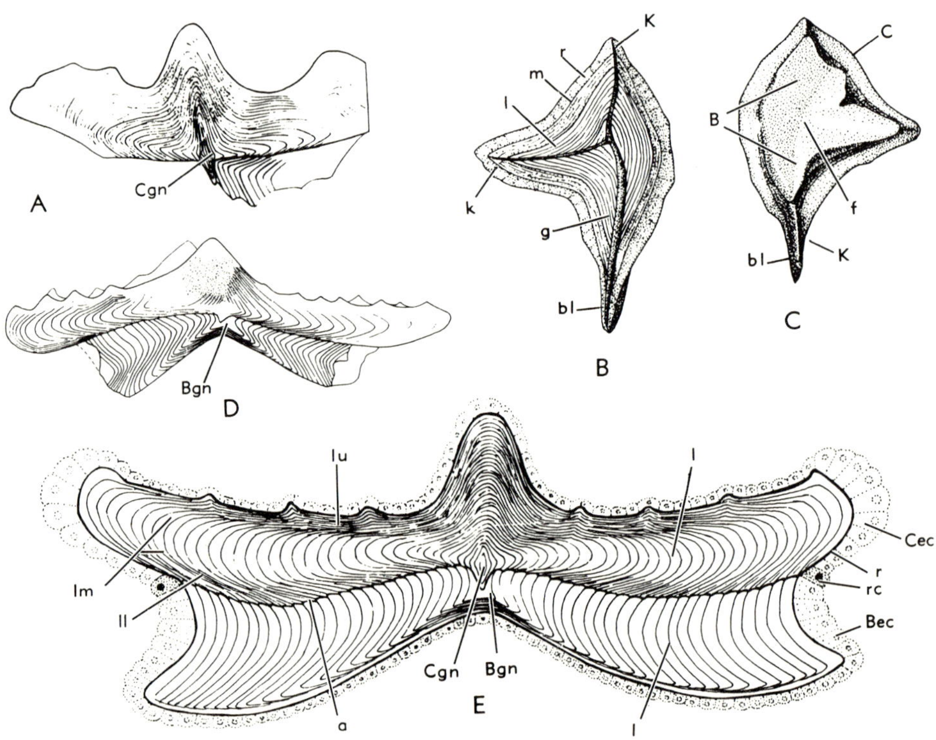

Fig. 48. Platelike conodonts showing features of conodont and basal plate (Gross, 1960).——*A. Palmatolepis foliacea* Youngquist; vertical section through portion of platelike conodont which shows growth nucleus of conodont *(Cgn),* ×125.——*B-D. Palmatolepis hassi* Müller & Müller; *B,* lower side of conodont with attachment area of basal plate, primary keel, and secondary keel, ×35; *C,* lower side of platelike conodont showing basal plate, ×35; *D,* vertical section through platelike conodont crossing azygous node, ×125 (3).——*E. Palmatolepis,* hypothetical diagram showing vertical section with cells that formed it, ×200 (3). [Explanation: *a,* attachment area of conodont and basal plate; *B,* basal plate; *Bec,* basal plate epithelial cell; *Bgn,* basal plate growth nucleus; *bl,* blade; *C,* conodont; *Cec,* conodont epithelial cell; *Cgn,* conodont growth nucleus; *f,* furrow on underside of basal plate; *g,* growth line on attachment area; *K,* keel, primary; *k,* keel, secondary; *l,* lamella; *ll,* lamella (lower); *lm,* lamella (median); *lu,* lamella (upper); *m,* margin of attachment area; *r,* reverse-curved surface of conodont; *rc,* resorbing cell at edge of basal plate.]

lamellae, which, however, are harmonic with those of the conodont. They stand more or less perpendicular to the attachment plane and are closed on the lower surface of the organ.

The conodonts, as well as basal plates, thus increase in size by growth in all directions, except on the lower side of the conodonts and the upper side of the basal plates. The attachment plane between them is a distinct area, marked by resorption of the conodont. This characteristic mode of growth undoubtedly has considerable importance for comparison of conodonts with other groups of animals as regards their systematic relationship.

REFERENCES

Bischoff, Günther, & Sannemann, Dietrich

(1) 1958, *Unterdevonische Conodonten aus dem Frankenwald:* Hess. Landesamt Bodenf., Notizbl., Bd. 86, p. 87-110, pl. 12-15.

Ellison, S. P.

(2) 1946, *Conodonts as Paleozoic guide fossils:* Am. Assoc. Petroleum Geologists, Bull., v. 30, p. 93-110, fig. 1-3.

Gross, Walter

(3) 1960, *Über die Basis bei den Gattungen Palmatolepis und Polygnathus (Conodontida):* Paläont. Zeitschr., Bd. 34, p. 40-58, fig. 1-9.

Hass, W. H.

(4) 1959, *Conodonts from the Chappel limestone of Texas:* U.S. Geol. Survey, Prof. Paper 294-J, p. 365-399, pl. 46-50, 1 text fig., 2 charts.

Helms, Jochen

(5) 1961, *Zur Taxionomie und Phylogenese von Palmatolepis (Conodontida, Oberdevon):* Geologie, v. 11, in press.

Müller, K. J.

(6) 1956, *Zur Kenntnis der Conodonten-Fauna des europäischen Devons, 1. Die Gattung Palmatolepis:* Senckenberg. Naturf. Gesell., Abhandl. Bd. 494, p. 1-70, pl. 1-11, 3 text fig.

(7) 1959, *Kambrische Conodonten:* Deutsch. geol. Gesell., Zeitschr., Bd. 111, T. 2, p. 434-485, pl. 11-15, fig. 1-11.

(8) 1960, *Wert und Grenzen der Conodonten-Stratigraphie:* Geol. Rundschau, Bd. 49, H. 1, p. 83-92, pl. 1-2.

———, & Müller, E. M.

(9) 1957, *Early Upper Devonian (Independence) conodonts from Iowa, part I:* Jour. Paleontology, v. 31, p. 1069-1108, pl. 135-142, fig. 1-8.

Rexroad, C. B.

(10) 1958, *The conodont homeomorphs Taphrognathus and Streptognathodus:* Jour. Paleontology, v. 32, p. 1158-1159.

Rhodes, F. H. T.

(11) 1952, *A classification of Pennsylvanian conodont assemblages:* Jour. Paleontology, v. 26, p. 886-901, pl. 126-129, fig. 1-4.

Scott, A. J., & Collinson, C. W.

(12) 1959, *Intraspecific variability in conodonts: Palmatolepis glabra Ulrich & Bassler:* Jour. Paleontology, v. 33, p. 550-565, pl. 75-76, fig. 1-4.

Walliser, O. H.

(13) 1957, *Conodonten aus dem oberen Gotlandium Deutschlands und der Karnischen Alpen:* Hess. Landesamt Bodenf., Notizbl., Bd. 85, p. 28-52, pl. 1-3, fig. 1-3.

Ziegler, Willi

(14) 1960, *Conodonten aus dem Rheinischen Unterdevon (Gedinnium) des Remscheider Sattels (Rheinisches Schiefergebirge):* Paläont. Zeitschr., Bd. 34, p. 169-201, pl. 13-15, fig. 1-2.

CONODONT CLASSIFICATION AND NOMENCLATURE

By Raymond C. Moore
[University of Kansas]

INTRODUCTION

The subject of classification and nomenclature of conodonts has been dealt with rather fully by Hass, Rhodes, and Müller in preceding chapters of this volume. Also it has been discussed by various workers previously (6-11, 13, 15). Why add anything more? I venture to extend discussion because these authors have left essential questions unresolved. We are still faced with the problem of how to handle the conflicts that arise from an illegal dual sort of nomenclature which they favor, one set of names being applied to discrete conodonts and the other to assemblages of conodonts presumed to comprise the remains of individual animals. Definition of genera and species based on the assemblages is asserted to distinguish "natural" taxa, whereas genera and species based on discrete conodonts are explicitly or implicitly interpreted to be arbitrary, artificial, and "unnatural" units which are acceptable because of their practical value to stratigraphic paleontology. Some authors have employed the designations "form-genera" and "form-species" in referring to taxa defined on the basis of discrete conodonts, although the Rules recognize no such categories for classification of animals (10, 11, 15). Müller (9) has used the terms "partial-genera" and "partial-species" for classificatory arrangement of discrete conodonts. Rhodes classes conodont assemblages in terms of genera and species but refers to discrete conodonts in terms of "genera" and "species" (*Treatise,* this volume). This is not helpful, since any discrete conodont undeniably constitutes a fossil record of some "natural" conodont-bearing animal and as such is fully entitled to first-class treatment at hands of zoological taxonomists. The only admissible distinction between conodont assemblages and discrete conodonts is the degree of their completeness (or rather, incompleteness) as fossil remains of once-living creatures. No difference in application to them of zoological classificatory and nomenclatural procedures specified by the Rules is allowable. Of course, here we encounter the real dilemma.

Following this preamble, I draw attention to the seeming fact that difficulties are all or nearly all of our own making. If we can undo what has been done, on the grounds that it lacked acceptably authoritative basis, problems vanish. If we revise our approach to the questions introduced by fossil conodonts, both assemblages and discrete individuals, our supposedly urgent need for a system of dual classification and nomenclature disappears. I propose to explore the possibility of undoing what has been done and achieving the suggested reorientation of approach to conodont classification and nomenclature. I shall try to show that a dual system of procedures is quite unnecessary. Seemingly, the sole requisite for success in removing difficulties is readiness on the part of a majority of workers to reject inadequately supported taxonomic conclusions, that is, those which depend on doubtful assumptions.

FACTUAL FOUNDATION

Let us begin by constructing a foundation of facts. A few are very elementary but not to be overlooked on this account.

(1) Each known kind of animal has only a single valid zoological name, which is binominal in form and different from the name of any other animal.

(2) The first-published zoological name that meets stipulations of the Rules is the accepted valid name of an animal, taking precedence over all other names that may be proposed.

(3) Innumerable discrete, disjunct, individually well-separated conodonts are found widely distributed as fossils in marine or semimarine sedimentary deposits ranging in age from Cambrian to Triassic.

(4) Many of these discrete conodonts have been demonstrated to possess great

value for stratigraphic zonation and correlation and for age determination of the enclosing sediment.

(5) These discrete conodonts have been classified and named in terms of species (called form-species or "species" by some authors), genera (called form-genera or "genera" by some authors), and families.

(6) Classification and scientific nomenclature in manner compliant with international Rules are requisite as applied to discrete conodonts in order to serve practical needs of stratigraphic paleontology.

(7) Assemblages of several kinds of conodonts have been found in such association as reasonably to indicate their derivation from a single conodont-bearing animal.

(8) These conodont assemblages have been assigned generic and specific names intended to designate the conodont-bearing animal thus represented by the fossil remains.

(9) Component individual conodonts of conodont assemblages have been designated by some authors using generic or generic-and-specific names derived from discrete-conodont classification and nomenclature.

(10) The use of different zoological names for a conodont assemblage and its several components has been challenged on the ground that it clearly disregards zoological Rules if such nomenclature is maintained and that it gives rise to intolerable confusion if effort is made to comply with the Rules by synonymizing the names which are in competition.

(11) A proposal to allow a limited sort of dual classification under sanction of international Rules by establishing a category of parataxa which would be independent of natural taxa for purposes of the Law of Priority but not of the Law of Homonymy (7) was rejected by the 1958 Zoological Congress which met in London. Therefore, such classification applied to discrete conodonts and conodont assemblages has been and is now illegal.

ASSUMPTIONS

In relation to the subject here discussed, the following two statements must be classified as assertions that only express assumptions.

(1) Some, if not all, conodont assemblages which have been designated by generic and specific names are trustworthy of interpretation as the composite fossil remains of individual conodont-bearing animals. This is reasonable and now so well documented that few paleontologists are unwilling to give it at least qualified assent. Accordingly, names given to the assemblages in compliance with the Rules are acceptable, provided conflict between them and names of disjunct conodonts is removed.

(2) Many, if not all, individual discrete conodonts found to occur as components of conodont assemblages are reliably identifiable as belonging to named genera and species of disjunct conodonts distinguished on the basis of specimens not originally found in assemblages. This is a critical, far-reaching assumption which demands close scrutiny, because it touches the very heart of our problem. Even so, it has been so long taken for granted rather thoughtlessly by paleontologists that they have not recognized its status as a quite unproved—possibly unprovable—assumption. In whatever degree the premise is discredited, nomenclatural problems diminish and they can disappear entirely.

EXAMPLE OF DUBOISELLA TYPICA

Let us test the line of thinking suggested by consideration of a chosen example. Almost any of the described and named taxa based on conodont assemblages are suitable for inquiry, except for the fact that component discrete conodonts in some assemblages are identified only to the generic level, without discrimination of species. *Duboisella typica* RHODES, 1952, which is the type-species of *Duboisella,* is a preferred example because all but one of its component discrete conodonts have been identified to the specific level and two of these are type-species of discrete conodont genera (10). The entire assemblage is illustrated in Figures 42,*4* and 43,*6*. The identified discrete conodont constituents are *Ligonodina typa* (GUNNELL), 1933; *Metalonchodina bidentata* (GUNNELL), 1931, which is the type-species of *Metalonchodina* BRANSON & MEHL, 1941; *Neoprioniodus conjunctus* (GUNNELL), 1931, which is the type-species of *Neoprioniodus* RHODES & MÜLLER, 1956;

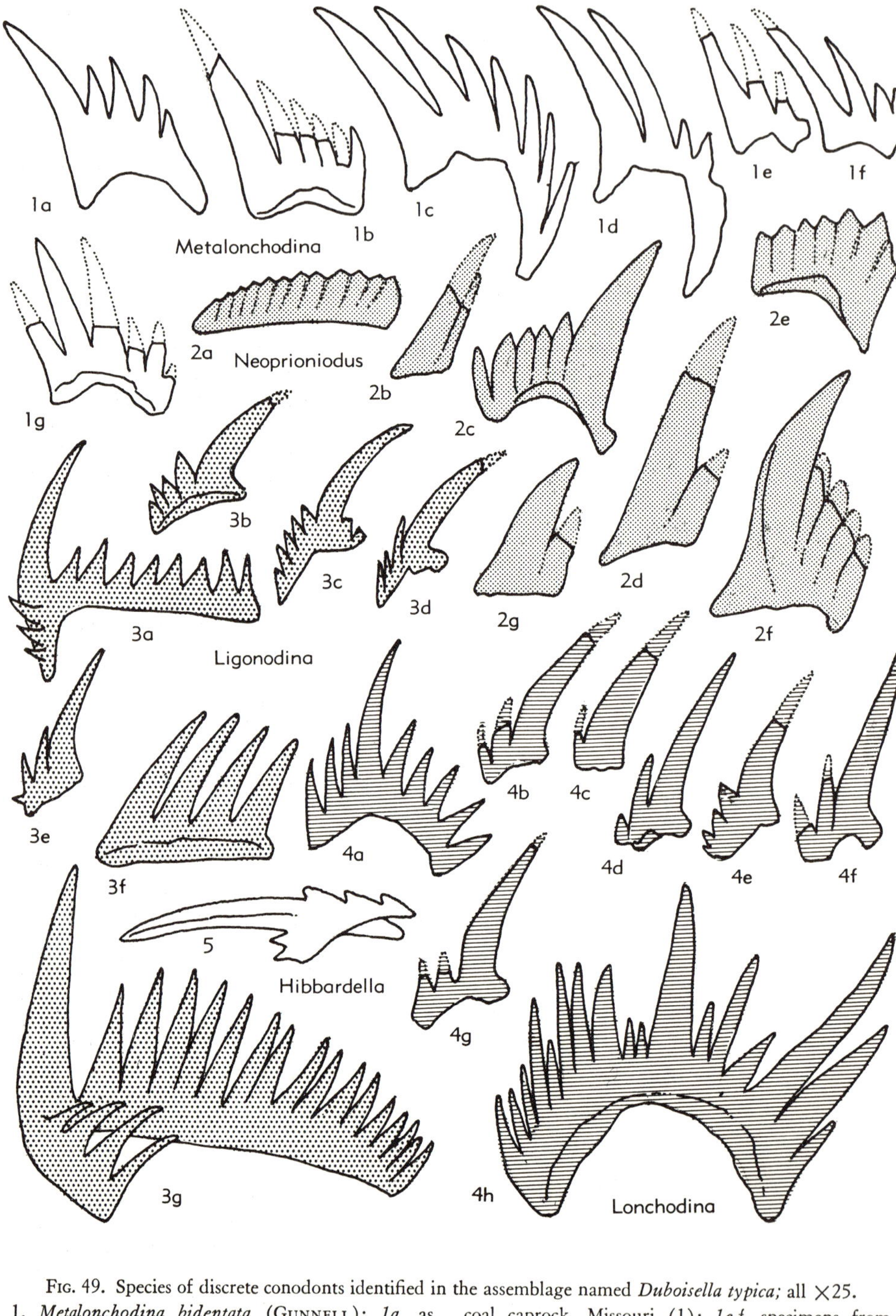

FIG. 49. Species of discrete conodonts identified in the assemblage named *Duboisella typica;* all ×25.

1. *Metalonchodina bidentata* (GUNNELL); *1a,* as identified in *D. typica* assemblage (10); *1b,* holotype from Fort Scott Limestone, Missouri (3); *1c,* specimen from Cherokee Shale, Missouri (2, mod. by HASS); 1d, specimen from Lexington coal caprock, Missouri (1); *1e,f,* specimens from Cherokee Group, Kansas (2); *1g,* specimen originally identified as *Prioniodus dactylodus* by GUNNELL (4).——2. *Neoprioniodus conjunctus* (GUNNELL); *2a,* specimen identified in *D. typica* as-

Lonchodina clarki (GUNNELL), 1931; and *Hibbardella* sp. Figure 49 furnishes illustrations of these individual conodonts, both as alleged to be found in the assemblage named *D. typica* and as recorded in isolated occurrence in various deposits; in order to facilitate comparisons, all are shown at the same magnification (×25).

It is unnecessary to pursue the nomenclatural complexities encountered in trying to apply the Rules if we should accept the stated identifications. The situation then becomes truly chaotic. Not only would the name *Duboisella typica* have to be replaced by one of its three 1931-dated components as the valid name of the conodontophorid animal represented by the assemblage, but all of the identified discrete components, being parts of the same animal, would be synonymous; their synonymy would be objective except for the fact that they have been subjectively identified, and therefore the synonymy is subjective. Further, because *Metalonchodina bidentata* and *Neoprioniodus conjunctus* are type-species of their respective genera, other species of these genera not belonging to the assemblage could be left without a generic name.

Complications of the sorts just noted are not in themselves important problems to be solved, since they are merely consequences of assumptions which may be unsound. The basic question is whether one nominal species (defined from a conodont assemblage) can be composed of or incorporate a number of other nominal species (defined from discrete conodonts). Of course, we must say "No," since such a concept is entirely inadmissible. We are sure that every sort of discrete conodont was derived from a once-living species of whole animal, but it does not follow that all conodonts having more or less similar shape are records of the same species of living animal. This is a crucial point. It is reasonable to suppose that quite different species of conodont-bearing animals possessed among their hard parts somewhat similar, or even exactly similar, discrete conodonts. Such components, naturally, would lack diagnostic value, whereas other components might be clearly distinguishable. For example, prioniodid, hindeodellid, or hibbardellid elements in conodont assemblage could well be less trustworthy indicators of specific distinctions among conodontophorid animals than associated platelike elements. In the same way, a group of horse teeth found together is likely to include incisors lacking in distinctive features along with highly diagnostic kinds of molars. Crinoid plates and columnals of generalized pattern, unidentifiable even as to family or order, are rather commonly found associated with some distinctive kinds of remains, although rarely in circumstances pointing to natural assemblages derived from single individuals. Unique types of echinoid spines may be trustworthy indicators of genera and species, whereas plates occurring with them are not similarly usable.

In addition, it is reasonable to postulate

---

semblage, illustrated by RHODES in pl. 128, fig. 5 (10); *2b,* holotype from Fort Scott Limestone, Missouri (3); *2c,* specimen from Cherokee Group, Missouri (2, mod. by HASS); *2d,* specimen from Cherokee Group, Kansas, identified as *Prioniodus cacti* by GUNNELL (4); *2e,* specimens of *N. conjunctus* termed senile by ELLISON, from Cherokee beds of Missouri (2); *2f,* specimen identified as *Prioniodus cacti* from early Pennsylvanian of Iowa by YOUNGQUIST & DOWNS (16); *2g,* specimen from Quivira Shale (Missourian) of Kansas City area (2).——— 3. *Ligonodina typa* (GUNNELL); *3a,* as identified in *D. typica* assemblage (10); *3b,* holotype from Winterset Limestone, Missouri (4); *3c,d,* specimens from Quivira Shale (Missourian) of Kansas City area (2); *3e,* specimen from Galesburg Shale (Missourian) of Kansas City area identified by GUNNELL as *Prioniodus? galesburgensis* (4); *3f,* specimen from early Pennsylvanian of Iowa identified as *Lonchodus?* sp. by YOUNGQUIST & HEEZEN (17); *3g, L. pectinata* ULRICH & BASSLER, type-species of *Ligonodina,* from Rhinestreet Shale, Upper Devonian, New York (HASS).———4. *Lonchodina clarki* (GUNNELL); *4a,* as identified in *D. typica* assemblage (10); *4b,* holotype from Fort Scott Limestone, Missouri (3); *4c,* specimen from East Mountain Shale (Desmoinesian) of Texas identified as *Prioniodus clarki* by STAUFFER & PLUMMER (14); *4d,e,* specimens from Quivira Shale (Missourian) of Kansas City area (2); *4f,* specimen from Hushpuckney Shale (Missourian) of Kansas City area (2); *4g,* specimen from Graford Formation (Missourian) of Texas identified as *Prioniodus cornutus* by STAUFFER & PLUMMER (14); *4h, L. typicalis* ULRICH & BASSLER, type-species of *Lonchodina,* from Rhinestreet Shale, Upper Devonian, of New York (HASS).———5. *Hibbardella* sp., as identified in *D. typica* assemblage (10).

that a given species of conodont-bearing animal may be represented not only by differently shaped discrete conodonts but that each of these components may exhibit variations within certain limits. Evidence supports this postulate (2, 5, 11). In view of variability judged to characterize many discrete conodonts that are classed as conspecific and the similarity of some discrete components of unlike conodont assemblages, it is hazardous to undertake firm identifications of the components of assemblages. Yet, if they are made, we must deal with their implications.

Turning again to *Duboisella*, if the assemblage component distinguished as *Metalonchodina bidentata* (Fig. 49,*1a*), for example, is considered to be unquestionably the same (on the basis of identity in form) as GUNNELL's type specimen of this species from the Fort Scott Limestone (Desmoinesian) of the Kansas-Missouri border (Fig. 49,*1b*), the species of conodont-bearing animal represented by the *Duboisella typica* assemblage (holotype) in black shale just below the La Salle Limestone (Missourian) of Illinois may be the same as the conodont-bearing animal that manufactured the Fort Scott specimen(s) of *M. bidentata.* It does not follow that the species represented respectively by the discrete conodont from Missouri and by the conodont assemblage from Illinois *must* be the same, because quite different animals may possess indistinguishable components of their hard parts. Thus, synonymization of *M. bidentata* and *D. typica* depends on unproved—and probably unprovable—assumptions. It is assumption of the validity of assumptions concerning identity (assumption multiplied by assumption) that makes trouble. The Rules demand that an author who accepts identity of the differently named genera and species shall abide by the Law or Priority, recognizing the first-published name and suppressing the junior synonym. Of course, other authors are not required to follow suit, and so may reject the synonymy.

If an assemblage component of *Duboisella* is only doubtfully considered to be equivalent to the discrete conodont from the Fort Scott Limestone named by GUNNELL, no conflict arises and both names may stand. Any paleontologist who questions identification of the *Duboisella* component as really an example of *M. bidentata* can accept both names as designations of species which are judged or assumed to be different. Examples of *M. bidentata* reported by ELLISON (1941) from the Cherokee Group (Fig. 49,*1c*) and by BRANSON & MEHL (1941) from the caprock of the Lexington coal in Missouri (Fig. 49,*1d*) if correctly identified, may (not must) represent occurrences of the *D. typica* conodont-bearing animal. Likewise, all correctly identified examples of *Ligonodina typa, Neoprioniodus conjunctus,* and *Lonchodina clarki* must constitute records of the presence of the *D. typica* animal, if we are certain beyond doubt that these various taxa (using the word advisedly) are really synonymous with *D. typica* (because some of their diagnostic remains are exactly equivalent to a part of the remains of *D. typica*). Doubt concerning the identity of one or more named discrete conodonts with *Duboisella* components would not help, if others should be accepted. Finally, it is evident that if we did not have to deal with assertedly definite identifications of the conodont-assemblage components, the whole problem would not exist.

DEDUCTIONS

If homeomorphic duplications of discrete conodonts exist, identity of form fails as indication of possible taxonomic identity. Among conodonts near-identity or unrecognizable homeomorphy of parts may relate to different species or even to different genera of animals. Little harm is done if isolated disjunct conodonts are incorrectly determined, whereas utmost trouble ensues from assertedly definite identifications of the discrete components of assemblages.

DUAL CLASSIFICATION AND NOMENCLATURE OF CONODONTS UNNECESSARY

The foregoing discussion indicates that dual classification and nomenclature are really unnecessary for application to conodonts. An individual isolated conodont is as truly the fossil representative of some species of animal as an assemblage of conodonts. Generic and specific names employed for these animals, whether based on disjunct conodonts or on assemblages, do not conflict unless and until effort is made to indicate the components of assemblages by names published for discrete conodonts which they may resemble. It is entirely appropriate and may be very useful to designate elements of an assemblage as *Hindeodella*-like, *Prioniodus*-like, and so on, or to employ such terms as hindeodellid, prioniodid, and others, for these are taxonomically noncommittal. Further, no valid objection could be offered to describing *Duboisella typica* as having components that closely resemble *Ligonodina typa, Metalonchodina bidentata,* and other mentioned species of disjunct conodonts. Characterization is as precise as though the respective components were explicitly affirmed to be specimens belonging to these species and the omission of definite identification is likely to prove more accurate from scientific viewpoints.

What about supposedly firm identifications of discrete conodont components of assemblages, as in *Duboisella typica,* which already have been published? Is it necessary that these should stand? By no means. Rhodes may agree that his identifications should all be modified by treating them as doubtful, or as indicative only of close resemblance. If he does not want to do this, he should spell out just what changes in zoological designations are needed for all affected genera and species. In any case, other paleontologists are free to reject the subjective synonymies which others may advocate. Accordingly, dual nomenclature is not only unacceptable and illegal, but it is unnecessary.

CONCLUSION

As summary, I point out that (1) all fossil remains are varyingly incomplete as records of the species which they represent; (2) with little doubt, "natural" assemblages of conodonts are more complete fossil records of conodont-bearing species of animals than individual discrete conodonts, but occurrences of both sorts are co-ordinate in taxonomic considerations; (3) the component discrete conodonts of assemblages should be discriminated only in terms of their resemblances to named discrete genera and species, and not as firmly identified individuals belonging to these taxa, because such identification is actually unprovable and because ramifying complexities in nomenclature can be avoided by omitting allegedly firm identifications which really depend on subjective assumptions.

Let us agree, then, on adopting a conservative, unassailable course which takes us around or away from conflict between names of genera and species respectively based on discrete conodonts and conodont assemblages. Bold workers who wish to proceed differently may do so, but then they are enjoined to tread carefully and follow through to ends that accord with the Rules. In my own view, the species, genera, and families distinguished on the basis of discrete conodonts, as described by Hass in this *Treatise,* are to be regarded as "natural" taxa, and the species and genera defined on the basis of conodont assemblages likewise.

REFERENCES

Branson, E. B., & Mehl, M. G.

(1) 1941, *New and little-known Carboniferous conodont genera:* Jour. Paleontology, v. 15, p. 97-106, pl. 19.

Ellison, Samuel

(2) 1941, *Revision of the Pennsylvanian conodonts:* Jour. Paleontology, v. 15, p. 107-143, pl. 20-23, fig. 1-4.

Gunnell, F. H.

(3) 1931, *Conodonts from the Fort Scott limestone of Missouri:* Jour. Paleontology, v. 5, p. 244-253, pl. 29.

(4) 1933, *Conodonts and fish remains from the Cherokee, Kansas City, and Wabaunsee groups of Missouri and Kansas:* Jour. Paleontology, v. 7, p. 261-297, pl. 31-33.

Hass, W. H.

(5) 1941, *Morphology of conodonts:* Jour. Paleontology, v. 15, p. 71-81.

(6) 1956, *Age and correlation of the Chattanooga Shale and the Maury Formation:* U.S. Geol. Survey, Prof. Paper 286, 47 p., 5 pl., 1 fig.

Moore, R. C., & Sylvester-Bradley, P. C.
(7) 1957a, *Suggested new article: Proposed recognition of the concept "parataxon" and the provision of rules for the nomenclature of units of this category:* Bull. Zool. Nomenclature, v. 15, pt. 1/4, p. 5-13.
(8) 1957b, *First supplemental application: application for a ruling of the International Commission directing that the classification and nomenclature of discrete conodonts be in terms of "parataxa":* Bull. Zool. Nomenclature, v. 15, pt. 1/4, p. 14-34.

Müller, K. J.
(9) 1956, *Taxonomy, nomenclature, orientation, and stratigraphic evaluation of conodonts:* Jour. Paleontology, v. 30, p. 1324-1340, pl. 145.

Rhodes, F. H. T.
(10) 1952, *A classification of Pennsylvanian conodont assemblages:* Jour. Paleontology, v. 26, p. 886-901, pl. 126-129, fig. 1-4.
(11) 1953, *Nomenclature of conodont assemblages:* Jour. Paleontology, v. 27, p. 610-612.

———, & Müller, K. J.
(12) 1956, *The conodont genus Prioniodus and related forms:* Jour. Paleontology, v. 30, p. 695-699.

Sinclair, G. W.
(13) 1953, *The naming of conodont assemblages:* Jour. Paleontology, v. 27, p. 489-490.

Stauffer, C. R., & Plummer, H. J.
(14) 1932, *Texas Pennsylvanian conodonts and their stratigraphic relations:* Univ. Texas, Bull. 3201, p. 13-59, pl. 1-4.

Sylvester-Bradley, P. C.
(15) 1954, *Form-genera in paleontology:* Jour. Paleontology, v. 28, p. 333-336.

Youngquist, W. L., & Downs, R. H.
(16) 1949, *Additional conodonts from the Pennsylvanian of Iowa:* Jour. Paleontology, v. 23, p. 161-171, pl. 30-31.

———, & Heezen, B. C.
(17) 1948, *Some Pennsylvanian conodonts from Iowa:* Jour. Paleontology, v. 22, p. 767-773, pl. 118.

SMALL CONOIDAL SHELLS OF UNCERTAIN AFFINITIES

By DONALD W. FISHER

[State Paleontologist of New York, New York State Museum, Albany. Published by permission of the Assistant Commissioner, New York State Museum and Science Service.]

CONTENTS

INTRODUCTION

Included in this section are the coleolids, cornulitids, hyolithellids, hyolithids, tentaculitids, and some miscellaneous conoidal shells which have sometimes been grouped with the aforementioned forms or with the "Pteropoda," or which show some similarity to them.

The creation of acceptable classifications of these fossils is gravely hampered by our lack of knowledge of the animals which inhabited the shells. The biologic importance of the existent structures is largely speculative; consequently, the taxonomic principles that can be applied to these long extinct groups are limited. Most skeletal structures must be appraised by circuitous methods. Such indirect avenues furnish the bases for most of the practical aspects of the classifications, but by themselves are apt to result in a high measure of artificiality. In selecting taxonomic criteria in fossils without living representatives, the value of a character usually is determined by its constancy in an aggregate of forms. Thus, a structure or characteristic that exists throughout a group of evidently different forms is considered to be a more significant taxonomic indicator than another structure or characteristic that is observed in only a few of the forms. The dependability of a character is increased if it is accompanied by other characters that exist in the identical grouping of specimens. It is frequently necessary to acknowledge one character as of prime significance and subordinate the others for the purpose of presenting a single taxonomic arrangement. However, purely subjective suppression of definitive taxonomic characters is assuredly not a valid basis for a natural classification.

Acknowledging that established biologic principles can be employed in problems of taxonomic appraisal, preferential decisions can be made with the expectation of approximating the natural grouping. For example, in the cornulitids study of external ornamentation contributes vital data to the understanding of ontogeny and polymorphism. Although *Cornulites* and *Tentaculites* possess homeomorphic casts of the body cavity, wall construction confirms that they are not even remotely related. This type of systematic approach coupled with distributional studies in time and space furnish data for the evaluation of more realistic concepts of taxa.

Regrettably, scarcity of adequate study material strongly handicaps application of the foregoing for many of the genera discussed here. Moreover, relative simplicity of form and paucity of characters have not inspired paleontologists to investigate these groups.

HISTORICAL REVIEW

It seems best to treat the probable biologic affinities of each group separately, for it is unlikely that any of the five major categories here discussed are related except at a phyletic level. However, so as to avoid repetition, whenever two or more of these groups have been united, they will be reviewed in these preliminary remarks.

SCHLOTHEIM (60), the nomenclator of *Tentaculites,* and EICHWALD (10), the nomenclator of *Hyolithes,* considered them to be crinoid arms and problematica, respectively. Together with *Conularia,* these two genera were grouped under the Pteropoda by BARRANDE (1) and HALL (17). Earlier, AUSTIN (1845) had placed *Tentaculites* in the Pteropoda. This disposition has been followed by most paleontologists. However, some have variously regarded tentaculitids as worms, spines of brachiopods, scaphopods, young cephalopods, spines of echinoids, or crinoid arms; whereas hyolithids have been considered to have also been worms or cephalopods. Many have avoided even tentative decisions and have assigned tentaculitids and hyolithids to the "waste basket," *incertae sedis.*

PELSENEER (54) was apparently the first openly to deny the pteropod assignment of the hyolithids and tentaculitids. He stated, "I . . . am firmly of the opinion that Pteropoda do not occur as fossils until the end of the Lower Tertiary." NEUMAYR earlier (1879) had considered hyolithids to be an extinct group, undoubtedly molluscan, but not related to pteropods. Were he to have stopped there, much of the later confusion might have been avoided. Regrettably, he united the hyolithids with the conulariids—two very different groups.

This denial of pteropod affinities prompted the setting up of a new niche for conulariids, hyolithids, and tentaculitids. Following MATTHEW (1889), most North American workers grouped them with the tube worms. Nevertheless, WALCOTT and many European workers continued the pteropod assignment, no doubt influenced largely by ZITTEL's widely used text (1913, 1937). HOLM's classic work (20) on the Swedish hyolithids and conulariids left them both without assignment but stated that they were not pteropods. *Tentaculites* was grouped under Annelida.

NAEF (48) was probably the first to realize the need for a separate name for these fossils without a "home" and, accordingly, he proposed the name Odontomorpha to include *Conularia, Hyolithes,* and *Styliolina.* No diagnosis of this assemblage was offered and *Tentaculites* and *Nowakia* were excluded. Similarly, HENRI and GENEVIÈVE TERMIER (66) proposed the name Eopteropoda to accommodate the conulariids and tentaculitids; the Eopteropoda were regarded as ancestral to the Pteropoda. In 1950 they removed the conulariids from the Eopteropoda, and in 1953 the name Eopteropoda seemingly went into oblivion when they placed tentaculitids (and hyolithids) in "Groupes d'affinités incertaines." Thus, Odontomorpha and Eopteropoda were essentially stillborn attempts to remove the cloak of obscurity from these "pteropod-like" fossils.

Several years earlier, KNIGHT (26) introduced evidence that *Conchopeltis* (a conulariid) was a coelenterate, rather than a mollusk. He further suggested that the Tentaculitidae bear some similarities to the coelenterates, possibly as part of a new class that would include the Conulariidae. In the same year, KIDERLEN convincingly demonstrated that conulariids were related to the Scyphozoa. Subsequently, KNIGHT (27) firmly rejected conulariids, hyolithids, and tentaculitids from classification with the gastropods, stating, ". . . all names of genera commonly referred to the Hyolithidae, Tentaculitidae, Torellellidae, and Conulariidae are omitted. These, in fact, I do not regard as even molluscan. Some, such as the Hyolithidae, may be mollusks, but they are more likely representatives of some phylum otherwise unknown and now wholly extinct; their resemblances to members of any living phyla are not impressive and may well be superficial. The Tentaculitidae, Torellellidae, and Conulariidae, with their radial and even four-fold symmetry, may possibly be related to the Scyphozoa, as suggested independently by KIDERLEN (1937) and by KNIGHT (1937A, p. 188)."

SHIMER & SHROCK (1944) placed *Coleoloides, Coleolus, Helenia, Hyolithellus, Hyolithes, Orthotheca, Tentaculites,* and *Styliolina* in Mollusca *incertae sedis.* In 1953, SHROCK did not regard the uniting of *Conularia, Hyolithes,* and *Tentaculites,* as the TERMIERS had done, to be an improvement and therefore he reverted to placing all three in *incertae sedis.* Following KNIGHT's proposal, MOORE (1953) questionably referred the Tentaculitidae to the conulariids, whereas the hyolithids were classed as doubtful pteropods. Conulariids have now been rejected from the hyolithids and tentaculitids and unequivocally ranked as a subclass (Conulata) in the Class Scyphozoa in the Phylum Coelenterata (*Treatise,* Part F, 1956).

G. P. LYASHENKO (31) made a noteworthy step forward by proposing the Class Coniconchia for the tentaculitids, questionably referring it to the Phylum Mollusca. Regrettably, no diagnosis was published until 1957, when according to my opinion a backward step was made by inclusion of the hyolithids in Coniconchia. Unification of two such basically different groups as hyolithids and tentaculitids in a single class is unfortunate and not supported by the evidence. It is disputable whether the two even belong to the same phylum. Whereas hyolithids were probably mollusks, they are distinct from any of the recognized molluscan classes. On the other hand, the occurrence of tiny pores in the tentaculitid shell wall and their lack of anything but radial symmetry casts some doubt on their molluscan affinities. However, like scaphopods, their shells may appear quite unlike those of mollusks, exhibiting only radial symmetry.

With increase in knowledge there has been a corresponding increase in the number of recognized major taxonomic categories. Following the discovery of a living monoplacophoran, *Neopilina,* the Monoplacophora were elevated to class rank within the Mollusca and have received wide acceptance in zoological texts and in this *Treatise* (Part I, 1960). Separately, the hyolithids and tentaculitids are no less distinct. Accordingly, it is here suggested that the hyolithids may be included in the new class Calyptoptomatida (with hyolithellids divorced from them) and that the tentaculitids and their allies may be grouped in the new class Cricoconarida. Each is considered to be an extinct class of the phylum Mollusca. This dichotomy emphasizes the uniqueness of each major group, a situation not implied by use of the inclusive names Coniconchia, Eopteropoda, Odontomorpha, or *incertae sedis.*

CLASSIFICATION

The tabular summary that follows shows the arrangement of taxa treated in this chapter. The numbers in parentheses indicate number of genera known in each taxon.

Divisions of Small Conoidal Shells of Uncertain Affinities

Mollusca *(phylum)*

Cricoconarida *(class)* (13). *L.Ord.-U.Dev.*
- Tentaculitida *(order)* (9). *L.Ord.-U.Dev.*
  - Tentaculitidae (2). *L.Sil.-U.Dev.*
  - Homoctenidae (3). *M.Dev.-U.Dev.*
  - Uniconidae (4). *L.Ord.-U.Dev.*
- Dacryoconarida *(order)* (4). *M.Sil.-U.Dev.*
  - Nowakiidae (3). *U.Sil.-U.Dev.*
  - Styliolinidae (4). *M.Sil.-U.Dev.*

Calyptoptomatida *(class)* (26). *L.Cam.-M.Perm.*
- Hyolithida *(order)* (18). *L.Cam.-M.Perm.*
  - Hyolithina *(suborder)* (17). *L.Cam.-M.Perm.*
    - Hyolithidae (2). *L.Cam.-M.Perm.*
    - Ceratothecidae (1). *U.Sil.-L.Dev.*
    - Orthothecidae (7). *L.Cam.-M.Dev.*
    - Sulcavitidae (5). *L.Cam.-Ord.*
    - Pterygothecidae (2). *Dev.*
  - Matthevina *(suborder)* (1). *U.Cam.*
    - Mattheviidae (1). *U.Cam.*
- Globorilida *(order)* (1). *M.Cam.*
  - Globorilidae (1). *M.Cam.*
- Camerothecida *(order)* (2). *Cam., Sil.*
  - Camerothecina *(suborder)* (1). *Cam., Sil.*
    - Camerothecidae (1). *Cam., Sil.*
  - Diplothecina *(suborder)* (1). *Cam.*
    - Diplothecidae (1). *Cam.*
- Order and Family Uncertain (5).

Phylum, Class Uncertain

- Hyolithelminthes *(order)* (6). *L.Cam.-Ord.*
  - Hyolithellidae (3). *L.Cam., M.Cam.?*
  - Torellellidae (3). *L.Cam.-Ord.*

Phylum, Class, Order Uncertain

- Coleolidae (7). *L.Cam.-Carb.*
- Cornulitidae (4). *M.Ord.-L.Carb.*

Phylum, Class, Order, Family Uncertain (3).

Supposed "Pteropoda" assigned to other groups (12).

Unrecognizable genera (6).

CRICOCONARIDS

Cricoconarids (tentaculitids, nowakiids, styliolinids) are small, narrow, straight, ringed true cones. As many unrelated animals have been given similar names (e.g., Tentaculata, Tentacularia, Tentaculatiana, Tentaculina), I believe that the use of a similar name for the supra-ordinal level of tentaculitids and its allies only magnifies confusion. To apply a name that implies the existence of tentacles, when their presence has not been demonstrated, is indefensible. However, the continuation of such names for subordinal taxa is mandated by previous wide usage.

Cricoconarids are exclusively Paleozoic, first encountered in the Early Ordovician (Lower Canadian=Tremadocian). The oldest known species is *Tentaculites (s.l.) lowdoni* FISHER & YOUNG (12), from the Lower Ordovician Chepultepec Limestone of Virginia. It has since been found in correlative strata in Pennsylvania and New York State. Cricoconarids are uncommon throughout the Ordovician but become increasingly abundant during the Silurian. They attained their maximum diversity and numbers during the Middle Devonian (Emsian-Eifelian-Givetian) only to become extinct during the Late Devonian (early Famennian). The last survivor is *Styliolina* sp., from the Gowanda Formation (lower Canadaway) of western New York. The youngest recorded form in the Eastern Hemisphere is *Uniconus livenensis* LYASHENKO, from the Liven Formation (upper Frasnian) of the Central Russian Platform. Their occurrence in post-Devonian strata is in cobbles of conglomerates derived from earlier Paleozoic formations.

MORPHOLOGY

Cricoconarids are gradually tapering, small, narrow cones with transverse rings, ringlets, and striae (Fig. 50). Either asymmetrical or symmetrically angulate, rounded, or ripple-like rings occur. Longitudinal striae or ridges may be present. The shell consists of calcium carbonate, except where secondarily replaced by silica. Cricoconarid shells range from less than 1 mm. to 80 mm. in length, with a maximum diameter of 6.5 mm. The smallest recorded species are *Styliolina domaniscense* LYASHENKO, from the Domanik beds (U.Dev., M.Frasn.) of southern Timan, with a length of 0.8 mm. and diameter of 0.17 mm., and *Homoctenus nanus* LYASHENKO, from the Semiluk beds (U.Dev., M.Frasn.) of the Russian Platform, wtih a length of 1 mm. and diameter of 0.12 mm.; the largest are *Tentaculites elongatus* HALL, from the Helderbergian (L.Dev.) of New York, with a length of 80 mm. and diameter of 6.5 mm., and *Tentaculites reedsi* VOKES, from the Shriver Chert (L.Dev.) of Pennsylvania, with a length of 75 mm. and diameter of 6 mm. Growth angles range from 2 to 18 degrees. Some reports of larger growth angles may be attributed to shell flattening.

The cricoconarid shell is morphologically divisible into four parts: (1) embryonic, (2) juvenile, (3) adult, and (4) apertural. The embryonic chamber, hollow in thin-shelled forms and hollow or solid in thick-shelled forms, either tapers to a blunt point or is expanded into a teardrop-like bulb. The juvenile portion shows extreme regularity of the rings, both in size and spacing, and is commonly septate, dividing this region into as many as nine camerae. Thick-shelled forms have more and thicker septa than thin-shelled forms. The adult region exhibits greater variation in kind and spacing of rings and has a growth angle of 2 to 7 degrees less than the adjacent juvenile portion. No notches or projections occur on the periphery of the aperture, which is at right angles to the shell axis. No operculum or siphon (connective passage between camerae) has been found.

Transverse and longitudinal thin sections and polished surfaces disclose that the walls consist of many laminae. These laminae are straight or gently undulating nearest to the internal cavity, but become more undulating within the shell wall and repeat the exterior sculpture in the outermost layers. The shell wall is prismatic. The wall interior may be smooth, or nearly so, ringed in manner repeating the exterior, or ringed differently from the exterior. The thicker-walled forms are pierced by a multiplicity of tiny radial canals which only penetrate to the internal cavity near the aperture. Constructionally,

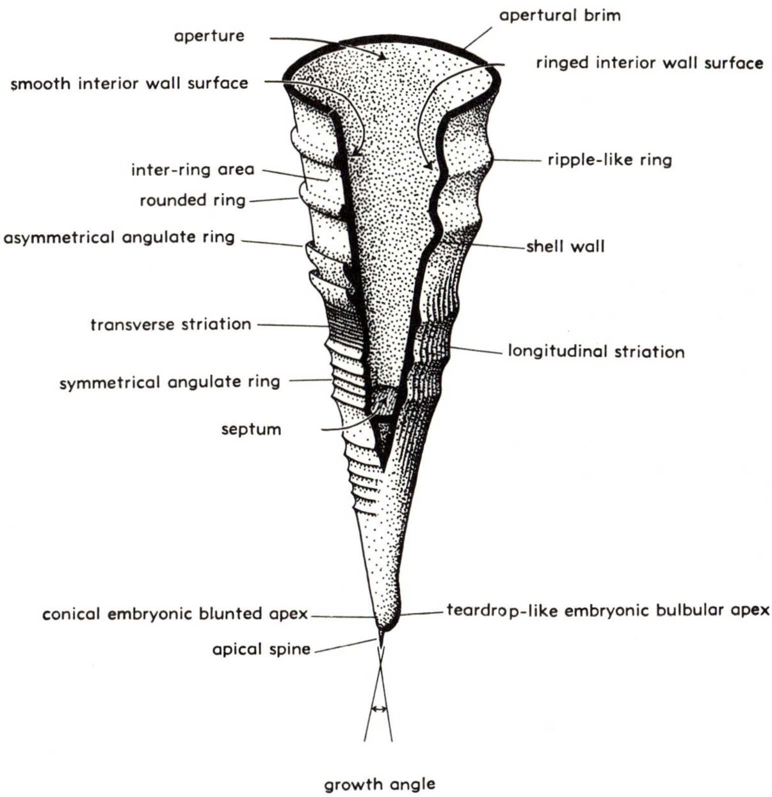

FIG. 50. Morphological features of cricoconarids (Fisher, n).

the shell wall resembles that of some brachiopods. It appears that the thickening of the shell takes place from the inside—the added layers having been produced by the mantle. As the animal tends to outgrow its living chamber, the mantle constructs additional shell material to accommodate the enlarged animal body. In response, the animal seeks to establish a new line of attachment. The old one is released and the animal slips forward. This leaves an empty space at the rear, which is soon closed off by construction of a septum. If the animal had communication through the shell, it is only evident in the most adapical mature region. Possibly the animal occupied only a relatively small portion of the large living chamber proximal to the aperture.

PALEOECOLOGY

Cricoconarids are found exclusively in marine rocks. They occur in all types of limestones but are especially prolific in lagoonal shallow-water deposits. They are common in all types of shales and argillaceous siltstones excepting red ones. They are less common in sandstones, occur sparingly in reef rock, graywacke, and dolomite, and are absent in saliferous and gypsiferous rocks. Generally speaking, cricoconarids were tolerant of many diverse environments. Careful study of the shell and manner of occurrence in the rock permits one to derive clues regarding their mode of life. Figure 51 illustrates several possible living habits, not all of which are equally plausible.

In general, cricoconarids occur in the rocks in two different ways: (1) extreme proliferation of complete specimens, commonly oriented similarly; (2) isolated specimens that commonly are incomplete apically or aperturally or both. This twofold manner of occurrence is inferred to signify a relatively quiet shallow-water environment with unidirectional oceanic currents in the former case and rough water in the latter case. It is curious that when bedding planes are replete with cricoconarids, little range in shell size is observed. Juveniles and adults are not haphazardly mixed. Mechanical size-sorting of dead shells is requisite to explain the phenomenon of three of four cones inserted within one another. When in immense numbers, relatively few species of other phyla are present, and these are customarily abundant and diminutive also. The usual faunal associates are: ostracodes, conodonts, small brachiopods, small pelecypods, and small bryozoan colonies. A purely mechanical distributional effect may explain this assortment, although this lack of diversity, associated with local abundance of individuals, is characteristic of waters with abnormal salinities, very muddy bottoms, or boreal environments. By contrast, whenever isolated cricoconarids are found, the faunal association is varied as to kind, size, and number of different representatives of many phyla.

Cricoconarid shells always lie parallel to bedding of strata that enclose them (Fig. 54). This implies a pelagic habit and strongly denies a fossorial one. Radial symmetry suggests a basic "up-and-down" differentiation of the animal. However, if the shell were upright, balancing on the bulbar or pointed apex, one would expect these comparatively fragile tips to be broken off. This is seldom the case. A possible reversal of this orientation, namely, with mouth directed downward and the organism hovering over the sea bottom, merits attention. A benthonic existence, with the long side of the cone in contact with the sea bottom, is refuted by the circular cross section of the fossils and lack of any wear of their prominent encircling rings. Moreover, the absence of an operculum, which would prohibit infiltration of mud and silt, would make a benthonic habit unfeasible.

A pelagic life is most compatible with accumulated evidence, which is insufficient, however, to resolve the question as to whether a nektonic or planktonic existence was more plausible. The multiplicity of shells and their preferred orientation in some strata might influence one to presume that some members of this class may have been distributed by oceanic currents. The relatively rapid world-wide dispersal of the dacryoconarids (nowakiids, styliolinids) lends credence to this view. There is no proof, however, that cricoconarids ever experienced an epinektonic or epiplanktonic existence.

Though a nektonic life appears most reasonable, it might be argued that the presence of rings on the shell exterior, especially rings of an angulate type, would impede swimming. It is noteworthy that the dacryoconarids possess ripple-like rounded rings, abortive ones, or none at all, thereby reducing surface friction. Probably the rings developed as a strengthening structure that served to combat the forces of agitated water. The development of thicker shells seems to have been a response to rough waters or elevated temperatures. It is difficult to comprehend how relatively heavy, thick-shelled cricoconarids (Tentaculitidae, Uniconidae) could have moved very far off the sea bottom. Perhaps these families were nektobenthonic, moving slowly about with the apical end upward and mouth directed downward, scavenging on the bottom. Coincidentally, these heavier types have the greatest number of camerae to compensate for their lesser buoyancy. These camerae must have functioned hydrostatically, permitting habitation in the pelagic realm. Since no connection existed between chambers, this hydrostatic capacity was fixed, prohibiting the versatility of rapid up-and-down movement—if they possessed any at all! The likelihood of cricoconarid bathymetric zonation is compelling. Since the septa are slightly concave toward the aperture, as in cephalopods, a moderate amount of reciprocal animal movement is presupposed. Lack of any operculum suggests the ability to move sufficiently fast to obviate the need for a protective lid for the soft parts.

Aside from the numerically superior nautiloid cephalopods, cricoconarid dominance of the pelagic realm went unchallenged until the Middle and Late Devonian, when the great development of goniatite cephalopods and fishes (acanthodians, arthrodires, osteichthyans) was introduced. Not being able to cope with the ecologic rivals, which were more active swimmers and predators,

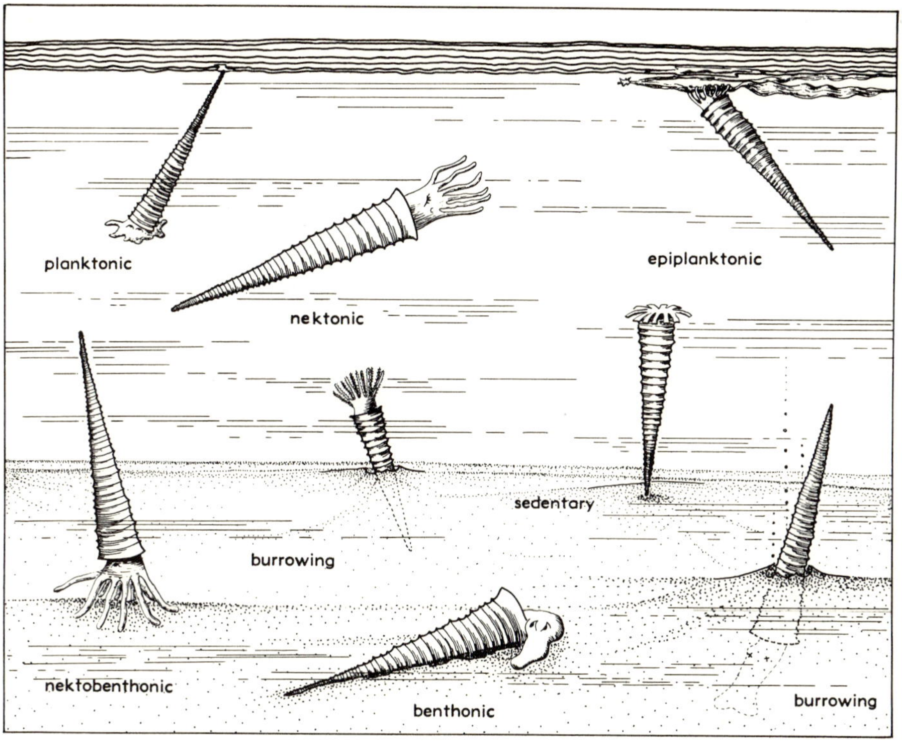

FIG. 51. Possible ecologic adaptations of cricoconarids (Fisher, n).

the cricoconarids diminished in an inverse ratio to the pronounced increase of the invaders. During the Silurian and Early Devonian, the incursion of bottom-dwelling fishes (ostracoderms, antiarchs) seemingly offered no serious competition to the cricoconarids' supremacy. Surely, if cricoconarids were benthonic, the effect of the co-existent fishes would have retarded their development. In contrast, cricoconarids experienced their optimum during the Silurian and Early and Middle Devonian.

In summary, it is suggested that the dacryoconarids were pelagic (principally planktonic) indigenous inhabitants of the upper reaches of the oceans, achieving relatively rapid world-wide dispersal via transoceanic currents. Had they been able to govern their movements, they might easily have escaped this distributive agent. Their nonseptate, thin shells suggest an inability to transgress bathyal zones or to live in areas of strong breaking waves. Among the tentaculitids *(sensu stricto)*, the multiseptate, thick-shelled Tentaculitidae and Uniconidae very likely were nektobenthonic scavengers in relatively warmer, more agitated waters, whereas a somewhat later stock, the Homoctenidae, with fewer septa and thinner walls, may have migrated to intermediate bathyal zones or more boreal environments.

CLASSIFICATORY STATUS

Ever since WALCH (69) first illustrated the fossils which SCHLOTHEIM (60) later named *Tentaculites* (Fig. 52), these curious fossils have defied taxonomic assignment. Since nothing is known of the organism which inhabited these shells, and since seemingly they have left no living descendants, cricoconarids cannot be placed with confidence in the scheme of zoological nomenclature.

VON BUCH (1830) thought that specimens of *Tentaculites* were spines of the brachiopod *Leptaena lata* (actually a chonetid). Failure to find chonetids or any other spine-bearing brachiopods in strata where cricoconarids are most prolific refutes such an assignment. Like reasoning may be applicable to consideration of *Tentaculites* as representing the spines of echinoids (EATON, 1832) or crinoid arms (60).

From gastropods, cricoconarids differ in possessing a straight septate calcitic shell with an untwisted embryonic chamber. Superficially, scaphopods resemble cricoconarids in that both display radial symmetry, and in this respect they are not mol-

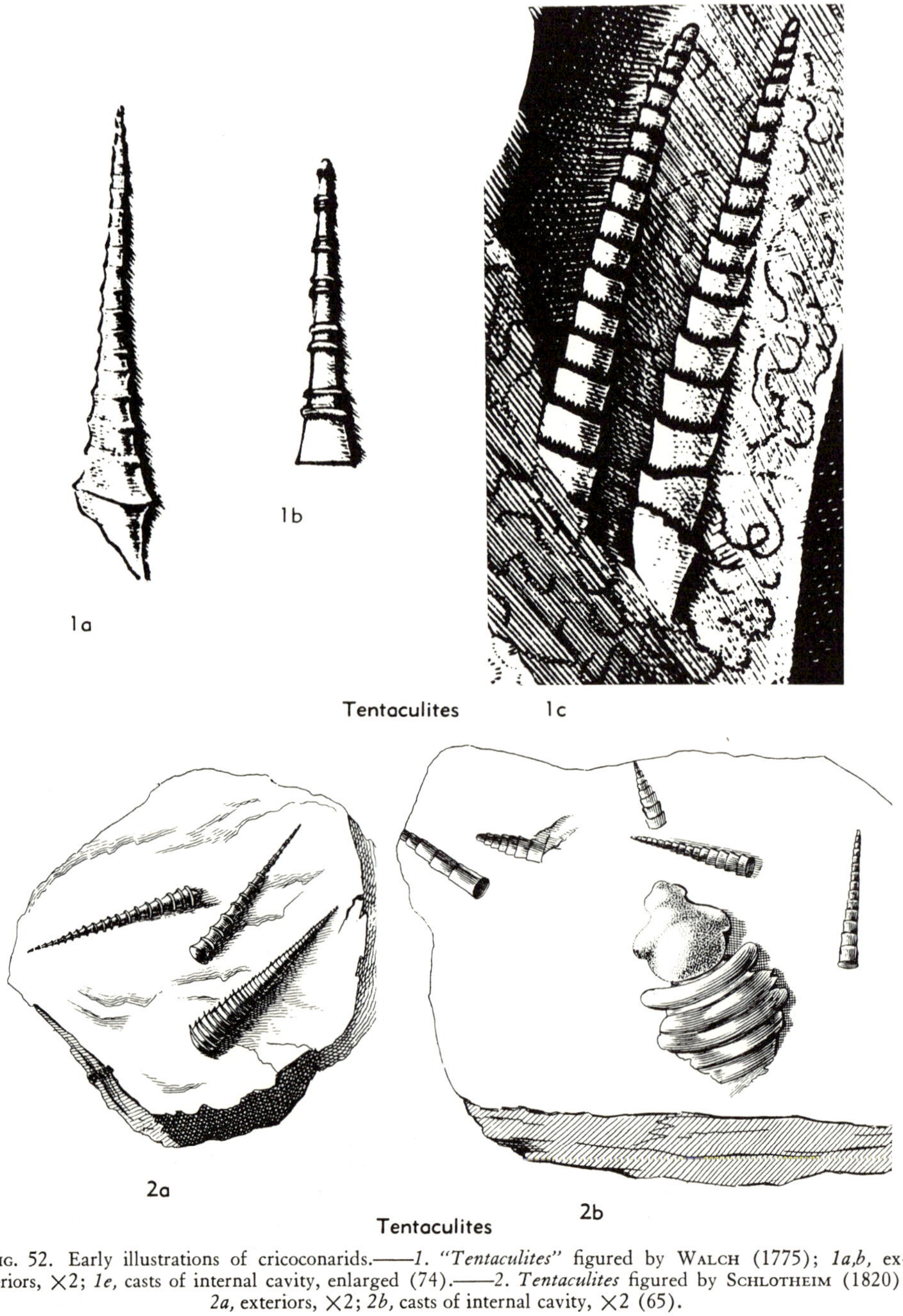

FIG. 52. Early illustrations of cricoconarids.——*1.* "*Tentaculites*" figured by WALCH (1775); *1a,b,* exteriors, ×2; *1e,* casts of internal cavity, enlarged (74).——*2. Tentaculites* figured by SCHLOTHEIM (1820); *2a,* exteriors, ×2; *2b,* casts of internal cavity, ×2 (65).

lusk-like. Basic differences are clearly marked, in that scaphopods have an apical aperture, well-defined longitudinal ribbing (usually), an absence of transverse rings, a nonseptate shell, and placement in living position at oblique angles to bedding. A considerable fundamental similarity to cephalopods may be seen in the mutual existence of a many-layered shell, presence of an embryonic bulb, and septate nature of the shell. Lack of a siphuncle and sutures in cricoconarids, however, reveals basic differences. Although it has been customary to group cricoconarids with the pteropods, cricoconarids lack certain fundamental characteristics of the latter group, namely, (1) an exceedingly thin shell, (2) a notch or projection on the apertural brim, and (3) presence of pteropodia or a swimming apparatus. Bilateral symmetry, a feature of true pteropods, cannot be demonstrated in the cricoconarids. A pseudobilateral symmetry is present in some forms (especially styliolinids) marked by a longitudinal depression caused by fracture of the thin shell during compaction. Thus, I reject the name Eopteropoda (66, 67) for tentaculitids, nowakiids, and styliolinids, because this name implies that the group was ancestral to living pteropods, which is an unconfirmed phylogenetic alliance. If such a relationship were real, it would be difficult to explain the long stratigraphic gap (Devonian to Tertiary) in which no fossil pteropods have been found.

Many paleontologists have identified cricoconarids as tubicolar worms. The presence of an embryonic chamber, multilayered wall, straight, septate shell, and free mode of living (tubicolar worms are usually curved and attached) seems to preclude any affinity with the worms.

Formerly it was customary to group cricoconarids with the conulariids. Now that the conulariids, with their quadrilateral radial symmetry and flexible chitinous wall, have become recognized as an extinct group of coelenterates, the basis for any supposed relationship vanishes. Nevertheless, the occurrence of tiny pores in cricoconarid walls, coupled with nothing but radial symmetry, does not preclude a coelenterate affiliation. Cricoconarids might be free-swimming hydroids.

Owing to their obscure biologic relationship, cricoconarids are frequently placed in *incertae sedis*. Such disposition masks the uniqueness of this fascinating group. Cricoconarids display many characteristics of the phylum Mollusca. Lyashenko (1955-1960) provisionally placed them here but denied their association with pteropods. I agree with Lyashenko and can see no better disposition than to give them separate phyletic rank, and that seems unwarranted. Accordingly Cricoconarida are here regarded as an extinct class of the phylum Mollusca.

CRICOCONARID CLASSIFICATION

To date, about 150 species of cricoconarids ranging from Early Ordovician to Late Devonian in age (Fig. 53) have been named, of which about a third have been described in detail by Lyashenko (1954-1959). Unfortunately, many earlier-named species supply inadequate data for modern generic assignment. Most of them will have to remain in *Tentaculites (sensu lato)* pending restudy, particularly of their internal structures. It is hoped that Lyashenko's recent excellent work will stimulate others to test the stratigraphic and paleoecologic usefulness of these fossils which have not received monographic treatment since the days of Barrande (1) and Hall (17, 18).

Inasmuch as nothing is known of the relationship of the animal to its shell, the sole recourse is to select a classification based on geometric configuration of the shell, with major features taken as a reflection of basic morphologic structures. The deficiencies of adopting such a scheme are obvious. The species concept becomes a typological one, unless variation within populations is carefully scrutinized and the modifying effects of diverse ecological factors are analyzed.

Gürich (15) was the first to attempt a division of the tentaculitids. He first considered both *Cornulites* and *Tentaculites* as members of the family Tentaculitidae under Vermes. *Tentaculites* was subdivided on the basis of the type of exterior ornamentation into four groups designated Clathrati, Annulati, Annulosi, and Coarctati. He further noted that the last

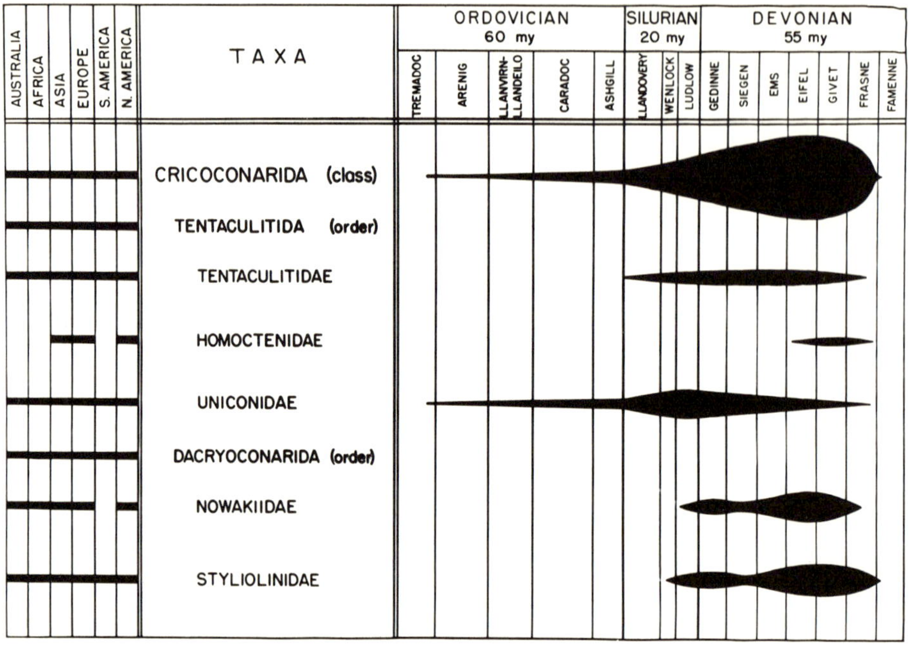

FIG. 53. Stratigraphic and geographic distribution of the Cricoconarida, with indication of comparative abundance (Fisher, n).

three groups stood close together and were linked by transitional forms. It is of interest that GÜRICH's four groups correspond approximately to four of my five recognized families of cricoconarids, namely, the Nowakiidae, Uniconidae, Homoctenidae, and Tentaculitidae, respectively. His *Styliolites* would constitute the fifth, the Styliolinidae. Whereas GÜRICH utilized exterior ornamentation for his subfamilial groups, I employ wall interior configuration for familial distinction, with surprisingly duplicated end results.

Some basic differences may be observed between LYASHENKO's (1954-1959) and my classification. The relative taxonomic value of some features is increased and that of others is decreased. Only time and usage will determine which (if either) will prove to be a practical workable arrangement. A summary of the criteria used and comparison of the two classifications follows.

The uniform shape of the fossils is of foremost importance. All are small (less than 80 mm. in length, averaging 20 mm.), narrow (averaging 1 to 3 mm.), tapering straight ringed cones (as they are true cones, they exhibit a circular cross section), terminating in a blunt point or expanded bulb. Here, then, is a major point of differentiation. Unquestionably, the type of apical termination, the embryonic stage, is of primary importance and, accordingly, this is the criterion for dividing the class into orders—(1) cricoconarids tapering to a bluntly pointed conical embryonic chamber forming the order Tentaculitida, and (2) those with expanded teardrop-like embryonic chambers constituting the new order Dacryoconarida.

It seems that the nature of the inner surface of the shell wall would, because of its proximity to the animal, naturally follow as a criterion of secondary importance. Therefore, I choose this feature for family differentiation. In the Tentaculitida, three types of wall interiors are known: (1) unpressed-ringed, but different from the exterior surface (Tentaculitidae); (2) depressed angulate-ringed, repeating features of the exterior surface (Homoctenidae); and (3) smooth, different from exterior

(Uniconidae). In the Dacryoconarida, two types of wall interiors are known: (1) undulatory ripple-like rings (Nowakiidae); and (2) smooth (Styliolinidae).

Genera are based on the type of exterior wall ornamentation, the most obvious character, though not necessarily the most basic. Species differentiation is based on minor details of the ornamentation and difference in growth angle of similar types.

Thickness of shell wall, uniformity or nonuniformity of rings and size differentiation of otherwise similar forms are considered ecological variants.

Class CRICOCONARIDA Fisher, n. class

[ety., *krikos*=ringed; *konarion*=small cones] [=Superorder Tentaculitoidea LYASHENKO, 1958 *(emend.)*]

Small, narrow, straight, ringed true cones belonging to various animals which possibly

Criteria Used in Classification of Cricoconarids (Tentaculitids, sensu lato)

G. P. LYASHENKO (1954-1959)		D. W. FISHER (herein)
No fundamentals given	CLASS	RINGED NARROW SMALL STRAIGHT CONES Radial symmetry (Cricoconarida)
Characteristics of tentaculitids and hyolithids lumped (Coniconchia)		Hyolithids *(sensu lato)* placed in separate class (Calyptoptomatida)
SYMMETRY Radial (Tentaculitoidea) Bilateral (Hyolithoidea)	SUPERORDER	No division
CHARACTER OF EXTERIOR Annulated (Tentaculitida) Swellings (Novakiida) Smooth (Styliolinida)	ORDER	SHAPE OF EMBRYONIC CHAMBER Blunted point (Tentaculitida) Teardrop-like (Dacryoconarida)
MANNER OF ARRANGEMENT OF RINGS ON EXTERIOR Uniform (Homoctenidae) Nonuniform (Tentaculitidae) Uniform (Novakiidae) No rings (Styliolinidae)	FAMILY	CHARACTER OF INTERIOR SIDE OF WALL Ringed, but different from exterior (Tentaculitidae) Ridged, repeats the exterior (Homoctenidae, *emend.*) Smooth, different from exterior (Uniconidae, *nom. transl.*) Smooth (Styliolinidae, *emend.*) Rippled, same as exterior (Nowakiidae, *nom. correct.*)
THICKNESS OF SHELL WALL and CHARACTER OF MARGIN OF INTERNAL CAVITY (Homocteninae, Uniconinae, Novakiinae, Crassilininae)	SUBFAMILY	No division
MINOR CHARACTERISTICS OF SCULPTURE and TYPE OF SHELL GROWTH	GENUS	TYPE OF EXTERIOR ORNAMENTATION, GROWTH ANGLE
DETAILS OF SCULPTURE, SIZE AND GROWTH ANGLE	SPECIES	FINE DETAILS OF ORNAMENTATION, DIFFERENCES IN GROWTH ANGLE
NONE?	ECOLOGICAL VARIANTS	THICKNESS OF SHELL, UNIFORMITY OR NONUNIFORMITY OF RINGS, SIZE OF OTHERWISE SIMILAR FORMS

are distantly related; presumably tentacle-bearing. [Includes tentaculitids, nowakiids and styliolinids.] *L.Ord.-U.Dev.*

Order TENTACULITIDA Lyashenko, 1955

[=Superfamily Tentaculitacea TERMIER & TERMIER, 1950 *(partim)*]

Cricoconarids with conical embryonal chamber terminating in a blunt pointed apex. Length, 1 to 80 mm., usually 15 to 30 mm.; shell wall thick or thin, with tiny radial canals piercing thick-walled forms; shell wall laminate, usually 2 to 5 layers. Juvenile portion of shell septate, forming several chambers distinct from large apertural cavity. Exterior covered by transverse rings of various size and spacing. Longitudinal striae rarely present. Interior wall surface ringed or smooth. About 110 species have been described. *L.Ord.(Tremadoc.)-U.Dev. (Up.U.Frasn.).*

Family TENTACULITIDAE Walcott, 1886

[=Coarctati GÜRICH, 1896 *(partim)*]
[*nom. correct.* MILLER, 1889 (*ex Tentaculidae* WALCOTT, 1886, *nom. imperf.*)]

Inner wall surface with depressed rings spaced at proportionately increasing intervals toward aperture; internal mold appearing as series of inverted invaginated cones. Walls thick, multilayered, and pierced by tiny radial canals. Juvenile portion septate, septa slightly concave toward aperture. Exterior rings more uniform in juvenile portion than mature region. *L.Sil.(L.Llandov.)-U.Dev. (Mid.M.Frasn.).*

Tentaculites SCHLOTHEIM, 1820 (p. 377) [**T. scalaris; non T. ornatus* SOWERBY, 1839 (*fide* LYASHENKO, 1955-1959)] [=*Dentalium (partim)* SCHROETER, 1784; *Lonchidium* EICHWALD, 1857; *Styliola* LUDWIG, 1864 *(partim)*]. Medium-size (15 to 30 mm.) cone, exterior with coarse transverse rings which are less uniform in spacing toward aperture; inter-ring area usually with transverse ringlets or striae noticeably developed only in adult region. Embryonic portion conical, hollow or solid. Growth angle, 7 to 12 degrees in juvenile portion, 3 to 7 degrees in mature portion. *?L.Ord., L.Sil.(Llandov.)-U.Dev.(Mid.M.Frasn.),* N.Am.-S.Am.-Eu.-Asia-Afr.-Austral.——FIGS. 54, *1*, 55,*1*. *T. bellulus* HALL, M.Dev.; 54,*1*, bedding surface of Arkona Sh., Arkona, Ont., with many well-preserved specimens, ×2; 55,*1a,b*, specimens from Menteth Ls., Canandaigua Lake, N.Y., ×7 (Fisher, n).——FIG. 54,*2*. *T. anglicus* SALTER, M.Ord.(Caradoc.), Eng.(Marshbrook); specimens subparallel in orientation, ×3 (Fisher, n).——FIGS. 54,*3*, 55,*7*. *T. gyracanthus* (EATON), L.Dev.; 54,*3*, bedding surface of Manlius Ls., Sharon, N.Y., with abundant nearly parallel specimens, ×2; 55, *7*, specimen from same horizon and locality attached to bryozoan, ×8 (Fisher, n).——FIG. 55, *6*. *T. arenosus* HALL, L.Dev.(Oriskany Ss.); *6a*, cast of interior (Cayuga, Ont.), ×4; *6b*, cast of interior within external mold (Glenerie, N.Y.), ×1 (Fisher, n).——FIG. 55,*8*. *T.* sp., Sil., Swed. (Gotl.), 10 specimens showing variation at a single locality (Klinteheim), ×4 (Fisher, n).——FIG. 56,*1*. *T.* sp., Dev., USSR; diagram. sec. showing chambers in apical region and external rings, ×6 (39, mod.).

Volynites LYASHENKO, 1957 (p. 87) [**V. russiensis*]. Medium-sized (10 to 15 mm.) cone with external various-sized rings, transverse striae usually present. Interior wall surface of adult region with irregularly spaced depressed rings different from exterior rings and spacing. No longitudinal striae. Growth angle, 10 to 13 degrees in juvenile portion, 6 or 7 degrees in mature portion. *U.Sil.-L.Dev.,* USSR-W.Eu.-N.Am. ——FIG. 56,*2*. **V. russiensis,* U.Sil.(Ludlov.), USSR; diagram. sec. showing apical chambers and external rings, ×7.5 (39, mod.).

Family HOMOCTENIDAE Lyashenko, 1955

[=Annulosi GÜRICH, 1896]

Inner wall surface with angulate depressed rings repeated on exterior surface as angulate crests; wall relatively thin, usually only 2 or 3 layers; no radial canals. Internal septa thin or absent or few in number (usually 1 or 2). *M.Dev.(Eifel.)-U.Dev. (Low.U.Frasn.).*

Homoctenus LYASHENKO, 1955 (p. 13) [**H. krestovnikovi*]. Small cone with exterior covered by angulate rings, size and spacing of which increase proportionately toward aperture; concave inter-ring areas wider than rings. Embryonic portion conical and separated from rest of internal cavity usually by a single septum; but 2 or 3 septa in some shells. No transverse or longitudinal striae. Growth angle 9 to 15 degrees in juvenile portion, 6 to 12 degrees in mature portion. *U.Dev.(L.Frasn.-Mid.M.Frasn.),* Eu.(USSR)-N.Am.(N.Y.) ——FIG. 57,*1*. **H. krestovnikovi,* USSR; diagram. sec., ×25 (39, mod.).

Denticulites LYASHENKO, 1957 (p. 87) [**Tentaculites lyashenkoi* LYASHENKO, 1957]. Small cone with fine and coarse rings, both rings and inter-ring areas covered with longitudinal furrows; wall thicker than in other homoctenids and consisting of many layers. Growth angle 7 to 10 degrees in juvenile portion, 3 to 6 degrees in mature por-

1 Tentaculites

2 Tentaculites

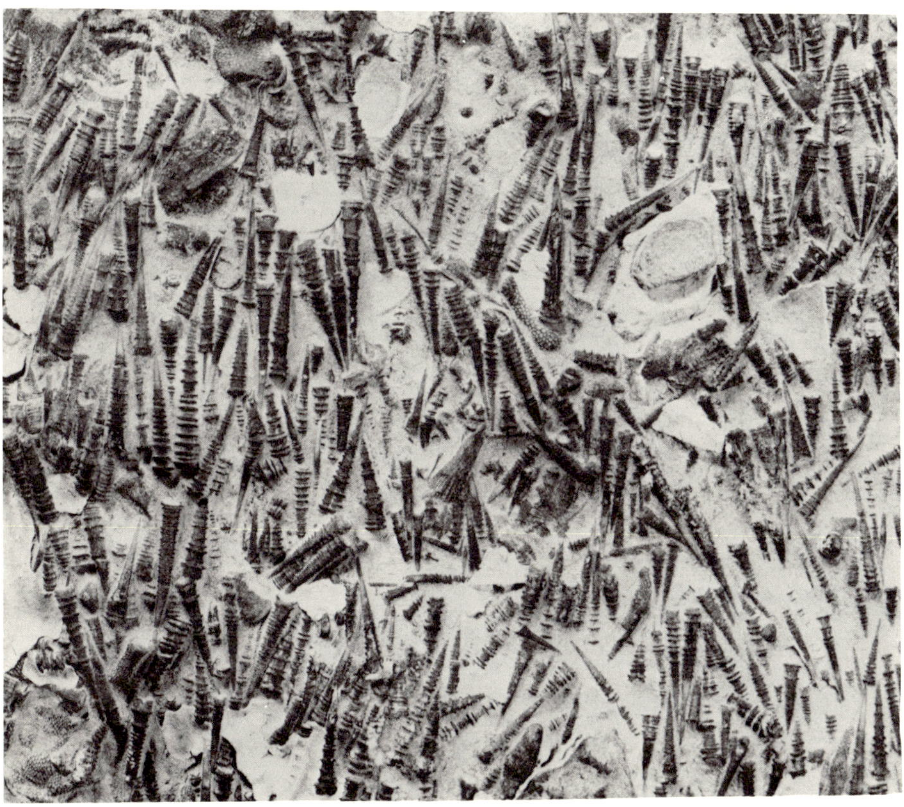

3 Tentaculites

Fig. 54. Species of *Tentaculites* (Tentaculitidae) (Fisher, n).

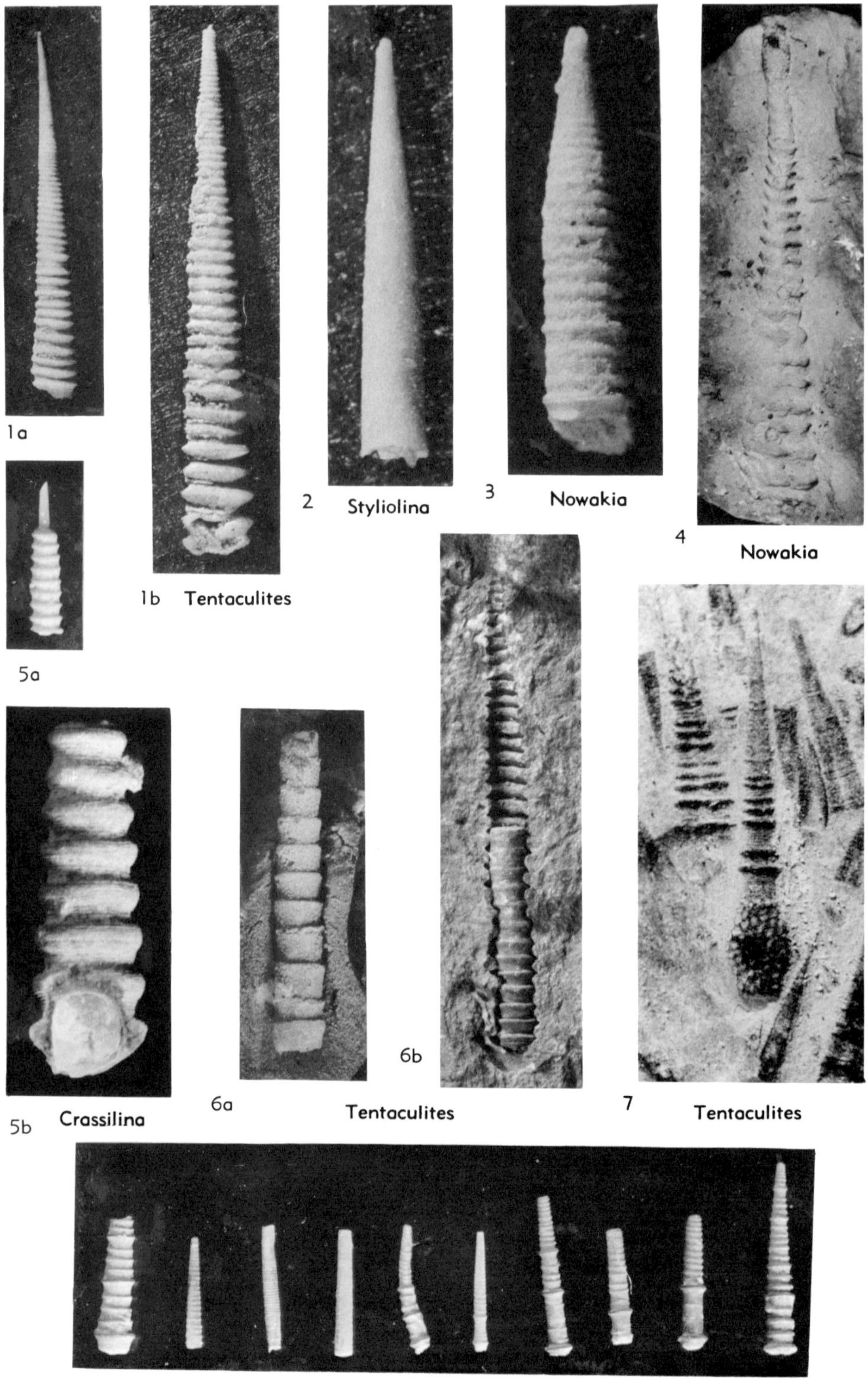

Fig. 55. Tentaculitidae, Nowakiidae, Styliolinidae (p. W110, W115-W116) (Fisher, n).

tion. *M.Dev.(U.Givet.)*, Eu.(USSR).——FIG. 57,*3*. **D. lyashenkoi* (LYASHENKO), diagram. sec., ×40 (39, mod.).

Polycylindrites LYASHENKO, 1955 (p. 13) [**Tentaculites nalivkini* LYASHENKO, 1954 (*non Polycylindrites nalivkini* LYASHENKO, 1955-1959)]. Small cone, rarely slightly curved, appearing to be series of ringed cylinders with common longitudinal axis, diameters of cylinders proportionately increasing toward aperture. Exterior composed of sharply elevated angulate rings similar to *Homoctenus* except that end of cylindrical portions of shell possess larger ring. Concave inter-ring areas wider than rings. No transverse or longitudinal striae. Usually only a single septum separating embryonic from mature portion, but septum rarely intact. Growth angle 12 to 15 degrees in juvenile portion, 7 to 11 degrees in mature portion. *U.Dev.(L.Frasn.-Low.U.Frasn.)*, Eu.(USSR).——FIG. 57,*2*. **P. nalivkini* (LYASHENKO); diagram. sec., ×20 (39, mod.).

FIG. 57. Homoctenidae (p. *W*110-*W*113) (Lyashenko, mod.).

Family UNICONIDAE Lyashenko, 1955

[=Annulati GÜRICH, 1896 (*partim*)]
[*nom. transl.* FISHER, herein (*ex* Uniconinae LYASHENKO, 1955)]

Interior wall surface smooth or nearly so; multi-layered thick wall pierced by tiny radial canals. Juvenile portion usually septate. Exterior with prominent transverse rings. *L.Ord.(Tremadoc.) - U.Dev.(Up.U. Frasn.)*.

Uniconus LYASHENKO, 1955 (p. 13) [**Tentaculites glaber* TRAUTSCHOLD, 1881]. Small cone with exterior covered by angulate rings which increase proportionately in size and spacing toward aperture. Rings asymmetrical; steeper part on apertural side. Juvenile portion septate. Growth angle 7 to 9 degrees in juvenile portion, 4 to 6 degrees in mature portion. *?L.Ord.-Sil.-U.Dev.(Up.U. Frasn.)*, Eu.(USSR).——FIG. 58,*1*. **U. glaber* (TRAUTSCHOLD), Frasn., USSR; diagram. sec., ×10 (39, mod.).

Contractenus LYASHENKO, 1959 (p. 81) [**C. markovskii*]. Medium-sized cones covered by coarse rounded rings separated by relatively wide inter-ring areas with smaller scattered wrinkles which are located rather haphazardly and at oblique angles to axis of shell; thin longitudinal wrinkles

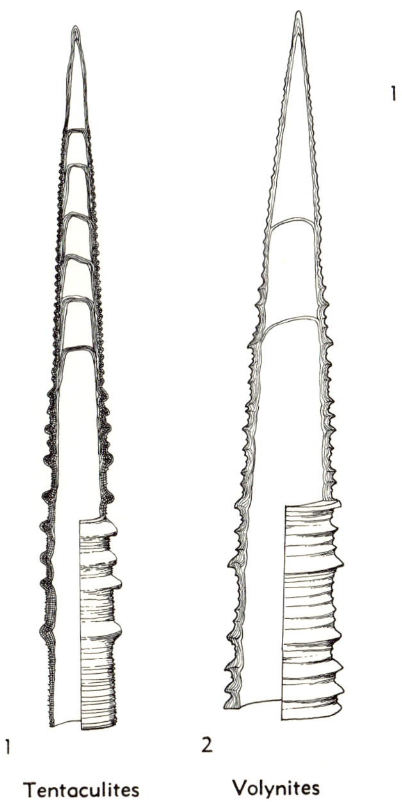

FIG. 56. Tentaculitidae (p. *W*110) (Lyashenko, mod.).

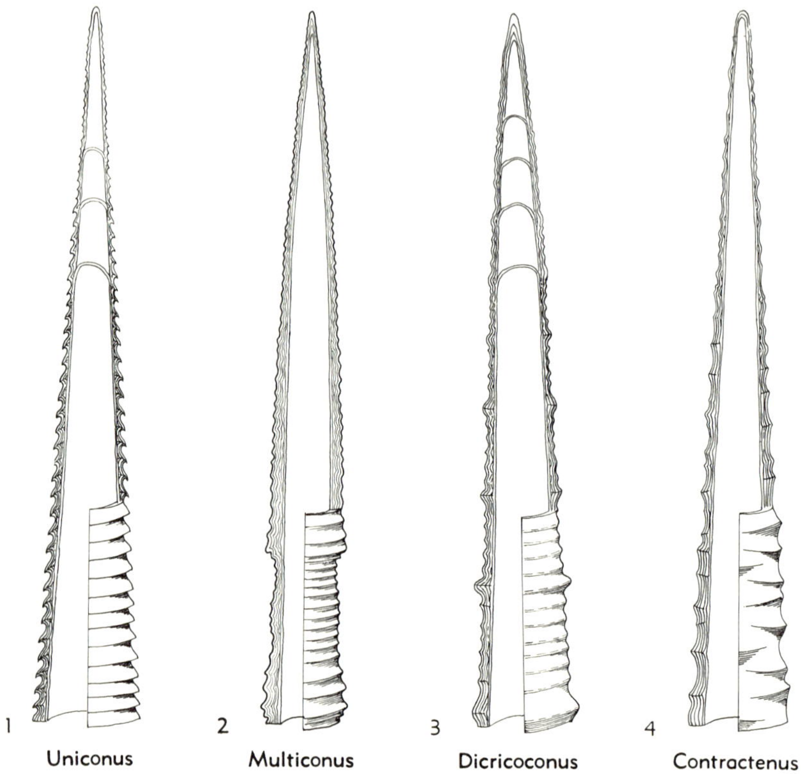

FIG. 58. Uniconidae (p. W113-W114) (Lyashenko, mod.).

may also occur. Growth angle 6 or 7 degrees in juvenile portion, 4 to 6 degrees in mature portion. *M.Dev.(U.Eifel.)*, Eu.(USSR).——FIG. 58,4. **C. markovskii;* diagram. sec., ×13 (39, mod.).

Dicricoconus FISHER, *nom. subst.* herein [*pro Heteroctenus* LYASHENKO, 1955, p. 12 (*non* POCOCK, 1893)] [**Tentaculites mesodevonicus* LYASHENKO, 1954]. Medium-sized cones with exterior covered by 2 types of rounded transverse rings: larger may be single or double, whereas smaller occur more uniformly spaced on inter-ring areas; transverse striae may appear on rings and inter-ring areas, and longitudinal striae also may occur. Juvenile portion septate with exterior rings of nearly uniform size and spacing. Growth angle 7 to 11 degrees in juvenile portion, 3 to 7 degrees in mature portion. *L.Sil.(M. Llandov.)-U.Dev.(Mid.U.Frasn.)*, N.Am.-S.Am.-Eu.-Asia-Austral.-Afr.——FIG. 58,3. **D. mesodevonicus* (LYASHENKO), Givet., USSR; diagram. sec., ×10 (39, mod.).

Multiconus LYASHENKO, 1955 (p. 14) [**M. schimanskii*]. Small to medium-sized cones having appearance of 2 or more invaginated ringed cones with common longitudinal axis. Exterior like *Uniconus* except that apertural end of invaginated cones has 3 to 6 larger rings. Septa and transverse and longitudinal striae not observed. Growth angle 7 to 11 degrees in juvenile portion, 4 to 6 degrees in mature portion. *U.Dev.(Up.M.Frasn.)*, Eu.(USSR).——FIG. 58,2. **M. schimanskii;* diagram. sec., ×13 (39, mod.).

Order DACRYOCONARIDA Fisher, n. order

[ety., *dakryon*=teardrop; *konarion*=very small cones] [=Novakiida LYASHENKO, 1955 (*nom. van.*) +Styliolinida LYASHENKO, 1955]

Small cricoconarids with pronounced teardrop-like embryonal bulb, which may have tiny apical spine emanating from it. Growth angle relatively greater than in Tentaculitida. Exterior smooth or covered by broad ripple-like rings with rounded crests and troughs. Longitudinal ornamentation usually present. Juvenile portion smooth or with weakly developed rings.

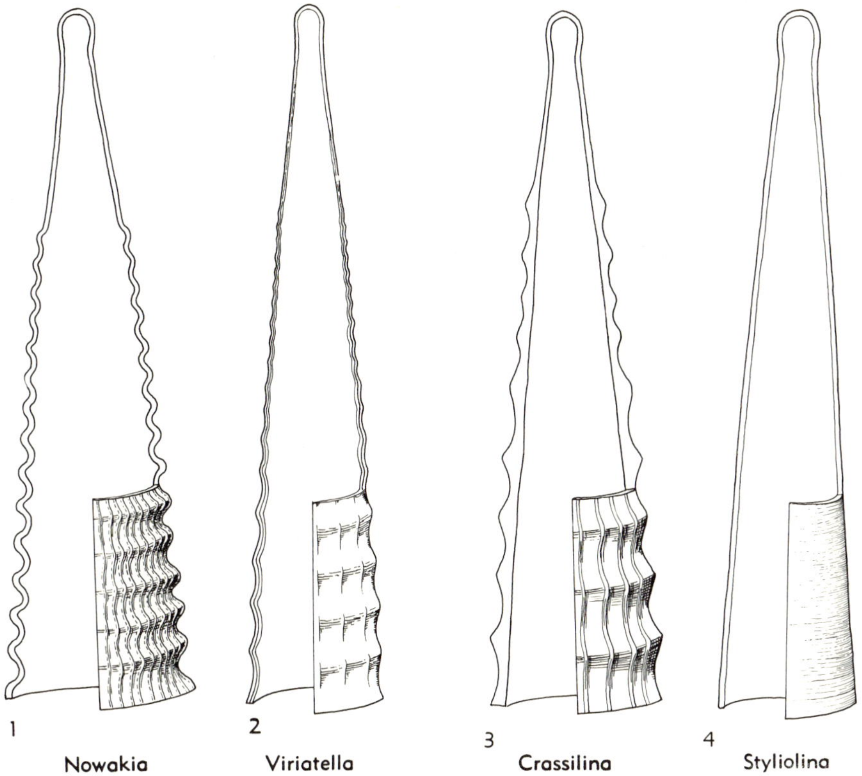

Fig. 59. Nowakiidae, Styliolinidae (Dacryoconarida) (p. *W*115-*W*116) (Lyashenko, mod.).

Shell wall thick or thin; no radial canals observed. Interior wall surface smooth or ringed. No evidence of septa, though Novak reported a septum between embryonic chamber and rest of interior. [About 40 species have been described; very abundant in Late Silurian and Early Devonian of Australia and Middle and early Late Devonian of North America.] *M.Sil.(Wenlock.)-U.Dev.(L.Famenn.).*

Family NOWAKIIDAE Bouček & Prantl, 1960

[=Clathrati Gürich, 1896] [=Novakiidae Lyashenko, 1955 *(partim) (nom. van.)*]

Dacryoconarids with an undulatory, ripple-like, ringed interior wall surface. *U.Sil.(L.Ludlov.)-U.Dev.(M.Frasn.).*

Nowakia Gürich, 1896 (p. 196) [**Tentaculites elegans* Barrande, 1852; SD Barrande, 1865] [=*Novakia* Tolmachov, 1926; (*non Novakia* Strobl, 1893]. Exterior covered by broad rounded ripple-like rings beginning about third of way from apex; transverse and longitudinal striae present. Wall thin; inner wall surface repeating exterior surface. Growth angle 13 to 18 degrees in juvenile portion, 10 to 13 degrees in mature portion. *U.Sil.(L.Ludlov.)-U.Dev.(M. Frasn.),* N. Am.-Eu.-Asia-Afr.-Austral. —— Fig. 55,*3*. *N. acuarius* (Richter), M.Dev. (Hlubocepy Ls.), Czech.; side view, ×10 (Fisher, n).——Fig. 55,*4*. *N.* sp., L.Dev.(Herdorfer Sh.), Belg.; specimen in matrix, ×1 (Fisher, n).——Fig. 59,*1*. *N.* sp., composite diagram. sec., ×25 (39, mod.).

Guerichina Bouček & Prantl, 1961 (p. 385) [**G. strangulata*]. Like *Nowakia,* but has wider and less pronounced ripplelike rings upon which are superimposed numerous narrow smooth delicate transverse ringlets; longitudinal striae absent. Wall thin. Growth angle 9-11° in juvenile portion. *L. Dev.(Dvorce & Prokop Ls.),* Eu.(Czech.).

Variatella Lyashenko, 1957 (p. 92) [**V. petrovi*]. Exterior consisting of rounded, ripplelike rings which increase in size and spacing proportionately from about third of distance from apex toward aperture; transverse and longitudinal striae present. Juvenile portion feebly ringed. Wall thin; interior wall surface repeating exterior surface. Growth angle 10 to 15 degrees in juvenile portion, 6 to 12 degrees in mature portion. *U.Sil.(Ludlov.)-U.Dev.(Up.L.Frasn.),* N.Am.-Eu.(USSR)-Afr.-Austral.——FIG. 59,*2*. **V. petrovi,* Frasn., USSR; diagram. sec., ×20 (39, mod.).

Family STYLIOLINIDAE Grabau, 1912

[Family cited by LYASHENKO, 1955, 1958, 1959, but unverifiable by me]

Dacryoconarids with a smooth interior wall surface. *M.Sil.(Wenlock.)-U.Dev.(L.Famenn.).*

Styliolina KARPINSKY, 1884 (p. 14) [**Styliola nucleata* KARPINSKY, 1884; SD LYASHENKO, 1958 (p. 184)] [=*Styliolites* GÜRICH, 1896]. [It is not clear from KARPINSKY's paper that he intended *Styliola nucleata* to be the type-species of his poorly described *Styliolina.* On the contrary, he seemed emphatic in not placing any of his species in *Styliolina!*]. Exterior smooth except for scattered transverse striae; a pseudo-longitudinal groove commonly present but this results from crushing of the thin shell; molds of internal cavity look like exterior except that embryonic bulb is less teardrop-like and may even be conical. Growth angle 8 to 14 degrees in juvenile portion, 3 to 11 degrees in mature portion. *M.Sil.(Wenlock.)-U.Dev.(L.Famenn.),* N.Am.-S.Am.-Eu.-Asia-Afr.-Austral.——FIG. 55,*2*. *S. clavulus* BARRANDE, M.Dev.(Hlubocepy Ls.), Czech.; side view, ×10 (Fisher, n).——FIG. 59,*4*. *S. grandis* LYASHENKO, Frasn., USSR, diagram. sec., ×25 (39, mod.).

Crassilina LYASHENKO, 1955 (p. 15) [**C. timanica;* SD LYASHENKO, 1957 (p. 97)]. Exterior covered with broad, undulatory rings and longitudinal and transverse striae. Wall thick. Molds of internal cavity almost identical to *Styliolina.* Growth angle 12 to 14 degrees in juvenile portion, 11 or 12 degrees in mature portion. *M.Dev.(Eifel.)-U.Dev.(Up.L.Frasn.),* N.Am.-Eu.(USSR).——FIG. 55,*5*. *C.* sp., L.Dev.(Camden Chert), Tenn.; *5a,b,* ×4, ×15 (Fisher, n).——FIG. 59,*3*. **C. timanica,* Dev., (Frasn.), USSR; diagram. sec., ×45 (39, mod.).

Metastyliolina BOUČEK & PRANTL, 1961 (p. 386) [**M. striatissima*]. Like *Styliolina* but with pronounced pseudo-longitudinal groove and numerous delicate longitudinal striae. Embryonic bulb not pronounced or clearly separated from rest of shell. Shell wall thin. Growth angle 9° in juvenile portion, 3° in mature portion. *M.Dev.(Couvin.) (Daleje F.)* Eu.(Czech.).

Striatostyliolina BOUČEK & PRANTL, 1961 (p. 386) [**Styliola strialula* NOVAK, 1882]. Like *Styliolina,* having a prominent pseudo-longitudinal groove, but with prominent longitudinal striations and sharply set off relatively larger embryonic bulb. Shell wall thin. *M.Dev.(Couvin.) (Daleje F.),* Eu.(Czech.).

CALYPTOPTOMATIDS

Defined here as an independent molluscan class under the name Calyptoptomatida are the long-known group of hyolithids and their allies. Their taxonomic placement has always been doubtful.

Hyolithids were first described by EICHWALD (10) from the Ordovician of Estonia. He regarded them as Problematica. Later investigators have variously referred to them as worms, pteropods, cephalopods, or have assigned them to *incertae sedis.* The principal studies have been undertaken by BARRANDE, COBBOLD, HALL, NOVAK, HOLM, MATTHEW, WALCOTT and lately SYSSOIEV, who has brought renewed interest to a group that has not received monographic treatment since the time of HOLM (20).

Calyptoptomatids are wholly Paleozoic forms. HOLM reported 178 described species, distributed as follows: Cambrian 55, Ordovician 68, Silurian 22, Devonian 30, Carboniferous 3, Permian 1. SINCLAIR (62) reported 363 described species, distributed as follows: Cambrian 179, Ordovician 111, Silurian 29, Devonian 51, Carboniferous 7, Permian 3, Triassic 1, and one dubious form from the Miocene. The presumed Triassic species has since been shown to be Permian. In 1958, SYSSOIEV stated that more than 400 species had been described. Undoubtedly, hyolithids were evolving for millions of years prior to the Cambrian, for they were already well diversified in the early Cambrian. Thereafter, they gradually declined, becoming extinct in the Middle Permian. More advanced mollusks and trilobites, in addition to rapidly evolved nautiloid cephalopods and "shell-cracking" placoderms, crowded the hyolithids out of existence.

MORPHOLOGY

The calyptoptomatids have bilaterally symmetrical, conoid, calcium carbonate shells that taper to a closed pointed or rounded apex and are open at their widest portion, the aperture. Shells are usually subtrigonal in cross-section but may also be circular, oval, elliptical, lenticular, pentagonal, or trapezohedral, with intermediate variations. In length, they range from 1 to 150 mm. Their growth angle is 10 to 40 degrees. They have medium to fairly thick laminated walls. In subtriangular forms, the two smaller sides join along a line located in a plane of bilateral symmetry. Despite compression, this juncture line is slightly raised, implying the existence of a median septum that resisted flattening. This septum may not have extended completely from the dorsal to the ventral side. A. R. PALMER, of the U.S. Geological Survey, has shown me some silicified specimens from the Cambrian of Nevada which have a median septum. The embryonic portion is conical, cylindrical, or globular. Thin sections and polished surfaces disclose chambers which are separated by imperforate septa in the juvenile portion of shell. The exterior surface is smooth or ornamented by fine growth lines or by transverse or longitudinal ridges (Fig. 60). In rare specimens, projections or "fins" extend along lateral sides of the shell. Commonly, a shelf or lip extends from the ventral side of the aperture. A swollen apertural brim may also be present.

In very rare individuals a pair of sliver-like broadly curved "supports" or "fins" has been found associated with *Hyolithes tricarinatus* WALCOTT and *H. carinatus* MATTHEW (from the Middle Cambrian Burgess Shale of British Columbia), *H. hathewayi* MATTHEW (from the Lower Cambrian of North Wales), *H. magnificus* BULMAN (from the Tremadoc), and unidentified hyolithids from the Lower Cambrian Kinzers Formation of Pennsylvania and the Middle Cambrian at Conception Bay, Newfoundland. Considering the extreme delicacy of these structures, it is likely that many, if not all, hyolithids possessed them.

The aperture (mouth) usually was closed by an operculum, the exterior of which had an eccentric summit with concentric growth lines. The opercula are generally subtrigonal or subquadrate. A pair of shelves radiate outward from the eccentric summit. Paired muscle scars occur on the underside of the operculum.

PALEOECOLOGY

Hyolithids, most common in argillaceous rocks, are found in all types of marine sedimentary rocks except those of hypersaline, dolomitic, or reef origin. They are exceedingly rare in graywacke and micaceous siltstone. They are usually associated with trilobites, brachiopods, and primitive-type gastropods. Rare associates are corals and bryozoans. It is of interest that fossilized hyolithids exhibit two modes of occurrence: (1) fossils characterized by a slight rolling of the shells with the middle portion well preserved and the apex and aperture broken, such shells being commonly oriented, size-sorted, and many specimens found without opercula or supports; and (2) fossils characterized by completeness of hard parts (operculum and supports with complete shells), isolated specimens that exhibit no prevailing orientation or sorting by weight or size. The first condition is interpreted to signify shallow agitated water and the second, deeper quiet water.

No agreement on orientation of hyolithid shells can be reported. SALTER, MATTHEW, and WALCOTT considered the longer side dorsal, whereas HALL and BILLINGS considered it ventral. ZAZVORKA (1930) called the longer surface "posterior." SYSSOIEV regarded the longer surface ventral, the position that I judge to be most likely. Obviously, any consideration of orientation necessarily depends upon the animals' living habit. At least three modes of life have been suggested: vagrant benthonic, pelagic, and sedentary. No evidence for a burrowing habit has been recorded.

Judging by their different shell configurations, calyptoptomatids assumed correspondingly different living habits (Fig. 61). Most can readily be divided dorsally and ventrally, implying a vagrant life. Those with marked flat ventral sides were undoubtedly benthonic. Some of these are curved upward, whereas others are straight.

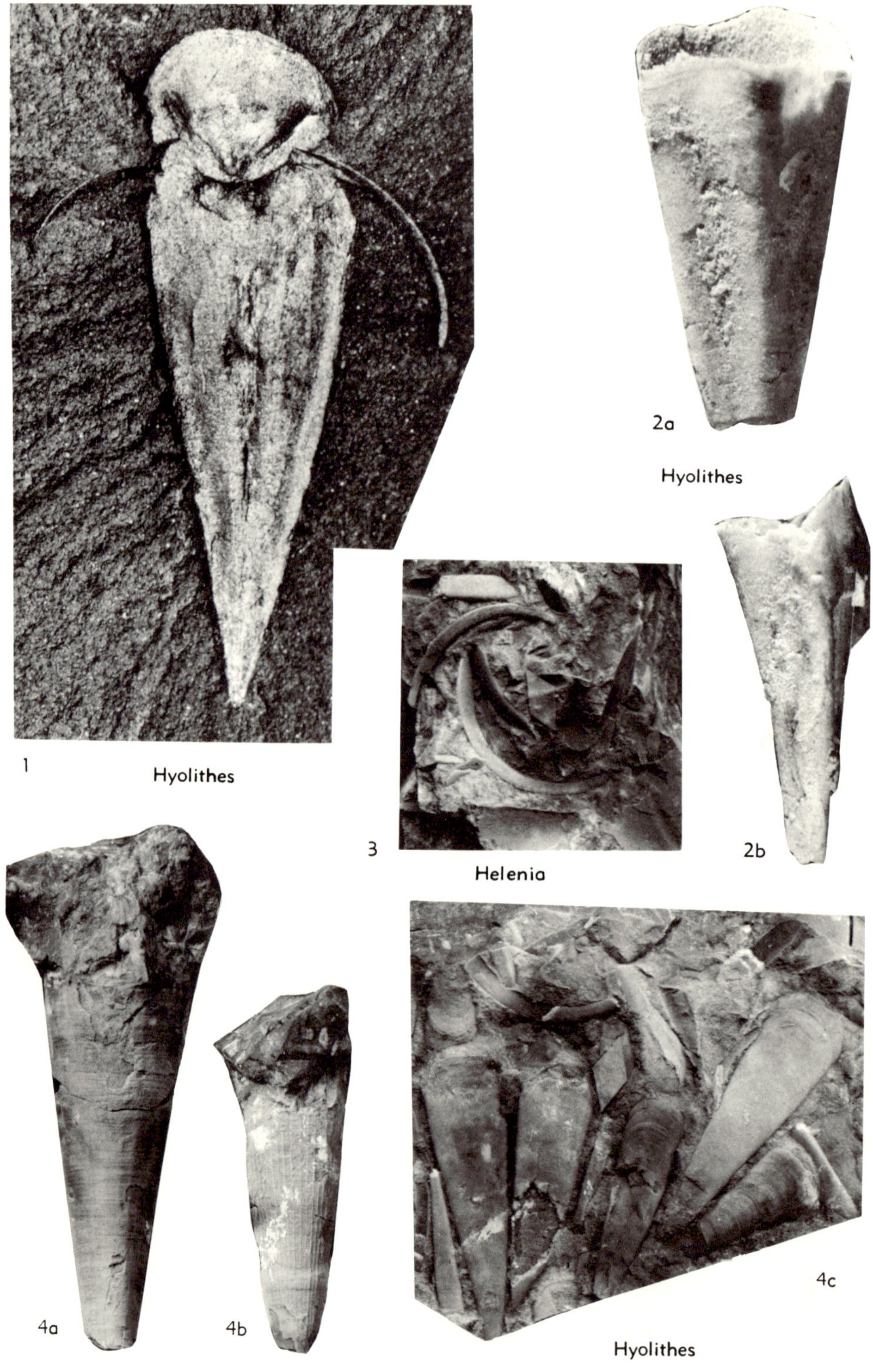

FIG. 60. Hyolithidae (p. W124) (Fisher, n).

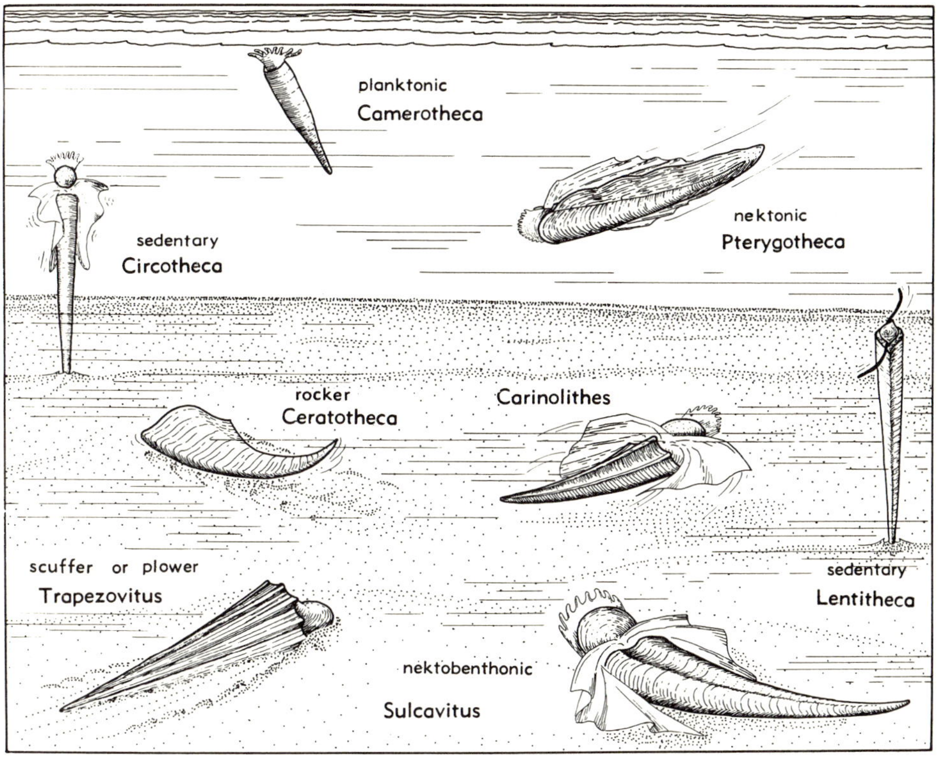

FIG. 61. Possible ecologic adaptations of hyolithids (Fisher, n).

The amount of curvature reflects the degree of bottom movement, the straight-shelled forms having greater surface friction to overcome and being less mobile. Strong longitudinal ribbing is highly developed on some straight hyolithids. Conceivably, this increased the durability of the shell in combating the effects of rough water and increased pressure. Many genera are imperfectly defined dorsally and ventrally. Such hyolithids with circular, elliptical, or quadrate cross sections may have been either pelagic or sedentary, with the apex thrust into the sea bottom. The thicker-shelled forms were assuredly incapable of a pelagic life unless they possessed a powerful locomotive mechanism, of which no evidence has been found.

Hyolithids with pronounced shelves or lips are interpreted as vagrant benthos. In addition to providing a lubricated platform for gliding of the animal into an extended feeding position, the shelf must simultaneously have inhibited influx of sediment into the living chamber. Calyptoptomatids without shelves are regarded as pelagic or upright sedentary forms. Conceivably, some may have maintained a pelagic habit throughout life, whereas others changed from a pelagic embryonic stage to a benthonic adult stage.

MATTHEW suggested that these organisms were sedentary, living with the point of the shell thrust into mud of the sea bottom. In contrast, RUEDEMANN (57) considered hyolithids as planktonic pteropods. While it is admissible that many hyolithids with circular or elliptical cross sections may have been sedentary, some possess several camerae, which would have increased buoyancy sufficiently to make probable a nektonic or planktonic existence. Heavy-shelled forms with shelves were unquestionably incapable of much movement and were quasi-sedentary or "scuffers" (e.g., *Trapezovitus*). Shelfless hyolithids, which cannot be differentiated dorsally and ventrally, probably were upright sedentary (e.g., Orthothecidae).

Some strongly curved shelfless hyolithids with convex ventral surfaces may have operated as "rockers." As in a rocking chair, these forms would experience limited mobility (e.g., *Ceratotheca*). Less curved, shelved hyolithids with convex ventral surfaces may likewise have accomplished some rocking action supplemental to their bottom swimming type of locomotion (e.g., *Carinolithes, Sulcavitus, Dorsolinevitus*). Among the other calyptoptomatids, camerothecids were assuredly pelagic. Of these, *Camerotheca,* with its large camerae and lack of top and bottom orientation, was probably planktonic. *Diplotheca,* with its unusually large dorsal cavity and chambered shell wall, possessed ample buoyancy to float and swim on the surface. The simple shell of *Globorilus* tells little, and therefore its living habit is vague. The supposed habitat of the anomalous *Matthevia* is worthy of special attention. Myriads of matthevinids occupy the flanking areas of *Cryptozoon* reefs in the Upper Cambrian Hoyt Limestone of New York State. Whether they represent a lag deposit or large numbers actually lived close to the reef base feeding on algal material and excreta from other reef dwellers cannot be stated convincingly. The latter condition is favored owing to the paucity of proof of mechanical abrasion on the shells. *Pterygotheca* and *Virgulaxonaria,* with their broad lateral extensions of the mantle, are visualized as nektonic dwellers of the open sea.

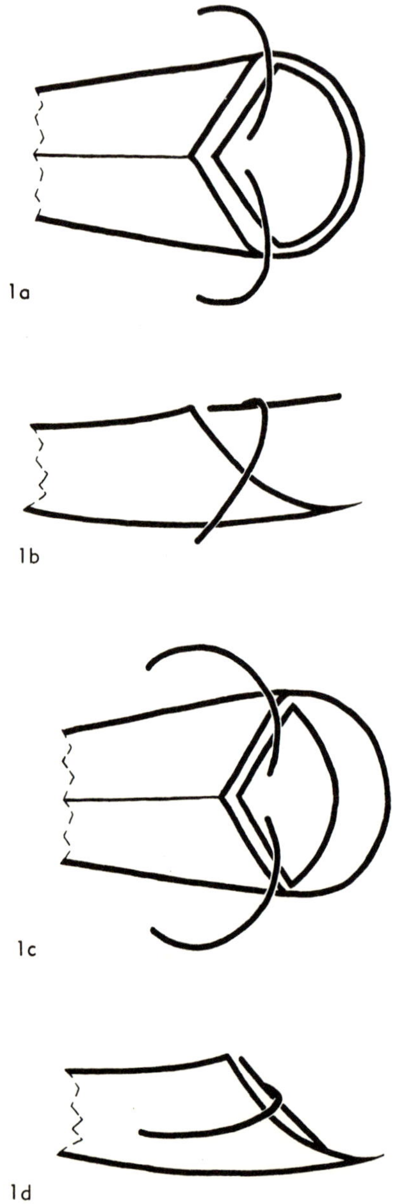

FIG. 62. Function of supports of hyolithids as interpreted by YOCHELSON; *1a,b,* shell with operculum open; *2a,b,* shell with operculum closed (Yochelson, mod.).

The purpose of the pair of "fins" or "supports" of calyptoptomatid organisms has long attracted attention. Some paleontologists have suggested that they were balancing structures for a hyolithid that presumably lived upright on its apex. HOWELL & STUBBLEFIELD (22) and YOCHELSON (personal communication) subscribe to this view, with the stipulation that it applies only to hyolithids which cannot be distinguished dorsally and ventrally. In this situation, supports would serve as a balancing mechanism. Others have suggested that they represent firm supports for a portion of the mantle usable for swimming. A novel proposal by YOCHELSON (76) interprets the supports as propping devices which held the operculum open while the animal was extended (Fig. 62). This hypothesis assumes that the supports were attached directly to the operculum, in which

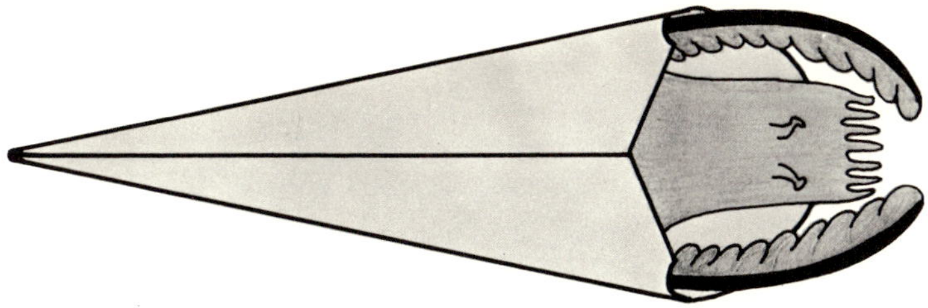

FIG. 63. Interpretation of hyolithid supports by SYSSOIEV (Syssoiev, mod.).

case they would be exposed and susceptible to breakage. But if the animal is extended, using the shelf as a gliding surface, there is no need for additional support to the operculum which already is held up by the animal! Furthermore, the role of an operculum, to afford protection to the vulnerable parts, would be negated if the supports were affixed to it. SYSSOIEV (65) has postulated that the supports were feeding arms which could be extended and withdrawn through narrow slits while the operculum remained closed or nearly closed (Fig. 63). It is noteworthy that many species have indentations at the apertural juncture of the dorsal and ventral sides which could have accommodated these "arms." Personally, I believe that the supports were retractable and am inclined to suppose that they served as stiff leading edges for "wings" which enabled the animal to move along the bottom like modern skates and rays (Figs. 64, 65).

CALYPTOPTOMATID CLASSIFICATION

For purposes of calyptoptomatid classification, the principal item of morphologic importance is shape of the shell's embryonic stage, either conical, globular, or cylindrical (Fig. 66). Accordingly, the class Calyp-

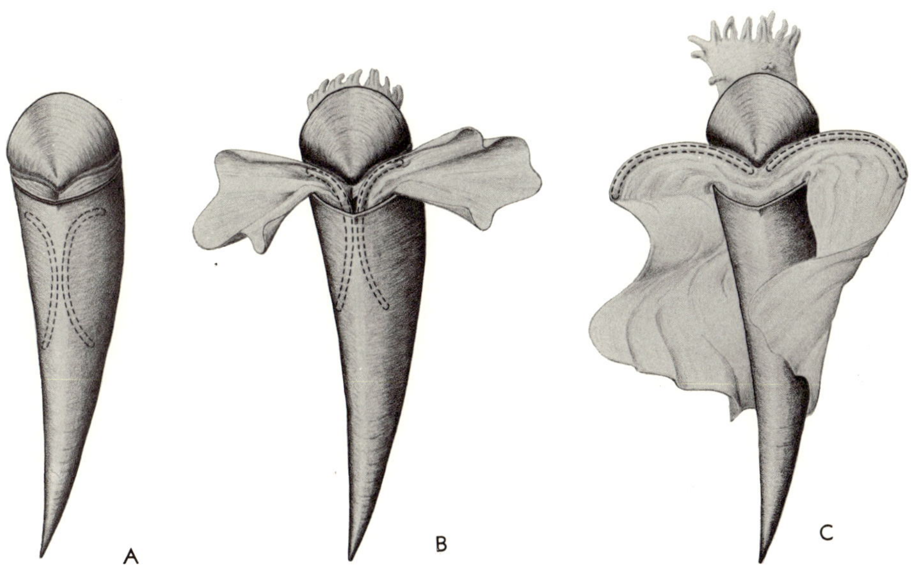

FIG. 64. Reconstruction of a hyolithid as it may have appeared in life; *A*, with operculum closed; *B*, with operculum open and animal partly extended; *C*, with operculum open and animal fully extended (supports indicated by broken lines) (Fisher, n).

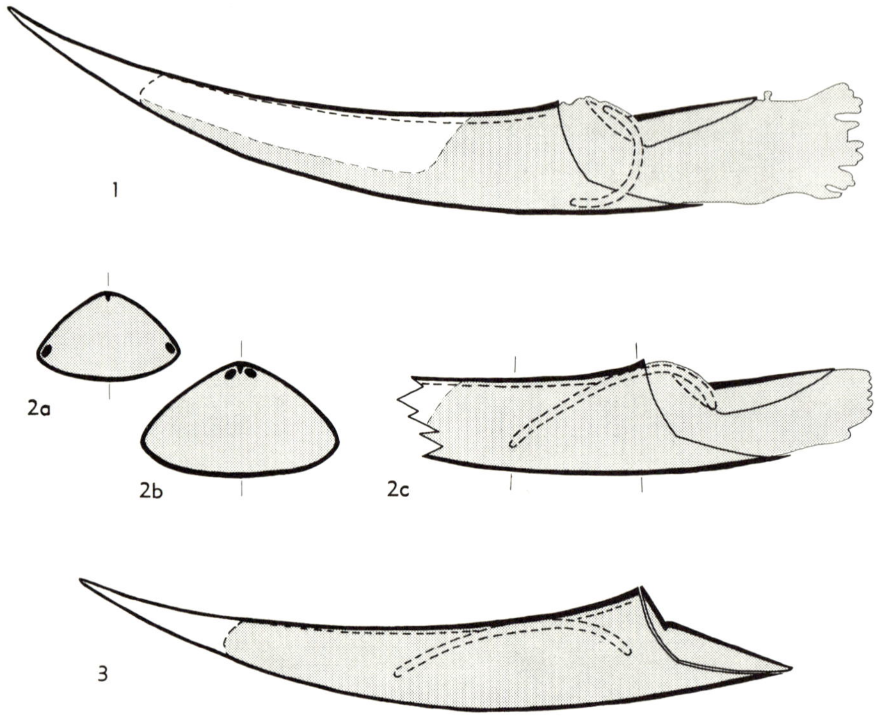

FIG. 65. Diagrams illustrating movement of hyolithid into shell, with indication of positions of operculum and its supports at different stages (Fisher, n).——*1*. Animal extended from shell.——*2a-c*. Animal partially drawn into shell, longitudinal and transverse sections.——*3*. Animal fully drawn into shell.

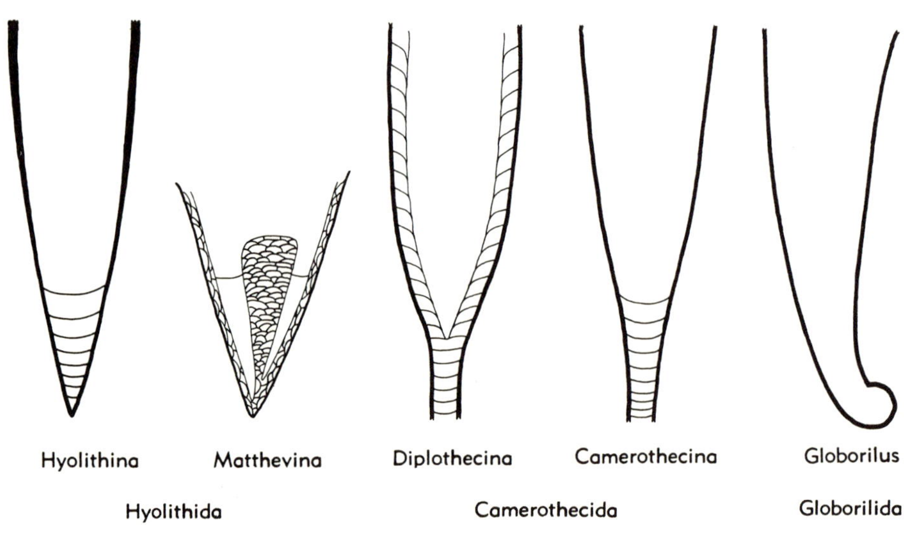

FIG. 66. Schematic longitudinal sections of major types of Calyptoptomatida (Fisher, n).

toptomatida is divided into three orders: Hyolithida (conical), Globorilida (globular), and Camerothecida (cylindrical). Further division of the first and last is possible, based on the nature of the embryonic portions. The Hyolithida are divisible into the suborders Hyolithina, with a multicamerate juvenile portion, and Matthevina, with a uni- or bicamerate juvenile portion. The Camerothecida are divisible into the suborders Camerothecina, with a noncamerate wall, and the Diplothecina, with a camerate wall. Family distinction depends on the character of the aperture. Generic and specific differentiation is based on shape of the shell cross section, kind and abundance of ornamentation, minor details of camerae, curvature, and size.

Class CALYPTOPTOMATIDA Fisher, n. class

[ety., *kalypta*, lid; *ptomatis*, cup that must be emptied at once because it will not stand upright] [=Superorder Hyolithoidea SYSSOIEV, 1957 *(partim)*]

Hyolithids and their allies herein are regarded as sufficiently unique to escape from the encompassing label of *incertae sedis*. A distinct grouping of these fossils is warranted with supposition that they constitute an extinct class of the phylum Mollusca. *L.Cam.-M.Perm.*

Order HYOLITHIDA Matthew, 1899

[*emend.* SYSSOIEV, 1957] [=Superfamily Hyolithacea TERMIER & TERMIER, 1950 *(partim)*]

Bilaterally symmetrical, pyramidal shells with conical embryonic chamber not differentiated outwardly or separated internally from remainder of shell. Operculum with one or two pairs of bilaterally symmetrical muscle scars. *L.Cam.-M.Perm.*

Suborder HYOLITHINA Matthew, 1899

[*nom. transl.* FISHER, herein]

Hyolithida with embryonic portion of interior not sharply separable from mature portion. More than one (usually 4 to 6) embryonic chambers, each extending across complete section of shell. *L.Cam.-M.Perm.*

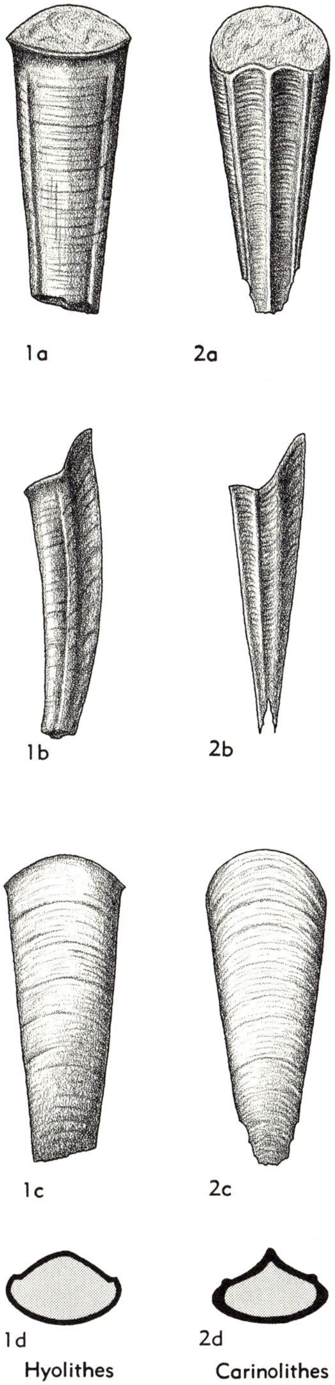

FIG. 67. Hyolithidae (p. W124) (Fisher, n).——*1. Hyolithes; a-d,* dorsal, lateral, ventral views, transverse section, ×0.5.——*2. Carinolithes; a-d,* dorsal, lateral, ventral views, transverse section, ×2.75.

Family HYOLITHIDAE Nicholson, 1872

Aperture with a strongly projecting lipped ventral side. *L.Cam.-M.Perm.*

Hyolithes Eichwald, 1840 (p. 97) [**H. acutus*] [=*Hyolithus* Hermannsen, 1847 *(nom. null.)*; *Hyolithis* Laseron, 1910*(nom. null.)*; *Orthoceras* Münster, 1840; *Theca* Sowerby, 1845; *Pugiunculus* Barrande, 1847; *Vaginella* d'Orbigny, 1850; *Cleodora* Ludwig, 1864; *Centrotheca* Salter, 1866; *Cleidotheca* Salter, 1866]. Cross section oval, subtriangular, trapezoidal, or subpentagonal. Exterior with growth lines only, no ribs. Dorsal side rounded, ventral side broad. Operculum subcircular to subquadrate to subtrigonal. Usually one pair of large muscle scars on underside of operculum. Rarely with 2 curved "arms" or "supports" found unattached to other hard parts. [Many species now assigned to *Hyolithes* probably belong to other hyolithid genera.] *L. Cam. - M. Perm.,* N. Am.-S. Am.-Eu.-Asia-Afr.-Austral.——Fig. 60,*1*. *H. carinatus* Matthew, M.Cam.(Burgess Sh.), Can.(B.C.); ventral side of specimen showing operculum and supports, ×4 (Fisher, n).——Fig. 60,*2*. *H.* sp., L.Perm. (Leonard, Bone Springs Ls.), W.Tex.; *2a,b,* dorsal and lateral views of silicified specimen, ×4 (Fisher, n).——Fig. 60,*4*. *H. terranovicus* Walcott, L.Cam., Manuels Brook, Newf.; *4a-c,* freed specimens and as found in matrix, ×1 (*4b* may represent a different hyolithid genus) (Fisher, n) [Also, compare Fig. 60,*3*, *Helenia,*].——Fig. 67,*1*. **H. acutus,* Ord., Est.; *1a-d,* dorsal, lateral, ventral views, transv. sec., ×0.5 (20).

Carinolithes Syssoiev, 1958 (p. 188) [**Hyolithes pennatulus* Holm, 1893]. Cross section subpentagonal to angulate oval, with broad ventral side; dorsum with 3 longitudinal keels, central one raised above flanking keels. Growth lines convex toward aperture on venter and interrupted by keels on dorsum so that 2 rows of concentric arcs occur between keels, which are convex toward aperture. [This and following genera attributed to Syssoiev, 1958 are given in the Russian *Fundamentals of Paleontology* (1958) as Syssoiev, 1957; I was unable to find a 1957 article by this author in which these generic names were proposed.] *M.Cam.-M.Ord.,* Swed.——Fig. 67,*2*. *C. pennatulus* (Holm), M.Cam.; *2a-d,* dorsal, lateral, ventral views, transv. sec., ×2.75 (20).

Helenia Walcott, 1889 (p. 39) [**H. bella*]. Elongate, narrow, flattened, curved tube, degree of curvature increasing toward closed end; cross section elongate-elliptical. Surface marked by irregular, transverse or concentric imbricating lines that vary in number and size. [I believe that fossils identified as *Helenia* are the supports of a relatively large hyolithid, *Hyolithes princeps,* with which it is associated.] *L.Cam.,* Newf.-Eng.——Fig. 60,*3*. **H. bella,* Manuels Brook, Newf.; fossils interpreted as "supports" of *Hyolithes princeps* or *H. terranovicus,* ×1 (Fisher, n).

Family CERATOTHECIDAE Fisher, n. fam.

Hyolithids with third of apical portion strongly curved obliquely upward and sideward, without lips or shelves; growth angle apparently increasing logarithmically. Aperture very wide compared to size of shell. *U.Sil.-L.Dev.*

Ceratotheca Novak, 1891 (p. 29) [**Hyolithes aduncus* Barrande, 1867]. Shell flattened, apical portion curved to right when viewed from above and with aperture oriented forward. Cross section broadly subtriangular, with wide convex ventral side. No operculum known. Two depressed furrows along median line of dorsum, 4 depressed furrows along median line of ventrum. Transverse striae on exterior. *U.Sil.-L.Dev.,* Eu.(Czech.-Eng.).

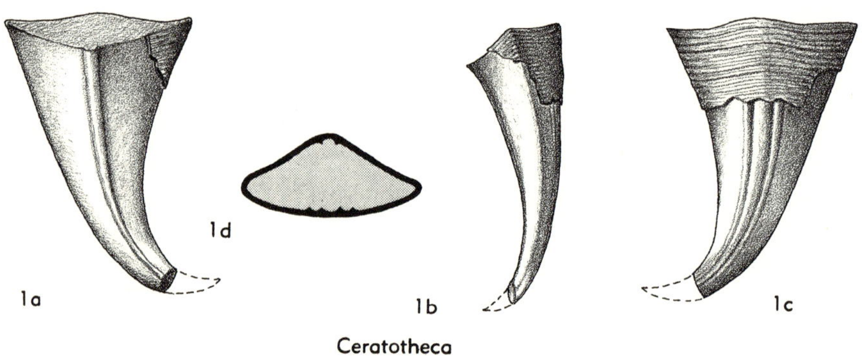

Fig. 68. Ceratothecidae (p. *W*124) (Fisher, n).——*1. Ceratotheca; a-d,* dorsal, lateral, ventral views, transverse section, ×2 (53).

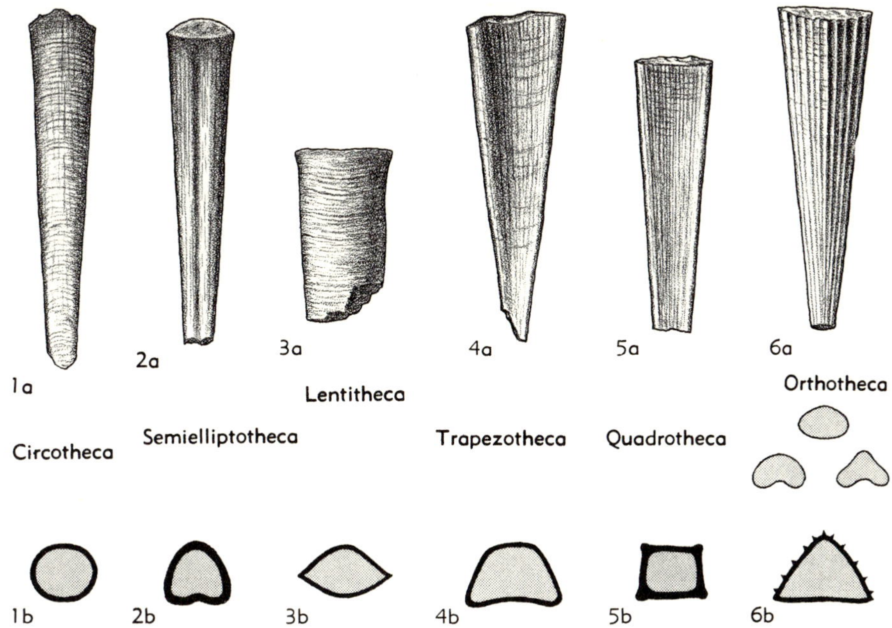

FIG. 69. Orthothecidae, lateral views and transverse sections (p. *W*125-*W*127) (Fisher, n).

——FIG. 68. **C. adunca* (BARRANDE), U.Sil. (Ludlov.), Eng.; *1a-d,* dorsal, lateral, ventral views, transv. sec., ×2 (53).

Family ORTHOTHECIDAE Syssoiev, 1958

Orthoconic hyolithids without noticeable lips or shelves. Aperture nearly at right angles to shell axis. *L.Cam.-M.Dev.*

Orthotheca NOVAK, 1886 (p. 36) [**O. intermedia*]. Cross section circular, subelliptical, subtrigonal, kidney- or heart-shaped. Aperture at right angles to shell axis. *L.Cam.-M.Dev.,* N.Am.-S.Am.(Bol.)-Eu.(Eng.-Swed.-Czech.-USSR)-Asia(China). —— FIG. 69,*6*. **O. intermedia,* Dev., Czech.; *6a,b,* lateral view, transv. secs., ×3 (53).

Bactrotheca NOVAK, 1891 (p. 34) [**Hyolithes teres* BARRANDE, 1867]. No apertural brim on dorsum. Shell thick, straight, elongate; cross section trapezohedral with rounded edges; dorsum with longitudinal and transverse ridges. Operculum thick-shelled, gently convex, quadrate in outline, with top parallel to bottom, sides convex, underside concave, with 3 triangular elevations which diverge from apex, center one most prominent, radial and concentric striae on operculum. Differs from *Orthotheca* in that ornamentation is on dorsum only and operculum has different shape and interior. *L.Ord.,* Czech.

Circotheca SYSSOIEV, 1958 (p. 187) [**Hyolithes stylus* HOLM, 1893]. Cross section circular or slightly elliptical. Aperture almost at right angles to shell axis. *L.Cam.,* N.Am.-Eu.(Eng.-Denm.); *L.Cam.-M.Cam.,* Swed.-USSR.——FIG. 69,*1*. **C. stylus* (HOLM), M.Cam., Swed.; *1a,b,* lateral view, transv. sec., ×1.8 (20).

Cryptocaris BARRANDE, 1872 (p. 459) [**C. suavis*]. May be operculum of *Orthotheca intermedia* NOVAK (*fide* NOVAK, 1886).

Lentitheca SYSSOIEV, 1958 (p. 187) [**Hyolithes peracutus* HOLM, 1893]. Cross section biconvex lens-shaped. Transverse parallel lines equidistant, resembling sutures of cephalopods. *U.Ord.-Sil.,* Eu.(Czech. - Norway - Swed.).——FIG. 69,*3*. **L. peracuta* (HOLM), Sil., Swed.; *3a,b,* lateral view, transv. sec., ×0.8 (20).

Quadrotheca SYSSOIEV, 1958 (p. 187) [**Hyolithes quadrangularis* HOLM, 1893]. Cross section square or trapezohedral, with 4 low thick longitudinal keels at corners (in this respect differing from *Bactrotheca*); sides flat or very slightly concave. Longitudinal striae on surface with cancellate pattern toward aperture. *L.Cam.,* N.Am.-Eu. (USSR); *L.Ord.,* Eu.(Swed.).——FIG. 69,*5*. **Q. quadrangularis* (HOLM), L.Ord., Swed.; *5a,b,* lateral view, transv. sec., ×3 (20).

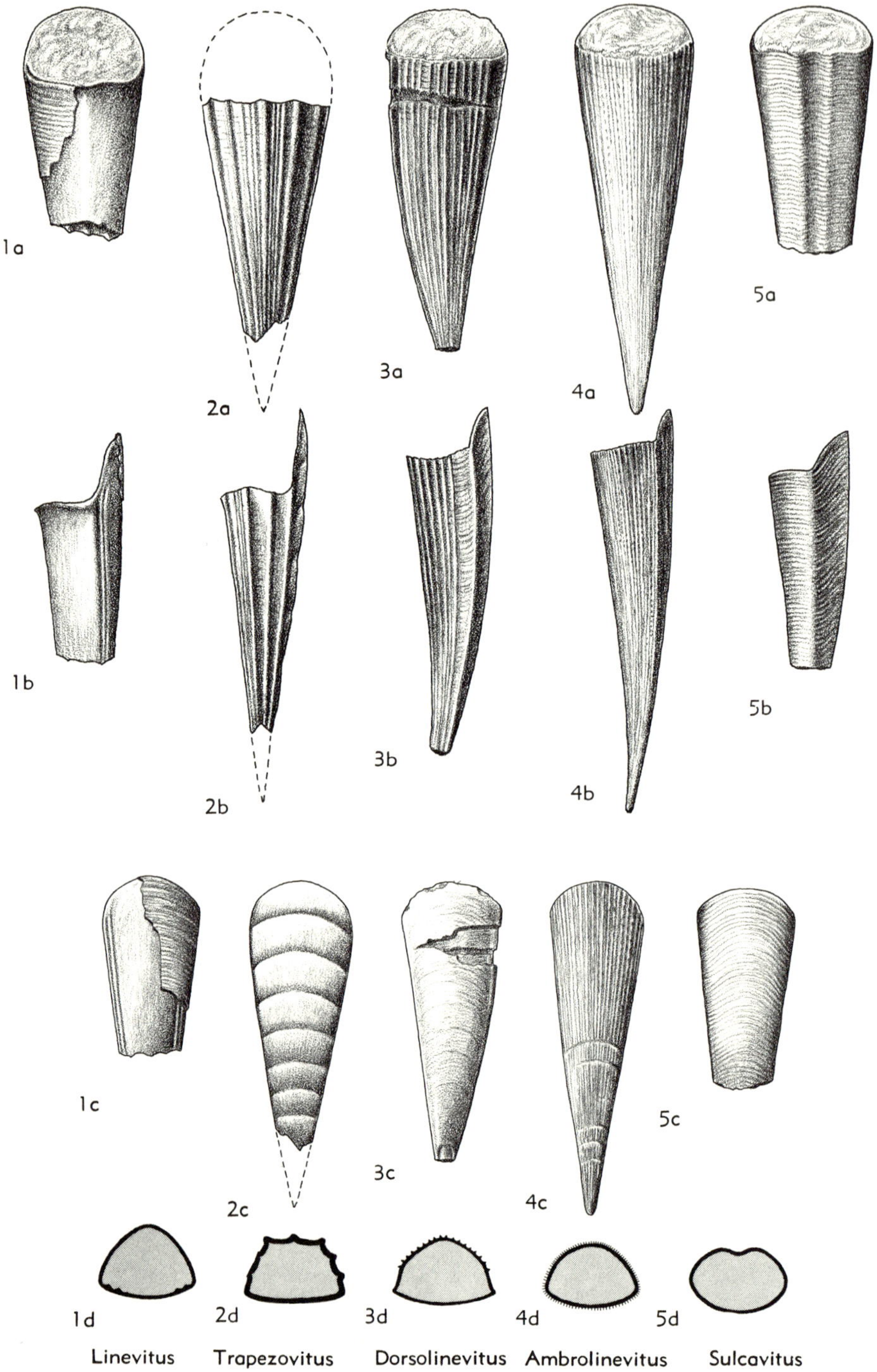

FIG. 70. Sulcavitidae (in order from top downward), dorsal, lateral, ventral views, transverse sections (p. *W*127) (Fisher, n).

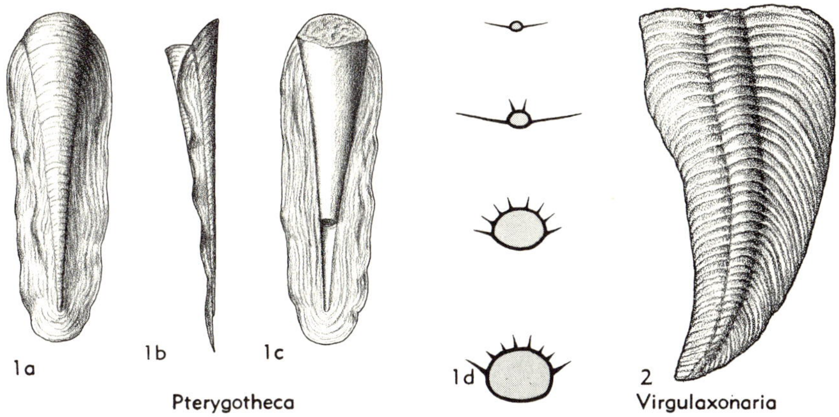

FIG. 71. Pterygothecidae (p. *W*127-*W*128) (Fisher, n).——*1. Pterygotheca; a-d,* dorsal, lateral, ventral views, transverse sections, ×1.5.——*2. Virgulaxonaria,* ?dorsal view, ×5.5.

Semielliptotheca SYSSOIEV, 1958 (p. 187) [**Hyolithes rosmarus* HOLM, 1893]. Cross section subtriangular to heart-shaped, ventrum slightly concave. Aperture at considerable angle to shell axis. *L.Ord.-M.Ord.,* Eu.(Swed.).——FIG. 69, *2.* **S. rosmarus* (HOLM); *2a,b,* lateral view, transv. sec., ×0.9 (20).

Trapezotheca SYSSOIEV, 1958 (p. 187) [**Hyolithes aemulus* HOLM, 1893]. Cross section trapezoidal with flat or slightly concave venter. Aperture almost at right angles to shell axis. *U.Cam.-L.Ord.,* N.Am.-Eu.(Swed.).——FIG. 69,*4.* **T. aemula* (HOLM), L.Ord., Swed.; *4a,b,* lateral view, transv. sec., ×1.6 (20).

Family SULCAVITIDAE Syssoiev, 1958

Apertural brim with strongly protruding lips, at base of which are small cuts along shell edge. *L.Cam.-Sil.*

Sulcavitus SYSSOIEV, 1958 (p. 188) [**Hyolithes caelatus* HOLM, 1893]. Cross section oval. Growth lines completely around shell, changing direction within dorsal concave depression along center occupying about third of shell width. *L.Ord.* Eu. (Czech.-Swed.).——FIG. 70,*5.* **S. caelatus* (HOLM), Swed.; *5a-d,* dorsal, lateral, ventral views, transv. sec., ×3 (20).

Ambrolinevitus SYSSOIEV, 1958 (p. 188) [**Hyolithes striatellus* HOLM, 1893]. Cross section triangular; venter and dorsum with longitudinal ribs. *L.Ord.-U.Ord.,* Eu.(Swed.).——FIG. 70,*4.* **A. striatellus* (HOLM), *4a-d,* dorsal, lateral, ventral views, transv. sec., ×0.85 (20).

Dorsolinevitus SYSSOIEV, 1958 (p. 188) [**Hyolithes dispar* HOLM, 1893]. Cross section like biconvex lens or subtriangular, with wide venter which bears growth lines convex toward aperture. Dorsum has longitudinal ribs in addition to growth lines. *L.Ord.-M.Ord.,* Eu.(Swed.).——FIG. 70,*3.* **D. dispar* (HOLM), *3a-d,* dorsal, lateral, ventral views, transv. sec., ×0.7 (20).

Linevitus SYSSOIEV, 1958 (p. 188) [**Hyolithes obscurus* HOLM, 1893]. Cross section subelliptical or subtriangular, with channel or groove on edges of venter paralleling long side of shell. Growth lines on all sides. *L.Cam.,* N.Am.-Eu.(Eng.); *L. Cam.-Sil.,* Eu.(Norway-Swed.-USSR).——FIG. 70, *1.* **L. obscurus* (HOLM), M.Cam., Swed.; *1a-d,* dorsal, lateral, ventral views, transv. sec., ×2.7 (20).

Trapezovitus SYSSOIEV, 1958 (p. 188) [**T. sinscus*]. Cross section trapezoidal. Venter with growth lines or smooth, gently sloping folds parallel to ventral portion of aperture. Dorsum has pronounced longitudinal ribs with intervening longitudinal striae. *L.Cam.,* Eu.(USSR).——FIG. 70,*2.* **T. sinscus; 2a-d,* dorsal, lateral, ventral views, transv. sec., ×6.5 (40).

Family PTERYGOTHECIDAE Syssoiev, 1958

Aperture with ventral lips bearing small notches along edges at base of lips. Dorsum has 1 to 4 pairs of "fins," one pair invariably along horizontal plane of shell (at juncture of dorsum and venter).

Pterygotheca NOVAK, 1891 (p. 45) [**P. barrandei*]. Cross section circular to suboval. Dorsum with few pairs of fins; their surfaces covered with growth lines. *Dev.,* Eu.(Czech.).——FIG. 71,*1.* **P. barrandei; 1a-d,* dorsal, lateral, ventral views, transv. secs., ×1.5 (53).

Virgulaxonaria Yin, 1937 (p. 290) [**V. elegans*]. Tapering conical shell about 8.5 mm. long, pointed at apical end and bent regularly in one plane. Surface ornamented with median rounded ridge and transverse lines; ridge occupying entire length of fossil; 2 flanking "wings" and central ridge entirely covered by fine growth lines, those on "wings," bending sharply toward apical side and in passing through lateral grooves on either side of central ridge showing slight curvature convex toward apical end. [Though specimens are flattened, they suggest comparison with the "wing"-bearing hyolithids (Pterygothecidae).] *L. Ord.*, China.——Fig. 71,*2*. **V. elegans; ?*dorsal surface, ×5.5 (Fisher, n).

Suborder MATTHEVINA Fisher, n. suborder

Bilaterally symmetrical, broadly conical shells with apical angle of 35 to 40 degrees; embryonic portion conical. Relatively thick-walled, with peculiar vesicular nature. Thickened imperforate median septum extending from apex toward aperture producing a double camerate juvenile portion. Oval to subquadrate cross section. Suboval to subquadrate operculum has concentric striae around an eccentric nucleus; radial creases may be present. Shell and operculum composed of calcium carbonate. *U.Cam.*

Family MATTHEVIIDAE Walcott, 1886

Characteristics of the suborder. Thickened median septum extending two-thirds length of shell, thickest at apertural end; single thin transverse septum dividing inner cavity into a large apertural cavity and 2 narrow apical chambers. *U.Cam.*

Matthevia Walcott, 1885 (p. 17) [**M. variabilis*]. Characteristics of the family. Inner transverse septum concave toward apertural chamber. Surface marked by undulating growth lines parallel to apertural margin. Fine papillae arranged in lines that cross at right angles in some specimens, whereas others show parallel papillae. Interior surface covered by a network of inosculating lines. *U.Cam.*, N.Am.(N.Y.-Que.)-Eu.——Fig. 72,*1*. **M. variabilis*, N.Y.; *1a,b*, incomplete shell, lateral views, ×3; *1c*, exterior of imperfect operculum, ×3; *1d*, dorsal view of shell, ×3; *1e*, interior surface of shell, ×6; *1f*, transv. sec. of shell showing 2 chambers and vesicular wall structure, ×3; *1g*, section of wall, ×6 (Fisher, n).

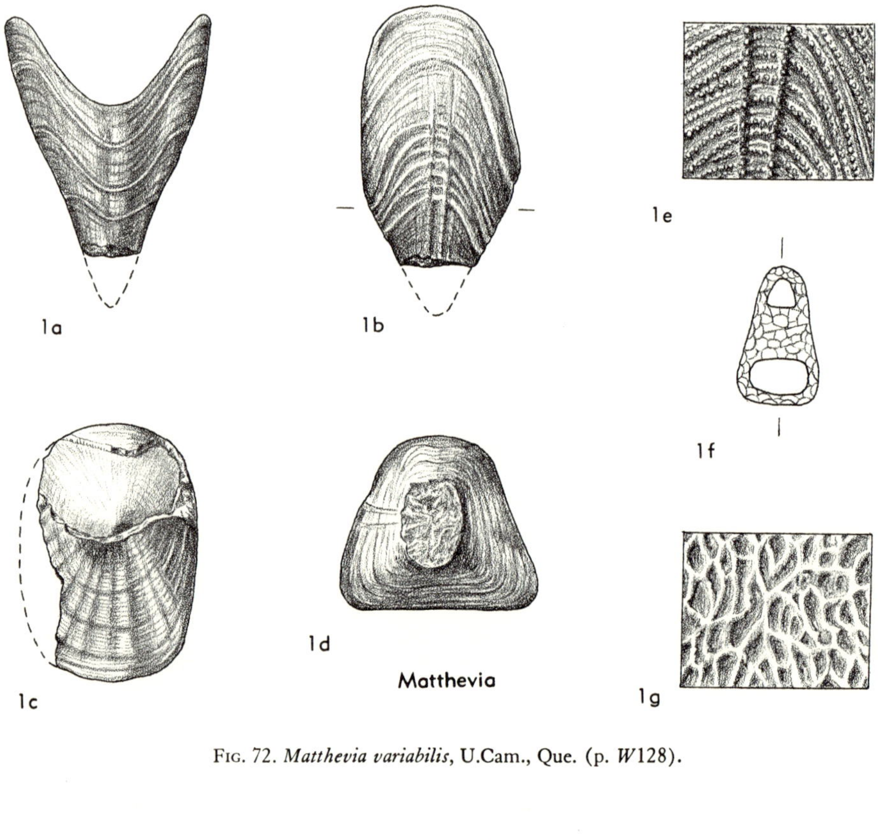

Fig. 72. *Matthevia variabilis*, U.Cam., Que. (p. *W*128).

Order GLOBORILIDA Syssoiev, 1957

Bilaterally symmetrical curved shell with curvature greater toward apex. Cross section circular to subtriangular; embryonic chamber globular. Uniformly curved very low conical operculum with subcircular to subquadrate outline. No external ornamentation visible. Interior unknown. *M.Cam.*

Family GLOBORILIDAE Syssoiev, 1958

Characteristics of the order. *M.Cam.*

Globorilus Syssoiev, 1958 (p. 189) [**Hyolithes globiger* Saito, 1936]. Characteristics of the family. *M.Cam.*, Korea.——Fig. 73,*3*. **G. globiger* (Saito); *3a,b*, lateral views of shell, ×2.4; *3c,d*, opercula, ×2.4 (40).

Order CAMEROTHECIDA Syssoiev, 1957

[*emend.* Fisher, herein]

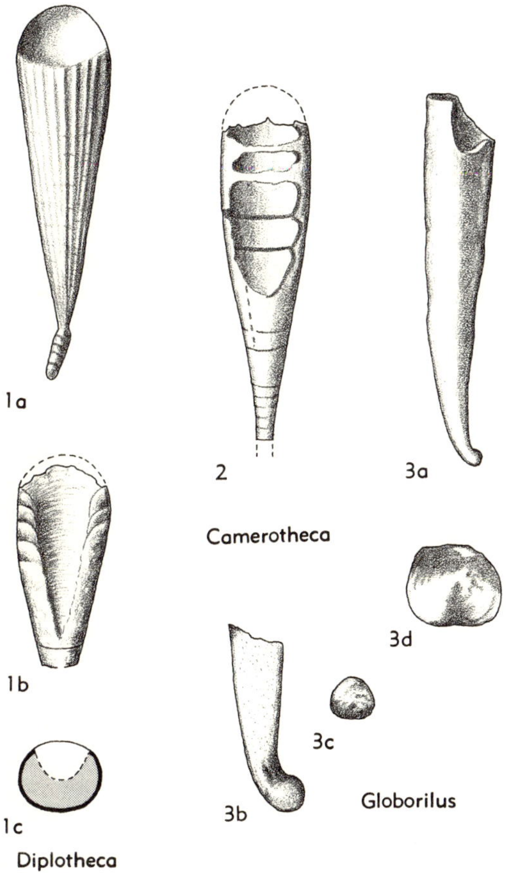

Fig. 73. Globorilida and Camerothecida (p. *W*129-*W*130) (40).

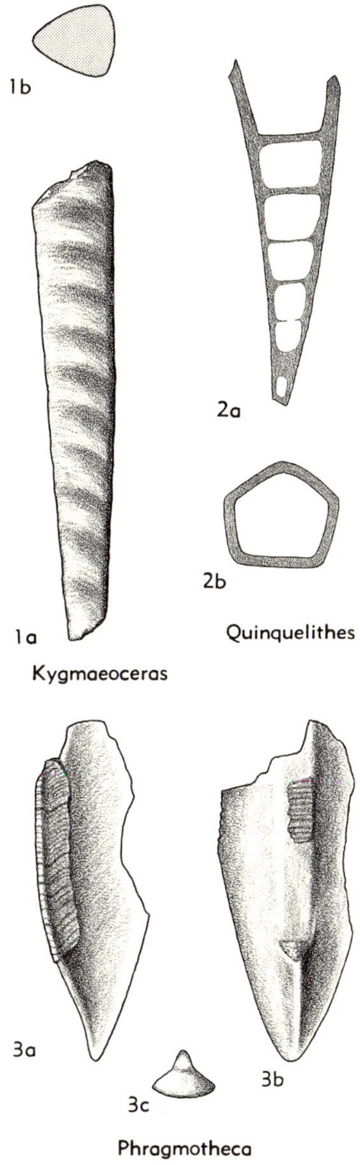

Fig. 74. Calyptoptomatida—Order and Family Uncertain (p. *W*130).

Bilaterally symmetrical shells with oval cross section; tubular embryonic stage without any appreciable angle of divergence. Small chambers separated by imperforate partitions in tubular portion, which progressively increases in size in mature stages. Angle of divergence increasing in adult stages. Side of shell nearly parallel near aperture. No apparent connection between chambers. Operculum unknown. *Cam.*

Suborder CAMEROTHECINA Fisher, n. suborder

Relatively thin, noncamerate wall. Camerae of body cavity large. *Cam.*

Family CAMEROTHECIDAE Syssoiev, 1958

Aperture incomplete, apparently at right angles to shell axis. *Cam.*

Camerotheca MATTHEW, 1885 (p. 149) [**C. gracilis*]. *Cam.*, N.Am.——FIG. 73,*2*. **C. gracilis*, L. Cam., Can.; lateral view of shell, ×1 (40).

Suborder DIPLOTHECINA, Syssoiev, 1957

[*nom. transl.* FISHER, herein (*ex* order Diplothecida SYSSOIEV, 1958)]

Bilaterally symmetrical shell with tubular embryonic stages. Relatively thick wall has transverse partitions dividing wall and tubular portion into small chambers. These partitions are usually curved, convex toward aperture. Longitudinal ribs on shell exterior. Aperture with lips. Operculum unknown. *Cam.*

Family DIPLOTHECIDAE Syssoiev, 1958

Characteristics of the order. *Cam.*

Diplotheca MATTHEW, 1885 (p. 149) [**D. acadica*]. Cross section subcircular to oval. Large central cavity occupies three-fourths volume of shell. *Cam.*, N.Am.(Can.).——FIG. 73,*1*. **D. acadica*, L.Cam.; *1a-c*, lateral and dorsal views, transv. sec., ×3.4 (40).

Order, Suborder, and Family UNCERTAIN

Kygmaeoceras FLOWER, 1954 (p. 31) [**K. perplexum*]. Straight, very slowly expanding shell with triangular cross section in form of narrow high isosceles triangle; slightly convex sides. Lateral surfaces with costae that slope toward apex as they approach flat (?ventral) base of triangle but disappear and do not cross it; costae continuous, though faint, over narrow dorsum in adult but not visible on dorsum of young stages; no trace of septa or other internal structure. Length about 26 mm., increasing from height of 5 mm. and width of 4 mm. to 6.5 mm. and 5 mm. respectively, at apertural end. [Genus may belong to the Orthothecidae.] *U.Cam.(Trempealeau)*, USA(Tex.-Nev.).——FIG. 74,*1*. **K. perplexum*, U.Cam., Tex.; *1a,b*, lateral view, transv. sec., ×2 (Flower, 1954, mod.).

Pharetrella HALL, 1888 (p. 7) [**P. tenebrosa*]. Cross section not known but seemingly like *Hyolithes* in lateral outline. Ornamentation consisting of imbricating transverse undulating striae. [May be synonymous with *Lentitheca*.] *U.Dev. (Geneseo)*, USA(N.Y.).

Phragmotheca BARRANDE, 1867 (p. 105) [**P. bohemica*]. Cross section triangular with acute middle keel on dorsal side. Poorly known. [May be a chiton.] *Sil.*, Czech.——FIG. 74,*3*. **P. bohemica*; *3a,b*, lateral and dorsal views, ×2; *3c*, transv. profile near tip, ×3 (1).

Quinquelithes SYSSOIEV, 1958 (p. 188) [**Q. pavonaceus*]. Cross section pentagonal, almost equilateral. Venter comparatively narrow. Relatively large camerae. [Systematic position of this genus unclear except that it is a calyptoptomatid.] *L. Cam.*, Eu.(USSR).——FIG. 74,*2*. **Q. pavonaceus*; *2a,b*, long. and transv. secs., ×5.5 (40).

OTHER SMALL CONOIDAL SHELLS

Phylum and Class UNCERTAIN

Order HYOLITHELMINTHES Fisher, n. order

[=Hyolithellida SYSSOIEV, 1957 (*partim*) *emend.*]

MORPHOLOGY

Small (length usually 5 to 15 mm.), almost cylindrical, brownish-black narrow, conical tubes (growth angle in adult region, 1 to 4 degrees); curved and irregular near closed apex but straightening toward aperture. Cross section circular to elliptical. Exterior surface of some fossils covered by small transverse ridgelets and striae; shell laminated, usually thick, with smooth interior surface. Body cavity elongate, devoid of septa or any other structure. Circular to subcircular operculum composed of many laminae, with eccentrically situated apex surrounded by closely spaced concentric growth lines; adult opercula thicker than translucent juvenile ones; on undersurface of operculum, paired muscle scars originate from point opposite apex and show bilateral symmetry, implying that soft parts of the animal likewise displayed bilateral symmetry. Tube and operculum composed of calcium-orthophosphate [$Ca_3(PO_4)_2$]; X-

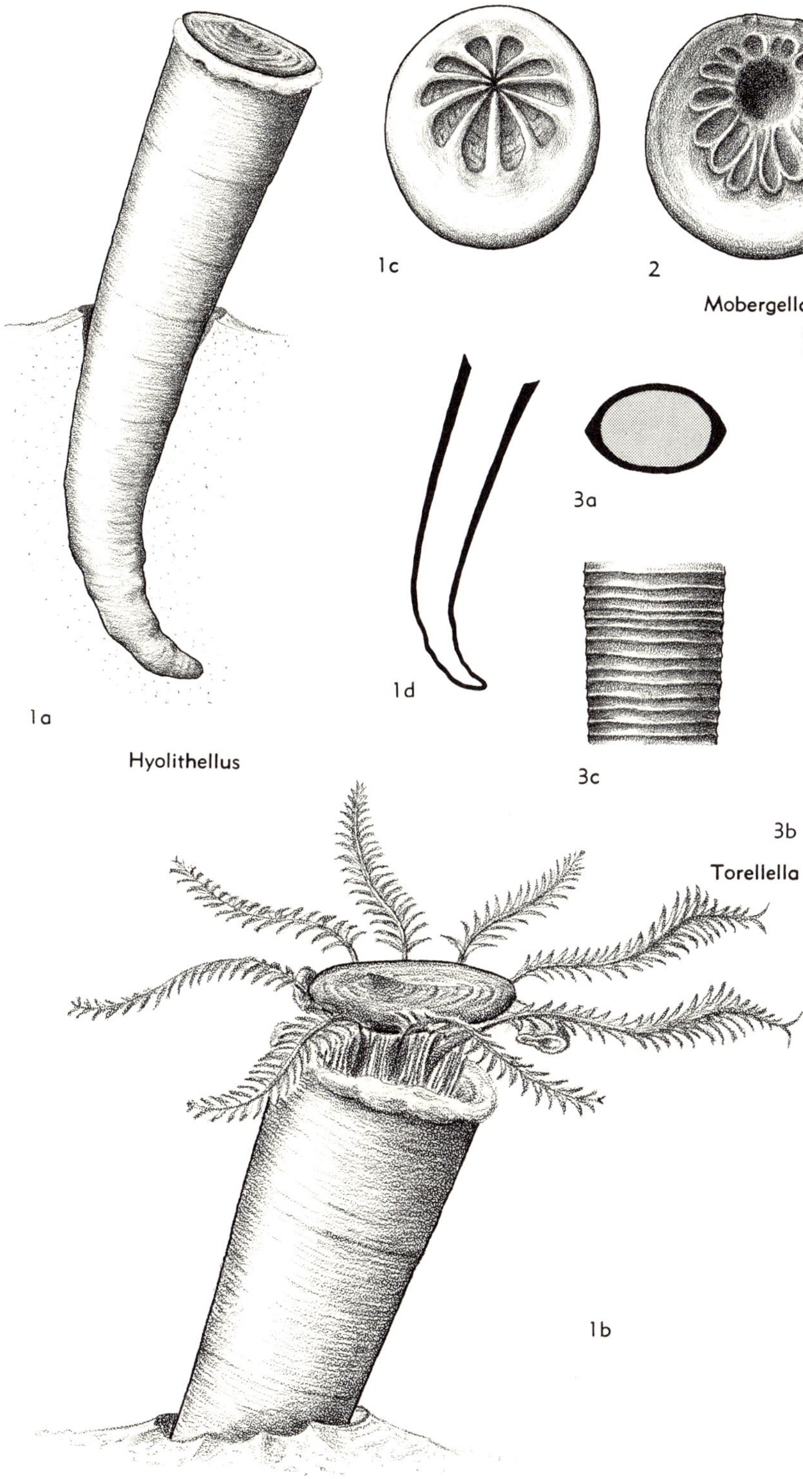

Fig. 75. Hyolithellidae (p. *W*132) (Fisher, n).

ray diffraction pictures serve to identify the mineral as fluorapatite.

PALEOECOLOGY

Hyolithelminths occur solely in marine rocks, including all types of limestones, black, gray or green calcareous or non-calcareous silty or nonsilty shales, less commonly sandstones or siltstones unless they are argillaceous; they are exceedingly rare in dolomites and graywackes. They are randomly arranged, with tubes always broken but opercula are complete and found parallel to bedding. Invariably, the curved and cylindrical portions occur separately, suggesting undue stress on the shell at this point. I suggest that hyolithelminths lived upright-sedentary lives with the cylindrical tube portion above sea bottom and the curved apical portion implanted on the soft bottom (Fig. 75,*1a,b*). The animal may have been able to expel itself partly and momentarily so as to ingest a larger amount of passing food. The eccentric arrangement of the muscle scars indicates that the operculum was susceptible of differential opening, which would have permitted access of the animal.

In Early Cambrian strata, where trilobites are wanting, poor, or nondiagnostic, etching with acetic or formic acids usually yields identifiable hyolithelminth opercula or shells, study of which in future promises to yield useful information on exact age relations. *Hyolithellus,* in particular, is a ubiquitous and prolific fossil in most Lower Cambrian faunas where diversity and multiplicity are unusual. The extensive geographic range of this form may be attributable to compounding its broad environmental tolerance in adult life with probable planktonic larval existence. About 35 species have been described. *L.Cam.-Ord.*

CLASSIFICATION

Hyolithellus and its allies have customarily been placed with the Annelida or some other worm phylum. SYSSOIEV (64, 65), however, classed *Hyolithellus, Coleoloides,* and *Coleolus* in the Hyolithellidae in the order Hyolithellida, on a par with the order Hyolithida. Both orders were placed in a superorder named Hyolithoidea, of the phylum Mollusca. I cannot agree with SYSSOIEV that *Hyolithellus* and *Hyolithes* are related, or even that they belong in the same phylum. Whereas attributes of hyolithids are decidedly molluscan, the hyolithelminths, because of their phosphatic shell composition, are more akin to some worm phylum, the entoproctids or phoronids. Similarly, the morphologically distinct calcium-carbonate shells of *Coleoloides* and *Coleolus* are rejected from placement with the hyolithelminths. They may be an aberrant branch of mollusks.

Family HYOLITHELLIDAE Walcott, 1886

[=Hyolithellida SYSSOIEV, 1957 (order)]

Hyolithelminths with a circular cross section and operculum with 4 to 7 paired muscle scars. *L.Cam.-Mid.M.Cam.*

Hyolithellus BILLINGS, 1871 (p. 240) [**Hyolithes micans* BILLINGS, 1871]. Rate of tapering for most of shell 1 or 2 degrees. Shell composed of very thin laminae which thicken progressively toward aperture. Some specimens show scattered oval pores. Apical portion curved and irregular in growth. Average tube size, 5 mm. long and 2 mm. in diameter. Opercula range from 0.25 to 2.5 mm. in diameter. Five pairs of teardrop-shaped muscle scars are arranged bilaterally on underside of operculum, those toward "ventral" side being longest, scars widest toward operculum periphery. *L.Cam.,* N.Am.-Eu.(Swed.-USSR); *L.Cam.-M.Cam.,* Eu.(G.Brit.-Norway)-S. Am.(Arg.).——FIG. 75,*1*. **H. micans* (BILLINGS), L.Cam., N.Y.; *1a,b,* shell in inferred living position, with operculum closed and open, ×6, ×8 (Fisher, n); *1c,* underside of operculum, showing 5 pairs of muscle scars, ×11; *1d,* diagram. long. sec. showing shell-wall thickening toward aperture ×3 (29).

Barella HEDSTRÖM, 1930 [*BARRANDE's "Opercule isole H" (1, pl. 9, figs. 16, 17) for which no name was given]. Originally described as a gastropod; undoubtedly a hyolithelminth operculum. Has 4 pairs of muscle scars. *Cam.,* Czech.

Discinella HALL, 1871 (p. 246) [*HALL designated no species. From description it is obvious that type is operculum of *Hyolithes micans* BILLINGS, 1871]. Originally described as a gastropod; assuredly the operculum of *Hyolithellus micans.* Therefore, *Discinella* is a senior synonym (March 1871) of *Hyolithellus* (Dec. 1871) but because of wide usage of latter name, the former should be suppressed.

Mobergella HEDSTRÖM, 1923 (p. 5) [**Discinella holsti* MOBERG, 1892]. Originally described as a gastropod; undoubtedly a hyolithelminth operculum. Has 7 pairs of muscle scars. Central area of underside of operculum relatively larger than in *Hyolithellus. L.Cam.,* Eu.(Norway-Swed.).——

FIG. 75,2. *M. holsti (MOBERG), Swed.; underside of operculum, showing 7 pairs of muscle scars, ×11 (19).

Family TORELLELLIDAE Holm, 1893

Hyolithelminths with an elliptical or biconvex lens-shaped cross section. In some forms 2 keels occur at "poles" of elongated cross section. Aperture at right angles to shell axis. Transverse striae and ribbing more pronounced than in Hyolithellidae. Opercula unknown. *L.Cam.-Ord.*

Torellella HOLM, 1893 (p. 146) [**Hyolithes laevigatus* LINNARSSON, 1871]. Narrow shell with elliptical or biconvex lens-shaped cross section, usually with 2 keels at "poles" of section. Transverse striae and rings generally present. *L.Cam.*, N.Am.-Eu.(Norway-Eng.); *L.Cam.-Ord.*, Eu. (Pol.-Swed.).——FIG. 75,*3*. **T. laevigata* (LINNARSSON), L.Cam., Swed.; *3a,b*, exterior of shell, ×3.5, ×8; *3c*, transv. sec., ×8 (20).

?**Pseudorthotheca** COBBOLD, 1935 (p. 27) [**P. acuticincta*]. HOWELL (p. *W*165) regards this genus as belonging to the Order Sedentarida.

?**Rushtonia** COBBOLD & POCOCK, 1934 (p. 323) [**R. lata*]. HOWELL (p. *W*165) regards this genus as belonging to the Order Sedentarida.

Phylum, Class, and Order UNCERTAIN

Family COLEOLIDAE Fisher, n.fam.

Tubuliform calcium-carbonate shells, extremely elongate-conical, almost cylindrical and commonly slightly curved toward apex; cross section circular to elliptical; comparatively thick-walled; laminated, interior surface smooth. Exterior surface smooth or with oblique or longitudinal ornamentation. Opercula and septa unknown. Length, 0.5 to 75 mm., diameter, 0.5 to 2.5 mm. at aperture. *L.Cam.-Carb.*

Coleolids are reported from limestones, black, gray and green shales and less commonly are found in sandstones. Their universal parallelism to bedding indicates a pelagic existence. In some places they occur in masses up to 3 shells thick, usually oriented. Invariably the apices and apertures are broken off. Their habitat is imperfectly known.

SANDBERGER (1852) and ROEMER (1853) compared *Coleoprion* to the living pteropod *Creseis*, and HALL (17, 18) placed *Coleolus* without question in the Pteropoda. SYSSOIEV (1958) placed *Coleoloides* and *Coleolus*, together with *Hyolithellus*, in the Mollusca. The nonmolluscan nature of *Hyolithellus* has already been discussed, but much may be said for retaining the coleolids in the Mollusca. Coleolids are possibly ancestral scaphopods but if so, they reveal no evidence of the burrowing habits which are so characteristic of living scaphopods. On the other hand, early scaphopods may have been pelagic. The pteropod assignment seems tenuous in view of the long stratigraphic gap (Carboniferous to Lower Tertiary) in which no fossil pteropods have been found and the thick nature of the shells, which are alien to living pteropods. The genera reviewed here are placed provisionally in the Mollusca, with full realization that they may prove to be, on further study, more closely allied to some phylum of worms. About 20 species of coleolids have been described.

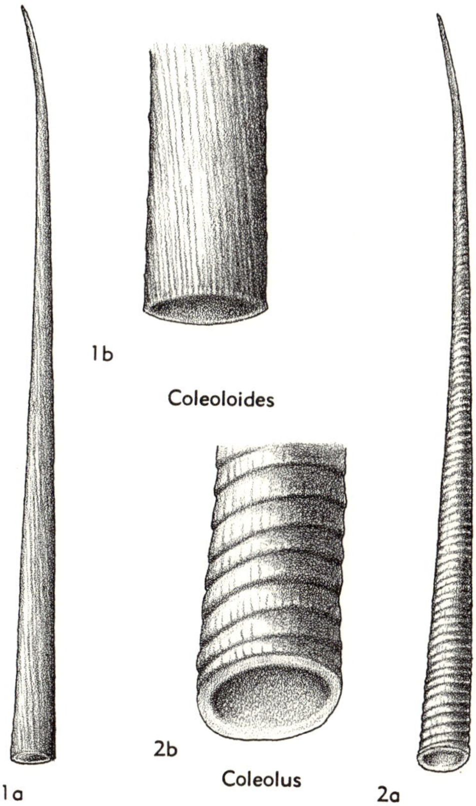

FIG. 76. Coleolidae (p. *W*134) (17, Fisher, n).

Coleolus HALL, 1879 (p. 184) [*nom. subst. pro Coleoprion* HALL, 1876 (*non* SANDBERGER, 1847)] [**Coleoprion tenuicinctum* HALL, 1876]. Surface with annulating striae or rings which are markedly oblique to shell axis; longitudinal striae may be present. *M.Sil.-Carb.*, N.Am.-Eu.-Austral.——FIG. 76,*2*. **C. tenuicinctus* (HALL), M.Dev., N.Y.; *2a,b,* exterior of shell showing aperture, ×2.5, ×7.5 (17).

Coleoloides WALCOTT, 1889 (p. 37) [**C. typicalis*]. Surface marked by parallel, longitudinal, slightly oblique striae. *L.Cam.*, N.Am.(Newf.-N.Y.).——FIG. 76,*1*. **C. typicalis,* Newf.; *1a,b,* exterior of shell showing aperture, ×2.5, ×7.5 (Fisher, n).

Coleoprion SANDBERGER, 1847 [**C. gracilis*]. Slightly undulating striae, oblique to shell axis, converging to distinct longitudinal groove that extends entire length of shell. *Sil.-U.Dev.*, N.Am. (Ont.)-Eu.

Paoshanella YIN, 1937 (p. 289) [**P. flexuosa*]. Slowly tapering, compressed; lenticular in cross section. Exterior with longitudinal striations. Though incomplete at both ends, shell measures 60 mm. [YIN classed genus in the Torellellidae but exterior ornamentation and composition suggests placement in the Coleolidae.] *L.Ord.*, China.

Polylopia CLARK, 1925 (p. 12) [**Salterella billingsi* SAFFORD, 1869]. Multilayered, narrow, straight conical shell, tapering uniformly and with marked longitudinal ribbing on outside of each layer. One to 5 walls, relatively thick, with light and dark layers alternating, latter invariably thicker. Cross section circular. No septa. [Shells found parallel to bedding and roughly oriented. No relation to *Salterella*. I previously (11) studied this genus and regarded it as an Ordovician scaphopod. This assignment depended mainly on whether apices of the specimens were complete or broken off. Excepting lack of curvature, all other characteristics point to scaphopod affinities. However, if such an assignment is rejected, then *Polylopia*, still deserving a molluscan assignment, should be placed with the coleolids.] *M.Ord.*, N.Am.(Tenn.).——FIG. 77,*1*. **P. billingsi* (SAFFORD); *1a,* part of slab showing several well-preserved specimens, ×1; *1b,c,* weathered specimen and polished rock section showing 3 layers of shell wall, ×4 (11).

Salopiella COBBOLD, 1921 (p. 362) [**S. obliqua*]. Relatively small (length about 3.5 mm., diameter, 0.6 mm.) coleolid with steplike ridges encircling tube obliquely to shell axis; ridges reproduced on interior, which is anomalous for coleolids. [Similar to *Torellella* but not phosphatic. Coleolid assignment uncertain.] *L.Cam.*, Eu.(Eng.).

Spirodentalium WALCOTT, 1890 (p. 271) [**S. osceola*]. Curved conical shell with exterior ornamented by spiral striae that pass around the shell

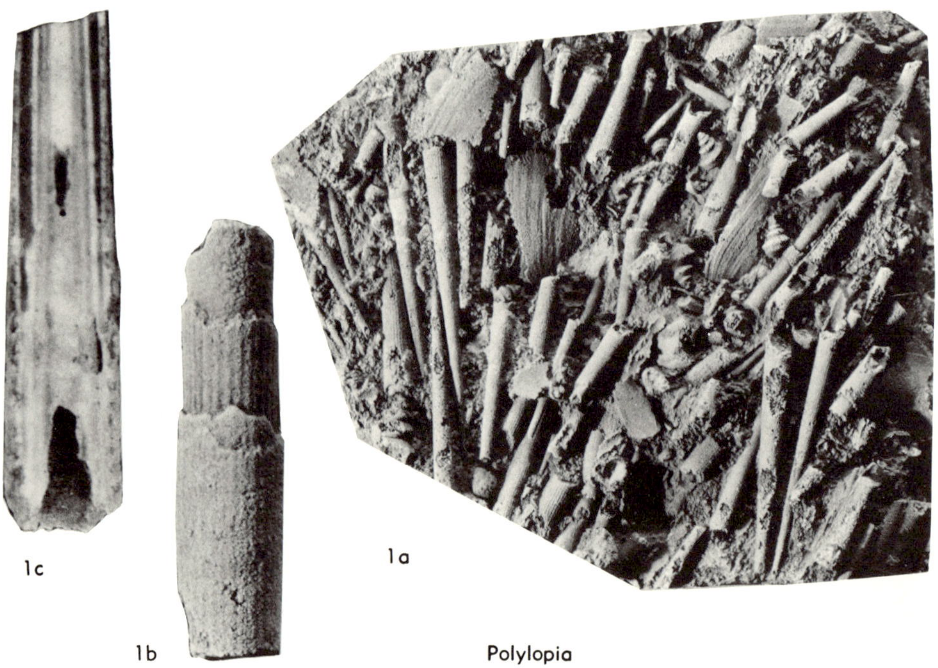

FIG. 77. *Polylopia billingsi* (SAFFORD), M.Ord., Tenn. (p. W134) (11).

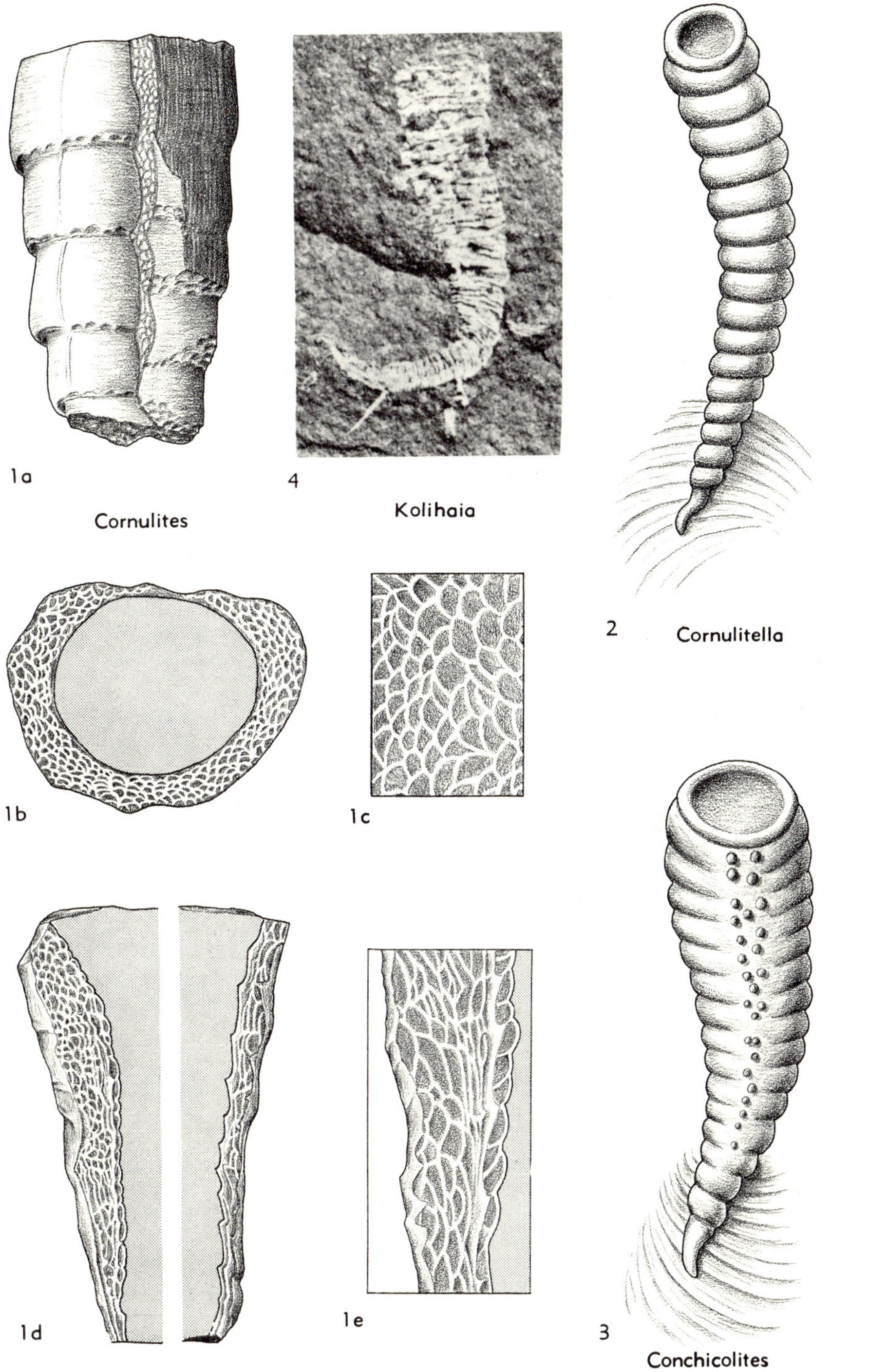

FIG. 78. Cornulitidae (p. *W*137-*W*138) (Fisher, n).

3 or 4 times in a length of 6 cm. *U.Cam.*, N.Am. (Wis.).

Family CORNULITIDAE Fisher, n. fam.

Flexuous, ringed, small to medium-sized tapering tubes of calcium carbonate, with circular cross section; rings dominant in specific stages of growth and longitudinal striae dominant in other stages. Relatively thick walls composed of large, rounded or oval cellular cavities, cells with thin walls and especially conspicuous in transverse rings which are narrower and closer together toward apex (Fig. 78). Mold of internal cavity consists of rodlike tapering body of invaginated cones with apices directed toward aperture and is thus homeomorphic with internal molds of Tentaculitidae. Length 5 to 80 mm., diameter at aperture, 2 to 20 mm. *M.Ord.-Miss.(L.Carb.)*, N.Am.-Eu.

Fossils classed as cornulitids are nearly smooth in early growth stages (Fig. 79) but become ringed in intermediate stages, and marked by predominant longitudinal striae in advanced stages; rings become obsolescent in gerontic stages. Free-swimming larvae guarantee dispersal. In very earliest stages, circular or suborbicular discs are mobile. As the tube tends to straighten, it affixes itself at the curved apical end to some object. Usual hosts are snails, brachiopods, or bryozoans. Certain cornulitid species seem to have selective preference for attachment to particular species of brachiopods or snails. The shell grows by absorption of calcium carbonate, in part, from the host. The vesicular tissue is rarely seen in early stages, but it develops rather rapidly during intermediate (juvenile) stages and reaches a maximum during advanced stages (adult and gerontic). Whereas the tube tends to straighten during juvenile and adult stages, it reverts to a flexuous condition in gerontic stages. During its embryonic existence, the tube consists essentially of inner and outer walls in contact with each other. The exterior wall develops rings at an early stage and areas between the rings produce sharp interior rings that commonly extend across the internal cavity so that when the wall is worn away or the fossil is sectioned, the appearance of a septate tube is obtained. Longitudinal striae become more pronounced with growth and continue to be a conspicuous feature after disappearance of the rings. Whenever gerontic cornulitids are found free, the ini-

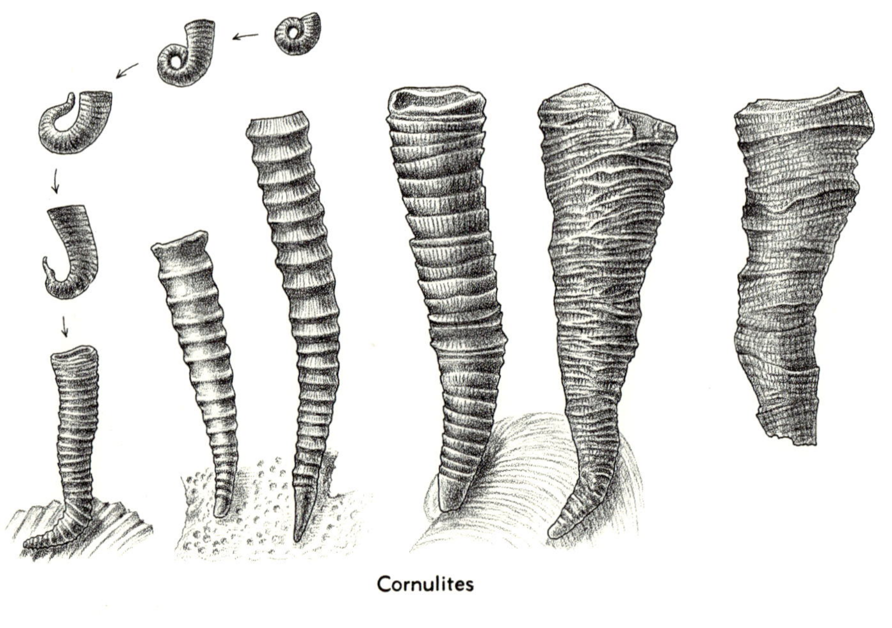

Cornulites

FIG. 79. Ontogeny of *Cornulites*, ×1 (20).

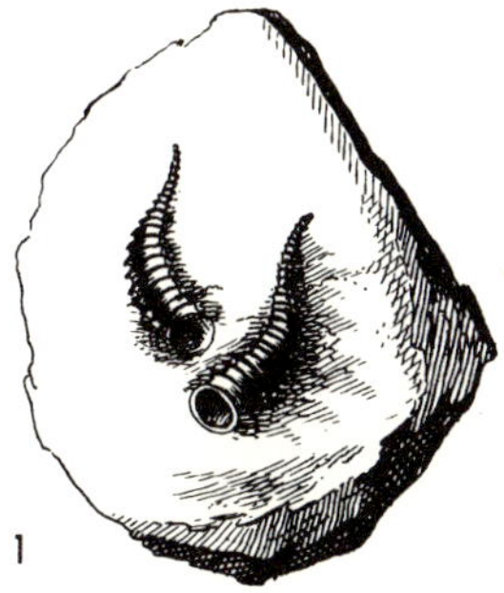

FIG. 80. *Cornulites serpularius,* Wenlock., Eng.; copy of SCHLOTHEIM's original figure (1820), ×2 (65).

tial point of the shell is missing and a still-visible fracture shows that it was broken off. Many gerontic tubes show evidence of damage and subsequent repair.

Cornulitids are generally found in calcareous strata. They are most abundant in limestones and calcareous shales and less common in calcareous sandstones. They are rare to absent in noncalcareous rocks. Seemingly, calcium carbonate is a necessary agent for their optimum growth. In juvenile and adult stages, cornulitids are exclusively epizoic. They may further be classed as epibenthonic, for they attach themselves to benthonic (sessile or slowly moving) forms that have calcium-carbonate-rich shells or skeletons. Whenever found, they are either isolated or in clusters on a single host.

SCHLOTHEIM (60) founded *Cornulites* (Fig. 80) to include certain fossils of somewhat doubtful affinities but allied most closely with tubicolar annelids. Although this is the usual assignment of cornulitids, the nature of their intricate cellular wall (alien to unequivocal tubicolar annelids) suggests the propriety of assigning them to some other group. This vesicular wall is similar to that of some coelenterates (stromatoporoids, calacareous hydroids), or some fusulines. Cornulitids are here unassigned taxonomically, pending further study. About 45 species have been described.

Cornulites SCHLOTHEIM, 1820 (p. 378) [**C. serpularius*]. Animal solitary, inhabiting a relatively long tube. Prominent external rings in adult stages and prominent longitudinal striae in gerontic forms. Cellular walls very thick. Tubes reach a length of 80 mm. and diameter of 20 mm. at aperture. *M.Ord.-M.Dev., ?Miss.,* N.Am.-Eu.——FIG. 78,*1*. **C. serpularius,* Wenlock, Eng.; *1a,* cast of internal cavity showing part of shell wall, ×4; *1b,c,* transv. sec. of shell wall, ×3.2, ×6.4; *1d,e,* long. sec. of shell wall, ×3.2, ×6.4 (18).——FIG. 81,*1*. *C. sterlingensis* (MEEK & WORTHEN), U.Ord., Baffin Land; *1a-c,* typical specimens, *1c* showing attachment to brachiopod shell, ×6 (Fisher, n).

Conchicolites NICHOLSON, 1872 (p. 203) [**C. gregarius*]. Animal social, tubes attached in clustered masses to a host shell. Tubes slightly curved, attached at smaller extremity; walls thin; exterior covered by short imbricated rings but devoid of longitudinal striae. Length 5 to 13 mm., diameter at aperture about 3 mm. *U.Ord. (Caradoc.)-L.Dev.,* N.Am.-Eu.——FIG. 78,*3*. **C. gregarius,* L.Sil., Eng.; exterior of shell showing aperture, ×6 (18).

Cornulitella HOWELL, 1952 (p. 37) [*nom. subst. pro Ortonia* NICHOLSON, 1872 (*non* WOOD, 1869)] [**Ortonia conica* NICHOLSON, 1872]. Animal solitary, inhabiting a ringed tube which is attached along whole of one side; conical tube slightly flexuous, somewhat flattened laterally; walls thick, cellular, markedly ringed on all sides except attached one. Longitudinal striae absent. Length 5 to 13 mm., diameter at aperture 2 or 3 mm. *M.Ord.-L.Carb.,* N.Am.——FIG. 78,*2*. **C. conica* (NICHOLSON), U.Ord., Ohio; exterior of shell showing aperture, ×6 (18).

FIG. 81. *Cornulites sterlingensis* MEEK & WORTHEN, U.Ord., Baffin Land, ×6 (p. *W*13) (Fisher, n).

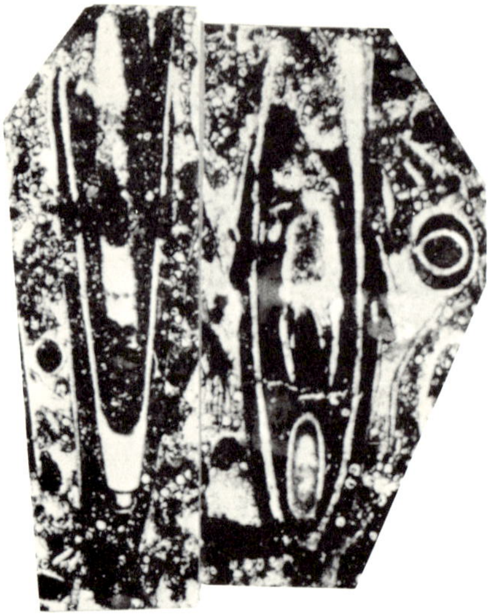

FIG. 82. *Biconulites grabaui,* L.Cam., China; specimens shown in polished rock sections, ×2 (p. W138) (15).

Kolihaia PRANTL, 1944 (p. 1) [**K. eremita*]. Animal gregarious with skeletons of conical tubes that curve abruptly at their proximal extremities, which show radical expansions that may bifurcate. Exterior with rings but no longitudinal striae. *M.Sil.(Wenlock.),* Eu.(Czech.).——FIG. 78,*4*. **K. eremita,* ×5 (Prantl, 1946).

Phylum, Class, Order, Family UNCERTAIN

Biconulites DE CHARDIN, 1931 (p. 184) [**B. grabaui*]. Narrow conical shell devoid of external markings. Internally consisting of 2 sets of inserted cones (direct cones and inverted cones) and central tube (possibly a fortuitous mechanical arrangement of shells, one within the other). Two types of direct cones are distinguished, a central cone with curved lower end and 3 or 4 septa, and 1 or 2 external direct cones concentrically around the central direct cones. No trace of fusion is seen between the central direct cones. Some shells show up to 4 inverted cones that act as "opercula." The geometric axis of the inverted cones does not correspond to that of the direct cones but runs obliquely to it. Inverted cones are always loose; central tube obscure. *L.Cam.,* China (Shansi).——FIG. 82,*1*. **B. grabaui;* specimens shown in polished rock sections, ×2 (5).

?**Lapworthella** COBBOLD, 1921 (p. 359) [**L. nigra*]. COBBOLD regarded this genus as intermediate between *Hyolithes (Orthotheca)* and *Salterella,* with some resemblance to *Tentaculites.* I believe that it is related to *Stenothecopsis.* It is considered by HOWELL (p. W164) to belong to Order Sedentarida.

Stenothecopsis COBBOLD, 1935 (p. 43) [**S. heraultensis*]. Shell consisting of small, slightly curved pyramid of calcium phosphate, oval or circular in cross section toward apex but becoming more quadrate toward aperture until in some, 2 longer sides may be as much as 3 times length of shorter sides; bilaterally symmetrical about plane containing longer axis of the section. Apex sharp, without evidence of attachment. Relatively thick, 3-layered wall composed of inner dull layer, intermediate nacreous layer, and outer black chitinous layer. Sides marked by strong, parallel transverse narrow ridges with flat interspaces containing 2 or more finer ridges, which may curve slightly toward apex. Cancellate pattern near aperture. Length 1 to 1.75 mm. [COBBOLD and POULSEN have questionably referred this genus to the Crustacea, though the shell is unlike any known primitive crustacean. A phoronid or entoproctid assignment seems to me not improbable, but similarity to conulariids is externally apparent and a molluscan assignment should not be dismissed, even though it is common practice to exclude phosphatic shells from this phylum. Like some brachiopods, some primitive mollusks may have had a phosphatic shell.] *L.Cam.,* USA(N.Y.)-Eu.(Fr.); *M.Cam.,* Eu.(S.Wales-Denm.).——FIG. 83,*1*. *S. schodackensis* LOCHMAN, L.Cam., N.Y.; lateral surface, ×50 (29).

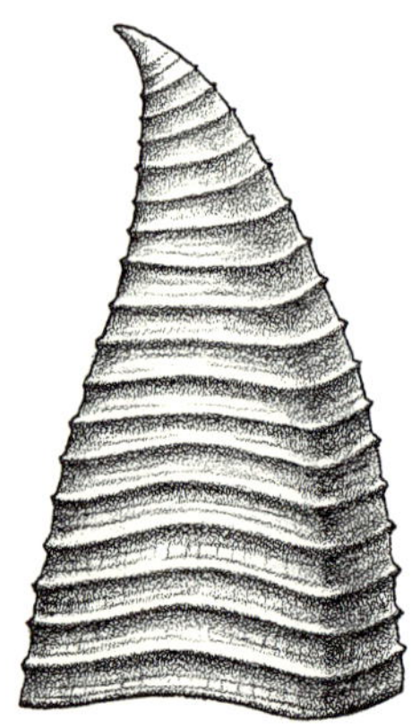

FIG. 83. *Stenothecopsis schodackensis* LOCHMAN, L. Cam., N.Y., ×50 (p. W138) (29).

PTEROPOD-LIKE GENERA PROVISIONALLY REFERRED TO OTHER GROUPS

Clathrocoelia Hall, 1879 (p. 203) [**C. eborica*]. Now recognized as the "wing" of the pelecypod *Actinopteria decussata. M.Dev.*, USA(N.Y.).

Clioderma Hall, 1861 (p. 96) [**C. saffordi* Hall, 1861; SD Whitfield & Hovey, 1898 (p. 59)]. Junior subjective synonym of *Pterotheca* Salter, 1853, a gastropod. *M.Ord.*, USA(N.Y.).

Enchostoma Miller & Gurley, 1896 (p. 29) [**Hyolithes lanceolatus* Miller, 1892 (=**H. milleri* Sinclair, 1946, p. 73)]. Elongate, lanceolate, straight cone of calcium phosphate, slightly curved apically, with circular or suboval cross section. Thin shell has longitudinal flutings. Length, 30 mm., width, 1.25 mm. [Probably a worm. Sinclair gave the new name *Hyolithes milleri* to *H. lanceolatus* Miller since the latter was preoccupied by *H. lanceolatus* (Morris), 1845 (as *Theca lanceolatus*).] *L. Miss.*, USA(Mo.).

Harttites Howell & Knight, 1936 [*nom. subst. pro Harttia* Walcott, 1884 (*non* Steindacher, 1877)] [**Harttia matthewi* Walcott, 1884]. Originally described as a gastropod, but rejected by Knight *et al.* (1960, *Treatise*, Part I, p. *I*324) as a mollusk; conceivably a brachiopod. *M.Cam.*, Can.(N.B.).

Latouchella Cobbold, 1921 (p. 366) [**L. costata*] [=*Oelandia* Westergård, 1936 (*fide* Knight *et al.*, 1960)]. Small, loosely coiled, broadly expanding calcareous cone with whorls not in contact; aperture oval, unnotched; sides with 6 to 8 elongate prominent ribs, swollen toward convex margin of shell which they do not cross, and narrowing to disappearance in opposite direction, ribs reproduced in interior. Dorsum narrowly rounded but not keeled. Length of shell, 4.5 mm., height, 2.5 mm.; aperture 2 by 4 mm. [Although assigned to gastropods by Cobbold, it was omitted from consideration by Knight (1941); in 1960, Knight *et al.* (*Treatise*, Part I, p. *I*172) included the genus in the Coreospiridae, a family of primitive Archaeogastropoda.]. *L.Cam.*, Eng.

Palaenigma Walcott, 1885 [*nom. subst. pro Tetradium* Schmidt, 1874 (*non* Dana, 1846, *nec* Safford, 1856)] [**Tetradium wrangeli* Schmidt, 1874]. Undoubtedly belongs with conulariids, as reported by Moore & Harrington (1956, *Treatise*, Part F, p. *F*62); the quadrilateral radial symmetry is diagnostic. *Ord.*, N.Am.-Eu.——Fig. 84, *1*. **P. wrangeli* (Schmidt), L.Ord., Eu.; *1a,b*, lateral view, transv. sec. showing quadriradial symmetry, ×1 (Fisher, n).

Pterotheca Salter, 1853 [**Atrypa transversa* Portlock, 1843; SD S. A. Miller, 1889] [=*Clioderma* Hall, 1861; *Aulacomerella* von Huene, 1900]. A primitive bellerophontid (Knight *et al.*, 1960, *Treatise*, Part I, p. *I*182). *M.Ord.-U.Ord.*, N.Am.-Eu.

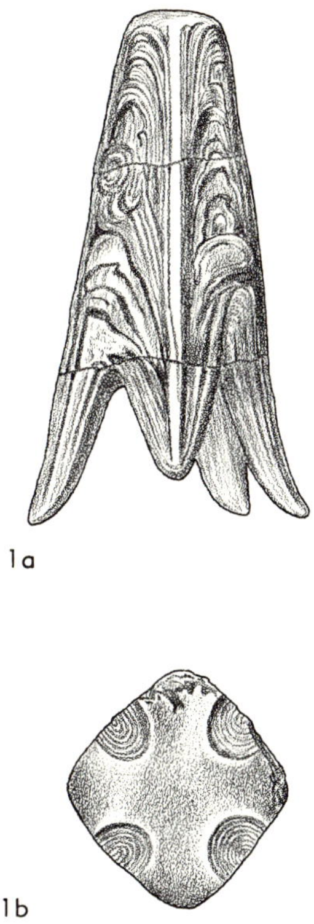

Fig. 84. *Palaenigma wrangeli* (Schmidt), L.Ord., Eu., ×1 (p. *W*139) (Fisher, n).

Randomia Matthew, 1899 [**R. aurorae*]. Height, 11 mm., aperture, 20 by 25 mm. Similar to the gastropod *Palaeacmaea* Hall & Whitfield, 1872 (U.Cam., N.Y.), but has strong radiating ridges that cross concentric ridges and prominent undulations of growth, whereas *Palaeacmaea* only has concentric lines. Considered to be a gastropod by Matthew but Knight (27) omitted mention of it. To me, it appears close, if not identical, to the gastropod. *Parmophorella* Matthew, 1886 (p. 59), according to Knight *et al.* (1960, *Treatise*, Part I, p. *I*331), *Randomia* is possibly a monoplacophora and belongs to the primitive archaeogastropod superfamily Helcionellacea. *M.Cam.*, Newfoundland. Erroneously cited as *Parmorphella* by Knight *et al.* (*Treatise*, Part I, p. *I*77); see following entry on *Scenella*. *L.Cam.*, Newfoundland.

Scenella Billings, 1872 [**S. reticulata*] [=*Parmophorella* Matthew, 1886]. Low cap-shaped shell with strong concentric wrinkles, now recognized as a tryblidioid gastropod (*Treatise,* Part I, p. *I*77). *Cam.,* N.Am.-NE.Asia.

Scenellopsis Resser, 1938 [**Scenella clotho* Walcott, 1905]. Small, limpet-like shell with eccentric apex and radiating and concentric lines. Muscular impression, evidently branching, produces folds on exterior; marginal outline flattened at broad end. Similar to *Scenella,* but not mentioned by Knight (27); considered probably not a mollusk (Knight *et al.,* 1960, *Treatise,* Part I, p. *I*324). *L.Cam.,* N.Korea; *M.Cam.,* China (Shansi-Shantung).

Stenotheca Hicks, 1872 (p. 180) [**S. cornucopia*]. Depressed conical univalve with oval aperture, apex curved toward end of shell. [Seems to have crustacean affinities; Knight *et al.* (*Treatise,* Part I, p. *I*324) class genus as a crustacean.] *Cam.,* N. Am. (Newf.-Mass.).

Stenothecoides Resser, 1938 [**Stenotheca elongata* Walcott, 1884]. Originally regarded as related to *Stenotheca* but has thick, impunctate, univalve shell with well-marked growth lines; suboval or subelliptical, elongated; internal surface with number of irregular, transverse ridges on each side of the mid-line. [Adapted for clinging to rocks, like limpets. Knight & Yochelson (1960, *Treatise,* Part I, p. *I*83) classed *Stenothecoides* definitely among monoplacophorans, assigning it to the family Cambridiidae, but in the same volume p. *I*324) Knight *et al.* doubtfully classify it as a crustacean; the latter entry probably is an oversight by these authors.] *L.Cam.,* N.Am.(Greenl.-Can.-N.Y.); *M.Cam.,* N.Am.(Can.-Nev.).

UNRECOGNIZABLE GENERA

Charruia Rusconi, 1955 [**C. annulata*]. Slightly curved shell, incomplete at both ends. Circular cross section; transverse rings. Composition and interior unknown. Length, 3.5 mm., diameter, 1.4 mm. [Classed as a gastropod by Rusconi and as a problematical organism by Knight *et al.* (*Treatise,* Part I, p. *I*324). A poor specimen is vaguely described.] *M.Cam.*(Isidreana F.) S.Am. (Arg.).

Cyrtotheca Hicks, 1872 [**C. hamula*]. Genus founded on a single deformed specimen. No description and poor illustration does not adequately diagnose the genus. *L.Cam.,* G.Brit.

Hyolithoconularia Termier & Termier, 1950 [**H. striata*]. Conoidal shell with growth lines and partitions as in *Hyolithes.* Apertural end has longitudinal bands with indented peristome forming 3 projections, of which one corresponds to a longitudinal band. Peristomal elevations recall those of the conulariids. Cross section and composition not reported. Originally classed with the Serpulitidae (*Coleolus, Coleoprion*) but seems to be a distorted conulariid. *L.Dev.,* Afr.(Morocco).

Macrotheca Waagen, 1880 (p. 178) [**M. wynnei*]. Conoidal, slowly tapering toward apex, with elliptical section and slightly flattened ?ventral side; apertural and apical ends not preserved. No trace of septa. Surface sculpture unknown. Shell comparatively thin. Estimated length, 320 mm. *Perm.,* India (Up. *Productus* Ls.)-Timor.

Pichynella Rusconi, 1954 (p. 42) [**P. annulata*]. Low cone with 7 "whorls" (helical nature cannot be proved from poor illustration). Composition of shell and nature of cross section not recorded. [Rusconi regarded *Pichynella* as a gastropod. This minute fossil (dia. 1 mm.) was considered by Knight *et al.* (1960, *Treatise,* Part I, p. *I*172) as possibly a protoconch of an archaeogastropod doubtfully assignable to *Helcionella* Grabau & Shimer.] *U.Cam.-L.Ord.,* S.Am.(Arg.).

Quilicanella Rusconi, 1952 (p. 86) [**Q. cuyana*]. Gently curved, rapidly expanding cone with 5 to 8 transverse rings. Composition of shell and cross section not indicated. [Rusconi regarded it as a gastropod. May be synonymous with *Lapworthella.*] *L.Cam.,* S.Am.(Arg.).

ACKNOWLEDGMENTS

I would be remiss if I failed to acknowledge the willing co-operation of those who have furnished specimens and foreign literature for study. Especial appreciation is extended Drs. G. A. Cooper, U.S. National Museum, Washington, D.C.; Edmund Gill, Victoria Museum, Melbourne, Australia; G. P. Lyashenko, Micropaleontological Laboratory, VNIGNI, Moscow, U.S.S.R.; Ferdinand Prantl, Narodni Museum, Praha, Czechoslovakia; Per Thorslund, Uppsala University, Uppsala, Sweden; and J. W. Wells, Cornell University, Ithaca, N.Y.

Dr. Ellis Yochelson, U.S. Geological Survey, freely provided unpublished information [now published, 76, 77] on *Hyolithes,* especially regarding the character and presumed function of the "supports."

In large measure this section is enhanced and made more useful by the line drawings which were executed by Mrs. John Winslow under my direction.

REFERENCES

Barrande, Joachim

(1) 1867, *Système Silurien du centre de la Bohême:* v. 3, Classe des Mollusques, Ordre des Ptéropodes, 179 p., 16 pl. (Prague & Paris).

(2) 1872, *Système Silurien du centre de la Bohême:* Supplément au Vol. I, Trilobites, Crustacés divers et Poissons, 647 p., 37 pl.

Billings, Elkanah

(3) 1872, *On some fossils from the primordial rocks of Newfoundland:* Canadian Naturalist, new ser., v. 6, p. 465-479.

Brastad, Johan

(4) 1915, *Discinella holsti faunaen ved Brastedelven nord for Gjøvik:* Norsk Geol. Tidsskr., v. 3, no. 5.

Chardin, P. T. de

(5) 1931, *On an enigmatic pteropod-like fossil from the Lower Cambrian of southern Shansi, Biconulites grabaui, nov. gen., nov. sp.:* Geol. Soc. China, Bull., v. 10, p. 179-188, pl. 1-2.

Clarke, J. M.

(6) 1913, *Fósseis devonianos do Paraná:* Serv. Geol. Mineral. Brasil, Mon., v. 1, xx+353 p., 27 pl. [Portuguese & Eng.]

Cobbold, E. S.

(7) 1919, *Cambrian Hyolithidae, etc., from Hartshill in the Nuneaton District, Warwickshire:* Geol. Mag., v. 56, p. 149-158, pl. 1-4.

(8) 1921, *The Cambrian horizons of Comley (Shropshire) and their Brachiopoda, Pteropoda, Gasteropoda, etc.:* Geol. Soc. London, Quart. Jour., v. 76, no. 304, p. 325-386, pl. 21-24.

(9) 1935, *Lower Cambrian faunas from Herault, France:* Annals & Mag. Nat. History, ser. 10, v. 16, p. 25-48, pl. 2-4.

Eichwald, Eduard von

(10) 1840, *Ueber das silurische Schichtensystem in Esthland:* Zeitschr. Nat.-Heilkunde Med.-chirur. K. Akad. St. Petersburg, H. 1 & 2.

Fisher, D. W.

(11) 1958, *Polylopia Clark, an Ordovician scaphopod:* Jour. Paleontology, v. 32, p. 144-146, pl. 13.

———, & Young, R. S.

(12) 1955, *The oldest known tentaculitid—from the Chepultapec limestone (Canadian) of Virginia:* Jour. Paleontology, v. 29, p. 871-875, pl. 88, fig. 1.

Flower, R. H.

(13) 1954, *Cambrian cephalopods:* New Mexico State Bur. Mines & Mineral Resources, Bull. 40, 51 p., pl. 1-3, fig. 1-7.

Gill, E. D.

(14) 1941, *The place of the genus Styliolina in the Palaeozoic palaeontology and stratigraphy of Victoria:* Royal Soc. Victoria, Proc., v. 53, pt. 1, p. 145-164, pl. 1.

Gürich, G.

(15) 1896, *Das Palaeozoicum im polnischen Mittelgebirge:* Russ.-Kais. mineral. Gesell. St. Petersburg, Verhandl., ser. 2, Bd. 32, p. 1-539, pl. 1-15.

(16) 1900, *Ueber Tentaculiten und Nowakien, fossile Röhrenthiere:* Schles. Gesell., Jahr.-ber., Bd. 77, T. 2, p. 32-35.

Hall, James

(17) 1879, *Containing descriptions of the Gasteropoda, Pteropoda and Cephalopoda of the Upper Helderberg, Hamilton, Portage and Chemung groups:* N.Y. State Geol. Survey, Palaeontology of New York, v. 5, pt. 2, x+492 p.; 113 pl. (sep.)

(18) 1888, *Supplement, Containing descriptions and illustrations of Pteropoda, Cephalopoda and Annelida:* N.Y. State Geol. Survey, Palaeontology of New York, v. 7 (Suppl. to v. 5, pt. 2), 42 p., 18 pl.

Hedström, Herman

(19) 1930, *Mobergella versus Discinella; Paterella versus Scapha & Archaeophiala (some questions on nomenclature):* Sver. Geol. Undersök., Afhandl., ser. C, no. 362, p. 1-8.

Holm, Gerhard

(20) 1893, *Sveriges Kambrisk-Siluriska Hyolithidae och Conulariidae:* Sver. Geol. Undersök., Afhandl., ser. C, no. 112, p. 1-172, pl. 1-6. [Eng. summary]

Howell, B. F.

(21) 1952, *New Carboniferous serpulid worm from Missouri:* Wagner Free Inst. Sci. Philadelphia, Bull., v. 27, no. 4, p. 37-40, pl. 1.

———, & Stubblefield, C. J.

(22) 1950, *A revision of the fauna of the north Welsh Conocoryphe viola beds implying a Lower Cambrian age:* Geol. Mag., v. 87, no. 1, p. 1-16, pl. 1-2.

Karpinsky, A. P.

(23) 1884, *Die fossilen Pteropoden am Ostabhange des Urals:* Acad. Impér. Sci. St. Pétersbourg, Mém., ser. 7, v. 32, no. 1, p. 1-21, pl. 1.

Kiaer, Johan

(24) 1916, *The Lower Cambrian Holmia fauna at Tømten in Norway:* K. Norske Vidensk. Skrift., no. 10, p. 1-134, pl. 1-14, fig. 1-15.

Kiderlen, Helmut

(25) 1937, *Die Conularien; ueber Bau und Leben der ersten Scyphozoa:* Neues Jahrb. Mineral.,

Beil.-Bd. 77, Abt. B, H. 1, p. 113-169, fig. 1-47.

Knight, J. B.

(26) 1937, *Conchopeltis Walcott, an Ordovician genus of the Conulariida:* Jour. Paleontology, v. 11, p. 186-188.

(27) 1941, *Paleozoic gastropod genotypes:* Geol. Soc. America, Special Paper 32, 510 p., 96 pl.

Koken, Ernst

(28) 1889, *Die Hyolithen der silurischen Gescheibe:* Deutch. geol. Gesell., Zeitschr., Bd. 41, p. 79-82, pl. 1-8.

Lochman, Christina

(29) 1956, *Stratigraphy, paleontology, and paleoecology of the Elliptocephela asaphoides strata in Cambridge and Hoosick quadrangles, New York:* Geol. Soc. America, Bull., v. 67, p. 1331-1396, pl. 1-10, fig. 1-2.

Lyashenko, G. P.

(30) 1953, *O stratigraficheskom znachenii tentakulitov:* Akad. Nauk SSSR Leningrad, Doklady, v. 91, no. 2, p. 371-374. [*Concerning the stratigraphic significance of tentaculitids.*]

(31) 1954, *Novye dannye o Devonskikh Tentakulitakh:* V.N.I.G.N.I., Paleontologyicheškiy sbornik, Vypusk 1, p. 31-43, pl. 7-12. [*New data concerning Devonian tentaculitids.*]

(32) 1955a, *Devonskie tentakulity, novakii i stylioliny tsentralnoy chasti Russkoy Platformy:* Akad. Nauk SSSR Leningrad, Geol. Inst., p. 1-18, pl. 1-2. [*Devonian Tentaculitidae, Novakiidae, and Styliolinidae from the central part of the Russian Platform.*]

(33) 1955b, *Novye dannye o sistematike tentakulitov, novakiy i stiliolii:* Moskov. Obshch. Ispyt. Prir., Bull., v. 30, no. 3, p. 94-95. [*New data concerning the systematics of tentaculitids, novakiids, and styliolinids.*]

(34) 1956, *Devonian Tentaculitida, Novakiida, and Styliolinida of the central part of the Russian Platform* (abs.): 20th Internat. Geol. Cong. Mexico City, p. 118-119.

(35) 1957, *Novye Klass iskopaemykh mollyuskov Coniconchia:* Akad. Nauk SSSR Leningrad, Doklady, Tom 117, no. 6, p. 1049-1052. [*Coniconchia, a new class of extinct Mollusca.*]

(36) 1957, *Novye roda Devonskikh tentakulitov:* Akad. Nauk SSSR Leningrad, Doklady, Tom 116, no. 1, p. 141-144, pl. 1. [*New families of Devonian tentaculitids.*]

(37) 1957, *Novye vidy srednedovonskikh tentakulitov i stiliolin tsentral'nykh oblastey Russkoy Platformy:* V.N.I.G.N.I., Vypusk 8, p. 212-233, pl. 1-3. [*New species of Middle Devonian tentaculitids and styliolinids from the central part of the Russian Platform.*]

(38) 1958, *Novye vidy Siluriyskikh tentakulitov Pribaltiki i Volyni:* V.N.I.G.N.I. no. 9, p. 19-26, pl. 1-5. [*Silurian tentaculitids from Podolia and the Baltic.*]

(39) 1959, *Konikonkhii Devona tsentral'nykh i vostochnykh oblastey Russkoy Platformy:* V.N.I.G.N.I., p. 1-220, pl. 1-31. [*Devonian Coniconchia in the central district of the Russian Platform.*]

———, & Syssoiev, V. A.

(40) 1958, *Tip Mollyuski? Klass konikonkhii:* Osnovy Paleontologii spravochnik Paleontologov Geologov SSSR (Orlov, Yu. A., Luppov, N. P. & Druschchits, V. V., eds.), v. 6, p. 179-191, pl. 1-7. [*Mollusca?; Class Coniconchia.*]

Matthew, G. F.

(41) 1886, *Illustrations of the fauna of the St. John Group, No. 3, descriptions of new genera and species:* Royal Soc. Canada, Trans., v. 3, sec. 4, p. 29-84, pl. 5-7.

(42) 1899, *Preliminary notice of the Etcheminian fauna of Newfoundland:* Nat. History Soc. New Brunswick, Bull., v. 4, p. 189-197, pl. 1-3.

(43) 1899, *The Etcheminian fauna of Smith Sound, Newfoundland:* Royal Soc. Canada, Trans., ser. 2, v. 5, p. 97-119, pl. 5-7.

(44) 1899, *Studies on Cambrian faunas No. 3—Upper Cambrian fauna of Mount Stephen, British Columbia—the trilobites and worms:* Royal Soc. Canada, Trans., ser. 2, v. 5, p. 39-66, pl. 1-3.

(45) 1901, *Acrothyra and Hyolithes, a comparison:* Royal Soc. Canada, Trans., ser. 2, v. 7, sec. 4, p. 93-107.

Miller, A. K.

(46) 1932, *The mixochoanitic cephalopods:* Univ. Iowa Studies, v. 14, no. 4, p. 1-67, pl. 1-9.

Miller, S. A.

(47) 1889, *North American geology and paleontology for the use of amateurs, students, and scientists:* 718 p., 1265 fig., Western Methodist Book Concern (Cincinnati, Ohio).

Naef, A.

(48) 1924(1926), *Studien zur generellen Morphologie der Mollusken:* Ergeb. & Fortschr. Zool., Bd. 6, H. 1.

Nicholson, H. A.

(49) 1872, *On the genera Cornulites and Tentaculites and on a new genus Conchicolites:* Am. Jour. Sci., v. 3, p. 202-206.

(50) 1872, *On Ortonia, a new genus of fossil tubicolar annelides, with notes on the genus Tentaculites:* Geol. Mag., v. 9, p. 446-449.

Novak, Ottomar

(51) 1882, *Über Böhmische, Thuringische, Greifensteiner und Harzer Tentaculiten:* Beiträge Paläont. & Geol. Österreich-Ungarns, p. 47-70, pl. 12-13.

(52) 1886, *Zur Kenntnis der Fauna der Etage F-f₁ in der Palaeozoischen Schichtengruppe Böhmens:* K. böhm. Gesell. Wiss., Sitzungsber., Jahrg. 1886, p. 660-685, pl. 1-2.

(53) 1891, *Revision der Palaeozoischen Hyolithiden Böhmens:* K. böhm. Gesell. Wiss., Abhandl., Bd. 7, pt. 4, p. 1-18, pl. 1-6, fig. 1-3.

Pelseneer, Paul

(54) 1888, *Report on the Pteropoda collected by HMS Challenger during the years 1873-76, pt. 3, Anatomy:* Rept. Sci. Results Explor. Voyage HMS Challenger, Zoology, v. 23, p. 1-97, pl. 1-5.

Pillet, Jean

(55) 1956, *Sur deux especes d'Annelides Tubicoles du Devonien armoricain:* Soc. Geol. France, Bull., v. 6, p. 839-846, text figs.

Richter, Rudolf

(56) 1854, *Thuringische Tentaculiten:* Deutsch. geol. Gesell., Zeitschr., Jahrg. 1854, Bd. 6, p. 275-290, pl. 1.

Ruedemann, Rudolf

(57) 1934, *Paleozoic plankton of North America:* Geol. Soc. America, Mem. 2, 141 p., 26 pl.

Rusconi, Carlos

(58) 1955, *Fosiles Cambricos y Ordovicias al oeste de San Isidro, Mendoza:* Museo Historia Nat. Mendoza, Revista, v. 8, p. 3-64, pl. 1-5.

Salter, J. W.

(59) 1848, *Mémoires of the Geological Survey of Great Britain and of the Museum of Practical Geology in London:* v. 2, pt. 1, p. 231-386, pl. 4-20.

(59a) 1866, *Mémoires of the Geological Survey of Great Britain and of the Museum of Practical Geology in London:* v. 3, p. 239-381, pl. 1-26.

Schlotheim, E. F. von

(60) 1820, *Die Petrefactenkunde auf ihrem jetzigen Standpunkte durch die Beschreibung seiner Sammlung versteinester und fossiler Uberreste des Thier- und Pflanzenreichs der Vorwelt erläutert:* 437 p., pl. 15-29.

Schmidt, Friedrich

(61) 1888, *Über eine neuendeckte Unter-Cambrische Fauna in Esthland:* Acad. Impér. Sci. St. Pétersbourg, Mém., ser. 7, v. 36, no. 2, p. 5, 25-26, pl. 1-2.

Sinclair, G. W.

(62) 1946, *Notes on the nomenclature of Hyolithes:* Jour. Paleontology, v. 20, p. 72-85.

Stubblefield, C. J., & Bulman, O. M.

(63) 1927, *The Shineton shales of the Wrekin District:* Geol. Soc. London, Quart. Jour., v. 83, p. 99-146.

Syssoiev, V. A.

(64) 1957, *K morfologii, sistematischeskomu polozhennyv i sistematike kholitov:* Akad. Nauk SSSR Leningrad, Doklady, v. 116, no. 2, p. 304-307. [*On the morphology, systematic position, and systematics of Hyolithoidea.*]

(65) 1959, *Ekologiya khiolitov:* Akad. Nauk SSSR Leningrad, Doklady, v. 127, no. 4, p. 892-895. [*Ecology of hyolithids.*]

Termier, Geneviève, & Termier, Henri

(66) 1947, *Généralités sur les invertébrés fossiles:* Notes & Mém. no. 69, Paléont. Marocaine, v. 1, p. 208-209, pl. 13.

(67) 1950, *Invertébrés de l'Ère Primaire, Pt. 4, annélides, arthropodes, échinodermes, conularides et graptolithes:* Paléont. Marocaine, v. 2, p. 108-117, pl. 232-234.

(68) 1953, *Groupes d'Affinités Incertaines:* in Traité de Paléontologie (Piveteau, J., ed.), v. 3, p. 1001-1018. (Les hyolithes, p. 1014-1015; Les tentaculites, p. 1016-1018.)

Walch, J. E. I.

(69) 1775, *Naturforscher:* pt. 7, p. 211-213, pl. 4, fig. 5 (Berlin, Tübingen).

Walcott, C. D.

(70) 1885, *Note on some Paleozoic pteropods:* Am. Jour. Sci., v. 30, p. 17-21, illus.

(71) 1886, *Second contribution to the studies on the Cambrian faunas of North America:* U.S. Geol. Survey, Bull. 30, 225 p., 33 pl.

(72) 1889, *Descriptive notes on new genera and species from the Lower Cambrian or Olenellus zone of North America:* U.S. Natl. Museum, Proc., v. 12, p. 33-46.

(73) 1890, *Description of new forms of Upper Cambrian fossils:* U.S. Natl. Museum, Proc., v. 13, p. 266-279, pl. 20-21.

(74) 1911, *Cambrian geology and paleontology II, no. 5, Middle Cambrian annelids:* Smithsonian Misc. Coll., v. 57, no. 5, p. 109-144, pl. 18-23.

Yochelson, E. L.

(75) 1957, *"Pteropods" of the Paleozoic:* Geol. Soc. America, Mem. 67, Treatise on marine ecology and paleoecology, pt. 2 (Ladd, ed.), 827 p.

(76) 1961, *The operculum and mode of life of Hyolithes:* Jour. Paleontology, v. 35, p. 152-161, pl. 33-34.

(77) 1961, *Notes on the class Coniconchia:* Jour. Paleontology, v. 35, p. 162-167.

Zazvorka, Vlastislav

(78) 1928, *Revision of the Hyolithi from dγ:* Palaeont. Bohemiae, no. 13, p. 1-22, pl. 1-3, fig. 1. [Czech. & Eng.]

WORMS

By B. F. Howell
[Princeton University]

CONTENTS

INTRODUCTION

The living animals commonly called worms are so varied in their anatomy that their classification is difficult. Today they are usually classified in several phyla which are based on features of the soft parts of their bodies. As fossil worms are seldom preserved so that even the *form* of the body, let alone the details of the anatomy, can be seen, their classification is even more difficult. For this reason paleontologists have tended to neglect most fossil worms except those which built tubes or distinctive kinds of burrows and those which had jaws composed of material that was easily preserved.

Nevertheless, worms have been abundant, especially in shallow areas, since Proterozoic times, and many genera and species of fossil worms have been described. Those genera and species that have been based on trails and burrows, and certain others whose relationships with the worms are doubtful, are dealt with in other parts of the present volume. Only the genera and species of which the tubes, jaws, or bodies have been preserved as fossils are considered here.

Because the jawed worms of the order Errantida, class Polychaetia, phylum Annelida, had jaw elements (scolecodonts) of various forms in one and the same species, whereas most genera are based on scattered scolecodonts, the arrangement of fossils representing this order in genera and families is as yet very tentative and subject to future revision.

DISTRIBUTION

A few worm trails have been found in Archeozoic rocks, and the trails and burrows of many kinds of worms are present in Proterozoic rocks, but no fossils of the bodies, jaws, or tubes of worms have been discovered in Precambrian strata, except for one tube from a Proterozoic formation. The bodies and tubes of several kinds of worms have been found in Cambrian beds, worm jaws (scolecodonts) occur in Ordovician marine sediments, and fossil worms of many kinds are common in Ordovician and later marine deposits and are occasionally found in fresh-water beds. Forty genera of worms have been reported from Cambrian rocks, 65 from the Ordovician, 54 from the Silurian, 45 from the Devonian, 29 from the Carboniferous, 18 from the Permian, 22 from the Triassic, 23 from the Jurassic, 35 from the Cretaceous, and 43 from the Cenozoic. Some of these genera, such as *Spirorbis,* which has been reported from rocks of all periods from Ordovician to Recent, have had very long-time ranges.

MORPHOLOGY AND ECOLOGY

All living worms are bilaterally symmetrical in the arrangement of their organs, most of them have a well-developed alimentary canal and ventral nervous system, and some have parapodia which they use in locomotion. The majority are crawlers, but some are burrowers and some free-swimmers. Most are adapted for life in the sea, but some live in fresh waters or on land, and not a few are parasitic. In their morphology they are adjusted to these different environments, some crawling forms having flattened or cylindrical bodies with parapodia, the burrowing forms having hairlike, wirelike, or larger cylindrical bodies, and the swimming forms having streamlined, fish-shaped bodies. They range in size from microscopic parasites to earthworms six feet long and large marine worms 90 feet in length. Some, such as the leeches, have sucking organs around the mouth; others possess a series of notched jaws; some bear a proboscis; some have external hairlike spines on their bodies; some are segmented, some not; some have eyes; some have gills; and some have palps around the mouth that are often gill-like in form. Some worms build calcareous tubes or ones composed of sand grains or other materials; they live in these tubes but are not attached to them. The worms of the past had these features also and most have lived in the same environments and had the same habits.

A discussion of the paleoecology of marine worms, with a bibliography, will be found in volume 2 of the *Treatise on Marine Ecology and Paleoecology,* published in 1957 as Memoir 67 of the Geological Society of America.

CLASSIFICATION

The proper classification of many extinct genera of worms is difficult, even when the body is preserved, because the soft parts needed for classification are not preserved. When the fossils are tubes or jaws their relationships with living worms can usually be determined; but even the jaws are difficult to classify generically, because it is known that jaws of several kinds were originally associated in a single species, whereas the fossil jaws are mostly widely scattered and found individually, so that their original associations cannot be learned. The classification of the genera of fossil worms adopted in the present work may therefore prove to be erroneous in part; however, it is the best that the author has been able to devise. The division of worms into phyla, classes, and lower-rank supra-generic taxa recognized in the *Treatise* follows. Numbers enclosed by parentheses indicate the number of known genera represented by fossils.

The adopted names of phyla and classes are those given in A. S. PEARSE's *Zoological names: a list of phyla, classes, and orders,* published by Section F of the American Association for the Advancement of Science in 1949.

Suprageneric Divisions of Worms

- Nemerta *(phylum)* (2). *Jur.-Rec.*
  - Class, order, and family uncertain (2). *Jur.-Rec.*
- Nematomorpha *(phylum)* (1). *Eoc.-Rec.*
  - Gordioida *(order)* (1). *Eoc.-Rec.*
    - Gordiidae (1). *Eoc.-Rec.*
- Nematoida *(phylum)* (4). *Oligo.-Rec.*
  - Phasmidia *(class)* (4). *Oligo.-Rec.*
    - Rhabditida *(order)* (4). *Oligo.-Rec.*
      - Ascaridatina *(suborder)* (4). *Oligo.-Rec.*
        - Plectidae (2). *Oligo.-Rec.*
        - Mermithidae (1). *Oligo.-Rec.*
        - Oxyuridae (1). *Quat.*
- Chaetognatha *(phylum)* (1). *Cam.-Rec.*
  - Class and order uncertain (1). *Cam.*
    - Amiskwiidae (1). *Cam.*
- Annelida *(phylum)* (147). *Prot.-Rec.*
  - Polychaetia *(class)* (143). *Cam.-Rec.*
    - Errantida *(order)* (44). *Ord.-Rec.*
      - Paulinitidae (4). *Ord.-Dev.*
      - Leodicidae (14). *Ord.-Rec.*
      - Aphroditidae (2). *Dev.-Rec.*
      - Lumbriconereidae (2). *Ord.-Jur.*
      - Nereidae (9). *Ord.-Rec.*
      - Onuphididae (2). *Ord.-Rec.*
      - Sigalionidae (1). *Ord.-Rec.*
      - Staurocephalitidae (2). *Ord.-Rec.*
      - Amphinomidae (1). *Jur.-Rec.*
      - Phyllodocidae (1). *Sil.-Rec.*
      - Glyceridae (3). *Ord.-Rec.*
      - Sprigginidae (1). *Cam.*
      - Family uncertain (2). *Ord.-Sil.*
    - Sedentarida *(order)* (90). *Cam.-Rec.*
      - Keiloritidae (1). *Ord.-Sil.*
      - Hermellidae (1). *Carb.-Rec.*
      - Sabellidae (2). *Cret.-Rec.*
      - Serpulidae (48). *Cam.-Rec.*
      - Terebellidae (12). *Cam.-Rec.*
      - Cirratulidae (1). *Mio.-Rec.*
      - Amphictenidae (2). *Perm.-Rec.*
      - Spionidae (1). *Mio.-Rec.*
      - Pikaiidae (2). *Cam.*
      - Arenicolidae (2). *Trias.-Rec.*
      - Chloraemidae (2). *Tert.-Rec.*
      - Family Uncertain (16). *Cam.-Sil.*
    - Miskoiida *(order)* (8). *Cam.-Ord.*
      - Miskoiidae (1). *Cam.*
      - Canadiidae (2). *Cam.*
      - Wiwaxiidae (3). *Cam.*
      - Family uncertain (2). *Ord.*
    - Order Uncertain (1).
  - Myzostomia *(class)* (1). *Ord.-Rec.*
    - Myzostomidae (1). *Ord.-Rec.*
  - Oligochaetia *(class)* (3). *Carb.-Rec.*
    - Plesiotheca *(order)* (1). *Carb.-Rec.*
      - Tubificidae (1). *Carb.-Rec.*
    - Prosotheca *(order)* (1). *Tert.-Rec.*
      - Enchytraeidae (1). *Tert.-Rec.*
    - Order Uncertain (1). *Ord.-Perm.*
      - Family Uncertain (1). *Ord.-Perm.*
- Sipunculoida *(phylum)* (6). *Cam.-Rec.*
  - Order Uncertain (6). *Cam.-Jur.*
    - Ottoiidae (1). *Cam.*
    - Family Uncertain (5). *Ord.-Jur.*
- Phylum Uncertain (10). *Cam.-Cret.*
  - Class Uncertain (10). *Cam.-Cret.*
    - Order Uncertain. *Cam.-Cret.*
      - Palaeoscolecidae (1). *Ord.*
      - Family Uncertain (9). *Cam.-Cret.*
- Phyla of worms (167). *Prot.-Rec.*

SYSTEMATIC DESCRIPTIONS

Phylum NEMERTA Delle Chiaje, 1841

Worms with elongate cylindrical, cord-like, or more or less flattened unsegmented bodies and a threadlike proboscis which can be withdrawn; most species marine, but some living in fresh waters or on land and some as parasites. *Jur.-Rec.*

Class, order, and family UNCERTAIN

Hirudella MÜNSTER, 1842 [**H. angusta*]. Body cylindrical, longer than that of *Legnodesmus,* tapering somewhat at anterior end; marine. *U. Jur.-Cret.,* Eu.——FIG. 85,*1*. **H. angusta,* U.Jur. (Solnhofen), Ger.; ×1 (103).

Legnodesmus EHLERS, 1869 [**L. ehlersi* HOWELL, 1958; SD HOWELL, 1958]. Body cylindrical, not

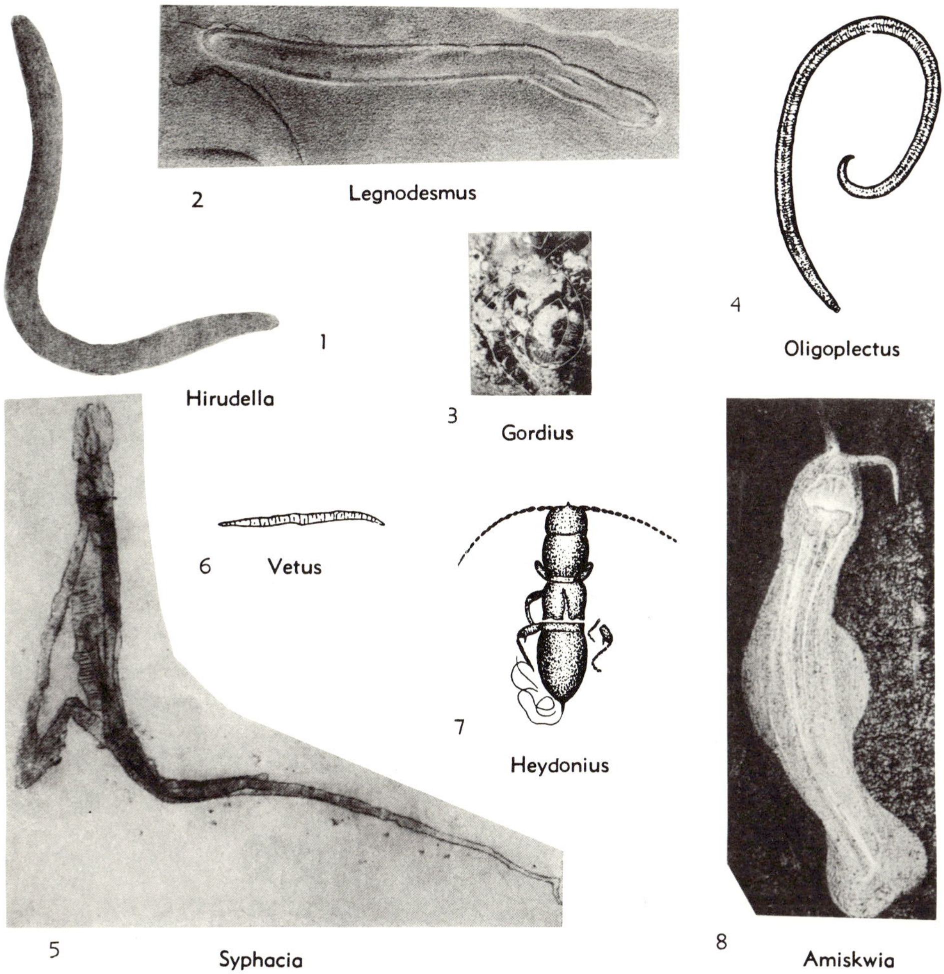

FIG. 85. Nemerta, Nematomorpha, Nematoida, Chaetognatha (p. W146-W148).

very long; marine. *U.Jur.*, Eu.——FIG. 85,*2*. **L. ehlersi* HOWELL, Solnhofen, Ger.; ×1 (34).

Phylum NEMATOMORPHA Vejovsky, 1886

Body hair-shaped, unsegmented, with reproductive organs dorsal to intestine. [These worms live in marine and fresh waters and as parasites.] *Eoc.-Rec.*

Order GORDIOIDA Ortlepp, 1924

Alimentary canal degenerate. *Eoc.-Rec.*

Family GORDIIDAE May, 1919

Ovaries and testes open at hind end of body. [Living in marine and fresh waters and as parasites.] *Eoc.-Rec.*

Gordius LINNÉ, 1758 [**G. aquaticus*]. Body filiform, parasitic in larval stages. [Adult living in fresh waters.] *Eoc.-Rec.*, Eu.——FIG. 85,*3*. *G. tenuifibrosis* VOIGT, Eoc.(Braunkohle), Ger.; ×7 (152).

Phylum NEMATOIDA Rudolphi, 1808

Body thread-shaped, with pointed ends. [These worms are usually minute and live in the soil, in fresh and marine waters, and as parasites in plants and animals.] *Oligo.-Rec.*

Class PHASMIDIA Chitwood, 1933

Phasmids present, caudal glands absent, sensory organs rarely setose. *Oligo.-Rec.*

Order RHABDITIDA Chitwood, 1933

Esophagus divisible into 3 regions: corpus, isthmus, and bulbar tract. *Oligo.-Rec.*

Suborder ASCARIDATINA Skrjabin, 1915

Lips 3 or 6 [Free-living or parasitic.] *Oligo.-Rec.*

Family PLECTIDAE Chitwood & Chitwood, 1937

Bulbar region of esophagus muscular. *Oligo.-Rec.*

Plectus BASTIAN, 1865 [**P. parientinus*]. *Rec.*

Oligoplectus TAYLOR, 1935 [**Auguillula succini* DUISBURG, 1862 *(partim)*]. Body annulated, tapering toward each end, with knoblike head and pointed tail. *Oligo.*, Eu.——FIG. 85,*4*. **O. succini* (DUISBURG), Ger.; ×80 (146).

Vetus TAYLOR, 1935 [**V. duisburgi*]. Body annulated, head rounded, tail bluntly pointed. *Oligo.*, Eu.——FIG. 85,*6*. **V. duisburgi*, Ger.; ×80 (146).

Family MERMITHIDAE Braun, 1883

Body up to 20 cm. or more in length; cuticle smooth, but containing criss-cross fibers; head rounded, with 4 submedian and 2 lateral papillae; spicules paired or single. *Oligo.-Rec.*

Mermis DUJARDIN, 1842 [**M. nigrescens*]. *Rec.*

Heydonius TAYLOR, 1935 [**Mermis matutina* MENGE, 1866]. Body 3.5 mm. long, 0.1 mm. wide, cylindrical, bluntly pointed at head and tail, with about 300 annules and 2 spicules. [Parasitic in insects.] *Oligo.*, Eu.——FIG. 85,*7*. **H. matitunus* (MENGE), Ger.; ×30 (146).

Family OXYURIDAE Cobbold, 1864

Body meromyarian, mouth with simple lips, male usually with one spicule but may be absent. *Quat.*

Oxyuris RUDOLPHI, 1803 [**O. curvula*]. *Rec.*

Syphacia SEURAT, 1916 [**Ascaris obvelata* RUDOLPHI, 1802]. Cuticle finely striated transversely; excretory pore very small and situated on median ventral line, behind esophageal bulb. [Parasitic in mammals.] *Pleist.*, Sib.——FIG. 85,*5*. *S.* cf. *obvelata* (RUDOLPHI); ×30 (31).

Phylum CHAETOGNATHA Leuckart, 1854

Swimming marine worms with fins on their tails, and in some on sides of body, and with rows of spines and hooks in mouth. *Cam.-Rec.*

Class and order UNCERTAIN

Family AMISKWIIDAE Walcott, 1911

[*nom. correct.* HOWELL, herein (*pro* Amiskwidae WALCOTT, 1911)]

Body with one pair of lateral fins; septum between head and trunk, but none between trunk and tail. *Cam.*

Amiskwia WALCOTT, 1911 [**A. sagittiformis*]. Body divided into broadly elongate oval head, cylindrical trunk, and expanded tail; head with bluntly pointed anterior end and bearing pair of tentacles. *M.Cam.*, Can.(B.C.).——FIG. 85,*8*. **A. sagittiformis*, Burgess Sh.; ×3 (154).

Phylum ANNELIDA Lamarck, 1809

Worms with distinct head, segmented trunk, and unsegmented pygidium. *Prot.-Rec.*

Class POLYCHAETIA Grube, 1850

Segments of trunk bearing lateral bundles of bristles called chaetae. [Mostly marine, but some live in brackish and fresh waters.] *Cam.-Rec.*

Order ERRANTIDA Audouin & Milne-Edwards, 1832

Segments of trunk all alike; mouth commonly bearing numerous pairs of notched jaws called scolecodonts. [Mobile.] *Ord.-Rec.*

Family PAULINITIDAE Lange, 1947

Mandibles inarticulate, shafts inwardly curved; maxillae in asymmetrical pairs; carriers short, smooth, slender, with curved margins; forceps asymmetrical, denticulated along entire inner margin, with large anterior hook; dental plates asymmetrical, denticulate, with shank on outer margin; unpaired piece denticulate, located on left side of apparatus; paragnaths asymmetrical, denticulate. *Ord.-Dev.*

Paulinites LANGE, 1947 [**P. paranaensis*]. One pair of long, conical, ventral mandibles; 2 short,

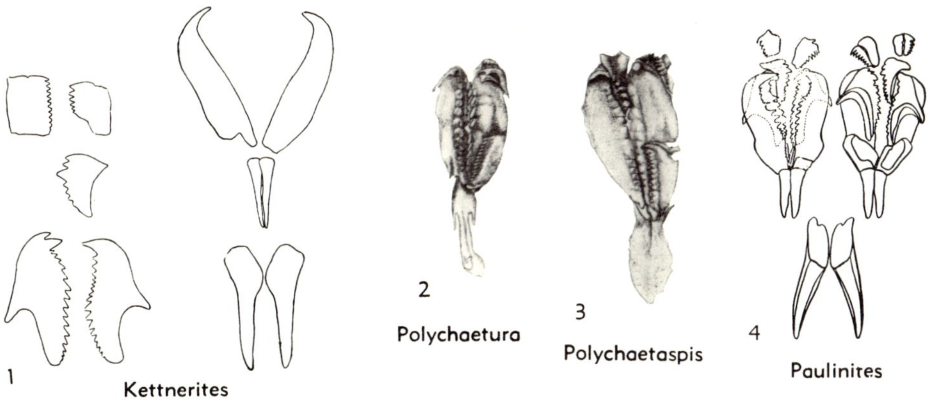

FIG. 86. Paulinitidae (p. *W*148-*W*149).

slender, posterior dorsal carriers; forceps asymmetrical, ending in stout hook, and with small, backward-directed denticles along whole length of inner margin; 2 small, irregularly dentate, asymmetrical, subtriangular dental plates; elongate, subtriangular unpaired piece; 2 small, irregularly oblong, asymmetrical paragnaths. *Dev.*, S.Am.——FIG. 86,*4*. **P. paranaensis,* Ponta Grossa F., Brazil; ×8 (85).

Kettnerites ŽEBERA, 1935 [**K. kosoviensis*]. Jaw apparatus composed of single pair of mandibles and 7 asymmetrical maxillary plates with pair of carriers; mandibles with broad anterior part which runs back into elongated, gradually narrowing basal shafts; carriers elongate, conical, with convex anterior margin; forceps asymmetrical, carinate, attached to carriers, with anterior hook; dental plates triangular, with shank at outer lateral margin; unpaired piece triangular, with unequally long denticles on inner lateral margin; paragnaths asymmetrical, approximately square, with row of minute denticles on inner margin of jaw. *Ord.-Dev.*, Eu.——FIG. 86,*1*. **K. kosoviensis,* Sil.(Budňany) Czech.; ×7 (139).

Polychaetaspis KOZLOWSKI, 1956 [**P. wyszogrodensis*]. Ten paired jaws and 2 or 3 unpaired jaws; forceps very asymmetrical, with denticles on entire length; denticulate basal piece; one pair of paragnaths. *Ord.*, Eu.——FIG. 86,*3*. **P. wyszogrodensis* (erratic boulder), Pol.; ×16 (80).

Polychaetura KOZLOWSKI, 1956 [**P. gracilis*]. One pair of denticulate forceps with basal plates; 2 pairs of denticulate maxillary plates. *Ord.*, Eu.——FIG. 86,*2*. **P. gracilis* (erratic boulder), Pol.; ×60 (80).

Family LEODICIDAE Treadwell, 1921

Body long, first 2 segments without parapodia, later segments with one branch or one and a half branches. [Marine.] *Ord.-Rec.*

Leodice LAMARCK, 1818. *Rec.*

Arabellites HINDE, 1879 [**A. hamatus*]. Forceps with very large hook and row of denticles on wide base; mandibles subquadrate in form with straight, denticulate, upper edge. *Ord.-Dev.*, N. Am.-Eu.——FIG. 87,*1*. **A. hamatus,* Ord.(Pulaski), Ont.; ×20 (66).

Diopatraites ELLER, 1938 [**D. conformis*]. Mandible consisting of 3-toothed frontal plate followed by tapering shaft with fine striae parallel to outer and posterior margins of plate; inner margin of plate straight; upper surface of shaft convex, lower side angular and concave. *Ord.-Dev.*, N.Am.——FIG. 87,*2*. **D. conformis,* Dev.(Potter Farm), Mich.; ×9 (41).

Eunicites EHLERS, 1868 [**E. avitus*]. Body long, with many spined parapodia, forceps in 2 parts; 3 to 5 pairs of mandibles; one unpaired plate. *Ord.-Rec.*, Eu.-N.Am.——FIG. 87,*13*. **E. avitus,* Jur.(Solnhofen), Ger.; ×0.5 (33).

Leodicites ELLER, 1940 [**L. variedentatus*]. Jaws of maxilla II triangular, without fang or primary denticle; inner margin bearing denticles which are variously shaped and not always uniform in arrangement; anterior margin round or slightly incurved to form blunt or acute shank; large indentation on outer margin just posterior to shank; fossa large, may occupy half to three-quarters of jaw length. *Ord.-Sil.*, N.Am.——FIG. 87,*4*. **L. variedentatus,* Sil.(Albion), USA(N.Y.); ×35 (42).

Marphysaites ELLER, 1945 [**M. aptus*]. Mandible consisting of 2 shafts joined or articulated at anterior end of inner margin; shafts of mandible elongate, wide anteriorly and tapering to pointed or blunt posterior end; thickened anterior margin straight or curved, shafts curving outward or

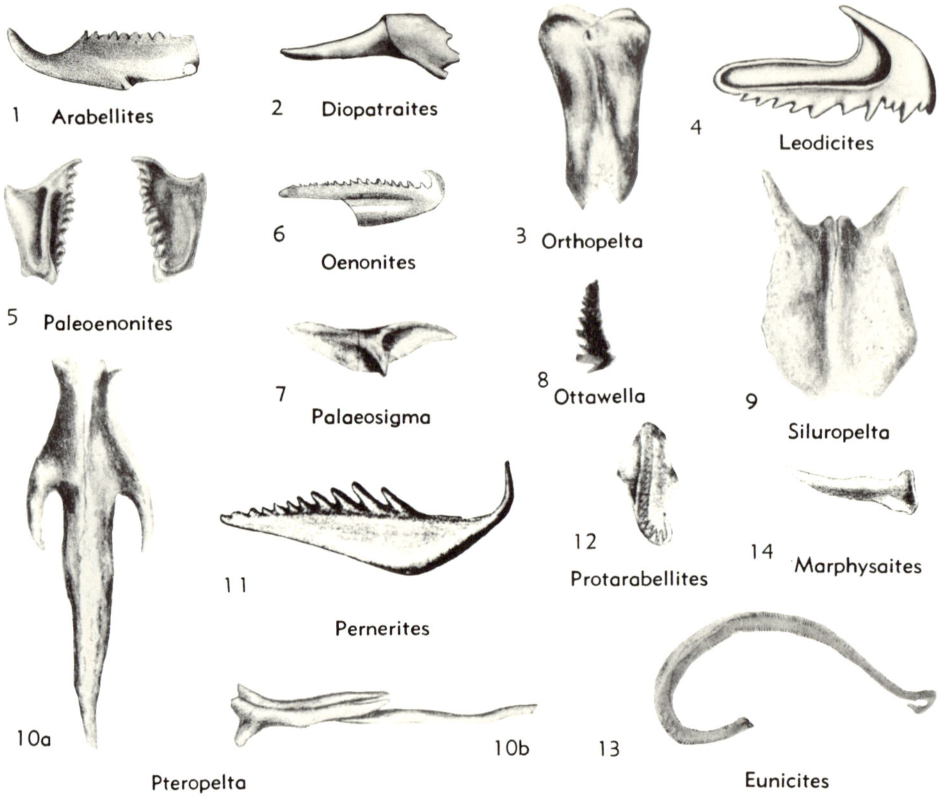

FIG. 87. Leodicidae (p. W149-W151).

nearly straight; surface of mandible convex or flattened. *M.Ord.*, Can.(Ont.).——FIG. 87,*14*. **M. aptus*, Coburg Ls.; ×28 (45).

Oenonites HINDE, 1879 [**O. curvidens*]. Jaws with more or less curved hook, followed by series of smaller teeth, similar in character to those of existing genus *Oenone*. *Ord.-Dev.*, Eu.-N.Am.——FIG. 87,*6*. **O. curvidens*, Ord.(Cinc.), Can.(Ont.); ×15 (66).

Orthopelta EISENACK, 1939 [**O. navis*]. Jaws without denticles, median knobs, or side wings; anterior end mostly short, vertical to median fissure or depressed or raised at side; median fissure extending to rear border. *Ord.-Sil.*, Eu.——FIG. 87,*3*. **O. navis*, Sil.(glacial boulder), Ger.; ×54 (38).

Ottawella WILSON, 1948 [**O. sinclairi*]. Jaw arched; base deep, laterally compressed, posterior and anterior margins convex, making acute angle with arched lower margin; abrupt lateral thickening at lowest anterior point; lower margin smooth, strengthened by ridgelike thickening; cusps pointed backward, irregular in size and shape, one con-considerably larger than others. *M.Ord.*, Can. (Ont.).——FIG. 87,*8*. **O. sinclairi*, Cobourg Ls.; ×10 (164).

Palaeosigma EISENACK, 1939 [**P. silurica*]. Jaw fan-shaped, with curved spike in middle. *Sil.*, Eu.——FIG. 87,*7*. **P. silurica*, Sil.(glacial boulder); ×54 (38).

Paleoenonites ELLER, 1942 [**P. accuratus*]. Jaws varying from triangular to rectangular in shape, with incurved anterior margin that ends in forward-directed shank; posterior ranging from acute extremity to broad truncated margin, which may be rounded, straight, incurved, or obliquely truncate; fossa ranging from narrow to broad, deep to shallow; inner and outer margins straight, incurved, or rounded; with series of sharply conical or short, blunt denticles on inner margin. *U.Ord.*, N.Am.——FIG. 87,*5*. **P. accuratus*, Erindale F., Can.(Ont.); ×25 (43).

Pernerites ŽEBERA, 1935 [**P. giganteus*]. Jaw subtriangular with long, curved, hook at anterior end, behind which is a series of denticles progressively smaller toward rear end of jaw and pointed backward at angle of 45 degrees to horizontal; shallow, longitudinal, depression separates upper portion of jaw from lower part. *Sil.*, Eu.——FIG. 87,*11*. **P. giganteus*, "e beta" Zone, Czech.; ×12 (169).

Protarabellites STAUFFER, 1933 [**P. humilis*]. Jaws and dental plates resembling those of *Arabellites*, but differing in having base much expanded laterally or flange that extends along most of both inner and outer sides; denticulate ridge usually crosses flattened base diagonally, bearing about 18 to 22 small teeth, and terminating anteriorly in prominent curved hook; base hollowed out along median line. *M.Ord.*, N.Am.——FIG. 87,*12*. **P. humilis*, Decorah Sh., USA(Minn.); ×30 (143).

Pteropelta EISENACK, 1939 [**P. gladiata*]. Jaws with small lateral spurs beside median portion and 2 backward-directed, winglike spines parallel to outer edges. *Sil.*, Eu.——FIG. 87,*10a,b*. **P. gladiata*, Sil.(glacial boulder), Ger.; ×70 (38).

Siluropelta EISENACK, 1939 [**S. lata*]. Jaws with strongly developed lateral spines and medial knobs; median cleft poorly developed toward rear and obsolete at end; no side wings. *Sil.*, Eu.——FIG. 87,*9*. **S. lata*, Sil.(glacial boulder), Ger.; ×54 (38).

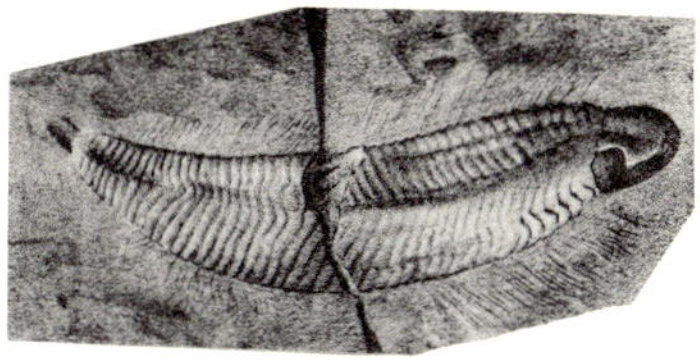

1 Protonympha

2 Sthenelaites

FIG. 88. Aphroditidae (p. *W*151).

Family APHRODITIDAE Savigny, 1820

Short worms with scales on their backs. [Marine.] *Dev.-Rec.*

Aphrodita LINNÉ, 1758 [**A. aculeata*]. *Rec.*

Protonympha CLARKE, 1903 [**P. salicifolia*]. Body tapering narrowly at anterior end and bluntly at posterior end, composed of about 50 segments, covered by overlapping plates, with narrow median elevation along axial line and long setae along sides. *U.Dev.*, N.Am.——FIG. 88,*1*. **P. salicifolia*, Portage Gr., USA(N.Y.); ×0.7 (17).

Sthenelaites ROVERTO, 1903 [**Nereites dasiaeformis* MASSALONGO, 1855]. Body long and narrow, with many thin chaetae, and with tiny scales on back. *Tert.*, Eu.——FIG. 88,*2*. **S. dasiaeformis* (MASSALONGO), Italy; ×10 (90).

Family LUMBRICONEREIDAE Schmarda, 1877

Body long, prostomium conical; cirri on upper surface rudimentary or lacking, no ventral cirri, bristles single or grouped or hook-shaped; lower jaw and 3 to 5 pairs of upper jaw elements without unpaired element. [Marine.] *Ord.-Rec.*

Lumbriconereis GRUBE, 1840 [**L. quadristriata*]. Upper jaws in 3 to 5 pairs; no unpaired element. *Plio.-Rec.*, cosmop.——FIG. 91,*2*. L. *ocellata* GRUBE, Rec., Philip.; ×20 (60).

Lumbriconereites EHLERS, 1869 [**L. deperditus*]. Jaws consisting of oblong or elongate, nearly straight to curved forms with denticulate ridge, in some specimens supported on margin of triangular basal flange, which may be broad and flat or concave and commonly terminating in rounded or sharp inner angle formed by inner, curved, posterior margins; anterior tooth or teeth usually elongate, bent backward, and followed by series of backward-directed teeth. [Marine.] *Ord.-Jur.*, Eu.-N.Am.——FIG. 89,*1*. **L. deperditus*, Jur.(Solnhofen), Ger.; ×0.25 (34).

Family NEREIDAE Savigny, 1820

Body long and many-segmented; 2 large jaws and usually paragnaths in 8 groups; first 2 pairs of parapodia simple, others double; with dorsal and ventral cirri; bristles with sickle-shaped ends; 2 anal cirri. [Marine.] *Ord.-Rec.*

Nereis LINNÉ, 1758 [**N. noctiluca*]. *Rec.*

Ctenoscolex EHLERS, 1869 [**C. procerus*]. Body long, poorly segmented along mid-line but better segmented at sides. *U.Jur.*, Eu.——FIG. 89,*2*. **C. procerus*, Solnhofen Ls., Ger.; ×1 (34).

Dinoscolites STAUFFER, 1933 [**D. mirabilis*]. Jaws massive, U- or V-shaped, limbs nearly circular in cross section, inner limb smooth, outer limb bearing irregular series of uneven teeth; anterior end of jaw bearing pair of large clawlike teeth, with several smaller teeth between. *M.Ord.*, N.Am. ——FIG. 91,*8*. **D. mirabilis*, Platteville Ls., USA (Minn.); ×30 (143).

Nawnites ROY, 1929 [**N. gilboensis*]. Body long and segmented; other characters not known. *M. Dev.*, N.Am.——FIG. 90,*1*. **N. gilboensis*, Ithaca F., USA(N.Y.); ×0.07 (123).

Nereidavus GRINNELL, 1877 [**N. varians*]. Jaws hollow, with more than 8 teeth, anterior one

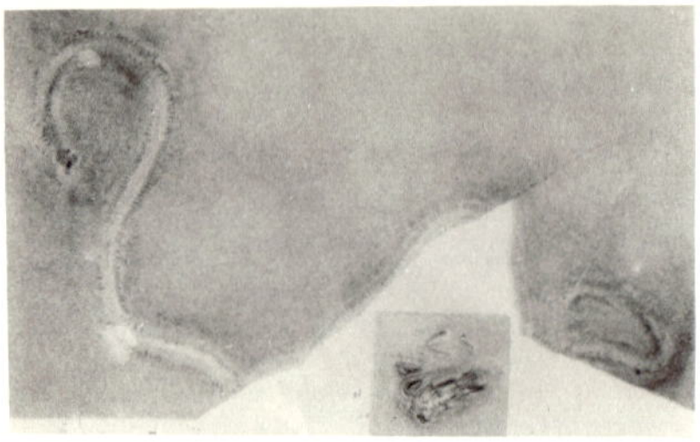

1 Lumbriconereites

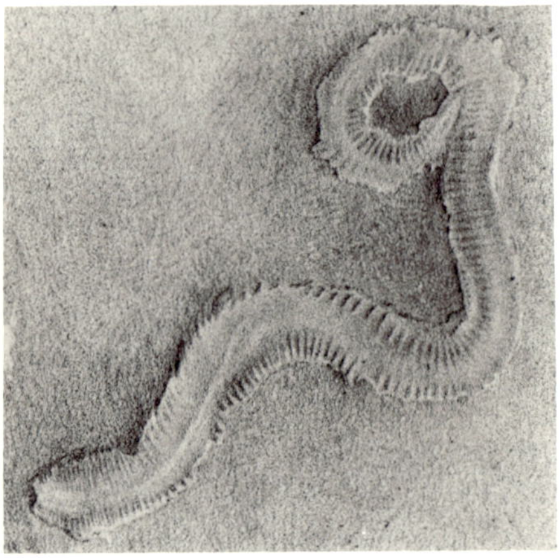

2 Ctenoscolex

FIG. 89. Lumbriconereidae, Nereidae (p. W151).

longest and twisted outward. *U.Ord.-U.Dev.*, N. Am.-Eu.——FIG. 91,*3*. **N. varians*, U.Ord. (Cinc.), USA(Ohio); ×5 (58).

Paleonereites STAUFFER, 1933 [**P. cervicornis*]. Jaws having thick base, with inner margin prolonged into broad, concave, flange which has arched margins; terminations of flange meeting in spinelike point; outer side continued downward from base of teeth, as convex, triangular, plate, lower margin of which passes into thin featheredges; anterior with strong, curved hook, usually with carinated inner edge, succeeded along denticulate outer edge by 6 or more long, slender, pointed teeth. *M.Ord.*, N.Am.——FIG. 91,*5*. **P. cervicornis*, Decorah Sh., USA(Minn.); ×30 (143).

Paranereites EISENACK, 1939 [**P. balticus*]. Jaws short and simple, with stout hook on end and no smaller teeth. *Sil.*, Eu.——FIG. 91,*6*. **P. balticus*, Sil.(glacial boulder); ×60 (38).

Pronereites STAUFFER, 1933 [**P. primus*]. Jaw stout, gently arched, base passing gradually into anterior hook, followed by 5 or more stout teeth; basal surface rounded, with slight carina on inner side running up into hook; slight flange also on inner side of jaw. *M.Ord.*, N.Am.——FIG. 91,*4*. **P. primus*, Glenwood F., USA(Minn.); ×30 (143).

Triadonereis MAYER, 1954 [**T. eckerti*]. Body segmented and shaped much like that of modern *Nereis*, tapering toward rear. *M.Trias.*, Eu.——FIG. 91,*7*. **T. eckerti*, Trochitenkalk, Ger.; ×2 (93).

Ungulites STAUFFER, 1933 [**U. bicuspidatus*]. Jaw clawlike, with prominent sharp tooth, or hook, succeeded on one or both sides by 1 or 2 (possibly more) similar but usually shorter teeth; base of jaw not distinctly separate from tooth or teeth and slightly arched from base to apex of tooth. *M.Ord.*, N.Am.——FIG. 91,*1a,b*. **U. bicuspidatus*, Decorah Sh., USA(Minn.); ×30 (143).

Family ONUPHIDIDAE McIntosh, 1910

Like Leodicidae, but with 2 tentacles and 2 palps. [Marine.] *Ord.-Rec.*

Onuphis AUDOUIN & MILNE-EDWARDS, 1833 [**O. eremita*]. *Rec.*

Hyalinaecites STAUFFER, 1933 [**H. typicalis*]. Maxillae large and nearly straight for 0.75 of length, then curved at 30-degree angle; bulge along upper surface and tip extended into point; second dental plate large, with 1, 2, or 3 larger anterior teeth curving out of line with cutting edge, having about 12 gradually diminishing, backward-pointing teeth. *M.Ord.*, N.Am.——FIG. 92,*4*. **H. typicalis*, Decorah Sh., USA(Minn.); ×30 (143).

Nothrites STAUFFER, 1933 [**N. elegans*]. Mandibles with long, tapering, rootlike shafts, triangular in cross section and denticulate, their 2 parts in contact or coalesced for short distance along sides at base of cutting edge and then diverging

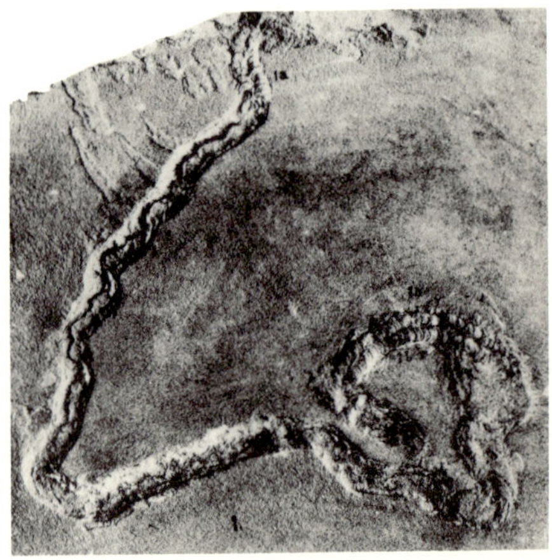

1 Nawnites

FIG. 90. Nereidae (p. W151).

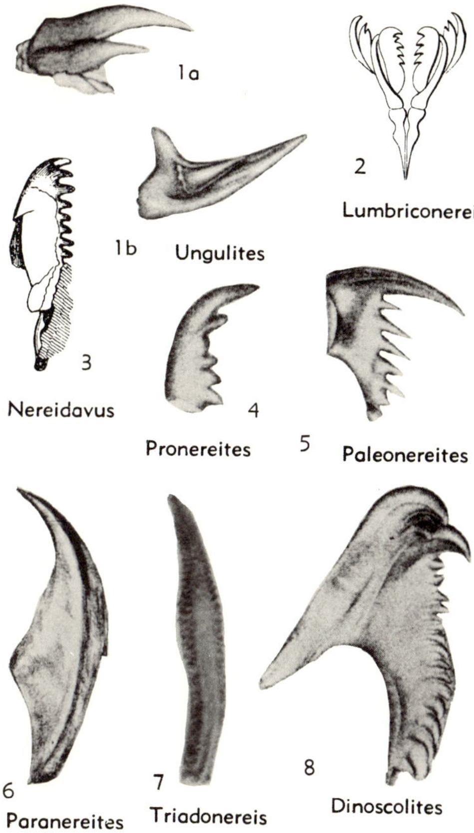

FIG. 91. Lumbriconereidae, Nereidae (p. *W*151-*W*152).

rather widely in shafts. *M.Ord.-Dev.*, N.Am.——FIG. 92,*3*. **N. elegans*, M.Ord.(Decorah Sh.), USA(Minn.); ×60 (143).

Family SIGALIONIDAE Kinberg, 1855

Body long and narrow; cephalic lobe rounded; feet on anterior segments bearing either an elytron or dorsal cirrus; feet on posterior segments bearing both elytra and dorsal cirri. [Marine.] *Ord.-Rec.*

Sigalion AUDOUIN & MILNE-EDWARDS, 1832 [**S. mathilde*]. *Rec.*

Thalenessites STAUFFER, 1933 [**T. lobatus*]. Thick polygonal chitinous dorsal scales or plates; jaws poorly developed. *M.Ord.*, N.Am.——FIG. 92,*6*. **T. lobatus*, Decorah Sh., USA(Minn.); ×35 (143).

Family STAUROCEPHALITIDAE Kinberg, 1865

[*nom. correct.* HOWELL, herein (*pro* Staurocephalidae KINBERG, 1865)]

Maxillae in 2 long rows on either side, very small and numerous, posterior one not a forcep; parapodia uniramous; setae simple and compound, 2 tentacles and 2 palps; dorsal and ventral cirri. *Ord.-Rec.*

Staurocephalites HINDE, 1879 [**S. niagarensis*]. Jaws elongate, compressed, denticulate, resembling those of *Staurocephalus*. GRUBE, 1855 (*non* BARRANDE, 1846). *Ord.-Dev.*, N.Am.-Eu.——FIG. 92, *1*. **S. niagarensis*, Sil.(Niagaran), Can.(Ont.); ×14 (66).

Marlenites ELLER, 1945 [**Oenonites marginatus* ELLER, 1944]. Jaw narrow, suboval in outline, usually short but may be elongate; with series of denticles extending full length of jaw or nearly to posterior end; denticles commonly uniform in size but usually larger at anterior end; anterior end of jaw usually broadly rounded, posterior end blunt or acute; underside, bearing denticles, may be straight, arched or curved, usually convex, but may be flattened or slightly concave near edges and at anterior end; entire upper side occupied by fossa with thickened and rounded margins. *M.Ord.-L.Sil.*, N.Am.——FIG. 92,*2*. **M. marginatus* (ELLER), Sil.(Manitoulin Ls.), Can(Ont.); ×30 (44).

Family AMPHINOMIDAE Savigny, 1820

Body long and cylindrical or flattened oval in cross section. Parapodia with 2 branches bearing simple bristles, and with 1 or 2 poorly developed appendages on pygidium. [Marine.] *Jur.-Rec.*

Amphinome BRUGUIÈRE, 1789 [**Aphrodita rostrata* PALLAS, 1766]. *Rec.*

Meringosoma EHLERS, 1869 [**M. curtum*]. Body short and broad, middle part of upper surface unsegmented, sides segmented, bristles on middle part short, longer on sides, and longest and hair-

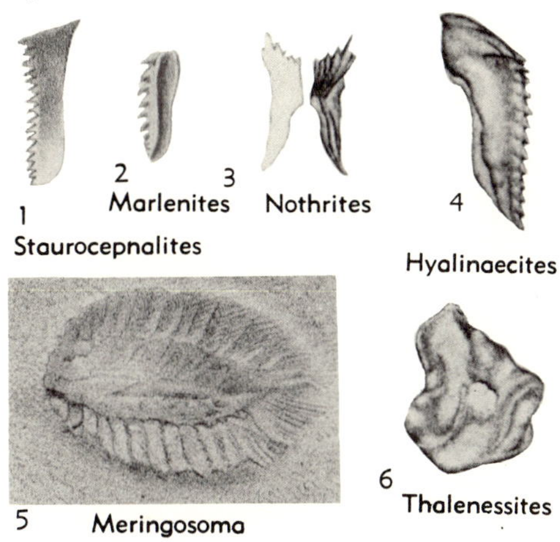

FIG. 92. Onuphididae, Sigalionidae, Staurocephalidae, Amphinomidae (p. *W*152-*W*153).

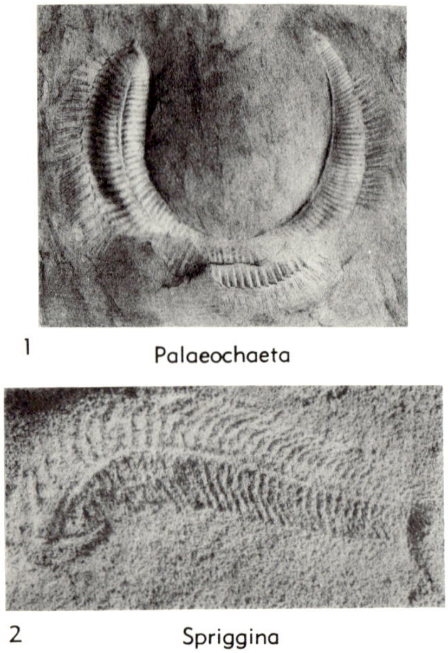

FIG. 93. Phyllodocidae, Sprigginidae (p. W154).

like on rear portions of sides. *U.Jur.*, Eu.——FIG. 92,*5*. **M. curtum*, Solnhofen Ls., Ger.; ×1 (34).

Family PHYLLODOCIDAE Grube, 1850

Body long and narrow, with many segments, which bear 1, 2, or more setae; dorsal and ventral cirri leaf-shaped; 2 anal cirri may be present. [Marine.] *Dev.-Rec.*

Phyllodoce RANZANI, 1817 [**P. maxillosa*]. *Rec.*

Palaeochaeta CLARKE, 1903 [**P. devonica*]. Anterior end blunt, rest of body tapering back to pointed rear; with more than 100 segments; body divided longitudinally by dorsal furrow; several setae on each parapodium. *Sil.-Dev.*, N.Am.-Eu. (Czech.).——FIG. 93,*1*. **P. devonica*, U.Dev. (Portage), USA(N.Y.); ×1 (17).

Family GLYCERIDAE Grube, 1850

Body cylindrical, tapering backward, segmented, segments generally bipartite or tripartite; prostomium cone-shaped; proboscis bearing papillae and jaws; parapodia with 1 or 2 setae; dorsal setae single, ventral setae in clusters; 2 anal cirri. [Marine.] *Ord.-Rec.*

Glycera SAVIGNY, 1818 [**G. unicornis*]. *Rec.*

Glycerites HINDE, 1879 [**G. sulcatus*]. Jaws consisting of simple curved hook with wide base, without smaller teeth, resembling those of modern *Glycera*. *Ord.-Dev.*, N.Am.——FIG. 94,*3*. **G. sulcatus*, U.Ord.(Cinc.), Can.; ×13 (66).

Ildraites ELLER, 1936 [**Arabellites bipennis* ELLER, 1934]. Anterior extremity of maxilla I with prominent pointed hook and row of several, usually acute, denticles along nearly straight inner lateral margin; posterior part sickle-shaped because of crescent-shaped bight. *Ord.-Dev.*, N.Am. ——FIG. 94,*1*. **I. bipennis* (ELLER), U.Dev. (Canadaway), USA(N.Y.); ×13 (39).

Paraglycerites EISENACK, 1939 [**P. necans*]. Hook-shaped jaws, without teeth except for large, dagger-shaped, spur at one side. *Jur.*, Eu.——FIG. 94,*2*. **P. necans*, U.Jur.(Kelloway, glacial boulder), Baltic; ×30 (38).

Family SPRIGGINIDAE Glaessner, 1958

Body rather flat; head without external segmentation, with lateral extensions which produce rough horseshoe shape; trunk consisting of very gently tapering series of segments, parapodia with acicular setae, pharynx well developed. [Marine.] *Cam.*

Spriggina GLAESSNER, 1958 [**S. floundersi*]. Characters of family; about 40 segments in trunk. *Cam.*, S.Austral.——FIG. 93,*2*. **S. floundersi*, ×1.3 (53).

Family UNCERTAIN

Ebetallites ŽEBERA, 1935 [**E. ancoraeformis*]. Jaw in form of semilunar plate with blunt spur on convex side; small, curved, semilunar, protuberance near concave border; small, curved, tooth near right obtuse point of jaw. [Marine.] *Sil.*,

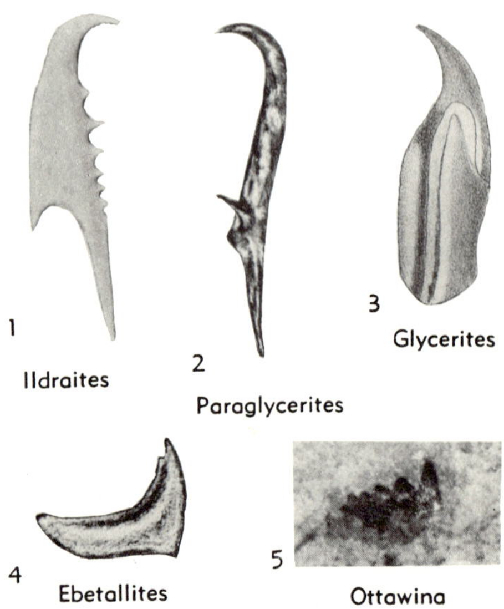

FIG. 94. Glyceridae, Family Uncertain (p. W154-W155).

Eu.——FIG. 94,*4*. **E. ancoraeformis*, Kosov F., Czech., ×35 (169).

Ottawina WILSON, 1948 [**O. trentonensis*]. Base of jaw deep and laterally compressed; denticulate margin straight, bearing more than 6 stout, blunt, upright cusps, anterior considerably larger than next one, others decreasing in size regularly toward posterior. *M.Ord.*, N.Am.——FIG. 94,*5*. **O. trentonensis*, Cobourg Ls., Can.(Ont.); ×6 (164).

Order SEDENTARIDA Lamarck, 1818

Worms that build tubes or burrows in which they live. [Marine.] *Cam.-Rec.*

Family KEILORITIDAE Allan, 1927

Worms that made perpendicular and diagonal burrows lined with membranous material, evenly rounded at lower end, irregularly constricted in some. *Ord.-Sil.*

Keilorites ALLAN, 1927 [*pro Trachyderma* PHILLIPS, 1848 (*non* LATREILLE, 1829)] [**Trachyderma crassituba* CHAPMAN, 1910]. Characters of family. *Ord.-Sil.*, Eu.-Austral.——FIG. 95,*1*. **K. crassituba* (CHAPMAN), Sil.(Melbourne), Austral. (Vict.); ×0.45 (15).

Family HERMELLIDAE Quatrefages, 1848

Body consisting of: (1) prostomium, with mouth surrounded by threadlike tentacles, (2) thorax, with first 2 segments bearing parapodia with hairlike bristles and 3 or 4 next segments bearing bifid parapodia with pin-shaped bristles, (3) abdomen, with comblike, hooked, bristles on dorsal surface and hairlike bristles on ventral surface, and (4) long, unsegmented, attenuate, posterior portion without parapodia or bristles. [Makes tubes of sand, many of which commonly occur together on ocean bottom.] *Carb.-Rec.*

Sabellaria LAMARCK, 1818 [**Sabella alveolata* LINNÉ, 1767] [=*Hermella* SAVIGNY, 1822]. Characters of family. *Carb.-Rec.*, cosmop.——FIG. 95, *4*. **S. alveolata* (LINNÉ), Rec., Eu.; ×1.3 (64).

Family SABELLIDAE Malmgren, 1867

Body long, somewhat flattened cylindrical; first 4 to 12 segments with dorsal bundles of hairlike bristles and ventral hooks; abdomen with many segments bearing dorsal hooks and ventral bristles; first segment with collar carrying tentacles; no operculum present; builds cylindrical tube of horny material or of sand grains or other small objects that are held together by sticky material. [Marine.] *Cret.-Rec.*

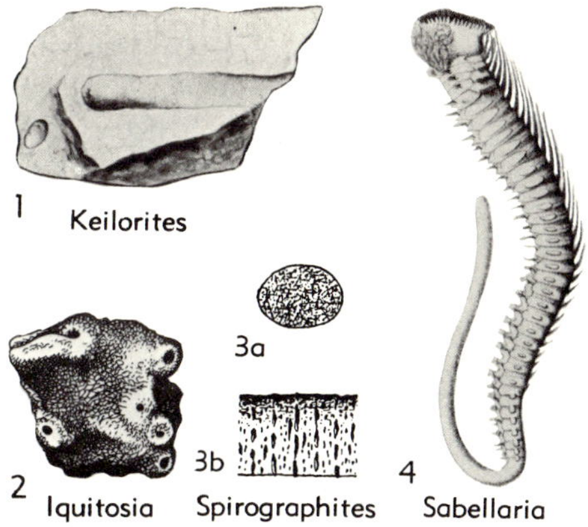

FIG. 95. Keiloritidae, Hermellidae, Sabellidae (p. *W*155).

Sabella LINNÉ, 1767 [**S. penicillus*]. *Rec.*

Iquitosia DE GREVE, 1938 [**I. bluntschlii*]. Built tiny smooth tubes in mud of ocean bottom; many tubes in group, each with mound around aperture. *Neog.*, S.Am.——FIG. 95,*2*. **I. bluntschlii*, Peru; ×6.7 (57).

Spirographites ASTRE, 1937 [**S. ellipticus*]. Subcylindrical membranous tube wtih numerous, concentric, discontinuous ridges on outside. *Cret.*, Eu. ——FIG. 95,*3*. **S. ellipticus*, Garumnian, Sp.; *3a,b*, ×0.7 (3).

Family SERPULIDAE Burmeister, 1837

Body cylindrical; thorax with 3 to 7 segments bearing bundles of hairlike bristles on dorsal side and hooks on ventral side; abdomen with many segments bearing hooks on dorsal side and bundles of hairlike bristles on ventral surface; numerous tentacles around mouth and calcareous or horny operculum; builds calcareous tube that is circular, polygonal, or triangular in cross section and may be ornamented on outside with concentric raised rings or longitudinal ridges or keels; usually attached for part or all of its length to substratum but some free throughout entire length. [Mostly marine but may occur in fresh waters.] *Cam.-Rec.*

A number of new subgenera, which should perhaps have generic rank (e.g., *Cycloserpula, Dorsoserpula, Tetraserpula, Pentaserpula, Hexaserpula*) are described by K.O.A. PARSCH from the Jurassic of

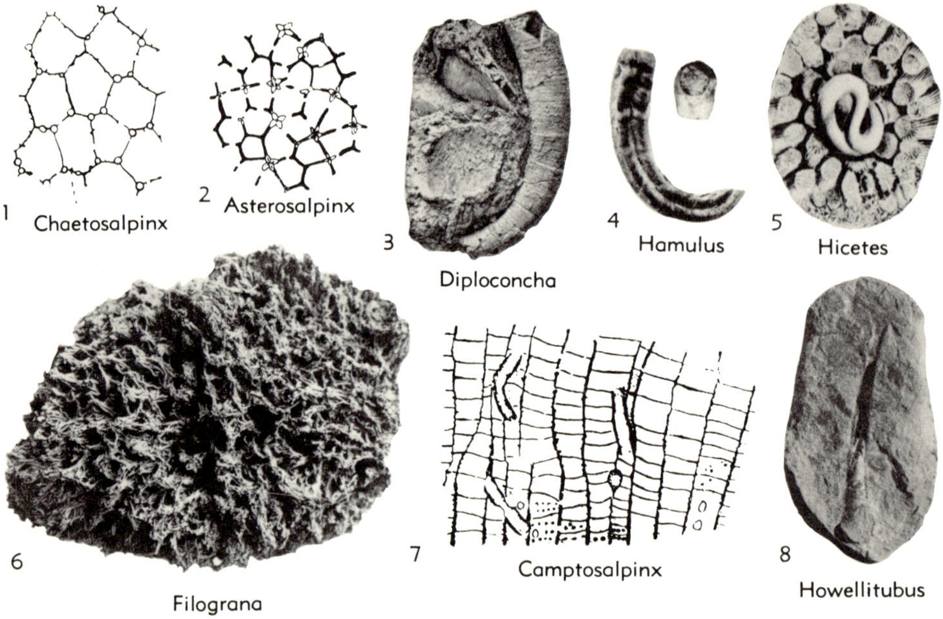

FIG. 96. Serpulidae (p. W156-W157).

Germany in Palaeontographica, Bd. 7, Abt. A, p. 211-240, 1956.

Serpula LINNÉ, 1768 [**Tubus vermicularis* ELLIS, 1755 (=*T. vermicularis* LINNÉ, 1768)]. Tube calcareous, tapering irregularly, coiled or contorted, lower end attached, remainder of tube more or less erect, surface bearing small concentric ridges; operculum horny. *Sil.-Rec.*, Eu.-N.Am.——FIG. 100,*3*. **S. vermicularis* (ELLIS), Rec., Fr.; ×0.7 (6).

Asterosalpinx SOKOLOV, 1948 [**A. asiaticus*]. Tube straight, with 4 or 5 longitudinal ridges on outer surface; commensal wtih favositid corals. *Dev.*, USSR.——FIG. 96,*2*. **A. asiaticus*, Novaya Zemlya; ×3.3 (140).

Camptosalpinx SOKOLOV, 1948 [**C. siberiensis*]. Tube curved, smooth; commensal with favositid corals. *U.Sil.*, USSR(Sib.).——FIG. 96,*7*. **C. siberiensis;* ×4 (140).

Cementula NIELSEN, 1931 [**C. sphaerica*]. Tube strongly coiled, with whorls cemented together by their outer layers, being then difficult to distinguish from outside, furrows between them smoothed down. *U.Cret.*, Eu.——FIG. 97,*6*. **C. sphaerica,* White Chalk, Denm.; ×4 (105).

Chaetosalpinx SOKOLOV, 1948 [**C. ferganensis*]. Tube straight, smooth; commensal with favositid corals. *U.Sil.*, USSR(Turkestan).——FIG. 96,*1*. **C. ferganensis;* ×4 (140).

Diploconcha CONRAD, 1875 [**D. cretacea*]. Sinuous calcareous tubes composed of numerous very thin concentric layers arranged as series of truncated cones, one within another, cones gradually increasing in size from apex of tube toward larger end; outer surface of tube bearing fine transverse growth lines and, distant from small end, more or less well-developed, coarser, irregular, transverse ridges which grow coarser away from apex. *U.Cret.*, E.N.Am.——FIG. 96,*3*. **D. cretacea,* Black Creek F., USA(N.Car.); ×0.7 (144).

Discouvermetulus ROVERTO, 1904 [**D. pissarroi*]. Tube attached, coiled in low spiral to form more or less regular disc, in center of which is small, smooth, globular nucleus. *Eoc.*, Eu.(Fr.).

Ditrupa BERKELEY, 1835 [**Dentalium corneum* LINNÉ, 1767 (=*Dentalium subulatum* DESHAYES, 1826)]. Tube calcareous, tapering, open at both ends; operculum thin and concentrically striate. *Cenoz.*, cosmop.——FIG. 97,*4*. **D. cornea* (LINNÉ), Plio., Italy; ×1 (119).

Ditrupula NIELSEN, 1931 [**Serpula canteriata* VON HAGENOW, 1840]. Tube free, curved, tapering, with 4 longitudinal rounded ridges on outer surface so that cross section is subquadrate. *U.Cret.*, Eu.——FIG. 97,*1*. **D. canteriata* (VON HAGENOW), Senon.; ×1 (105).

Filograna OKEN, 1815 [**Serpula filograna* LINNÉ] [=*Filogranula* NIELSEN, 1931 (*non* LANGERHANS, 1884)]. Shell smooth, very slender, filiform, gregarious; operculum obliquely truncate. *Cret.-Rec.*, Eu.-N.Am.——FIG. 96,*6*. *F. implexa* BERKELEY, Plio., Italy; ×0.7 (122).

Galeolaria LAMARCK, 1818 [*non* DE BLAINVILLE, 1830] [**G. caespitosa*]. Tube rather short, straight

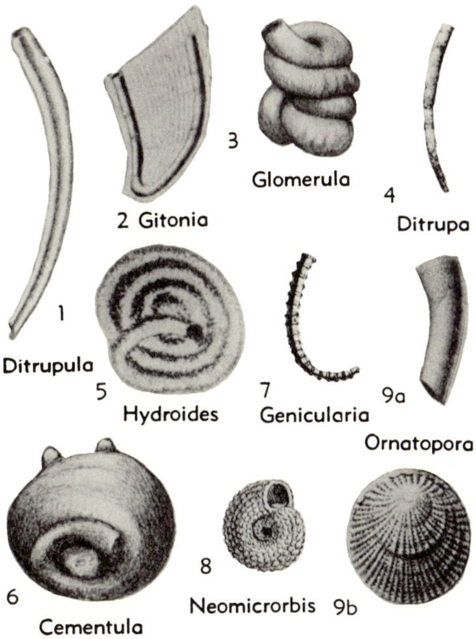

FIG. 97. Serpulidae (p. *W*156-*W*158).

or curved, with 4 equally spaced longitudinal ridges on outer surface, lower surface fixed to substratum; operculum orbicular, helmet-shaped. *Jur.-Rec.*, Eu.——FIG. 99,*2*. *G. prolifera* (GOLDFUSS), Jur., Ger.; ×4 (55).

Genicularia QUENSTEDT, 1858 [**G. ornata*]. Tube curved, with numerous concentric flanges throughout its entire length. *Jur.*, Eu.——FIG. 97,*7*. **G. ornata*, Brauner Zeta, Ger.; ×1 (114).

Gitonia CLARKE, 1908 [**G. corallophila*]. Tubes straight or curved, smooth; built within corals and stromatoporoids. *Sil.-Dev.*, N.Am.——FIG. 97,*2*. **G. corallophila*, M.Dev.(Onondaga), USA (N.Y.); ×1 (18).

Glomerula NIELSEN, 1931 [**Serpulites gordialis* VON SCHLOTHEIM, 1820]. Tube labyrinthically coiled, same diameter throughout, separate coils free, not cemented together. *Cret.*, Eu.——FIG. 97,*3*. **G. gordialis* (VON SCHLOTHEIM), U.Cret. (Senon.); ×4 (105).

Hamulus MORTON, 1834 [**H. onyx*] [=*Falcula* CONRAD, 1870]. Tube with from 3 to 7 axial ribs; early stages attached, usually broken away and solitary in adult; operculum calcareous, consisting of interior disc with three-cornered, elongate, posterior process. *Cret.*, N.Am.-Trinidad-Eu.-Palest.——FIG. 96,*4*. **H. onyx*, U.Cret.(Eutaw), USA(Ala.); ×1.3 (68).

Hicetes CLARKE, 1908 [**H. innexus*]. Tube irregularly coiled and same in diameter throughout; built within the coral, *Pleurodictyum*. *Dev.*, Eu.-N.Am.——FIG. 96,*5*. **H. innexus*, U.Dev.(Hamilton), USA(N.Y.); ×0.7 (18).

Howellitubus RICHARDSON, 1956 [**H. whitfieldorum*]. Tube straight or somewhat curved, composed of many layers arranged as cones, one within another, circular in cross section, tapering, with increasing flare at apertural end; wall thick apically, thin aperturally. *Penn.*, N.Am.——FIG. 96,*8*. **H. whitfieldorum*, Penn.(Francis Creek), USA(Ill.); ×0.7 (116).

Hydroides GUNNERUS, 1768 [**H. norvegica*]. Tube long, slender, curved, subquadrangular, about same in diameter throughout, adherent almost all of its length; operculum chitinous, with crenulate margin, funnel-shaped. *Eoc.-Rec.*, cosmop.——FIG. 97,*5*. **H. norvegica*, Rec., Norway; ×2 (119).

Jereminella LUGEON [**J. pfenderae*]. Tubes straight or slightly curved, up to 6 inches long; no important ornamentation on outer surface. *U.Cret.*, Eu.——FIG. 98,*1*. **J. pfenderae*, MAASTRICHT., Fr.; ×0.7 (99).

Josephella CAULLERY & MESNIL, 1896 [**J. marenzelleri*]. Tube solitary, cylindrical, small, orna-

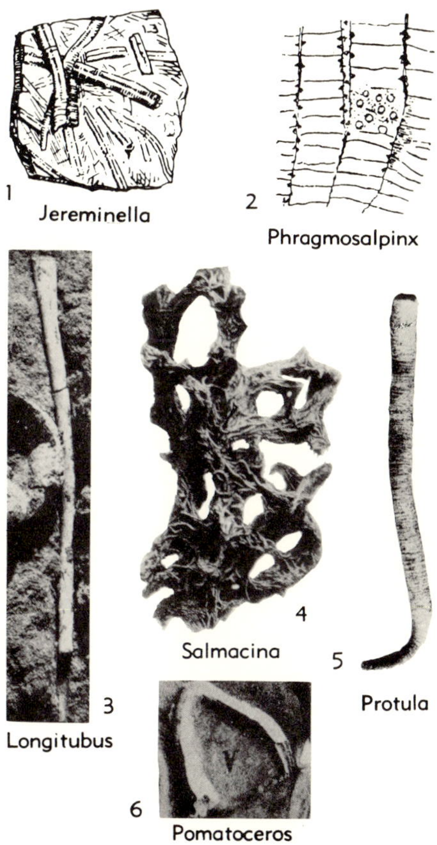

FIG. 98. Serpulidae (p. *W*157-*W*159).

mented with fine concentric ridges, some more prominent than others; operculum calcareous, conical, with denticulate upper edge. *Mio.-Rec.,* Eu. ——FIG. 99,*3*. **J. marenzelleri,* Rec., Fr.; ×140 (14).

Longitubus HOWELL, 1943 [**Hamulus lineatus* WELLER, 1907]. Tube calcareous, straight, unornamented, except for fine, closely spaced, concentric ridges; nearly same in diameter throughout length. *U.Cret.,* E.N.Am.——FIG. 98,*3*. **L. lineatus* (WELLER), Merchantville F., USA(N.J.); ×0.7 (68).

Mercierella FAUVEL, 1923 [**M. enigmatica*]. Operculum vesicular, with many simple chitinous spines; tube calcareous, round, with fine, closely spaced, concentric ridges and widely spaced, larger, concentric flaring, ridges or flanges; aperture of tube flaring. *Mio.-Rec.,* Eu.——FIG. 99, *7a,b*. **M. enigmatica,* Rec., Fr.; ×3, ×15 (47).

Neomicrorbis ROVERTO, 1903 [**Serpula granulata* SOWERBY, 1829]. Tube coiled almost in single plane, with about 2 coils; surface covered with longitudinal rows of prominent granules. *Cret.-Eoc.,* Eu.——FIG. 97,*8*. **N. granulata* (SOWERBY), Cret., Eng.; ×4 (141).

Ornatoporta GARDNER, 1916 [**O. marylandica*]. Tube arcuate, tapering; surface with fine radial lirae diverging in all directions from strongly eccentric nucleus, number more than doubled near margin by intercalation and bifurcation; with concentric lirae, in part incremental, and 2 to 5 prominent growth stages, as well as fine, crowded

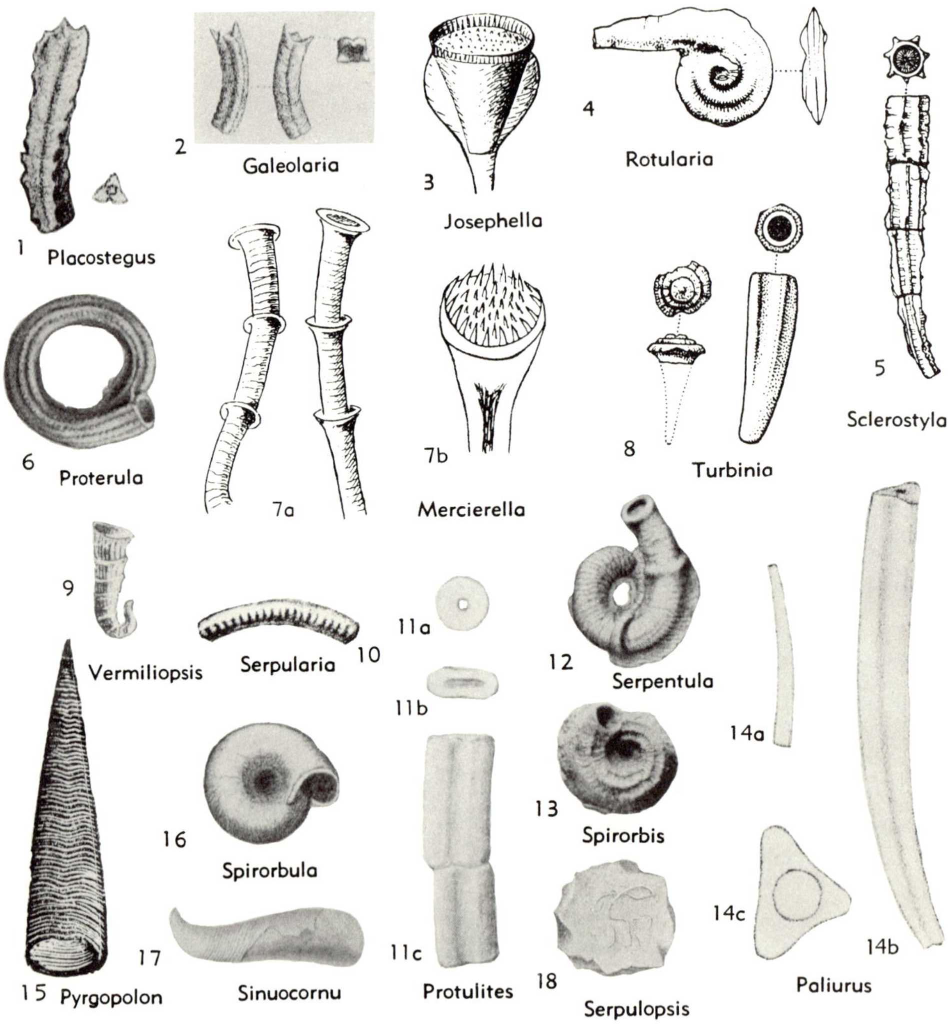

FIG. 99. Serpulidae (p. W156-W161).

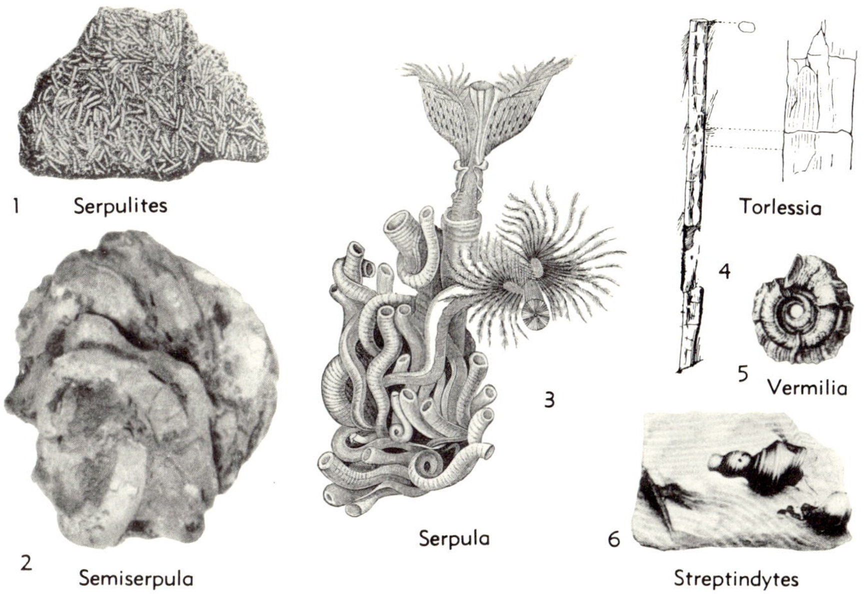

FIG. 100. Serpulidae (p. *W*156, *W*160-*W*161).

threadlets that do not override radial lirae but closely dissect interradial ones; operculum with reticulate sculpture. *U.Cret.*, N.Am.——FIG. 97, *9a,b*. **O. marylandica,* Monmouth F., USA(Md.); ×1, ×4 (51).

Paliurus GABB [**P. triangularis*]. Tube straight, slightly twisted, or bent, triangular in cross section, circular internally. *Eoc.*, N.Am.——FIG. 99, *14a-c*. **P. triangularis,* Vincentown F., USA(N.J.); ×1, ×2, ×6 (50).

Phragmosalpinx SOKOLOV, 1948 [**P. australiensis*]. Tube straight, smooth, with horizontal tabulae; commensal with favositid corals. *Dev.*, Austral.-USSR(Sib.).——FIG. 98,*2*. **P. australiensis,* Austral.; ×4 (140).

Placostegus PHILIPPI, 1844 [**Serpula tridentatus* FABRICIUS, 1779]. Tube triangular in cross section, with the 3 edges extended as serrate keels; operculum calcareous. *Plio.-Rec.*, cosmop.——FIG. 99,*1*. **P. tridentatus* (FABRICIUS), Pleist., Italy; ×1 (119).

Pomatoceros PHILIPPI, 1844 [**Serpula triquetra* LINNÉ, 1758]. Tube triangular, commonly curved, upper keel slightly serrate; tube attached by lower surface; operculum a truncated cone with 1 to 3 spines. *Mio.-Rec.*, cosmop.——FIG. 98,*6*. **P. triqueter* (LINNÉ), Plio., Italy; ×0.67 (119).

Proterula NIELSEN, 1931 [**P. costata*]. Tube elongate, more or less coiled, curved from side to side, adherent by nearly whole length, even in diameter throughout. *Paleoc.*, Eu.——FIG. 99,*6*. **P. costata,* Denm.; ×3 (105).

Protula RISSO, 1826 [**P. rudolphi*]. Base of gradually tapering, smooth, cylindrical tube attached, anterior part of tube free and erect. *Eoc.-Rec.*, cosmop.——FIG. 98,*5*. *P. canavarii* ROVERTO, Plio., Italy; ×0.7 (119).

Protulites JASKÓ, 1940 [**P. segmentata*]. Tube small, probably calcareous, with rather thick wall; circular in cross section, not tapering. *Oligo.*, Eu. ——FIG. 99,*11a-c*. **P. segmentata,* Hung.; ×5 (76).

Pyrgopolon MONTFORT, 1808 [**P. mosae*] [=*Entalium* DEFRANCE, 1819; *Pharetrium* KÖNIG, 1825]. Shell free, conical, with internal conical compartment walls through which runs small longitudinal tube to small apical aperture; outer surface ornamented with fine concentric lines. *Cret.-Eoc.*, Eu.——FIG. 99,*15*. **P. mosae,* Maastricht.; ×4 (97).

Rotularia DEFRANCE, 1827 [**Serpula spirulaea* LAMARCK, 1818]. [=*Spirulaea* BROWN, 1828; *Tubulostium* STOLICZKA, 1869]. Tube helically coiled, with same diameter throughout most of length but ending at apertural end in restricted tube of smaller diameter; restricted portion (and in some part of unrestricted tube posterior to it) extending tangentially from coiled portion; posterior end of tube usually attached to substratum; outer surface of tube smooth or concentrically wrinkled; one or 2 longitudinal keels present in some species. *U.Cret.-Eoc.*, cosmop.——FIG. 99,*4*. **R. spirulaea* (LAMARCK), Eoc., Fr.; ×1 (165).

Salmacina CLAPARÈDE, 1870 [**S. incrustans*]. Tube

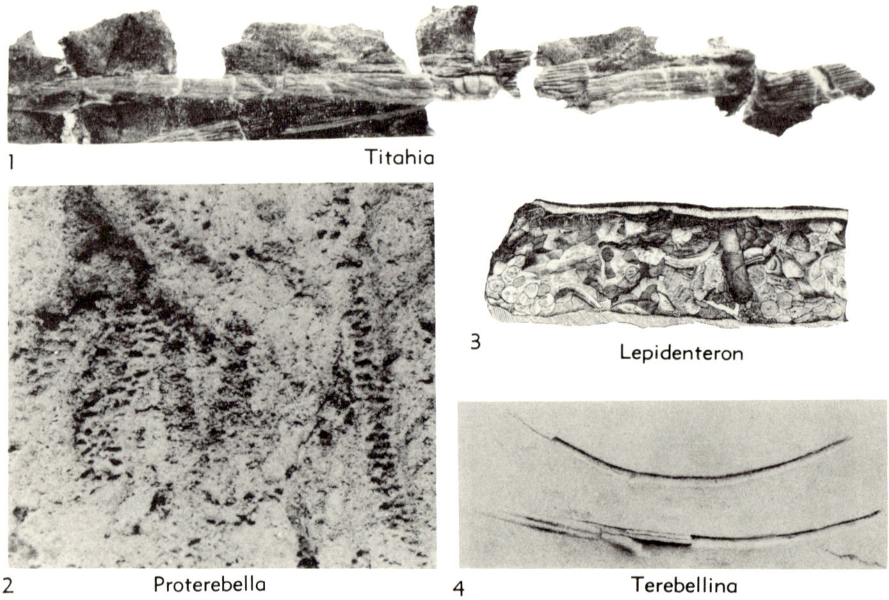

FIG. 101. Terebellidae (p. W161-W162).

flexuous, diameter small as in *Filograna* but without operculum. *Trias.-Rec.*, Eu.——FIG. 98,*4*. *S. aedificatrix* CLAPARÈDE, Rec., Italy; ×0.7 (119).

Sclerostyla MØRCH, 1863 [**Serpula (Sclerostyla) ctenactis* MØRCH]. Tube curved, tapering, with 5 to 7 longitudinal external flanges; tube wall composed of parabolic layers with rims pointing outward and forming fine concentric lines on outer surface of tube; operculum calcareous, stalk bearing 2 incised grooves which repeatedly branch upon cone to make network of incised reticulations. *Eoc.-Rec.*, cosmop.——FIG. 99,*5*. *S. mellevillei* (NYST & LEHON), Eoc.(Barton.), Eng.; ×1 (166).

Semiserpula WETZEL, 1957 [**S. chilensis*]. Tube partly phosphatic, cylindrical, not tapering, irregularly coiled spirally, surface smooth. *Paleog.*, S. Am.——FIG. 100,*2*. **S. chilensis*, Chile; ×0.45 (160).

Serpentula NIELSEN, 1931 [**Serpula ampullacea* SOWERBY, 1829]. Tube comparatively short, more or less coiled from side to side, cemented along most of its length to some foreign object, thick--ness rapidly increasing from apex toward aperture. *U.Cret.-Paleoc.*, Eu.——FIG. 99,*12*. **S. ampullacea* (SOWERBY), U.Cret., Denm.; ×1 (105).

Serpularia MÜNSTER, 1840 [**S. crenata*]. Tube same in diameter throughout, bearing parallel concentric ridges on one side. *Ord.*, Eu.——FIG. 99,*10*. **S. crenata*, Ord. (Orthoceratite Ls.), S. Ger.; ×3 (102).

Serpulites BLUMENBACH, 1803 [**S. coacervatus*]. Tubes small, short, nearly straight, outer surface bearing concentric striations. *Mesoz.*, Eu.——FIG. 100,*1*. **S. coacervatus*, Ger.; ×0.7 (70).

Serpulopsis GIRTY, 1912 [**Serpula insita* WHITE, 1879]. Tube very small, free or attached, tortuous. *Penn.*, N.Am.——FIG. 99,*18*. **S. insita* (WHITE), USA(Ind.); ×1 (162).

Sinuocornu HOWELL, 1959 [**Serpulites curtus* SALTER, 1848]. Tube short, rather rapidly tapering, sinistrally curved, with moderately strong, oblique growth lines on outer surface. *M.Sil.*, Eu.——99,*17*. **S. curtum* (SALTER), Wenlock., Eng.; ×1 (131).

Spirorbis DAUDIN, 1800 [**Serpula spirorbis* LINNÉ, 1758] [=*Spirillum* OKEN, 1807; *Microconchus* MURCHISON, 1839; *Gyromices* GOEPPERT, 1853; *Palaeorbis* BENEDEN & COEMANS, 1867]. Tube small, coiled in flat spiral, some shells bearing concentric ridges, attached to substratum. [Marine and fresh water.] *Ord.-Rec.*, cosmop.——FIG. 99, *13*. **S. catagraphus* ROVERTO, Plio., Italy; ×12 (122).

Spirorbula NIELSEN, 1931 [**Serpula aspera* VON HAGENOW, 1840]. Tube wound in spiral whorls, either lying in same plane or forming spirally enrolled, inversely conical, or cylindrical bodies. *U.Cret.-Paleoc.*, Eu.——FIG. 99,*16*. **S. aspera* (VON HAGENOW), *U.Cret.(Senon.)*, Ger.; ×6 (105).

Streptindytes CALVIN, 1888 [**S. acervulariae*]. Tube

coiled spirally, with concentric growth annulations; grew inside corals and stromatoporoids. *Sil.-Carb.*, N.Am.-Eu.——FIG. 100,*6*. **S. acervulariae*, M.Dev.(Hamilton.), Iowa; ×1 (18).

Torlessia BATHER, 1905 [**T. mackayi*]. Tube straight, slightly tapering, with stout walls. *L. Mesoz.*, N.Z.——FIG. 100,*4*. **T. mackayi*; ×0.7 (4).

Turbinia MICHELIN, 1845 [**T. graciosa*]. Tube gently curved, heptagonal, with 7 longitudinal ridges and irregular concentric growth lines; operculum cone-shaped, with upper face bearing radiating ridges and convex in center; ridges on upper face continued onto lower surface of upper part of cone; lower end of cone bifid. *Eoc.*, Eu. ——FIG. 99,*8*. *T. abbreviata* (DESHAYES), Eoc. (Lutetian), Eng.; ×8 (166).

Vermilia LAMARCK, 1818 [**V. triquetra*]. Tube variously curved, attached by its side, one or more teeth on edge of aperture, longitudinal and concentric ridges on outer surface. *Carb.-Rec.*, Eu. ——FIG. 100,*5*. *V. manicata* (REUSS), Neog. (Torton.), Aus.; ×1.3 (137).

Vermiliopsis SAINT-JOSEPH, 1906 [**Vermilia infundibulum* LANGERHANS, 1884]. Tube curved, with longitudinal ridges and concentric flanges which make it appear to be made up of a series of nested tubes. *Neog.*, Eu.——FIG. 99,*9*. *V. elegantula* (ROVERTO), Neog.(Torton.), Aus.; ×2 (137).

Family TEREBELLIDAE Grube, 1850

Tubes usually straight, formed of fragments of rocks, shells, and other small objects, cemented by worm to make solid cylinder. [Marine.] *Cam.-Rec.*

Terebella LINNÉ, 1767 [**T. lapidaria*]. Tube irregular in form, composed of fragments of many kinds of materials. *Jur.-Rec.*, cosmop.——FIG. 102,*7*. *T. conchligea* (PALLAS), Rec., Eu.; ×1 (6).

Cryptosiphon PRANTL, 1948 [**C. terebelloides*]. Tubes composed of shells of brachiopods, gastropods, and ostracodes, or of tests of small trilobites. *Ord.*, Eu.——FIG. 102,*4*. **C. terebelloides*, Llanvirn., Czech.; ×1 (113).

Lepidenteron FRITSCH, 1878 [**L. longissimum*]. Tube formed of scales and bones of fishes. *Cret.*, Eu.——FIG. 101,*3*. **L. longissimum*, Czech.; ×0.7 (49).

Paraterebella HOWELL, 1955 [**Terebellopsis scotti* HOWELL, 1953] [=*Terebellopsis* HOWELL, 1953 (*non* LEYMERIE, 1844)]. Tube composed of compactly, solidly built small fragments of shells and segments of crinoid stems. *Penn.*, N.Am.——FIG. 102,*1*. **P. scotti* (HOWELL), USA(Tex.); ×4 (71).

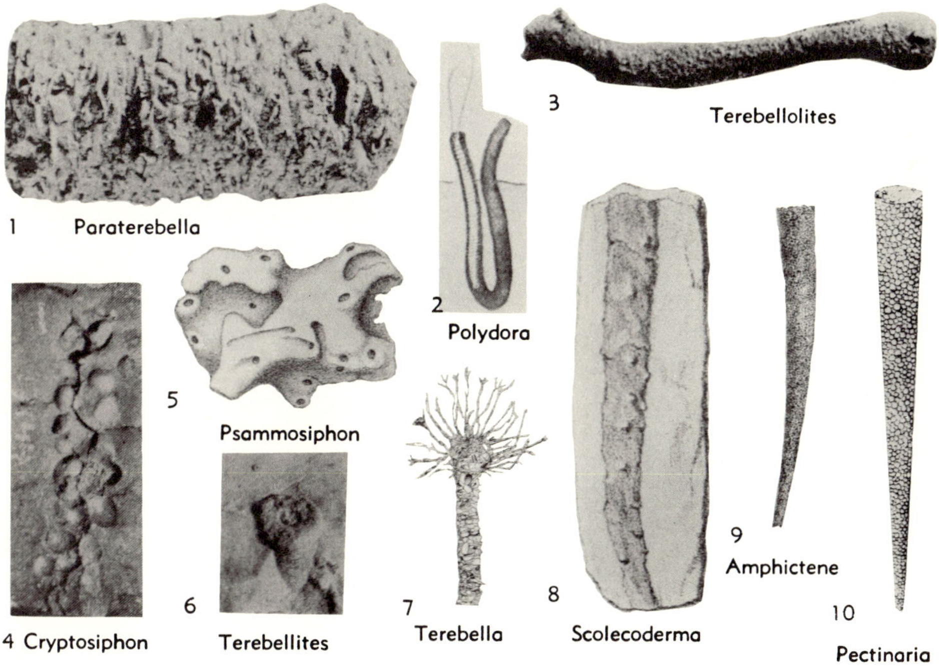

FIG. 102. Terebellidae, Amphictenidae, Spionidae (p. *W*161-*W*163).

Proterebella HOWELL, 1953 [*P. permiana]. Tubes curved or sinuous, composed of small sand grains. *M.Perm.*, N.Am.——FIG. 101,*2*. *P. permiana,* Kaibab Ls., USA(Ariz.); ×27 (72).

Psammosiphon VINE, 1882 [*P. amplexus]. Tubes very small, attached singly or in clusters, composed of minute sand grains. *Sil.-Carb.*, Br.I.——FIG. 102,*5*. *P. amplexus,* M.Sil.(Wenlock.), Eng.; ×7 (151).

Scolecoderma SALTER, 1855 [*S. antiquissima]. Tubes membranous. *Cam.*, Br.I.——FIG. 102,*8*. *S. tuberculata* SALTER, M.Cam., Wales; ×1 (133).

Streblosoma SARS, 1872 [*S. cochleatum]. Tube free, smooth, composed of sand or mud, irregularly curved or coiled in regular convolutions. *Eoc.-Rec.*, Eu.

Terebellina ULRICH, 1910 [*T. palachei]. Tubes long, subcylindrical, gently curved, acuminate at lower end, with rather thick walls composed of cemented minute siliceous grains and with surface obscurely striated transversely. *Jur.*, N.Am.——FIG. 101,*4*. *T. palachei,* Yakutat F., Alaska; ×0.7 (150).

Terebellites HOWELL, 1943 [*T. franklini]. Tube with thick walls composed of compactly cemented small sand grains, walls thicker than diameter of dwelling cavity within them. *M.Cam.*, Newf.——FIG. 102,*6*. *T. franklini,* Cloud Rapids F.; ×3 (69).

Terebellolites DESIO, 1940 [*T. fezzanensis]. Tube gently curved, some bifurcate, composed of small particles of sand. *Dev.*, N.Afr.——FIG. 102,*3*. *T. fezzanensis,* Libya; ×2 (30).

Titahia WEBBY, 1958 [*T. corrugata]. Tube large, slightly tapering, with prominent longitudinal ribs; wall siliceous, composed of cemented aggregation of sand grains. *L.Mesoz.*, N.Z.——FIG. 101, *1*. *T. corrugata;* ×0.7 (157).

Family CIRRATULIDAE Carus, 1863

Body segmented throughout, with capillary chaetae on each side in 2 bundles,

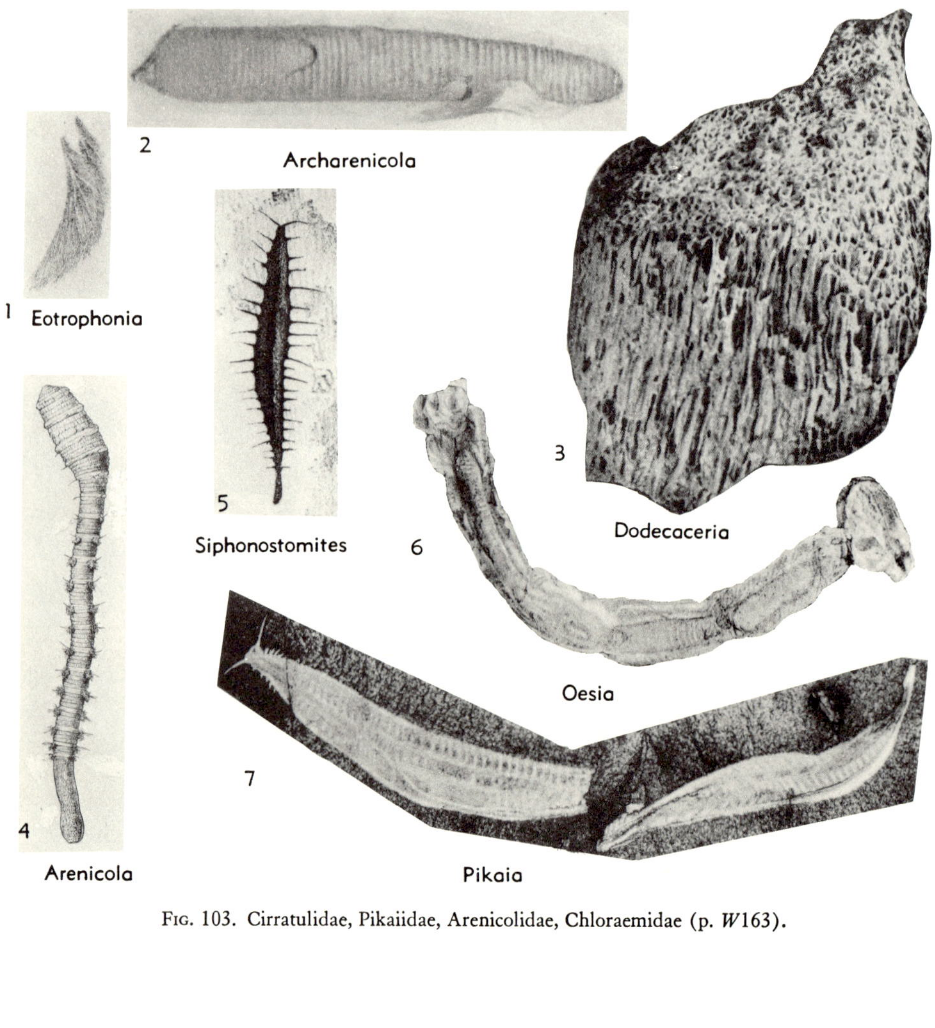

FIG. 103. Cirratulidae, Pikaiidae, Arenicolidae, Chloraemidae (p. W163).

carried by small papillae; live in burrows or calcareous tubes. [Marine.] *Mio.-Rec.*

Cirratulus LAMARCK, 1801 [**Aphrodite cirrata* MÜLLER, 1776]. *Rec.*

Dodecaceria ÖRSTED, 1843 [**D. concharum*]. Tubes calcareous, somewhat sinuous, crowded together in masses. *Mio.-Rec.*, cosmop.——FIG. 103,*3*. *D. fistulicola* EHLERS, Rec., USA(Calif.); ×1 (115).

Family AMPHICTENIDAE Malmgren, 1867

Tube straight or slightly curved, composed of sand grains or other material cemented together, nearly cylindrical, but somewhat tapering and open at both ends. [Marine.] *Perm.-Rec.*

Amphictene SAVIGNY, 1820 [**Amphitrite auricoma* MÜLLER, 1788]. Tube slightly curved. *Perm.-Rec.*, cosmop.——FIG. 102,*9*. **A. auricoma* (MÜLLER). Rec., Eu.; ×1 (6).

Pectinaria LAMARCK, 1818 [**Nereis cylindraria belgica* PALLAS, 1766]. Tube straight. *Mio.-Rec.*, cosmop.——FIG. 102,*10*. **P. belgica* (PALLAS), Rec., USA(Mass.); ×1 (56).

Family SPIONIDAE Sars, 1861

Worms with dorsal chaetae comprising fringed capillaries and ventral chaetae consisting of crochets. They build long, flexible tubes of mud or sand. [Marine.] *Mio.-Rec.*

Spione ÖRSTED, 1844 [**S. trioculata*]. *Rec.*

Polydora BOSC, 1801 [**Polydora cornuta*]. Tube composed of mud, forming U-shaped burrow in chalk, limestone, a shell, or shale. *Mio.-Rec.*, cosmop.——FIG. 102,*2*. *P. ciliata* (JOHNSTON), Rec., Br.I.; ×1 (1).

Family PIKAIIDAE Walcott, 1911

[*nom. correct.* HOWELL, herein (*pro* Pikaidae WALCOTT, 1911)]

Body slender, many-segmented; small head with well-developed eyes; parapodia on anterior segments. [Marine.] *Cam.*

Pikaia WALCOTT, 1911 [**P. gracilens*]. Body tapering at each end; 2 tentacles on head. *M.Cam.*, N. Am.——FIG. 103,*7*. *P. gracilens*, Burgess Sh., Can.(B.C.); ×2 (154).

Oesia WALCOTT, 1911 [**O. disjuncta*]. Head wider than rest of body, very small hooks on anterior part of body. *M.Cam.*, N.Am.——FIG. 103,*6*. **O. disjuncta*, Burgess Sh., Can.(B.C.); ×1 (154).

Family ARENICOLIDAE Audouin & Edwards, 1833

Burrowing worms with cylindrical body larger at anterior end and indistinctly segmented, capillary chaetae on dorsal surface and short crochets on ventral surface. *Trias.-Rec.*

Arenicola LAMARCK, 1801 [**Lumbricus marinus* LINNÉ, 1758]. Burrows in mud or muddy sand; no chaetae on 2 anterior segments. *Cret.-Rec.*, cosmop.——FIG. 103,*4*. **A. marina* (LINNÉ), Rec., Br.I.; ×0.5 (6).

Archarenicola HORWOOD, 1912 [**A. rhaetica*]. Body annulate, annuli of 2 sizes, possibly forming segments; appendages paired, on alternate annuli and consisting of capillary notopodial setae; head lacking appendages, but having frilled prostomium. *U. Trias.*, Eu.——FIG. 103,*2*. **A. rhaetica*, Rhaet., Eng.; ×1 (67).

Family CHLORAEMIDAE Malmgren, 1867

Body cylindrical or spindle-shaped; segments short, equipped with papillae and bristles, bristles on first few segments longer than those on later segments and directed forward; parapodia widely bifid; back covered with small hairlike bristles, ventral surface bearing small S-shaped or sickle-shaped bristles. *Ord.-Rec.*

Chloraema DUJARDIN, 1838 [**C. edwardsii*]. *Rec.*

Eotrophonia ULRICH, 1878 [**E. setigera*]. Segments with tuft of setae on each side at junction with adjacent segment and tuft on upper surface; tufts composed of 20 to 40 or more setae, directed obliquely outward. *U.Ord.*, N.Am.——FIG. 103,*1*. **E. setigera*, USA(Ohio); ×18 (149).

Siphonostomites ROVERTO, 1904 [**Nereites hesionoides* MASSALONGO, 1855]. Body subcylindrical, fusiform, with bristles along entire length, those on middle of body longer than near ends. *Tert.*, Eu.——FIG. 103,*5*. **S. hesionoides* (MASSALONGO), Italy; ×3 (90).

Family UNCERTAIN

Byronia MATTHEW, 1899 [**B. annulata*]. Tube curved, horny, wall thin, outer surface bearing concentric annulations. *M.Cam.*, N.Am.——FIG. 104,*1*. **B. annulata*, Stephen F., Can.(B.C.); ×4 (91).

Campylites EICHWALD, 1856 [**Serpulites longissimus* SOWERBY, 1839]. Tube large, curved, composed of numerous thin layers, tapering very gradually. *Sil.*, Eu.——FIG. 105,*1*. **C. longissimus* (SOWERBY), U.Sil.(U.Ludlov.), Br.I., ×0.3 (142).

Hammatopsis HADDING, 1913 [**H. scanicus*]. Body finely segmented, ends of segments forming triangular points. *Ord.*, Eu.——FIG. 104,*5*. **H. scanicus*, Swed.; ×1 (62).

Khemisina TERMIER & TERMIER, 1951 [**K. annulata*]. Tube calcareous, tapering to point, with well-defined median longitudinal furrow; covered on outside with arenaceous granules and bearing diagonal ridges, running concentrically from longitudinal furrow around tube. *Ord.*, N.Afr.——Fig. 104,*7*. **K. annulata*, Morocco; ×0.7 (147).

?**Lapworthella** Cobbold, 1921 [**L. nigra*]. Tube tapering, circular or subpolygonal in cross section, consisting of 2 layers, outer chitinous and thinner than inner, which is calcareous; ornamented externally with pronounced concentric raised ridges (considered by Fisher to be related to *Stenothecopsis*). *L.Cam.*, Eu.——Fig. 104,*9*. **L. nigra*, Eng.; ×10 (20).

Lockportia Howell, 1959 [*pro Dactylethra* Ruedemann, 1925 (*non* Cuvier, 1829; *nec* Meyrick, 1906)] [**Dactylethra conspicua* Ruedemann, 1925]. Body shaped like finger of glove; smooth, leathery test. *M.Sil.*, N.Am.——Fig. 105,*2*. **L. conspicua* (Ruedemann), Lockport Dol., USA (N.Y.); ×0.7 (128).

Melanostrophus Öpik, 1930 [**M. fokini*]. Tube chitinous, long, smooth, irregularly curved and crooked. *Ord.*, Eu.——Fig. 104,*14*. *M. signum* Öpik, M.Ord.(Kuckers), Est.; ×2.5 (108).

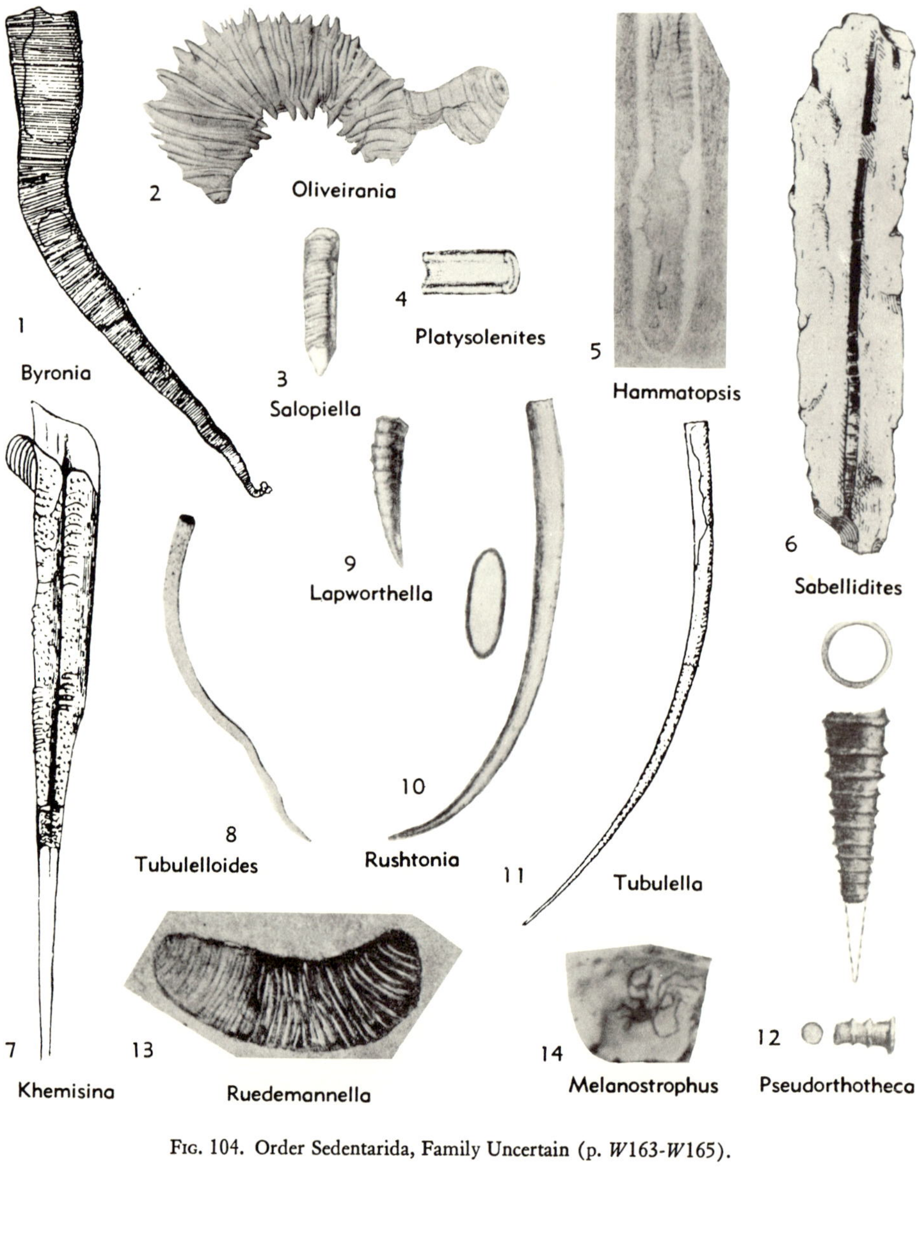

Fig. 104. Order Sedentarida, Family Uncertain (p. *W*163-*W*165).

Oliveirania MAURY, 1927 [**O. santacatharinae*]. Body segmented, curved, having biramous parapodia except at extremities and 2 short hooklike jaws. *Sil.*, S.Am.——FIG. 104,*2*. **O. santacatharinae*, Brazil; ×1 (92).

Platysolenites EICHWALD, 1860 [**P. antiquissimus*]. Tube calcareo-siliceous, flattened, with sides bent upward. *L.Cam.*, Eu.——FIG. 104,*4*. **P. antiquissimus*, USSR; ×8 (37).

?Pseudorthotheca COBBOLD, 1935 [**P. acuticincta*]. Tube phosphatic, with concentric annulations marked by incised striae or raised engirdling ribs which may be sharply defined or obsolescent. [Considered by FISHER to belong to family Torellellidae of Order Hyolithelminthes.] *Cam.*, Eu.-N.Am.——FIG. 104,*12*. **P. acuticincta*, L.Cam. (*Heraultia* Ls.), Fr.; ×5 (21).

Ruedemannella HOWELL, 1959 [**Bertiella obesa* RUEDEMANN, 1925] [=*Bertiella* RUEDEMANN, 1925 (*non* STILES & HASSALL, 1902)]. Body plump, segmented, smooth except for circular nodes on some segments, arranged in pairs; has parapodia and slender, subtriangular jaws. *Sil.*, N.Am.——FIG. 104,*13*. **R. obesa* (RUEDEMANN), U.Sil.(Bertie), USA(N.Y.); ×1 (128).

?Rushtonia COBBOLD & POCOCK, 1934 [**R. lata*]. Tube phosphatic, tapering, curved in one plane, cross section elliptical; minute concentric striae on outer surface. [Considered by FISHER (p. *W*133) to belong to Family Torellellidae of Order Hyolithelminthes.]. *L.Cam.*, Eu.——FIG. 104,*10*. **R. lata*, Eng.; ×7 (22).

Sabellidites YANICHEVSKY, 1926 [**S. cambriensis*]. Tubes long, straight, curved, or twisted, circular in cross section, outer surface with fine concentric striae which are not continuous around entire tube; concentric striae less prominent than in *Platysolenites*. *L.Cam.*, Eu.-N.Am.(Va.).——FIG. 104,*6*. **S. cambriensis*, Blue Clay, USSR; ×1 (168).

Salopiella COBBOLD, 1921 [**S. obliqua*]. Tube elliptical in cross section, margins subparallel, wall thin, possibly calcareous, formed of closely set, concentric, steplike annulations, which are reproduced on inside of tube. *L.Cam.*, Eu.——FIG. 104,*3*. **S. obliqua*, *Strenuella* Ls., Eng.; ×4 (20).

Tubulella HOWELL, 1949 [**Urotheca flagellum* MATTHEW, 1899] [=*Urotheca* MATTHEW, 1899 (*non* COCTEAU & BIBRON, 1843)]. Tube long and curved, chitinous, smooth or having fine concentric growth lines. *M.Cam.*, N.Am.(Newf.-B.C.).——FIG. 104,*11*. **T. flagellum* (MATTHEW), Stephen F., B.C.; ×5 (91).

Tubulelloides HOWELL, 1949 [**Serpulites gracilis* RUEDEMANN, 1916]. Tube flexuous, chitinous, smooth, with longitudinal marginal welt on each side and basal disc. *Ord.*, N.Am.——FIG. 104,*8*. **T. gracilis* (RUEDEMANN), M.Ord.(Canajoharie), USA(N.Y.); ×1 (127).

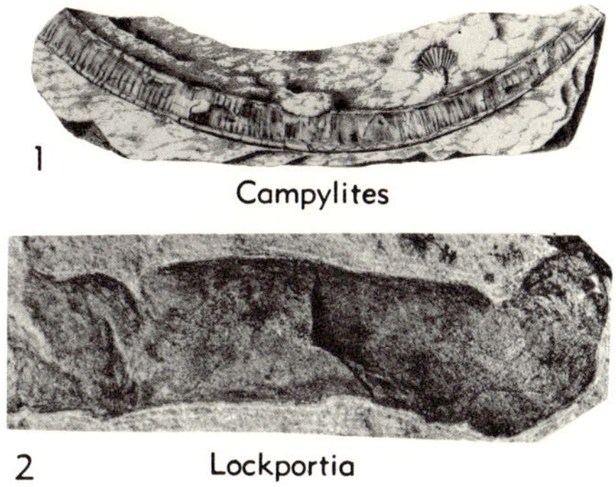

FIG. 105. Order Sedentarida, Family Uncertain (p. *W*163-*W*164).

Order MISKOIIDA Walcott, 1911

[*nom. correct.* HOWELL, herein (*pro* Miskoida WALCOTT, 1911)]

Segments and parapodia similar throughout length of body; proboscis retractile; enteric canal straight; body not distinctly specialized into sections. [Marine.] *Cam.-Ord.*

Family MISKOIIDAE Walcott, 1911

[*nom. correct.* HOWELL, herein (*pro* Miskoidae WALCOTT, 1911)]

Body elongate, slender, with numerous uniform segments; anterior end with rows of strong setae around mouth; surface of anterior portion with numerous papillae; parapodia abundant, branched; proboscis elongate, retractile; enteric canal straight, with enlargement in anterior portion. *M. Cam.*

Miskoia WALCOTT, 1911 [**M. preciosa*]. Body roughly divided into 3 areas, anterior, central, which is more or less expanded, and posterior, which is slender; setae around mouth; anterior and central parts segmented, posterior part smooth, but bearing setae; elongated papillae on anterior part, arranged in longitudinal rows. *M.Cam.*, W. N.Am.——FIG. 106,*7a,b*. **M. preciosa*, Burgess Sh., Can.(B.C.); ×2 (154).

Family CANADIIDAE Walcott, 1911

[*nom. correct.* HOWELL, herein (*pro* Canadidae WALCOTT, 1911)]

Body slender, formed of long segments bearing setiferous parapodia, with dorsal and ventral bundles of setae; head small, with 2 large tentacles. [Marine.] *Cam.*

Canadia WALCOTT, 1911 [**C. spinosa*]. Pair of parapodia, with dorsal and ventral bundles of strong, nonjointed, setae on each segment. *M.Cam.*,

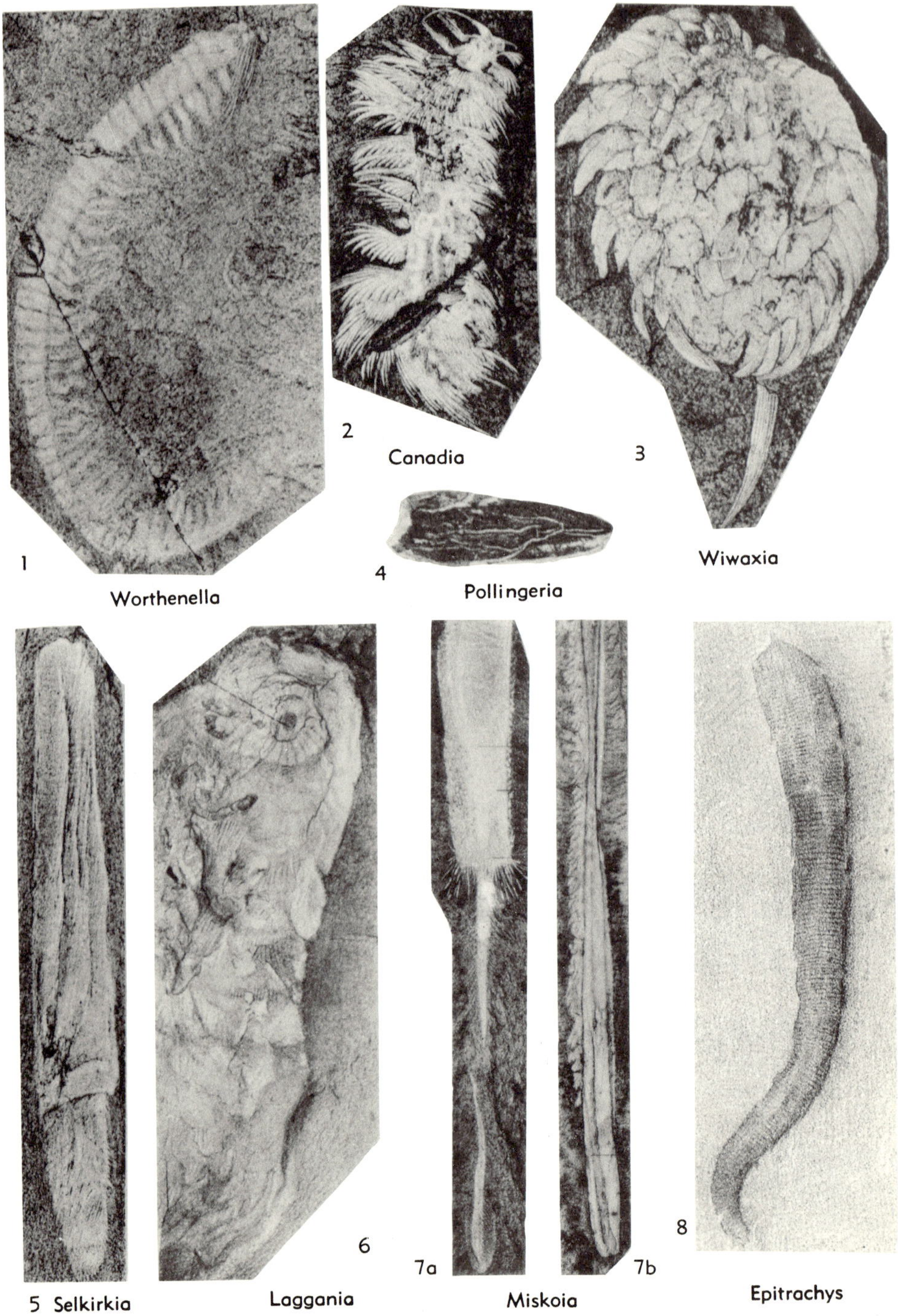

FIG. 106. Miskoiidae, Canadiidae, Wiwaxiidae, Polychaetia—Order Uncertain, Oligochaetia—Order and Family Uncertain, Sipunculoida—Family Uncertain (p. W165-W170).

W.N.Am.——FIG. 106,*2*. *C. spinosa*, Burgess Sh., Can.(B.C.); ×2 (154).

Selkirkia WALCOTT, 1911 [*Orthotheca major* WALCOTT, 1908]. Body segmented and bearing short spines at base of anterior portion; lived in tube which was probably chitinous. *M.Cam.*, W.N. Am.——FIG. 106,*5*. *S. major* (WALCOTT), Burgess Sh., Can.(B.C.); ×3 (154).

Family WIWAXIIDAE Walcott, 1911

[*nom. correct.* HOWELL, nerein (*pro* Wiwaxidae WALCOTT, 1911)]

Body oval, covered with dorsal ribbed scales and strong, elongate spines. [Marine.] *Cam.*

Wiwaxia WALCOTT, 1911 [*Orthotheca corrugata* MATTHEW, 1899]. Entire dorsal surface covered by long scales. *M.Cam.*, W.N.Am.——FIG. 106,*3*. *W. corrugata* (MATTHEW), Burgess Sh., Can. (B.C.); ×2 (154).

Pollingeria WALCOTT, 1911 [*P. grandis*]. Body covered with thin, smooth, elongate, dorsal scales. *M.Cam.*, W.N.Am.——FIG. 106,*4*. *P. grandis*, Burgess Sh., Can.(B.C.); ×2 (154).

Worthenella WALCOTT, 1911 [*W. cambria*]. Body slender, elongate, formed of 46 or more segments and small head; each segment with annular median furrow that divides it into rings; head formed of 2 or 3 segments, probably with eye and one or more pairs of short, jointed tentacles and pair of long, filament-like palps; anterior 34 segments with strong parapodia, each divided into 2 filamentous branches; parapodia of next 8 segments longer and more compact. *M.Cam.*, W.N. Am.——FIG. 106,*1*. *W. cambria*, Burgess Sh., Can.(B.C.); ×2 (154).

Family UNCERTAIN

Eopolychaetus RUEDEMANN, 1901 [*E. albaniensis*]. Body long, slender, segmented, each segment bearing 5 to 8 annulations and long, untufted, setae on ?dorsal side. [Marine.]. *U.Ord.*, N.Am. ——FIG. 108,*11*. *E. albaniensis*, Canajoharie Sh., USA(N.Y.); ×4 (126).

Pontobdellopsis RUEDEMANN, 1901 [*P. cometa*]. Body cylindrical or rather long conical, regularly tapering, and terminating abruptly in flat disc at ?anterior end; segmented; segments smooth. *U. Ord.*, N.Am.——FIG. 108,*10*. *P. cometa*, Canajoharie Sh., USA(N.Y.); ×3 (126).

Order UNCERTAIN

Laggania WALCOTT, 1911 [*L. cambria*]. Body elongate, pear-shaped, slightly flattened on ventral surface; mouth ventral, near anterior end, surrounded by ring of plates; surface marked by longitudinal radiating lines of chaetae. [Marine.] *M.Cam.*, W.N.Am.——FIG. 106,*6*. *L. cambria*, Burgess Sh., Can.(B.C.); ×1 (155).

Class MYZOSTOMIA Graff, 1884

Body flat, disc-shaped, not segmented, with an external chitinous cuticle, 5 pairs of parapodia, each with hook and supporting rod. [Parasitic on and in crinoids; marine.] *Ord.-Rec.*

Family MYZOSTOMIDAE Graff, 1884

Alimentary canal ramified, parapodia connected by muscles which converge to a central muscular mass; body divided into paired chambers by incomplete septa. *Ord.-Rec.*

Myzostomum LEUCKART, 1827 [*M. costatum*]. *Rec.*

Myzostomites CLARKE, 1921 [*M. clarkei*; SD HOWELL, herein]. Formed small gall-like protuberances, with central perforation, on columns of crinoids. *Ord.-Jur.*, cosmop.——FIG. 108,*14*. *Myzostomites* sp. CLARKE, Carb., locality unknown; ×1 (19).

Class OLIGOCHAETIA Grube, 1850

Body segmented, with setae but no parapodia. [Mostly terrestrial, but a few live in fresh and marine waters.] *Carb.-Rec.*

Order PLESIOTHECA Michaelsen, 1930

Setae in bundles, each with indeterminate number of setae; male ducts opening to exterior one segment behind their funnels. *Carb.-Rec.*

Family TUBIFICIDAE Vejdovsky, 1884

No asexual reproduction; spermathecae situated not far from gonads. *Carb.-Rec.*

Tubifex LAMARCK, 1816. *Rec.*

Pronaidites KUŠTA, 1888 [*P. carbonarius*]. Body long, thin, and segmented. *Carb.*, Eu.——FIG. 107,*8*. *P. carbonarius*, Czech.; ×1 (81).

Order PROSOTHECA Michaelsen, 1930

Male ducts on the segment which follows the testicular segment. *Tert.-Rec.*

Family ENCHYTRAEIDAE Vejdovsky, 1879

Setae needle-like or hook-shaped, without distinct nodulus, body straight or doubly recurved in S-shape. *Oligo.-Rec.*

Enchytraeus Henle, 1837 [**E. albidus*]. Setae in 2 ventral and 2 lateral bundles. *Oligo.-Rec.*, cosmop.——Fig. 108,*6*. **E. albidus*, Rec., Ger.; ×1 (65).

Order and Family UNCERTAIN

Lumbricopsis Fritsch, 1907 [**L. permicus*]. Body long, with many segments, each bearing pair of wartlike markings which probably indicate loca-

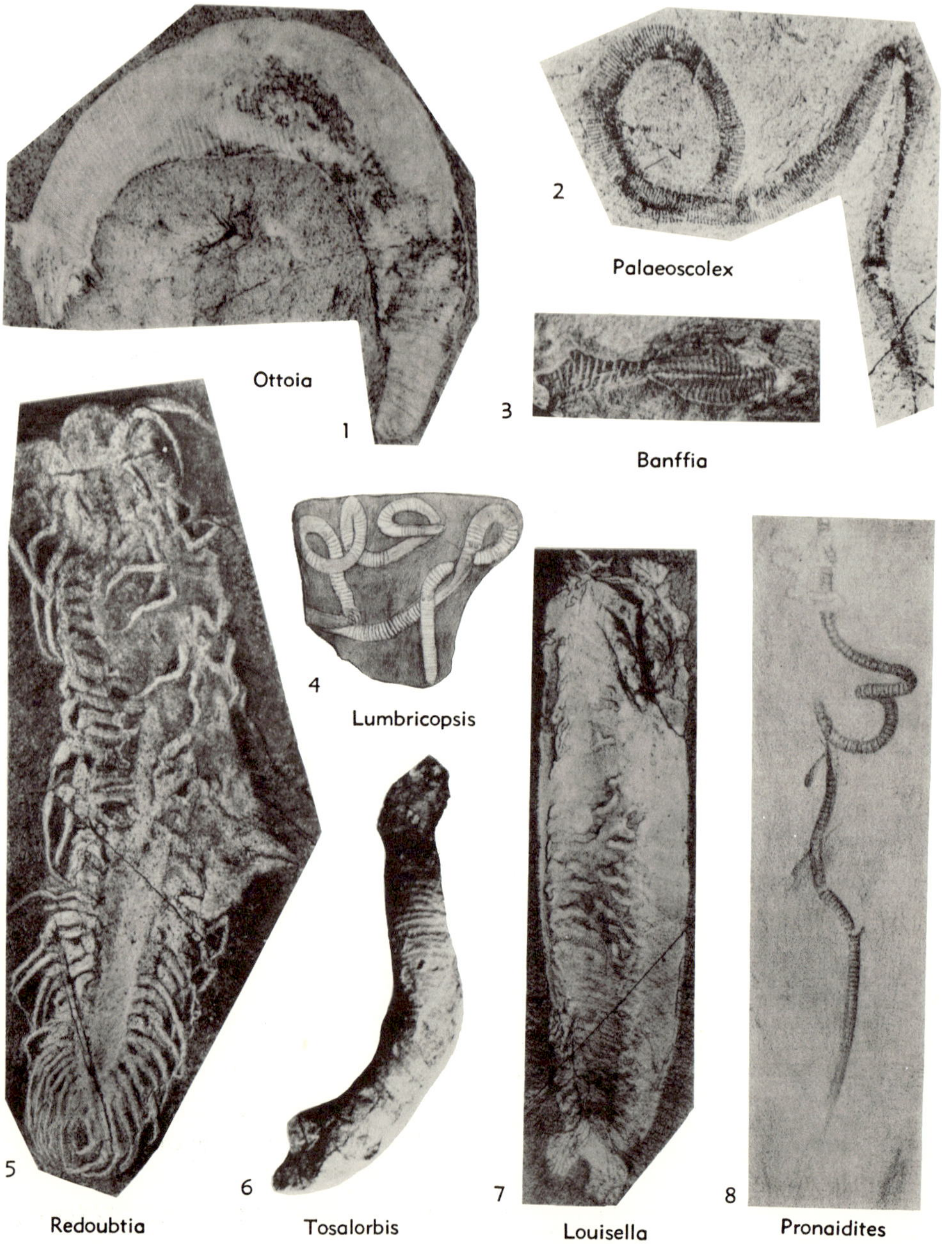

Fig. 107. Tubificidae, Ottoiidae, Palaeoscolecidae, Oligochaetia—Family Uncertain, Sipunculoida—Family Uncertain (p. *W*167-*W*170).

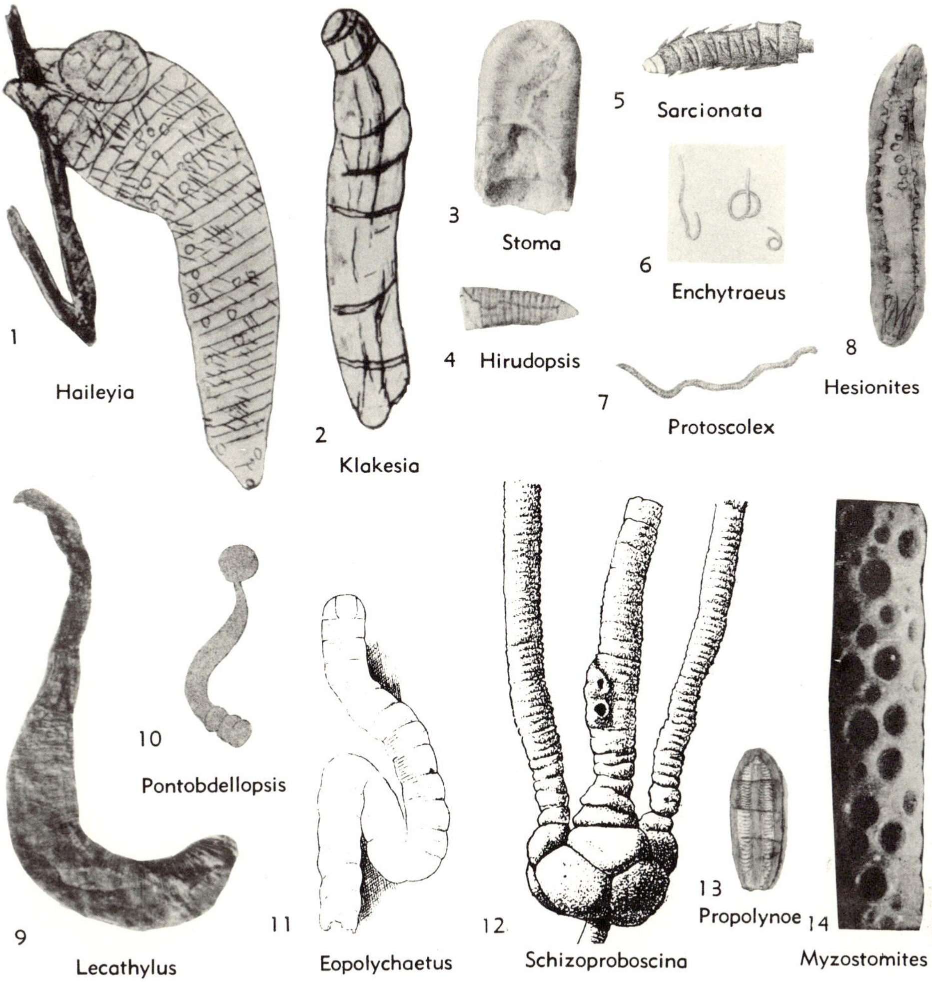

FIG. 108. Myzostomidae, Tubificidae, Enchytraeidae, Miskoiida—Family Uncertain, Sipunculoida—Family Uncertain, Phylum Uncertain—Family Uncertain (p. W167-W170).

tion of clitellum. *Perm.*, Eu.——FIG. 107,*4*. **L. permicus*, Czech.; ×0.25 (49).

Phylum SIPUNCULOIDA Sedgwick, 1898

Body unsegmented or poorly segmented, cylindrical or subcylindrical, with retractile introvert at anterior end, which may be armed with chitinous hooks. [Marine.] *Cam.-Rec.*

Order UNCERTAIN

Family OTTOIIDAE Walcott, 1911

[*nom. correct.* HOWELL, herein (*pro* Ottoidae WALCOTT, 1911)]

Body cylindrical, elongate, with numerous segments that vary in width posteriorly, with hooks around mouth and at posterior end; with papillose introvert or proboscis. *Cam.*

Ottoia WALCOTT, 1911 [**O. prolifica*]. Body elongate, tapering at each end, with many segments; minute hooks arranged in 5 or 6 concentric rings at anterior end and concentric row of stronger hooks at posterior end. *M.Cam.*, W.N. Am.——FIG. 107,*1*. **O. prolifica*, Burgess Sh., Can.(B.C.); ×1 (154).

Banffia WALCOTT, 1911 [**B. constricta*]. Body elongate, constricted medially; anterior section larger, elongate-spatulate in outline, with surface marked by fine, transverse, slightly imbricating lines which define narrow segments; posterior section narrowly elliptical, truncated at ends, surface with stronger lines than on anterior section. *M.Cam.*, W.N.Am. ——FIG. 107,*3*. **B. constricta*, Burgess Sh., Can. (B.C.); ×1 (154).

Family UNCERTAIN

Epitrachys EHLERS, 1869 [**E. rugosus*]. Body cylindrical, tapering, with many short segments that are more or less granulose on their outer surfaces. [Marine.] Jur., Eu.——FIG. 106,*8*. **E. rugosus*, Solnhofen Ls., Ger.; ×1 (34).

Lecathylus WELLER, 1925 [**L. gregarius*]. Body flask-shaped, bulblike in front, tapering backward to elongate, slender, tubular portion; anterior end attached to substratum; surface of anterior region crossed by lines about 0.2 mm. apart, lines becoming fainter, more irregular, and wrinkled toward rear. *Sil.*, N.Am.——FIG. 108,*9*. **L. gregarius*, Racine Dol., USA(Ill.); ×1 (124).

Louisella WALCOTT, 1911 [**L. pedunculata*]. Body elongate, tapering toward both ends, flattened on ventral surface, which bears longitudinal rows of podia, and 2 peltate extensions at posterior end. *M.Cam.*, W.N.Am.——FIG. 107,*7*. **L. pedunculata*, Burgess Sh., Can.(B.C.); ×1 (155).

Schizoproboscina YAKOVLEV, 1939 [**S. ivanovi*]. Living in curved calcareous tubes, open at both ends. [Ectoparasitic on crinoids.] *Carb.*, Eu.——FIG. 108,*12*. **S. ivanovi*, USSR; ×1 (167).

Stoma HADDING, 1913 [**S. hians*]. Body shaped like glove finger, crossed by fine striations. *Ord.*, Eu.——FIG. 108,*3*. **S. hians*, *Dicellograptus* Z. Swed., ×2 (62).

Phylum, Class, and Order UNCERTAIN

Family PALAEOSCOLECIDAE Whittard, 1953

Characters those of only known genus, *Palaeoscolex*. *Ord.*

Palaeoscolex WHITTARD, 1953 [**P. piscatorum*]. Body with many annulations marked by papillae, numbering about 60 to 80 in a ring, that were probably provided with very delicate chaetae; each metamere shows 2 bands, one without ornamentation, the other with innumerable minute chaetae, pores, or papillae; jaw apparatus consisting of pair of mandibles. [Marine.] *L.Ord.*, Eu.——FIG. 107, *2*. **P. piscatorum*, Tremadoc., Eng.; ×3 (163).

Family UNCERTAIN

Haileyia RUEDEMANN, 1934 [**H. adhaerens*]. Body stout, composed of 30 or more narrow segments, separated by sharp, hairlike sutures, and bearing irregularly distributed minute papillae and delicate setae; no parapodia except 2 large posterior ones for attachment; anterior part of body abruptly contracted to short, subtriangular, cephalic region. *Ord.*, W.N.Am.——FIG. 108,*1*. **H. adhaerens*, USA(Idaho); ×8 (129).

Hesionites FRITSCH, 1907 [**H. bioculata*]. Head with 2 eyes, pair of tactile organs, and jaws; body with about 20 segments, each with pair of parapodia, except for head and last posterior segments; each parapodium bearing about 5 setae. *Penn.*, N.Am.——FIG. 108,*8*. **H. bioculata*, Carbondale F., USA(Ill.); ×0.16 (49).

Hirudopsis MOBERG & SEGERBERG, 1906 [**H. koepingensis*]. Body apparently segmented, in short chitinous tube that is bluntly pointed at posterior end. *Ord.*, Eu.——FIG. 108,*4*. **H. koepingensis*, *Shumardia* Z., Swed.; ×3 (96).

Klakesia RUEDEMANN, 1934 [**K. simplex*]. Body smooth, cylindrical, composed of few large segments; anterior end rounded or provided with short proboscis-like lobe; posterior end terminated by plate or segment; 2 converging subtriangular plates (jaws?) at front extremity. *Sil.*, Alaska.——FIG. 108,*2*. **K. simplex*; ×8 (129).

Propolynoe FRITSCH, 1907 [**P. laccoei*]. Body short and wide, composed of about 40 segments, with 2 ?eyes on rectangular head, each segment having pair of parapodia, which bears bundles of setae. *Penn.*, N.Am.——FIG. 108,*13*. **P. laccoei*, Carbondale F., USA(Ill.); ×0.3 (49).

Protoscolex ULRICH, 1878 [**P. covingtonensis*]. Body long and very slender, uniform in width throughout, with rather thick test, many short segments. *U.Ord.*, USA.——FIG. 108,*7*. **P. covingtonensis*, Economy Sh., USA(Ky.); ×1 (149).

Redoubtia WALCOTT, 1918 [**R. polypodia*]. Body of moderate length, with many long parapodia. M.Cam., W.N.Am.——FIG. 107,*5*. **R. polypodia*, Burgess Sh., Can.(B.C.); ×2 (156).

Sarcionata COSTA, 1856 [**S. proboscidata*]. Body short, cylindrical, segmented, with proboscis and scattered thin, tapering appendages. *Cret.*, Eu.——FIG. 108,*5*. **S. proboscidata*, Italy; ×1 (25).

Tosalorbis KATTO, 1960 [**T. hanzawai*]. Body elongate, cylindrical, composed of many narrow segments; anterior end unknown. *Oligo.-Eoc.*, Japan.——FIG. 107,*6*. **T. hanzawai*, Eoc. (Muroto); ×1 (78).

REFERENCES

Abel, Othenio

(1) 1935, *Vorzeitliche Lebensspuren:* 644 p., 530 text fig., G. Fisher (Jena).

Allan, R. S.

(2) 1927, *Keilorites (a new generic name for a Silurian annelid from Australia):* Geol. Mag., v. 64, p. 240.

Astre, Gaston

(3) 1937, *Un annélide sabellien dans le Garumnien de Saldés:* Soc. Histoire Nat. Toulouse, Bull., v. 71, p. 192-194.

Bather, F. A.

(4) 1905, *The Mount Torlesse annelid:* Geol. Mag., new ser., decade 5, v. 2, p. 532-541.

Beneden, P. J. Van, & Coemans, E.

(5) 1867, *Note sur un insecte (Omalia macroptera) et un gastéropode pulmoné (Palaeorbis ammonis) du terrain houiller:* Acad. Royale Sci., Lettres & Beaux-Arts Belgique, Bull., new ser., v. 23, p. 384-401.

Benham, W. B.

(6) 1896, *Archannelida, Polychaeta, and Myzostomaria:* The Cambridge Natural History, v. 2, p. 239-344, Macmillan Co. (London, New York).

Berkeley, M. J.

(7) 1835, *Observations upon the Dentalium subulatum of Deshayes:* Zool. Jour., v. 5, p. 424-427.

Blainville, M. H. D. de

(8) 1830, *Dictionnaire des Sciences Naturelles:* Cuvier, F. G., ed., v. 60, p. 113 (Strasbourg, Paris).

Blumenbach, J. F.

(9) 1803, *Specimen archaeologiae telluris, terrarum imprimis Hannoveranarum:* Soc. Reg. Sci. Gottingensis, Comment., v. 15, p. 132-156, pl. 1-2.

Bosc, L. A. G.

(10) 1801, *Histoire naturelle des vers, contenant leur description et leur moeurs:* v. 1, p. 150, illus., Roret (Paris).

Bronn, H. G.

(11) 1828, *Verzeichnis der bei dem Heidelberger Mineralien-Komptoir verkäuflichen Konchylien-, Pflanzenthier- und anderen Versteinerungen:* Zeitschr. Mineral., v. 2, p. 544.

Bush, K. J.

(12) 1910, *Tubicolous annelids of the tribes Sabellides and Serpulides from the Pacific Ocean:* Smithsonian Inst., Harriman Alaska Ser., v. 12, p. 167-246, pl. 21-44.

Calvin, Samuel

(13) 1888, *On a new genus and new species of tubicolar Annelida:* Am. Geologist, v. 1, p. 24-28.

Caullery, Maurice, & Mesnil, Félix

(14) 1896, *Note sur deux Serpuliens nouveaux, Oriopsis metchnikowi, n.g., et Josephella marenzelleri, n.g., n.sp.:* Zool. Anzeiger, v. 19, p. 482-486.

Chapman, Frederick

(15) 1910, *New and little-known Victorian fossils in the National Museum, Part 10. Some Paleozoic worms and Crustacea:* Royal Soc. Victoria, Proc., v. 22, pt. 2, p. 101-112, pl. 27, 29.

Claparède, Edouard

(16) 1870, *Les annélides chétopodes du Golfe de Naples, Part 2:* Soc. Phys. & Histoire Nat. Genève, Mém., v. 20, pt. 1, p. 1-225, pl. 17-31.

Clarke, J. M.

(17) 1903, *Some Devonic worms:* N.Y. State Museum, Bull. 69, p. 1234-1238, pl. 27-28.

(18) 1908, *Beginnings of dependent life:* N.Y. State Museum, Bull. 121, p. 146-196, pl. 1-12.

(19) 1921, *Organic dependence and disease; their origin and significance:* 114 p., illus., Yale Univ. Press (New Haven).

Cobbold, E. S.

(20) 1921, *The Cambrian horizons of Comley (Shropshire) and their Brachiopoda, Pteropoda, Gasteropoda, etc.:* Geol. Soc. London, Quart. Jour., v. 76, p. 325-386, pl. 21-24.

(21) 1935, *Lower Cambrian faunas from Hérault, France:* Annals & Mag. Nat. History, ser. 10, v. 16, p. 25-48, pl. 2-4.

———, & Pocock, R. W.

(22) 1934, *The Cambrian area of Rushton (Shropshire):* Royal Soc. London, Philos. Trans., Ser. B, v. 223, p. 305-409, pl. 38-45.

Conrad, T. A.

(23) 1870, *Notes on Recent and fossil shells, with descriptions of new species:* Am. Jour. Conchology, v. 6, p. 71-78.

(24) 1875, *Descriptions of new genera and species of fossil shells of North Carolina:* in Kerr, W. C., North Carolina Geol. Survey, Rept. 1, Append. A, p. 1-28, pl. 1-2.

Costa, O. G.

(25) 1856, *Paleontologia del regno di Napoli:* Accad. Pontaniana, Atti, v. 7, p. 113-378, pl. 11-28.

Daudin, F. M.
(26) 1800, *Recueil de mémoires et notes sur des espèces inédites ou peu connues de mollusques, de vers et de zoophytes:* 48 p., illus., Fuchs (Paris).

Defrance, D. F.
(27) 1827, *Dictionnaire des sciences naturelles:* CUVIER, F. G., ed., v. 46, p. 321 (Strasbourg, Paris).

Defrance, M. J. L.
(28) 1819, *Dictionnaire des sciences naturelles:* CUVIER, F. G., ed., v. 14, p. 517 (Strasbourg, Paris).

Deshayes, G. P.
(29) 1826, *Anatomie et monographie du genre Dentale:* Soc. Histoire Nat. Paris, Mém., v. 2, p. 321-378, pl. 15-18.

Desio, Ardito
(30) 1940, *Vestigia problematiche paleozoiche della Libia:* Museu Libico Storia Nat., Ann., v. 2, p. 47-92, pl. 4-13.

Dubinin, V. B.
(31) 1948, *Poux (Anoploura) et nematodes pléistocènes trouvés lors de l'examen de Spermophiles fossiles du bassin du fleuve Indiguirka (Sibérie Nord-orientale):* Acad. Nauk SSSR Leningrad, Doklady, v. 62, p. 417-420, text fig. 1-3.

Duisburg, H. von
(32) 1862, *Beitrag zur Bernstein-Fauna:* K. Phys.-Oekon. Gesell. Königsberg, Schrift., Bd. 3, p. 31-36, pl. 1.

Ehlers, Ernst
(33) 1868, *Ueber eine fossile Eunicee aus Solenhofen (Eunicites avitus), nebst Bemerkungen über fossile Würmer überhaupt:* Zeitschr. Wiss. Zool., Bd. 18, p. 421-443, pl. 29, fig. 1-2.
(34) 1869, *Ueber fossile Würmer aus dem lithographischen Schiefer in Bayern:* Palaeontographica, v. 17, p. 145-175, pl. 31-37.
(35) 1901, *Die Anneliden der Sammlung Plate, Fauna Chiliens:* Zool. Jahrbücher, Suppl., v. 5, p. 251-272.

Eichwald, Eduard von
(36) 1856, *Beitrag zur geographischen Verbreitung der fossilen Thiere Russlands, Alte Periode:* Soc. Impér. Nat. Moscou, Bull., v. 29, p. 88-127, 406-453, 555-608.
(37) 1860, *Lethaea rossica ou paléontologie de la Russie:* v. 1, 681 p., 59 pl., figs.

Eisenack, Alfred
(38) 1939, *Einige neue Annelidenreste aus dem Silur und Jura des Baltikums:* Zeitschr. Geschiebeforsch., v. 15, p. 153-176, pl. A-C.

Eller, E. R.
(39) 1934, *Annelid jaws from the Upper Devonian of New York:* Carnegie Museum, Ann., v. 22, p. 303-316, pl. 22-23.
(40) 1936, *A new scolecodont genus, Ildraites, from the Upper Devonian of New York:* Carnegie Museum, Ann., v. 25, p. 73-76, pl. 11.
(41) 1938, *Scolecodonts from the Potter Farm Formation of the Devonian of Michigan:* Carnegie Museum, Ann., v. 27, p. 275-286, pl. 28-29.
(42) 1940, *New Silurian scolecodonts from the Albion beds of the Niagara Gorge, New York:* Carnegie Museum, Ann., v. 28, art. 2, p. 9-46, pl. 1-7.
(43) 1942, *Scolecodonts from the Erindale, Upper Ordovician, at Streetville, Ontario:* Carnegie Museum, Ann., v. 28, art. 11, p. 241-270, pl. 1-4.
(44) 1944, *Scolecodonts of the Silurian Manitoulin dolomite of New York and Ontario:* Am. Midland Naturalist, v. 32, p. 732-755, pl. 2-4.
(45) 1945, *Scolecodonts from the Trenton Series (Ordovician) of Ontario, Quebec, and New York:* Carnegie Museum, Ann., v. 30, art. 12, p. 119-212, pl. 1-7.

Fabricius, J. C.
(46) 1779, *Reise nach Norwegen, mit Bemerkungen aus der Naturhistoirie und Oekonomie:* p. 385 (Hamburg).

Fauvel, Pierre
(47) 1923, *Un nouveau serpulien d'eau saumâtre, Mercierella, n.g., enigmatica, n.sp.:* Soc. Zool. France, Bull., v. 47, p. 424-430, fig. 1a-o.

Fritsch, Anton
(48) 1878, *Die Reptilien und Fische der böhmischen Kreideformation:* 46 p., 10 pl., F. Rivnáč (Prague).
(49) 1907, *Miscellanea Palaeontologica, I. Palaeozoica:* 23 p., 12 pl., 4 fig. (Ebendort).

Gabb, W. M.
(50) 1876, *Notes on American Cretaceous fossils, with descriptions of some new species:* Acad. Nat. Sci. Philadelphia, Proc., p. 276-324, pl. 17.

Gardner, J. A.
(51) 1916, *Systematic Paleontology, Upper Cretaceous. Vermes:* Maryland Geol. Survey, Upper Cretaceous vol., p. 745-749, pl. 47.

Girty, G. H.
(52) 1912, *On some new genera and species of Pennsylvanian fossils from the Wewoka Formation of Oklahoma:* N.Y. Acad. Sci., Ann., v. 21, p. 119-156.

Glaessner, M. F.
(53) 1958, *New fossils from the base of the Cambrian in South Australia:* Royal Soc.

South Australia, Trans., v. 81, p. 185-188, pl. 1.

Goeppert, H. R.

(54) 1853, in Germar, E. F., *Die Versteinerungen des Steinkohlengebirges von Wettin und Löbejün im Saalkreise (Petrificata stratorum Lithanthracum Wettini):* H. 8, p. 111, illus., G. Schwetschke (Halle).

Goldfuss, G. A.

(55) 1833, *Petrefacta Germaniae:* T. 1, div. 3, p. 222-242, pl. 68, fig. 11f, Arnz & Co. (Düsseldorf).

Gould, A. A.

(56) 1841, *Report on the Invertebrata of Massachusetts, comprising the Mollusca, Crustacea, Annelida, and Radiata:* xiii+373 p., 15 pl., Folsom, Wells & Thurston (Cambridge).

Greve, L. de

(57) 1938, *Eine Molluskenfauna aus dem Neogen von Iquitos am oberon Amazonas in Peru:* Schweiz. Palaeont. Gesell. Zurich, Abhandl., v. 71, p. 1-133, pl. 1-10.

Grinnell, G. B.

(58) 1877, *Notice of new genus of annelids from the Lower Silurian:* Am. Jour. Sci., ser. 3, v. 14, no. 81, p. 229-230.

Grube, Eduard

(59) 1840, *Actinien, Echinodermen und Würmer des Adriatischen—und Mittelmeers, nach eigenen Sammlungen beschrieben:* p. 79 (Königsberg).

(60) 1878, *Annulata Semperiana, Beiträge zur Kenntniss der Annelidenfauna der Philippinen:* Akad. Impér. Sci. St. Pétersbourg, Mém., ser. 7, v. 25, no. 8, pl. 8, fig. 6a.

Gunnerus, J. E.

(61) 1768, *Om nogle Norske coraller:* K. Norske Vidensk., Skrift., v. 4, p. 38-73, pl. 2, fig. 11-13.

Hadding, Assar

(62) 1913, *Undre Dicellograptusskiffern i Skåne jämte nagra därmed ekvivalenta bildningar:* Lund Univ., Årsskr., new ser., sec. 2, v. 9, no. 15, p. 29, pl. 1.

Hagenow, Friederich von

(63) 1840, *Monographie der rügen'schen Kreide-Versteinerungen, Abth. 2: Radiolaren und Annulaten:* Neues Jahrb. Mineral., Jahrg. 1840, p. 631-672, pl. 9.

Hempelmann, Friedrich

(64) 1931, *Erste und zweite Klasse der Vermes Polymera (Annelida), Archiannelida=Ur-Ringel-Würmer und Polychaeta=Borsten-Würmer:* Handbuch der Zoologie, Bd. 2, pt. 7, p. 198, text fig. 238, de Gruyter (Berlin).

Henle, F. G. J.

(65) 1837, *Ueber Enchytraeus, eine neue Anneliden-Gattung:* Arch. Anat., Physiol. & Wiss. Med., Jahrg. 1837, p. 74-90, pl. 6, fig. 1-9.

Hinde, G. J.

(66) 1879, *On annelid jaws from the Cambro-Silurian, Silurian, and Devonian formations in Canada and from the Lower Carboniferous in Scotland:* Geol. Soc. London, Quart. Jour., v. 35, p. 370-389, pl. 18-20.

Horwood, A. R.

(67) 1912, *On Archarenicola rhaetica, sp. nov..* Geol. Mag., new ser., dec. 5, v. 9, p. 395-399, pl. 21, fig. 1-4.

Howell, B. F.

(68) 1943, *Hamulus, "Falcula," and other Cretaceous Tubicola of New Jersey:* Acad. Nat. Sci. Philadelphia, Proc., v. 95, p. 139-166, pl. 19-20.

(69) 1943, *Faunas of the Cambrian Cloud Rapids and Treytown Pond formations of northern Newfoundland:* Jour. Paleontology, v. 17, p. 236-247, pl. 36-39.

(70) 1949, *New hydrozoan and brachiopod and new genus of worms from the Ordovician Schenectady formation of New York:* Wagner Free Inst. Sci. Philadelphia, Bull., v. 24, no. 1, p. 1-10, pl. 1-2.

(71) 1953, *A new terebellid worm from the Carboniferous of Texas:* Wagner Free Inst. Sci. Philadelphia, Bull., v. 28, no. 1, p. 1-4, pl. 1.

(72) 1953, *New Permian terebellid worm from Arizona:* Wagner Free Inst. Sci. Philadelphia, Bull., v. 28, no. 4, p. 25-28, pl. 1.

(73) 1955, *Paraterebella, new name for Terebellopsis Howell, 1955:* Jour. Paleontology, v. 29, p. 189.

(74) 1958, *Type species for the nemertan worm genus, Legnodesmus:* Jour. Paleontology, v. 32, p. 247.

(75) 1959, *Three notes on Silurian worm genera:* Jour. Paleontology, v. 33, p. 487.

Jaskó, Sándor

(76) 1940, *A Rima és Tarna közenek oligocén rétegei és Kövületei:* Földtani Közlöny, v. 70, p. 294-317, pl. 9.

Johnston, George

(77) 1837, *The natural history of British zoophytes:* Mag. Zool. & Botany, v. 1, p. 64-81.

Katto, Jiro

(78) 1960, *Some problematica from the so-called unknown Mesozoic strata of the southern part of Shikoku, Japan:* Tohoku Univ., Sci. Repts., ser. 2 (Geol.), spec. v. 4, p. 323-334, pl. 34-35.

König, C. D. E.
(79) 1825, *Icones fossilium sectiles, centuria prima:* (London).

Kozlowski, Roman
(80) 1956, *Sur quelques appareils masticateurs des annélides polychètes ordoviciens:* Acta Palaeont. Polonica, v. 1, p. 165-210.

Kušta, Jan
(81) 1888, *Beitrag zur Kenntniss der Steinkohlenfauna bei Rakonitz:* Böhm. Gesell. Wiss., Math-Naturwiss. Cl., Sitzungsber., Jahrg. 1887, p. 561-564, 1 pl., fig. 1.

Lamarck, J. B. de
(82) 1801, *Système des animaux sans vertèbres:* p. 324 (Paris).
(83) 1818, *Histoire Naturelle des Animaux sans Vertèbres:* v. 5, p. 348, 359, 371 (Paris).

Lange, F. W.
(84) 1947, *Anelídeos poliquetos dos folhelhos devonianos de Paraná:* Arquivos Museu Paranaense, v. 6, p. 161-230, pl. 17-32.
(85) 1949, *Polychaete annelids from the Devonian of Paraná, Brazil:* Bull. Am. Paleontology, v. 33, no. 134, p. 1-102, pl. 1-16.

Langerhans, P.
(86) 1884, *Die Würmfauna von Madeira:* Zeitschr. Wiss. Zool., v. 40, p. 247-285, pl. 15-17.

Linné, Carl
(87) 1758, 1767, *Systema naturae:* ed. 10 (1758), p. 647-648, 787; ed. 12 (1767), p. 1092, 1263, 1265, 1268.

Lugeon, Maurice
(88) 1916, *Sur l'inexistence de la nappe du Augsmatthorn:* Soc. Vaudoise Sci. Nat., Bull., v. 51, Procès-Verbaux, p. 55-57.

Madsden, F. J.
(89) 1957, *On Walcott's supposed Cambrian holothurians:* Jour. Paleontology, v. 31, p. 281-282.

Massalongo, A. B.
(90) 1855, *Monografia delle nereidi fossili del M. Bolca:* 55 p., 6 pl. (Verona).

Matthew, G. F.
(91) 1899, *Upper Cambrian of Mount Stephen, British Columbia: the trilobites and worms:* Roy. Soc. Canada, Trans., ser. 2, v. 5, sec. 4, p. 39-66, pl. 1-2.

Maury, C. J.
(92) 1927, *Fósseis Silurianos de Santa Catharina:* Serv. Geol. & Mineral. Brasil, Bull. 23, p. 8, pls., fig. 1.

Mayer, Gaston
(93) 1954, *Neue Beobachtungen an Lebensspuren aus dem Unteren Hauptmuschelkalk (Trochitenkalk) von Wiesloch:* Neues Jahrb. Mineral., Abhandl., Bd. 99, p. 223-229, pl. 16-18.

Menge, F. A.
(94) 1866, *Ueber ein Rhipidopteron (Triaena tertiaria) und einige Helminthen im Bernstein:* Naturforsch. Gesell. Danzig, Schrift., Bd. 1, T. 3.

Michelin, J. L. H.
(95) 1840-47, *Iconographie zoophytologique, description par localités et terrains des polypiers fossiles de France et pays environnants:* 348 p., 79 pl. (Paris).

Moberg, J. C., & Segerberg, C. O.
(96) 1906, *Bidrag till kännedomen om Ceratopygeregionen med särskild hänsyn till dess utveckling i Fogelsångstrakten:* K. Fysiog. Sällsk., Handl., new ser., v. 17, p. 1-113, pl. 1-7.

Montfort, P. D. de
(97) 1808, *Conchyliologie systématique et classification méthodique des coquilles:* p. 395-396, pl. 99 (Paris).

Mørch, O. A. L.
(98) 1863, *Revisio critica Serpulidarum:* Naturhist. Tidsskr., ser. 3, v. 1, p. 347-370, 1 pl.

Moret, Léon
(99) 1948, *Manuel de paléontologie animale:* ed. 2, p. 300, text fig. 112D, Masson et Cie. (Paris).

Morton, S. G.
(100) 1834, *Synopsis of the organic remains of the Cretaceous Group of the United States:* 88 p., pl. 2, 16 (Philadelphia).

Müller, O. F.
(101) 1788, *Zoologia danica, seu animalium Daniae et Norwegiae rariorum ac minus notorum, descriptiones et historia:* v. 1, p. 26, pl. 26 (Havniae).

Münster, George von
(102) 1840, Beiträge zur Petrefacten-kunde, v. 3, p. 115, pl. 9, fig. 4.
(103) 1842, Beiträge zur Petrefacten-kunde, v. 5, p. 98, pl. 1.

Murchison, R. I.
(104) 1839, *The Silurian System:* 768 p., 27 pl., John Murray (London).

Nielson, K. B.
(105) 1931, *Serpulidae from the Senonian and Danian deposits of Denmark:* Dansk Geol. Foren., Medd., v. 8, p. 71-113, pl. 1-3.

Oken, Lorenz
(106) 1807, *Gottingsche lehrte Anzeigen:* v. 2, p. 1168.
(107) 1815, *Okens Lehrbuch der Naturgeschichte:* T. 3, Abt. 1, p. 381.

Öpik, A. A.
(108) 1930, *Beiträge zur Kenntnis der Kukruse-(C_2-C_3-) Stufe in Eesti:* Univ. Tartu (Dorpat.), Acta & Comment., sec. A, v. 29, art. 4, p. 11, pl. 1.

Örsted, A. S.

(109) 1843, *Annulatorum danicorum conspectus:* fasc. 1, p. 44.

Pallas, P. S.

(110) 1766, *Miscellanea zoologica:* p. 118, pl. 9.

Philippi, R. A.

(111) 1844, *Einige Bemerkungen über die Gattung Serpula, nebst Aufzählung die von mir im Mittelmeer mit dem Thier beobachten Arten:* Archiv Naturgesch., Jahrg. 10, Bd. 1, p. 192.

Phillips, John, & Salter, J. W.

(112) 1848, *Palaeontological appendix to Prof. John Phillips' memoir on the Malvern Hills compared with the Palaeozoic districts of Abberley, etc.:* Geol. Survey Grt. Britain & Museum Pract. Geol. London, Mem., v. 2, pt. 1, p. 331-386, pl. 4-30.

Prantl, Ferdinand

(113) 1948, *Some terebelloid remains from the Ordovician of Bohemia:* K. České Společnosti Nauk, Trída Mathematico-Přírodorvědecką, Věstnik, Ročník 1948, Číslo 8, p. 1-8, pl. 1-2.

Quenstedt, F. A.

(114) 1858, *Der Jura:* fasc. 3, p. 521, pl. 69.

Reish, D. J.

(115) 1952, *Discussion of the colonial tube-building polychaetous annelid, Dodecaceria fistulicola Ehlers:* So. California Acad. Sci., Bull., v. 51, p. 103-107, pl. 20.

Richardson, E. S.

(116) 1956, *Pennsylvanian invertebrates of the Mazon Creek area, Illinois:* Fieldiana, Geol., v. 12, p. 1-76.

Risso, Antoine

(117) 1826, *Histoire naturelle des principales productions de l'Europe méridionale et particulièrement de celles des environs de Nice et des Alpes Maritimes:* v. 4.

Roger, Jean

(118) 1943, *Les polypiers du gisement pliocene ancien de Dar bel Hamri (Maroc):* Muséum Natl. Histoire Nat., Bull., ser. 2, v. 15, no. 6.

Roverto, Gaetano

(119) 1899, *Serpulidae del Terziario e del Quaternario in Italia:* Palaeontographia Italica, v. 4, p. 47-91, pl. 6-7.

(120) 1903, *Annellidi del Terziario:* Revista Italiana Paleont., v. 9, p. 103-104.

(121) 1904, *Contributo allo studio dei vermeti fossili:* Soc. Geol. Italiana, Bull., v. 23, p. 67-69.

(122) 1904, *Studi monografici sugli annellidi fossili, I, Terziario:* Palaeontographica Italica, v. 10, p. 1-73, pl. 1-4.

Roy, S. K.

(123) 1929, *Contributions to Paleontology:* Field Museum Nat. History, Pub. 254, Geol. Ser., v. 4, no. 5, p. 201-228, pl. 32-40.

———, & Croneis, Carey

(124) 1931, *A Silurian worm and associated fauna:* Field Museum Nat. History, Pub. 298, Geol. Ser., v. 4, p. 229-247, pl. 43-45.

Rudolphi, C. A.

(125) 1802, *Beobachtungen über die Eingeweidwürmer:* in Wiedemann, C. R. W., Archiv Zool. & Zootomie, v. 2, p. 18.

Ruedemann, Rudolf

(126) 1901, *Hudson River beds near Albany and their taxonomic equivalents:* N. Y. State Museum, Bull. 42, p. 489-596, pl. 1-2.

(127) 1916, *Account of some new or little-known species of fossils:* N. Y. State Museum, Bull. 189, p. 7-97, pl. 1-33.

(128) 1925, *Some Silurian (Ontarian) faunas of New York:* N. Y. State Museum, Bull. 265, p. 43-45, pl. 14-15.

(129) 1934, *Paleozoic plankton of North America:* Geol. Soc. America, Mem. 2, p. 86, 89, pl. 21.

Saint-Joseph, Antoine de

(130) 1906, *Les Annélides polychètes des côtes de France (océan et côtes de Provence):* Sci. Nat., Ann., ser. 9, v. 3, p. 145-260, pl. 1-5.

Salter, J. W.

(131) [*See* Phillips & Salter]

(132) 1855, Malvern Club, Trans. pt. 1, p. ?, *fide* Murchison, R. I., in Siluria, ed. 4 (1867), p. 514.

(133) 1881, *On the fossils of North Wales:* Geol. Survey Grt. Britain & Museum Pract. Geol., Mem., v. 3, ed. 2, append. by Etheridge, R., p. 331-611, pl. 1-26.

Sars, G. O.

(134) 1872, *Diagnoser af nye annelider fra Christianiafjorden:* Vidensk.-Selsk. Khristiania, Forhandl., Year 1871, p. 406-417.

Savigny, J. C.

(135) 1820, *Système des annélids, principalement de celles des côtes de l'Égypt et de la Syrie, . . . , Description de l'Égypt:* Histoire Nat., v. 21, p. 88.

Schlotheim, E. F. von

(136) 1820, *Die Petrefactenkunde auf ihrem jetzigen Standpunkte durch die Beschreibung seiner Sammlung . . . erläutert:* T. 1, p. 96, Becker (Gotha).

Schmidt, W. J.

(137) 1955, *Die tertiären Würmer Österreichs:* Österreich. Akad. Wiss., Math.-Naturwiss. Kl., Denkschr., Bd. 109, T. 7, pl. 7.

Seurat, L. G.
(138) 1916, *Sur les Oxyures des manifères:* Soc. Biol., Hebdomadaires Séances & Mém., Comptes Rendus, v. 79, p. 64-68.

Šnajdr, Milan
(139) 1951, *On errant Polychaeta from the Lower Paleozoic of Bohemia:* Geol. Survey Czech., Bull., v. 18, Paleontology, p. 241-296, pl. 27-36.

Sokolov, B. S.
(140) 1948, *Commensalism in favositids:* Akad. Nauk SSSR, Izvestia, Biol. Ser., p. 101-110, fig. 1-10.

Sowerby, J. de C.
(141) 1829, *The mineral conchology of Great Britain:* v. 6, p. 200, pl. 579.
(142) 1839, *Fossil shells of the upper Ludlow rock:* in MURCHISON, R. I., The Silurian System, pt. 2, p. 579-734, pl. 1-27, John Murray (London).

Stauffer, C. R.
(143) 1933, *Middle Ordovician Polychaeta from Minnesota:* Geol. Soc. America, Bull., v. 44, p. 1173-1218, pl. 59-61.

Stephenson, L. W.
(144) 1923, *Invertebrate fossils of the Upper Cretaceous formations:* North Carolina Geol. Survey, v. 5, pt. 1, p. 1-402, pl. 1-100.

Stoliczka, F.
(145) 1868, *Cretaceous fauna of southern India, II, The Gastropoda:* Geol. Survey India, Mem., (Palaeont. Indica), p. 236-237.

Taylor, A. L.
(146) 1935, *A review of the fossil nematodes:* Helminthol,. Soc. Washington, Proc., v. 2, p. 47-49, fig. 6.

Termier, Henri, & Termier, Geneviéve
(147) 1951, *Sur deux formes énigmatiques de l'Ordovicien marocain: Leckwickia et Khemisina:* Serv. Geol. Maroc, Div. Mines & Geol., Notes, v. 5, p. 187-198.

Torell, O. M.
(148) 1870, *Petrificata suecana Formationis cambricae:* Lund Univ., Årsskr., v. 6, no. 8.

Ulrich, E. O.
(149) 1878, *Observations on fossil annelids, and descriptions of some new forms:* Cincinnati Soc. Nat. History, Jour., v. 1, p. 87-91, pl. 4.
(150) 1910, *Fossils and age of the Yakutat Formation: description of collections made chiefly near Kadiak, Alaska:* Smithsonian Inst., Harriman Alaska Ser., v. 4, Geol. & Paleont., p. 125-146, pl. 11-21.

Vine, G. R.
(151) 1882, *Notes on the annelid Tubicola of the Wenlock shales from the washings of Mr. George Maw,* F.G.S.: Geol. Soc. London, Quart. Jour., v. 38, p. 377-393, pl. 15.

Voigt, Ehrhard
(152) 1938, *Ein fossiler Saitenwürm (Gordius tenuifibrosus, n.sp.) aus der eozäner Braunkohle des Geiseltales:* Nova Acta Leopoldina, new ser., v. 5, pl. 351-360, pl. 55-56.

Walcott, C. D.
(153) 1908, *Mount Stephen rocks and fossils:* Canadian Alpine Jour., v. 1, p. 232-248, pl. 1.
(154) 1911, *Middle Cambrian annelids:* Smithsonian Misc. Coll., v. 57, no. 5, p. 109-142, pl. 18-23.
(155) 1911, *Middle Cambrian holothurians and medusae:* Smithsonian Misc. Coll., v. 57, no. 3, p. 41-68, pl. 8-13.
(156) 1918, *Geological explorations in the Canadian Rockies, Explorations and fieldwork of the Smithsonian Institution in 1917:* Smithsonian Misc. Coll., v. 68, no. 12, p. 4-20, text fig. 3-19.

Webby, B. D.
(157) 1958, *A Lower Mesozoic annelid from Rock Point, southwestern Wellington, New Zealand:* New Zealand Jour. Geol. & Geophys., v. 1, p. 509-513.

Weller, Stuart
(158) 1907, *A report on the Cretaceous paleontology of New Jersey, based on the stratigraphic studies of George N. Knapp:* Geol. Survey New Jersey, Paleont., v. 4, p. 310, pl. 19.
(159) 1925, *A new type of Silurian worm:* Jour. Geol., v. 33, p. 540-544, fig. 1.

Wetzel, Walter
(160) 1957, *Semiserpula, eine neue Röhrenwurm-Gattung aus dem Alt-Tertiär Chiles:* Senckenbergiana Lethaea, Bd. 38, p. 29-35, pl. 1.

White, C. A.
(161) 1879, *Descriptions of new species of invertebrate fossils from the Carboniferous and Upper Silurian rocks of Illinois and Indiana:* Acad. Nat. Sci. Philadelphia, Proc. for 1878, p. 29-37.
(162) 1883, *Contributions to invertebrate paleontology, No. 8; Fossils from the Carboniferous rocks of the interior states:* U.S. Geol. & Geog. Survey Territories, Ann. Rept. 12, pt. 1, p. 155-171, pl. 39-42.

Whittard, W. F.
(163) 1953, *Palaeoscolex piscatorum, gen. et sp. nov., a worm from the Tremadocian of Shropshire:* Geol. Soc. London, Quart. Jour., v. 109, pt. 2, p. 125-133, pl. 4-5.

Wilson, A. E.
(164) 1948, *Miscellaneous classes of fossils, Ottawa*

Formation, Ottawa-St. Lawrence Valley Canada: Geol. Survey Canada, Dept. Mines & Res., Mines & Geol. Br., Bull. 11, p. 55-56, pl. 24-25.

Wrigley, Arthur

(165) 1950, *Les opercules de serpulidés de l'Éocène du Bassin de Paris:* Soc. Géol. France, Bull., ser. 5, v. 19, p. 499-505.

(166) 1951, *Some Eocene serpulids:* Geologist's Assoc., Proc., v. 62, p. 177-202.

Yakovlev, N. N.

(167) 1939, *Sur la découverte d'un parasite original des crinoïdes marins du Carbonifère:* Acad. Sci. USSR, Proc., v. 22, p. 146-148, fig. 1-7.

Yanichevsky, M.

(168) 1926, *Ob ostatkakh Trubchatykh chervey iz Kembriyskoy Siney Gliny:* Paleont. Soc. Russia, Jour., v. 4, p. 99-111. [On fossil Tubicola in the Cambrian Blue Clay]

Žebera, K.

(169) 1935, *Les conodontes et les scolecodontes du Barrandien:* Acad. Tchèque Sci., Bull. Internatl., Cl. Sci. Math., Nat. & Méd., v. 36, p. 88-96, pl. 1-2.

TRACE FOSSILS AND PROBLEMATICA

By Walter Häntzschel

[Geologisches Staatsinstitut, Hamburg, Germany]

CONTENTS

INTRODUCTION

A vast majority of the fossils dealt with in this part of the *Treatise* consist of tracks, trails, and burrows; these collectively may be designated as *trace fossils.* Only a small minority are *body fossils,* all of them difficult to interpret and doubtful as to classificatory status. Actually, a considerable number of the so-called body fossils are not fossils at all, since they have come to be recognized as inorganic structures. Those that formerly were given a generic name (e.g., *Eozoon, Eophyton*) are summarized under a separate heading at the end of the text. Trace fossils and doubtful sorts of body fossils are often termed *Problematica,* because of their more or less obscure nature.

Although a large part of the trails and burrows were made by worms, they are treated here rather than in the chapter on worms, which is restricted to such body fossils as are normally attributed to annelids and other phyla of worms.

Owing to lack of a satisfactory classificatory system for the fossils here dealt with, their names have been arranged alphabetically within the several parts, because this is most convenient for reference.

Tracks, trails, and burrows are often hard to interpret, leaving considerable uncertainty as to origin, and thus they never have been very popular with paleontologists. Their importance for paleozoology is small and they can only rarely be used as index fossils. In almost all textbooks on paleontology a detailed chapter on trails and burrows *(Lebensspuren)* is missing. The present treatment is the first one which tries to cover all forms so far described and to deal with them uniformly. In the future it will certainly be necessary to make corrections, especially in nomenclature, for it is often extremely difficult or impossible to assign type-specimens in this special field of paleontology. Here, "genera" and "species" do not have the usual taxonomic meaning, but are only supposed to indicate a certain formal assemblage. The naming of fossils, their delimitation, and thus their synonymy is nowhere in paleontology so dependent on personal opinion as in this group of fossils.

DEFINITIONS

hieroglyphs (FUCHS, 1895). "Problematical fossils. . . . , which . . . in their appearance are suggestive of drawings, ornaments or even letters."

ichnocoenosis (LESSERTISSEUR, 1955). An association of *Lebensspuren,* corresponding to "biocoenosis."

ichnofossil (German, *Spuren-Fossil,* SEILACHER, 1953). Trace fossil.

Ichnolites (E. HITCHCOCK, 1841). Name proposed for a "class" including all sorts of tracks.

ichnolithology (E. HITCHCOCK, 1841). Same as ichnology, but term not widely adopted.

ichnology (BUCKLAND, about 1830). The entire field of *Lebensspuren* (all tracks, trails, and burrows); in the fossil state, paleoichnology or palichnology; Recent, neoichnology.

Lebensspur (ABEL, 1912) (French, *trace d'activité* LESSERTISSEUR, 1955). Trace fossil or ichnofossil; also used for Recent tracks and burrows; introduced into English literature by Japanese authors. ABEL included also pathological phenomena, etc. Shortest definition (given by HAAS, 1954, and improved by SEILACHER, 1955): "Lebensspuren are structures in the sediment left by living organisms."

Spreite. German noun literally translated as "spread," meaning something spread between two supports, as the web of a duck's foot.

trace fossil (S. SIMPSON, 1956, originally published without definition but the following one given by SIMPSON in 1957). Sedimentary structure resulting from the activity of an animal moving on or in the sediment at the time of its accumulation; includes tracks, burrows, feeding marks, and other traces.

tracks (PETTIJOHN, 1957). Impressions left in soft material by the feet of animals.

trails (PETTIJOHN, 1957). More or less continuous markings left by an organism as it moved over the bottom.

vestigiofossil (MOORE, 1956). Unpublished suggestion to replace the term "ichnofossil" because of its bilinguistic derivation from both Latin and Greek.

GEOLOGICAL OCCURRENCE

Trace fossils and problematic fossils occur in the minerogenic and biogenic, marine, limnic, and continental sedimentary rocks of all geologic systems from the Precambrian to the Recent. In a number of rocks they are so characteristic and numerous as to furnish the name of the stratigraphic unit, e.g., *Scolithus* [*recte Skolithos*] Sandstone, Fucoid Sandstone, Grès à *Harlania,* etc. In this type of sediment contemporaneous body fossils are usually absent, but the trace fossils inform us of the existence of large numbers of bottom-dwelling animals. Trace fossils can thus give much information about the area of sedimentation. A prerequisite to this use of trace fossils is sufficient knowledge of the Recent *Lebensspuren* of all biotopes, a goal which neoichnology has yet to attain.

PRESERVATION

Lebensspuren are very transient structures as compared with shells, skeletons, or other hard parts, and in general they have little chance of being preserved as fossils. Nevertheless, the large number of trace fossils in many marine, limnic, and even continental sediments shows that their fossilization is possible. A favorable consistency and rapid solidification of the sediment are prerequisites. Among other factors, grain size of the sediment is correlated with distinctness of preservation.

Trace fossils occur as deformations of the sediment or its bedding planes. Burrows may be preserved as hollows or fillings of hollows. Often animal activity is indicated merely by the disturbance of the lamination as seen in vertical sections (*Wühl-Gefüge* or *Fossi-Textur* of RUDOLF RICHTER).

Only a part of the *Lebensspuren* become imprinted on the surface of the sediment. Surface tracks and trails, especially various footprints, are likely to be exposed to destruction by flooding or by shifting of the topmost strata at the bottom of shallow seas and only a small fraction of them are preserved.

SEILACHER emphasized recently that many *Lebensspuren* are made *within* deposits of sediment *(Innen-Spuren)*, mainly at the level of the boundary between a sand and underlying clay layer. This holds good for complete forms and especially for reliefs ("demireliefs" of earlier literature). The latter are found at the lower surface of sandstone beds and represent sand fillings of impressions made on the surface of clay within the sediment. They are distinguished from surface tracks on sand by their much more distinct preservation, as verified by experiments with Recent animals. SEILACHER calls this type of preservation "positive endogenic hyporelief" and depending on the mode of formation, he distinguishes epi- and hyporeliefs; either of them may be exogenic or endogenic and positive or negative.

Occasionally, a repetition of like forms close by or on top of each other may be observed (horizontal and vertical repetition of SEILACHER). In the fossil state this phenomenon may represent the work of starfishes and ophiuroids (e.g., *Asteriacites* SCHLOTHEIM). The animal that produced the trails causes repetition of them by changing its resting place. Impressions of this type are certain indications of trails, as opposed to body fossils.

HISTORICAL REVIEW

No complete history of palichnological investigations has yet been written. WINKLER (47) published a chronologically arranged annotated bibliography covering palichnological publications for the period 1828 to 1886. The following section will describe only a few stages of the rather intermittent development of this branch of paleontology. Among the very numerous (more than 1,000) publications on palichnology since about 1820, very few have dealt with the entire subject; the more complete works include FUCHS' (10) discussion of the *Fucoiden und Hieroglyphen,* ABEL'S (1) *Vorzeitliche Lebensspuren,* LESSERTISSEUR'S (23) *Traces d'activité,* and SEILACHER'S (38-42) not yet completed series on *Studien zur Palichnologie.*

In the early years of paleontology many fossils, especially cylindrical and U-shaped burrows, which have now been identified as *Lebensspuren,* were considered to be remains of marine algae. This is apparent in names like *Fucoides, Algacites, Chondrites,* and the many generic names ending with *-phycus.* It was the ramification of the burrows which was considered most conclusive evidence for their interpretation as plants. In publications on these "algae" Recent Thallophyta were commonly figured in order to show the identity or relationship of the fossil form with them. Occasionally, even the drawings of the fossils were modified so as to make them look more like algae.

In 1881 NATHORST (28), on the basis of systematic neoichnological observations, pointed out the striking similarity of many Problematica with the tracks and trails of marine invertebrates. This evidence, together with previously published (1873) information that animal trails may ramify, permitted NATHORST to challenge the doctrine of plant origin for these fossils. Simultaneously, but independently, JAMES (19) in the United States, expressed the same conviction as the Swedish scientist, although his papers have often been overlooked. He showed most of the algal interpretations to be untenable and cautioned urgently against the widespread practice of merely describing and naming isolated and often insufficiently preserved findings. It is indeed proper that this habit of referring every mark found in the rocks of the earth to some sort of organism, and calling all uncertain marks marine plants, should be protested. If the practice is not discontinued, the nomenclature of the science will be so encumbered with useless names that chaos will result. As JAMES put it, when "every turn made by a worm or shell, and every print left by the claw of a Crustacean is described as a new addition to science, it is time to call halt and eliminate some of the old before making any more new species."

Only gradually did NATHORST's interpretation of many fossil "algae" as *Lebensspuren* become generally accepted. Several authors, led by DE SAPORTA, opposed him, and between the years 1880 and 1885 violent discussions between the representatives of the two schools of thought took place, as can be shown by the many publications of DE SAPORTA, DELGADO, NATHORST and others. Even today several "genera" of *Lebensspuren* (e.g., *Chondrites, Fucoides*), are sometimes interpreted as algae. Recent Canadian and Indian papers refer typical fucoids to "algae." FUCINI, in extensive publications, described Problematica, mainly inorganic markings, as plant fossils. Other trace fossils, such as the strange "spreadburrows" *(Spreiten-Bauten)* are not yet generally accepted as such (e.g., some French authors), and sometimes they are still considered body fossils. The discussion about these problematical fossils has lasted for decades, modern counterparts having been found only quite recently. FUCHS (10), after an extensive study of the vast material kept in European collections, on the whole confirmed NATHORST's interpretations. A number of especially peculiar forms, such as the "genera" *Paleodictyon, Hercorhaphe,* and *Spirorhaphe,* summarized under the term hieroglyphs *(Graphoglypten),* were assumed by FUCHS to be spawn, presumably of gastropods. Similar interpretations are still being discussed for similar forms (e.g., *Spirodesmos*).

After tens of years of stagnation following the turn of the century, substantial progress was made in *Lebensspuren* studies by ABEL and his pupils and especially in the course of "actuopaleontologic" investigations in marine biology of the North Sea tidal flats by RUDOLF RICHTER. His studies included a survey of Recent and fossil worm trails and burrows, an elucidation of general questions of palichnology and their utilization for paleogeography, an interpretation of many Problematica, as well as an analysis of numerous arthropod trails and Recent and fossil U-shaped burrows. The efforts and results of RICHTER's collaborators at the marine-geologic Forschungsanstalt "Senckenberg" in Wilhelmshaven (HÄNTZSCHEL, SCHÄFER, SCHWARZ, TRUSHEIM) are to be found in the same general direction. The most recent contributions are those of SEILACHER, which are also based on thorough investigations of Recent and fossil materials. He merits consideration for his realization that the majority of fossil *Lebensspuren* are made inside the sediment (interior traces), and especially for his well-thought-out, ecologically founded classification of all *Lebensspuren*.

CLASSIFICATION

The possible diversity of *Lebensspuren* made by an individual animal, dependent on its activity (crawling, eating, running, burrowing, swimming), and the dependence of traces on fortuitous properties or conditions of the sediment, make it impossible to classify *Lebensspuren* in a manner corresponding to a zoological pattern. One can not arrange all trace fossils according to their producers, because the makers of *Lebensspuren* can only rarely be detected unequivocally. In early stages of paleontological research, when trace fossils were mostly interpreted as marine algae, they were arranged exclusively according to morphological characters. The shape of the "thallus" was regarded as a determining factor; fucoid species were distinguished according to the angle of divergence of their branches.

FUCHS (10), admitting them to be trace fossils, tried to arrange them in family-like groups, mainly after morphological criteria: crawling trails and burrows, hieroglyphs, fucoids, and *Spreiten*-burrows classed as Alectoruridae ("the darkest and most enigmatical area in the kingdom of the problematical fossils"). The difficulties of classification, partly due to many homeomorphies, caused RICHTER to express caution against too narrow a grouping of trace fossils. He gave good examples of a possible simple classification in his distinction of U-shaped burrows with or without *Spreite* (Rhizocorallidae, Arenicolitidae) and in his division of worm trails and burrows according to "architectural basic forms" *(bauliche Grund-Formen)* on a mechanical and biological basis. KREJCI-GRAF (22) suggested a very detailed classification on a genetic basis, distinguishing superordinal units consisting of tracks (organic and inorganic), trails, burrows, hieroglyphs, and fucoids. Chiefly, he gave definitions of these units and an extremely detailed subdivision with isolated examples, but no real classification of the trace fossil genera.

Classification according to ecological principles proposed by SEILACHER (1953) is based on the fact that different groups of animals with similar life habits produce trails with similar basic characters, even though the animals themselves have quite different body shapes. Working out these common basic characters he recognized five groups: dwelling burrows (Domichnia); feeding burrows (Fodinichnia); feeding trails (Pascichnia); resting trails (Cubichnia) (="repose imprints" KUENEN, 1957); and crawling trails (Repichnia). For each of these groups characteristic features may be noted. Thus many *Lebensspuren*, provided that they are well preserved, become attributable to one of these groups, which may be characterized as follows:

(1) DOMICHNIA, simple or U-shaped tunnels made in the sediment at right angles to the surface, representing the permanent domicile of hemisessile anglers or whirlers;

(2) FODINICHNIA, extensive tunnels and tunnel systems with good utilization of space and surface, used by hemisessile sediment-eaters simultaneously as domicile and "mine" or hunting-ground;

(3) PASCICHNIA, highly winding bands or tunnels, not crossing each other, with intense utilization of the surface available for grazing or feeding that often results in surface ornamentation such as meanders ("parqueting," *Parketierung*);

(4) CUBICHNIA, isolated impressions with outlines corresponding roughly to shapes of their producers, often arranged parallel to each other as a result of like orientation (rheotactic rectification) toward currents (vertical and horizontal repetition possible);

(5) REPICHNIA, furrows, longitudinal swellings, and crawling tunnels of variable direction, ramified or unramified, smooth or sculptured.

SEILACHER's system has the advantage of collecting ecologically similar groups of *Lebensspuren*. Questions as to identity of their producers may be disregarded here, for these can only rarely be answered unequivocally from morphological criteria. The characterization of groups is independent of time; for example, the assemblage termed Cubichnia is equally valid for extinct arthropods of the Paleozoic (e.g., trilobites), as for Recent arthropods that have a corresponding way of life.

The classification suggested by LESSERTISSEUR (23) is based mainly on morphological features and distinguishes *traces exogènes* (simple bilobate, and trilobate crawling trails, meanders, spirals, starlike trails, etc.) and *traces endogènes* (burrows and tunnels of various forms, fucoids, resting trails, U-shaped burrows with or without *Spreite* and screw-shaped burrows).

NOMENCLATURE OF TRACE FOSSILS

It has become customary to use binary nomenclature for trace fossils in the same way as that used for body fossils. With trace fossils, however, "genera" and "species" have a different meaning from that which is applicable to body fossils. As may be understood from the history of palichnology, the number of names created is much greater than is necessary. Much too finely differentiated genera and species have been distinguished as a result of their having been thought to be plant fossils. This is especially true for the host of fucoids, as evidenced by description of the "genus" *Fucoides* by JAMES (1894). The numerous isolated descriptions scattered throughout world literature in paleobotanical, paleozoological, faunistic, stratigraphical, regional geological, and strictly palichnological papers have led to an excessive number of described genera and species. Owing to the world-wide distribution of numerous trace fossils and to their frequently considerable vertical ranges the "new" forms were often published without knowledge or consideration of earlier literature. As early as 1884, JAMES (19) pronounced an unheeded warning against the ballast of useless names.

Binary nomenclature has not been accepted universally for problematical fossils. Many authors have declined to give formal names to trace fossils, an understandable and justified procedure, especially with poorly preserved forms. However, experience shows that these unnamed forms usually escape notice in later literature. Furthermore, several attempts have been made to name trails simply as "species" of one genus, *Ichnium,* including trails of vertebrates as well as invertebrates. It is also possible to give purely descriptive names to morphologically corresponding assemblages (e.g., *Ichnia catenaria, Ichnia spicea*), if an individual description or name is to be avoided.

FAUL's (1951) suggestion that vertebrate tracks be designated by formulas has little chance of becoming generally accepted; in any case this is not applicable to trails of invertebrates.

Despite its drawbacks, the binary system appears to be the most suitable one. It is unavoidable that trace fossils, which were formerly assumed to be bodily preserved plants or animals and were named accordingly, now carry inconsistent names and must retain them (e.g., *Fucoides*, for feeding burrows of marine animals, and *Spongia ottoi* GEINITZ, for the feeding trail of an arthropod or worm). It is inadmissible to name the trail by the neozoological name of the presumed producer, as was done for a beaded trail from the German Jurassic called *Corophium* by PUTZER.

Addition of the suffix *-ichnus* to the generic name of trace fossils is recommended in order to render them immediately recognizable. If, in addition, the name offers a morphological mark of distinction (e.g., *Sagittichnus* SEILACHER) or if the producer can clearly be indicated (e.g., *Pelecypodichnus* SEILACHER), such a name will convey a clear conception of the named object. When giving new names this aim ought to be kept in mind.

IMPORTANCE OF TRACE FOSSILS FOR STRATIGRAPHY, PALEOGEOGRAPHY, AND PALEOECOLOGY

Lebensspuren usually have little importance in stratigraphy. In restricted areas, however, they may attain the rank of index fossils for the field geologist. A burrow (e.g., *Arenicolites franconicus* TRUSHEIM, from the Muschelkalk of Southern Germany) may serve as an example; this fossil occurs abundantly in a layer only 3 to 4 cm. in thickness and may be followed for a horizontal distance of 26 km.

The vast majority of trace fossils remain unchanged through the geologic eras. This is true for nondescript, smooth furrow-like crawling trails and cylindrical burrows, as well as for more distinctive U-shaped burrows with *Spreite* and even for the honeycomb-like networks named *Paleodictyon* which are known from Silurian to Tertiary time. Only occasionally do trace fossils turn out to be true index fossils; one example is the beaded coprolite, *Tomaculum* GROOM, which so far has been found only in Ordovician strata of England, France, Czechoslovakia, and Germany.

In structurally complicated areas where inverted beds must be reckoned with, burrows and trails may be useful for distinguishing the top and bottom of strata. Especially well suited to this purpose are U-shaped burrows, which are invariably built either horizontally or with the curved part toward the bottom. Similarly, a sequence of strata may be correctly oriented with the help of footprints of vertebrates or of *Lebensspuren* preserved in relief at the bottom of a sandstone bed overlying clay.

Trace fossils may be useful for paleoecologic and paleogeographic investigation of ancient habitats, if their ecological significance is understood. Whereas faunas mostly represent thanatocoenoses, the trace fossils are important as examples of autochthonous life within or on the sediment.

Our insufficient knowledge of Recent trails and traces makes difficult the use of trace fossils for reconstruction of areas of sedimentation. Recent *Lebensspuren* have been thoroughly investigated in only a few biotopes, especially on the shore and in shallow seas. Besides, we know mainly surface traces. Therefore, *Lebensspuren* often have been considered to be evidence of sedimentation in shallow seas, especially if other assumed criteria for shallow seas, such as ripple marks, seemed to confirm this deduction. Such assumptions are dangerous, since digging and burrowing animals and ripple marks have been identified in submarine photographs taken at considerable depths. Drilling cores taken by the Albatros Expedition from depths of more than 5,000 m. are reported to be shot with burrows of unknown animals. Similarly, the presence of *Lebensspuren* must not be taken as certain indication that the sediments concerned were laid down under the influence of the tide, that is, in areas of sedimentation corresponding to tidal flats of the Dutch and German North Sea coast. Trails and traces offer no certain criteria for the recognition of fossil *Watten* (HÄNTZSCHEL, 1953, 1955).

Perhaps one may draw conclusions concerning the area of sedimentation from "ichnofacies" features comprising associations of traces, which so far have received little notice. SEILACHER (1954) found conspicuous differences between the ichnofacies of the Flysch and that of the Molasse, as well as in older sediments of corresponding types. However, further investigations are needed before ichnofacies may be safely used in paleogeographic studies; present conclusions are highly tentative.

TRACE FOSSILS

The definition of "trace fossils" in the introduction shows which fossils are to be dealt with in this chapter. They might have been arranged according to the system based on ecological principles suggested by SEILACHER, which does justice to their peculiarities as compared with body fossils. However, inasmuch as many genera cannot at present be placed within this system, the genera are arranged alphabetically here for simplicity and easy reference. Besides, the individual descriptions generally indicate the position of the trace fossil within SEILACHER's system.

The author wants to emphasize again that with the trace fossils the meaning of "genus" differs much from that applicable to body fossils. In a number of instances it has been impossible to prepare clear, exact and unequivocal descriptions because of unsatisfactory original illustrations, unavailable or lost type material, and varying opinions as to the limits of the "genera." Likewise, great diversity of the fossils themselves, and the generally unsatisfactory status of trace-fossil research have adversely affected the preparation of meaningful descriptions. The writer is well aware of the shortcomings of this work.[1]

SYSTEMATIC DESCRIPTIONS

Acanthichnus HITCHCOCK, 1858. Linear trails of insects, in 2 parallel rows; each impression turned slightly outward, generally quite numerous; tracks opposite; "genus" including widely different "species" (17). *Trias.*, USA(Mass.).

Aglaopheniolites. According to SEILACHER (personal communication, 1956) name used in Italian paleontological collections for trace fossil from Italian Flysch. [Very probably manuscript name.]

Annelidichnium KUHN, 1937 [**A. triassicum*]. Tunnel fillings with irregular sculpture; ornamented with sharp or rounded longitudinal ridges or blunt tubercles (16). *U.Trias.*, Ger.

Archaeonassa FENTON & FENTON, 1937 [**A. fossulata*]. Trails made by Gastropoda; see "group" *Scolicia* DEQUATREFAGES, 1849 (p. *W*215) (23). *L.Cam.*, Can.(B.C.).

Arenicolites SALTER, 1857 [**Arenicola carbonarius* BINNEY, 1852]. U-shaped, thick or thin, rounded

[1] It should be noted that in some of the generic descriptions that follow, designation of the author as well as the date of the type-species is lacking. In every such case the pertinent information is unobtainable, owing to the rather casual way that trace fossils have commonly been treated in the literature.

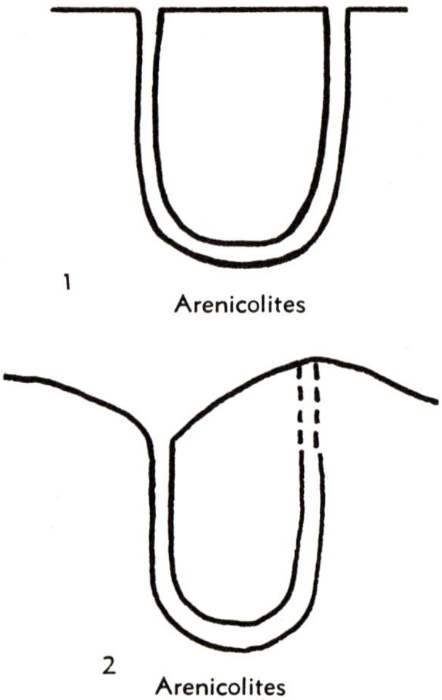

FIG. 109. Trace fossils (p. W183).

or compressed burrows without *Spreite;* walls unsculptured, sculptured, or lined; perpendicular to bedding plane (1,23). *Cam.-Rec.,* Eu.-N.Am.——FIG. 109,*1. A.* sp. SALTER; *schematic* (130).——109,*2. A. franconicus* TRUSHEIM; schem. cross sec. of burrow, ×0.8 (130).

Arthraria BILLINGS, 1874 [**A. antiquata*]. Bars on bedding surfaces with spheroidal expansions at each end, similar to pair of dumbbells; related to *Bifungites* DESIO (6). [The specimens from the Cincinnatian of Ohio, according to CASTER's and the author's investigations, are certainly U-shaped burrows with *Spreite,* similar to and possibly identical with *Corophioides* or *Diplocraterion* (33).]. *Cam.-Dev.,* N.Am.

Arthrophycus HALL, 1852 [**Fucoides alleghaniensis* HARLAN, 1831 (=*Fucoides harlani* CONRAD, 1838; ?*Rauffella palmipes* ULRICH, 1889)] [=*Harlania* GOEPPERT, 1852; *Arthrophicus* PACHECO, 1908 (obj.); *Arthrichnites* (author, date unknown)]. Curving "stems," simple or usually in bunches; surface showing regularly spaced transverse ridges, commonly bearing median depression; "branches" may reach length of 2 feet (60 cm.) or more; variously regarded as inorganic, seaweed, burrows, or trails produced by arthropods or worms; for most probable explanation as feeding burrows see SARLE, 1906, and SEILACHER (41). [According to SEILACHER, the "genus" can be regarded as a synonym of *Phycodes* RICHTER, 1850 (1,23).] *Sil., ?Dev.,* N.Am.-S.Am.-N.Afr.-?Eu.——FIG. 111,*3.* **A. alleghaniensis* (HARLAN), L.Sil., N.Y.; ×0.3 (Häntzschel, n).

Asabellarifex KLÄHN, 1932. Poorly founded, rather superfluous "genus" proposed for vertical burrows resembling *Sabellarifex,* but believed to be burrowed in downward direction, not built upward as tubes like *Sabellarifex* (16). *L.Cam.(Pleist.* drift), Ger.-Swed.

Asaphoidichnus MILLER, 1880 [**A. trifidus*]. Trackways with 2 rows of impressions comprising separate and distinct tracks, anterior third of each track trifid and thrown outward, middle toe a little longer than others; distance between rows about 1.5 inches (37.5 mm.) (23). [Probably made by trilobites.] *Ord.,* USA(Ohio).——FIG. 110,*3. A. dyeri* MILLER; ×0.5 (100).

Asteriacites SCHLOTHEIM, 1820 [*non* SCHLOTHEIM, 1822] [**A. lumbricalis*] [=*Heliophycus* MILLER & DYER, 1878; *Spongaster* FRITSCH, 1908 (*non* EHRENBERG, 1860)]. Impressions in form of starfishes with transversely sculptured arms, commonly showing horizontal or vertical repetition (39). [Interpreted as resting tracks of brittle stars, auluroids, and asteroids.] *Sil.-Tert.,* Eu.-N.Am.——FIG. 110,*2.* **A. lumbricalis,* L.Jur., Ger.; ×0.5 (39). (See also FIG. 130,*4.*)

Asterichnites BROWN & VOKES, 1944 [**A. octoradiatus*]. Rows of stellate imprints, each consisting of unmarked central disc and 8 radiating grooves 13 to 18 mm. long; arranged in rows on stratification planes (16). [Probably produced by tentacles of dibranchiate cephalopod.]. *L.Cret.,* N.Am.——FIG. 110,*1.* **A. octoradiatus,* Mowry Sh., USA (Mont.); *1a,b,* ×0.08, ×0.6 (55).

Asterophycus LESQUEREUX, 1876 [**A. coxii*]. "Front" expanded at its base in 5 flattened starlike oblong or obovate divisions; surface wrinkled longitudinally; diameter 6 to 12 cm. (16). [Interpreted by DAWSON (5) as burrows of ?worms.] *Carb.,* N.Am.——FIG. 111,*1.* **A. coxii,* Ky.; ×0.3 (93).

Asterosoma VON OTTO, 1854 [*non* GRUBE, 1867] [**A. radiciforme*]. Big stars (diameter ca. 20 cm.) with elevated center; rays bulbous, tapering toward ends, longitudinally wrinkled (16). [Very probably burrows with radiating (?feeding) trails.] *U.Cret.(Turon.),* Eu.——FIG. 111,*2.* **A. radiciforme,* Ger.; ×0.3 (103).

Astropolithon DAWSON, 1878 [**A. hindii*]. Oval depressions surrounded by raised ridge from which several raised lines radiate, in some forms bifurcating; rays poorly developed, especially in smaller specimens. [Originally explained as "fucoids with radiating fronds" but later (DAWSON, 5) as mouths of large burrows with radiating trails. No type or other specimen could be located.] *L.Cam.,* Can.(Nova Scotia).——FIG. 112, *5.* **A. hindii;* ?×0.7 (5).

Atollites MAAS, 1902 [*non* ZUBER, 1910] [**A. zitteli*]. Starlike fossils, formerly regarded mostly as belonging to Scyphozoa (*Treatise,* p. *F73,* Fig.

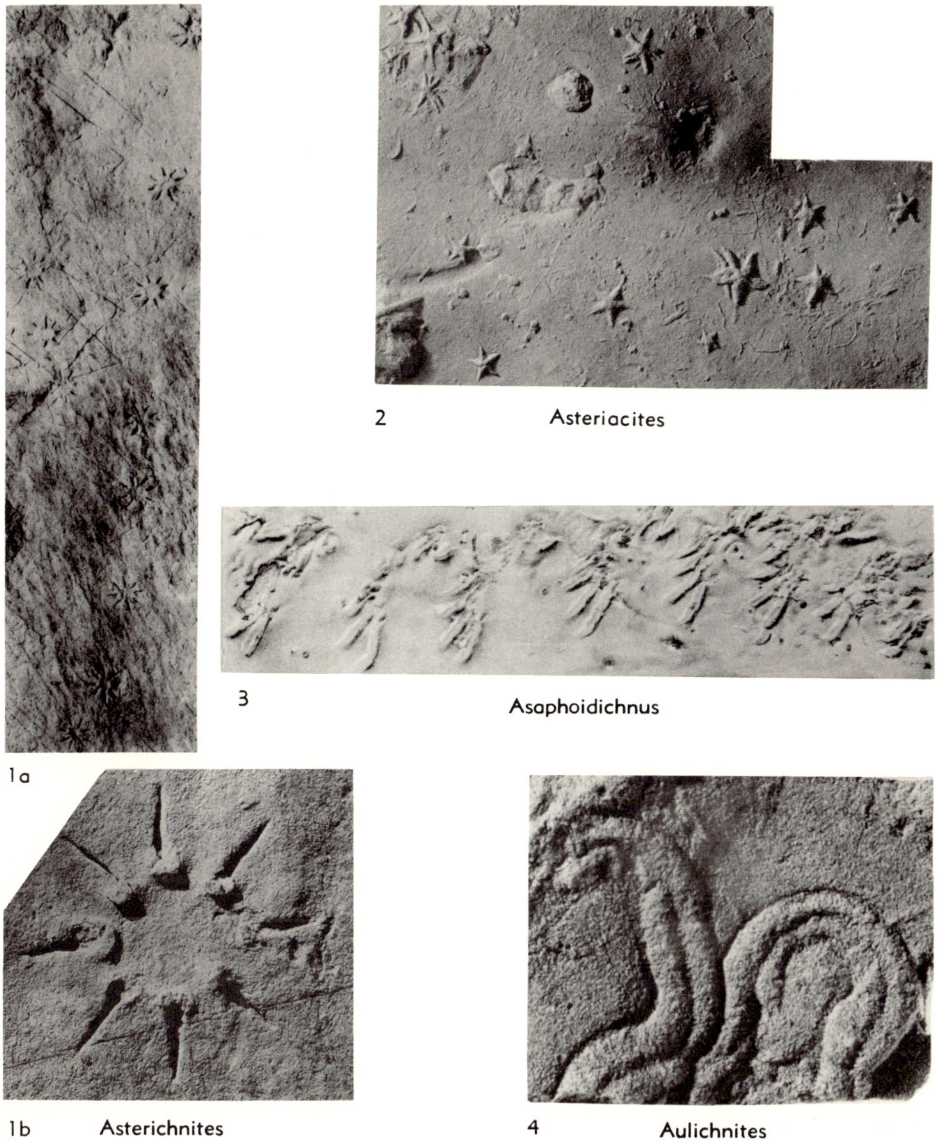

FIG. 110. Trace fossils (p. *W*184-*W*185).

59). [Very probably starlike feeding trails; resembling some forms of *Lorenzinia.*] *L.Cret.,* Czech.

Aulichnites FENTON & FENTON, 1937 [**A. parkerensis*]. Very similar to *Scolicia* DE QUATREFAGES, 1849 (3) but without elevated mesial ridge (23). *Penn.,* N.Am.——FIG. 110,*4*. **A. parkerensis,* USA(Tex.); ×1 (Howell, n).

Balanoglossites MÄGDEFRAU, 1932 [**B. triadicus*] [=*Unculiferus* HUNDT, 1941]. Burrows, 1 to 3 cm. wide and up to 15 cm. deep, irregularly branching, with several openings that do not taper; may bear transverse ridges and delicate longitudinal striations on wall (23). [Traces of the bristle rows of a polychaete(?).] *Ord., M.Trias. (Muschelkalk),* Ger.

Bassaenia RENZ, 1925 [**Lorenzinia (B.) moreae*]. Sometimes regarded as subgenus of *Lorenzinia* GABELLI, 1900; more probably feeding trail than scyphozoan (*Treatise,* Part F, p. *F*43, Fig. 32,*6*). *U.Cret.,* Greece.

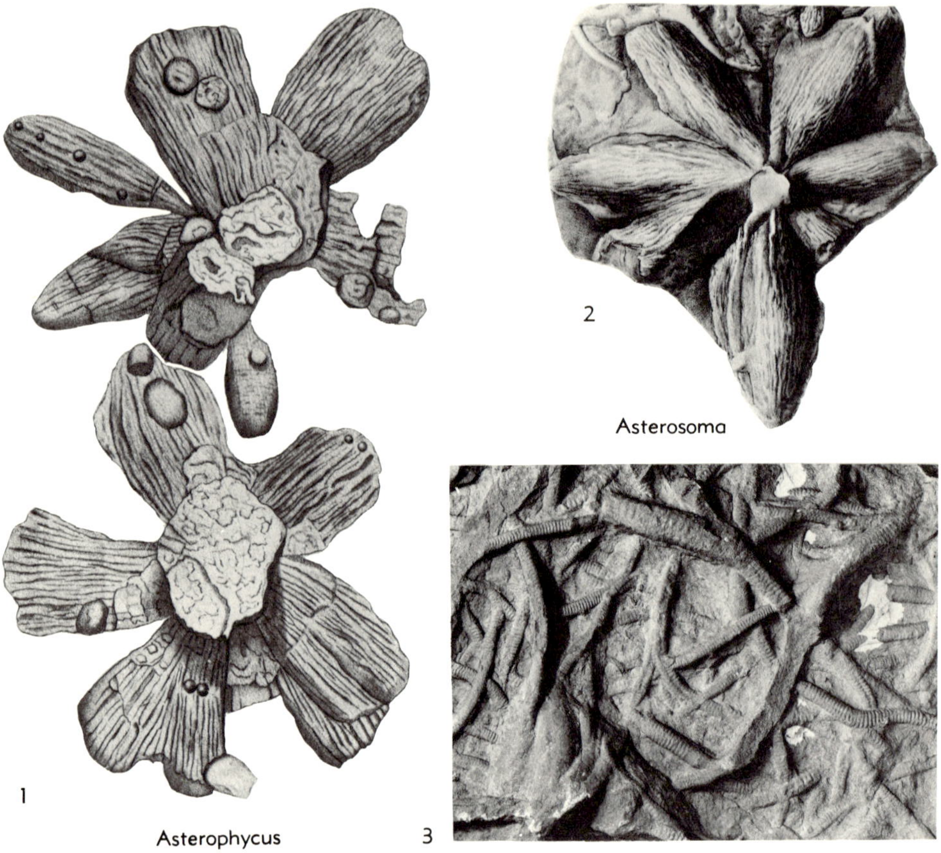

FIG. 111. Trace fossils (p. W184).

Belorhaphe FUCHS, 1895 [**Cylindrites zickzack* HEER, 1877] [=*Beloraphe* AUCTT.]. Sharply zig-zag-shaped locomotion traces; commonly with short protrusions at corners (10). *Cret.-L.Tert. (Flysch)*, Eu.——FIG. 112,*1*. *B.* sp. FUCHS, Aus.; ×0.4 (10).

Bergaueria PRANTL, 1946 [**B. perata*]. Cylindrical protrusions with smoth walls; length and diameter subequal, about 40 mm.; lower end rounded, with shallow trough; at lower surface of bedding planes in sandstones (23). [?Resting trails of actinians.] *Ord.*, Czech.——FIG. 112,*4*. **B. perata; 4a,* casts in overlying sandstone; *4b,* original burrow-cavities in underlying shale, ×0.3 (107).

Bifasciculus VOLK, 1960 [**B. radiatus*]. Feeding burrow, consisting of many (up to 40) tunnels, 2 to 3 cm. long, ending blindly, radiating from a center, bent slightly upward and downward (16). *Ord.(Griffel-Schiefer)*, Ger.——FIG. 113,*2*. **B. radiatus;* ×1 (133).

Biformites LINCK, 1949 [**B. insolitus*]. Bimorphous form, consisting of narrow section, partly divided by longitudinal furrows, continuing into wider section with prominent transverse ribs; resembles shafted hand grenade; fillings visible at lower surface of layers (24). [According to SEILACHER, comprises dwelling burrows *(Domichnia)*.] *U. Trias.(M.Keuper)*, Ger.——FIG. 114,*1*. **B. insolitus; 1a,* ×0.8; *1b,* schematic, ×1 (24).

Bifungites DESIO, 1940 [**B. fezzanensis* (?=*Buthotrephis impudica* HALL, 1852)]. Structures like dumbbells within bedding planes, 2 to 3 cm. long; ends spherical or more commonly hemispherical; diameter up to 1 cm.; somewhat resembling *Arthraria* (41). *L.Cam.*, Pakist.; *Ord.*, Czech.; *?Sil.*, USA(N.Y.); *Dev.*, N.Afr.——FIG. 113,*1*. **B. fezzanensis,* M.Dev.-U.Dev., N.Afr.; ×0.7 (6).

Bifurculapes HITCHCOCK, 1858 [**B. laqueatus;* SD HÄNTZSCHEL, herein] [=*Bifurculipes, Biferculipes, Bifurcalipes* HITCHCOCK, 1865 *(errore)*]. Four regular rows of tracks, resembling small forks when united at base, as happens commonly; may have 2 additional rows; pairs of tracks opposite

(17). [Made by insects or crustaceans.] *Trias.*, USA(Mass.).——FIG. 112,*3*. **B. laqueatus;* ×0.7 (96).

Blastophycus MILLER & DYER, 1878 [**B. diadematus*]. According to JAMES (1884), identical to *Trichophycus.* SEILACHER (personal communication, 1955) regards these fossils as worm burrows projecting above surface of sediment (19). *Ord.*, USA(Ohio).

Bostricophyton SQUINABOL, 1890 [**B. pantanellii;* SD ANDREWS, 1955] [=*Bostrichophyton* ANDREWS 1955 *(errore)*]. Spiral threads, similar to *Chondrites intricatus* or *C. arbulescens* (16). *Cret.-L. Tert.*, Eu.——FIG. 112,*2*. **B. pantanellii,* Tert. (Flysch), Italy; ×0.7 (10).

Boteillites. According to SEILACHER (personal communication 1956) name used in Italian paleontological collections for trace fossil from the Italian Flysch. [Probably manuscript name.]

Camptocladus FENTON & FENTON, 1937 [**C. intertextus*]. Nondescript, branched, flexuous, intertwined burrows; "genus" proposed on assumption that burrows are of crustacean origin (23). *Penn.*, USA(Tex.).

Caridolites NICHOLSON, 1873 [**C. wilsoni*]. Tracks, not described in detail; thought to be made by *Ceratiocaris* (16). *L.Paleoz.*, Eng.

Caulerpites STERNBERG, 1833 [**Fucoides lycopodioides* BRONGNIART, 1828; SD ANDREWS, 1955] [=*Caulerpides* SCHIMPER, 1869]. Very heterogeneous "genus" including plants (even conifers, according to SCHIMPER) as well as trails (e.g., *C. marginatus* LESQUEREUX, 1869=aff. *"Taonurus"; C. annulatus* ETTINGHAM, 1863=aff. *Keckia;* other "species" partly also classified with Recent genus *Caulerpa* LAMOUROUX, 1809; trails like *Nereites, Phymatoderma* and similar "genera").

Chondrites STERNBERG, 1833 [*non* M'COY, 1848] [**Fucoides targionii* BRONGNIART, 1828; SD ANDREWS, 1955] [=*Algacites* SCHLOTHEIM, 1822 *(partim); Fucoides* BRONGNIART, 1823 *(partim); "Gigartinites"* BRONGNIART, 1823 (not used as genus); *Caulerpites, Sphaerococcites* STERNBERG, 1833 *(partim); Buthotrephis* HALL, 1847; *Phymatoderma* BRONGNIART, 1849; *?Trevisania* ZIGNO, 1856; *Phycopsis* FISCHER-OOSTER, 1858; *Bythotrephis* EICHWALD, 1860; *Nulliporites* HEER, 1865; *Chondrides, Leptochondrides* SCHIMPER, 1869; *?Theobaldia* HEER, 1877 *(partim); Palaeochondrites* DE SAPORTA, 1882; *Chondropogon* SQUINABOL, 1890; *?Prochondrites* FRITSCH, 1908; *Labyrinthochorda* WEISSENBACH, 1931; *Clematischnia* WILSON, 1948]. Very plantlike, regularly ramifying tunnel structures which neither cross each other nor anastomose; should be interpreted as dwelling burrows or feeding burrows; width of tunnels remaining equal within a system, otherwise varying from large (e.g., *Buthotrephis*) to small (e.g., *Chondrites*); very common trace fossil, usually named "fucoid"; some with transverse-

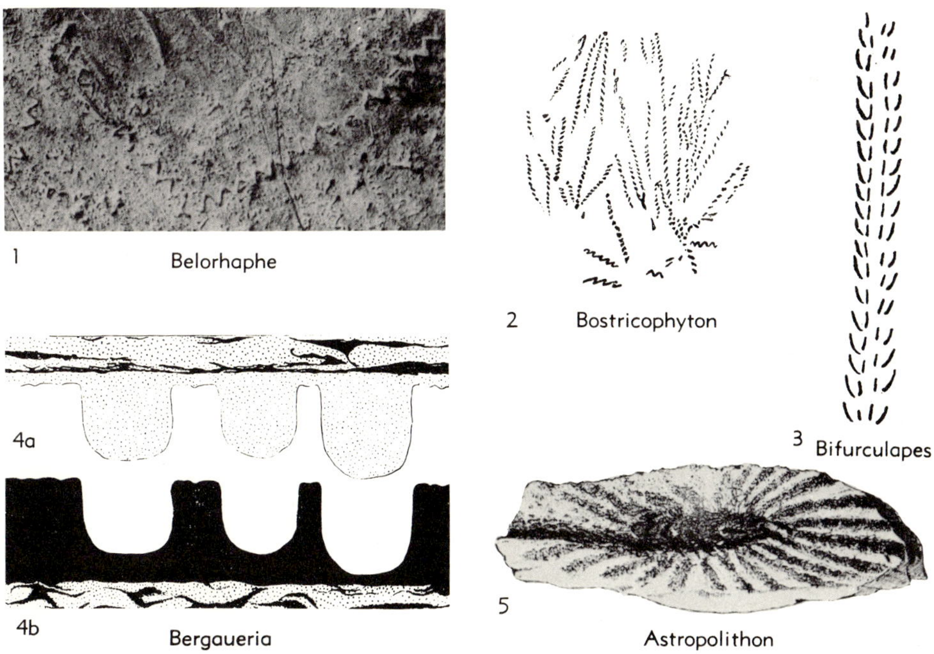

FIG. 112. Trace fossils (p. W184-W187).

1 

Bifungites

2 

Bifasciculus

FIG. 113. Trace fossils (p. W186).

ly built-in ellipsoidal excrement pills; surface pattern commonly very regular, effected by phobotaxis (43). [Probably made by marine worms.] *Cam.-Tert.*, cosmop.——FIG. 115,*1*. *C.* sp., U.Cret.; *1a*, small form, Maastr., Sp.; ×1 (12); *1b*, large form, Aus.; ×1 (15); *1c*, reconstr. of tunnel system (43).

Climacodichnus HITCHCOCK, 1865 [**C. corrugatus*]. Small, ladderlike rows of impressions, resembling steps of *Acanthichnus* (18). [Possibly made by arthropods.] *Trias.*, N.Am.——FIG. 117,*3*. **C. corrugatus*, USA(Mass.); ×0.2 (18).

Climactichnites LOGAN, 1860 [**C. wilsoni*] [=*Climachtichnites* MILLER, 1877 *(errore)*; *Climactichnides* CHAPMAN, 1878]. Very large crawling tracks (width about 15 cm., maximum length 3 to 4 m.), with prominent, slightly arched or V-shaped transverse ridges and very delicate, closely spaced arched rills; dishlike impressions, oval, distinctly bounded at the end (?or beginning) of tracks (1,23). [Probably formed by mollusks, arthropods, or worms.] *U.Cam.*, N.Am.——FIG. 115,*2*. **C. wilsoni*, Potsdam Ss., USA(N.Y.); ×0.02 (134).——FIG. 116,*5*. *C. youngi* (CHAMBERLIN), St. Croix, USA(Wis.); ×0.3 (134).

Cochlea HITCHCOCK, 1858 [*non* MARTYN, 1784; *nec* GRAY, 1847)] [**C. archimeda*]. Trackway resembling an archimedean screw (17). [Junior homonym.] *Trias.*, USA(Mass.).

Cochlichnus HITCHCOCK, 1858 [**C. anguineus* (=*Palaeophycus kochi* LUDWIG, 1869)] [*Sinusia* KRESTEW, 1928 (*non* CARADJA, 1916); *Sinusites* RENIER, 1938 *(errore)*]. Regularly meandering trails, resembling sine curve (17). *Carb.*, Eu.-N. Am.; *Trias.*, N.Am.——FIG. 116,*1*. **C. anguineus*, Trias., USA(Mass.); ×0.7 (17).

Confervites BRONGNIART, 1828 [**C. thoreaeformis;* SD ANDREWS, 1955] [=*Confervides* SCHIMPER, 1869]. Most forms placed here, especially those from Tertiary beds, are remains of threadlike algae (PIA, 1927), or tissue residues of higher plants. [According to NATHORST (1881), some "species," such as *C. padellae* HEER, 1877, are probably trace fossils resembling *Chondrites* (*Treatise*, p. E104).] *Jur.-Tert.*, Eu.

Conopsoides HITCHCOCK, 1858 [**C. larvalis*]. Tracks in 3 (?4) rows, divergent from median line; foot linear, blunt anteriorly; track terminated, usually in front, by slight mound of mud (17). *Trias.*, USA(Mass.).

Copeza HITCHCOCK, 1858 [**C. triremis*]. Resembling *Lithographus* HITCHCOCK, but having oblique markings within longitudinal ones (17). *Trias.*, N.Am.——FIG. 116,*4*. **C. triremis*, USA(Mass.); ×0.7 (96).

Coprinisphaera SAUER, 1955 [**C. ecuadoriensis*]. Balls with a hole; about 6 cm. in diameter; walls about 1 cm. thick; mostly hollow or filled with consolidated mass similar to argillaceous excrement; found in loess-like tuffs *(cangagua)*. [Probably breeding places of scarabaeid beetles.] *Pleist.* (guide fossil of 3rd interglacial stage), S. Am.(Ecuad.-Colomb.).

Corophioides SMITH, 1893 [**C. polyupsilon*] [=*Arenicoloides* BLANCKENHORN, 1916; *Arenicolithes* HILDEBRAND, 1924 *(errore)*, *Corophyoides* ÖPIK, 1956 *(errore)*. U-shaped *Spreiten* burrows similar to *Rhizocorallium*, but shorter and always perpendicular to bedding plane (33). [*Arenicoloides* comprises crescent-shaped grooves in bedding planes produced by erosion of burrows to their basal ends.] *Cam.-U.Cret.*, Eu.-Asia.——FIG. 117,*1*. *C. luniformis* (BLANCKENHORN), L.Trias., Ger.; *1a*, side, ×0.6; *1b*, side (somewhat schematic), ×1; *1c*, lower ends of U-shaped burrows with *Spreite*, ×0.6 (1).——FIG. 117,*2*. *C.* sp. cf. *C. rosei* DAHMER, L.Cam., Pak.; ×0.6 (41).

Corophites ABEL, 1935 [*nom. nud.*]. Suggested as name for burrows made by Recent amphipod *Corophium*, especially for (rare) simple shafts with sidewise branchings. *Rec.*

Cosmorhaphe FUCHS, 1895 [=*Cosmoraphe* FUCHS, 1895]. "Free meanders" of extraordinarily regular form in 2 size orders; windings not close to each other; form reminiscent of some spawn-strings of gastropods (10). *Cret.-L.Tert.(Flysch)*, Eu.——FIG. 118,*3*. *C.* sp., Low.M.Eoc.(Flysch), Pol.; ×0.6 (89).

Crossopodia M'COY, 1851 [**C. scotica;* SD HÄNTZSCHEL, herein] [=*Crassopodia* TATE, 1859 *(errore)*; *Crossochorda* SCHIMPER, 1879; *Chrossocorda, Chrossochorda, Chrossocarda* WILLIAMSON, 1887 *(errore)*]. Meandering trails resembling *Nereites, Myrianites,* and partly *Psammichnites;* width about 1 cm.; "segments" forming a broad dense fringe on each side; mostly with median furrow (23). [Concerning *C. henrici* (GEINITZ) see *Dictyodora*]. *Ord.-Carb.*, Eu.-Asia(Pak.).——FIG. 118,*2*. **C. scotica* (M'COY), Ord., Fr.; ×0.5 (116).——FIG. 118,*4*. *C. tuberculata* (WILLIAMSON), Carb., Eng.; ×0.3 (140).

Cruziana D'ORBIGNY, 1842 [**C. furcata;* SD SEILACHER, 1955] [=*Bilobites* D'ORBIGNY, 1839 (*non* DEKAY, 1824; *nec* RAFINESQUE, 1831); *Bilobichnium* KREJCI-GRAF, 1932]. Shallow pocket-like pits, passages, or pocket burrows shoveled or scratched by trilobites; cross ribs obliquely placed but more regularly distributed and set at more acute angle than in *Rusophycus,* which should be interpreted as resting-trails (41). *Cam.-Sil.*, cosmop.——FIG. 119,*5*. *C. furcifera* D'ORBIGNY, Ord., Port.; *5a, 5b,* both ×0.3 (63).——FIG. 119, *6*. *C.* sp., Ord., Bol.; ×0.4 (83).

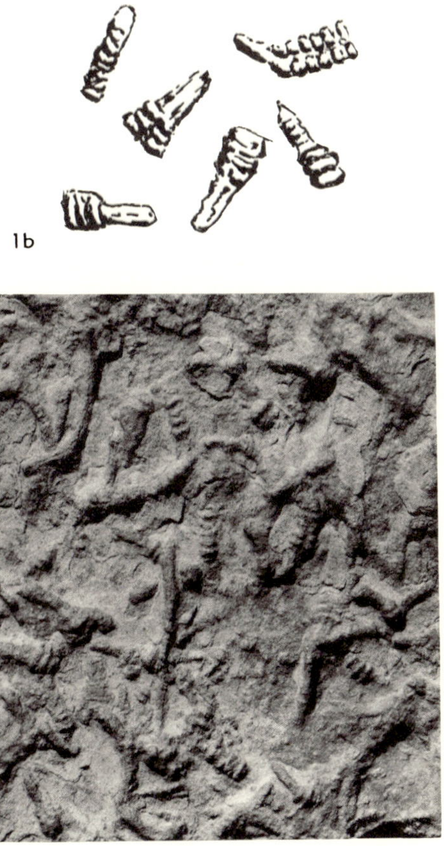

FIG. 114. Trace fossils (p. W186).

Cunicularius HITCHCOCK, 1858 [**C. retrahens*]. Nondescript ramified tunnel tracks (17). *Trias.*, USA(Conn.-Mass.).

Curvolithus FRITSCH, 1908 [**C. multiplex;* SD HÄNTZSCHEL, herein]. Interior tracks, ribbon-shaped, flat, consisting of 3 parts, with broad, usually smooth central stripe (1). [Produced by gastropods.] *Ord.*, Czech.; *Jur.*, Ger.——FIG. 119, *3*. *C.* sp., M.Jur., Ger.; ×0.7 (41).

Cylindricum LINCK, 1949 [**C. gregarium* (=*Tubifex antiquus* PLIENINGER, 1845)]. Plugs (fillings of tubes) shaped like test tubes, rounded at lower end, with smooth walls, not pointed; present in groups at lower surface of sandstone beds, perpendicular to bedding-plane; diameter up to 5 mm., length up to several cm.; dwelling burrow (24). *U.Trias.(M.Keuper)*, *?M.Jur.*, Eu.——FIG.

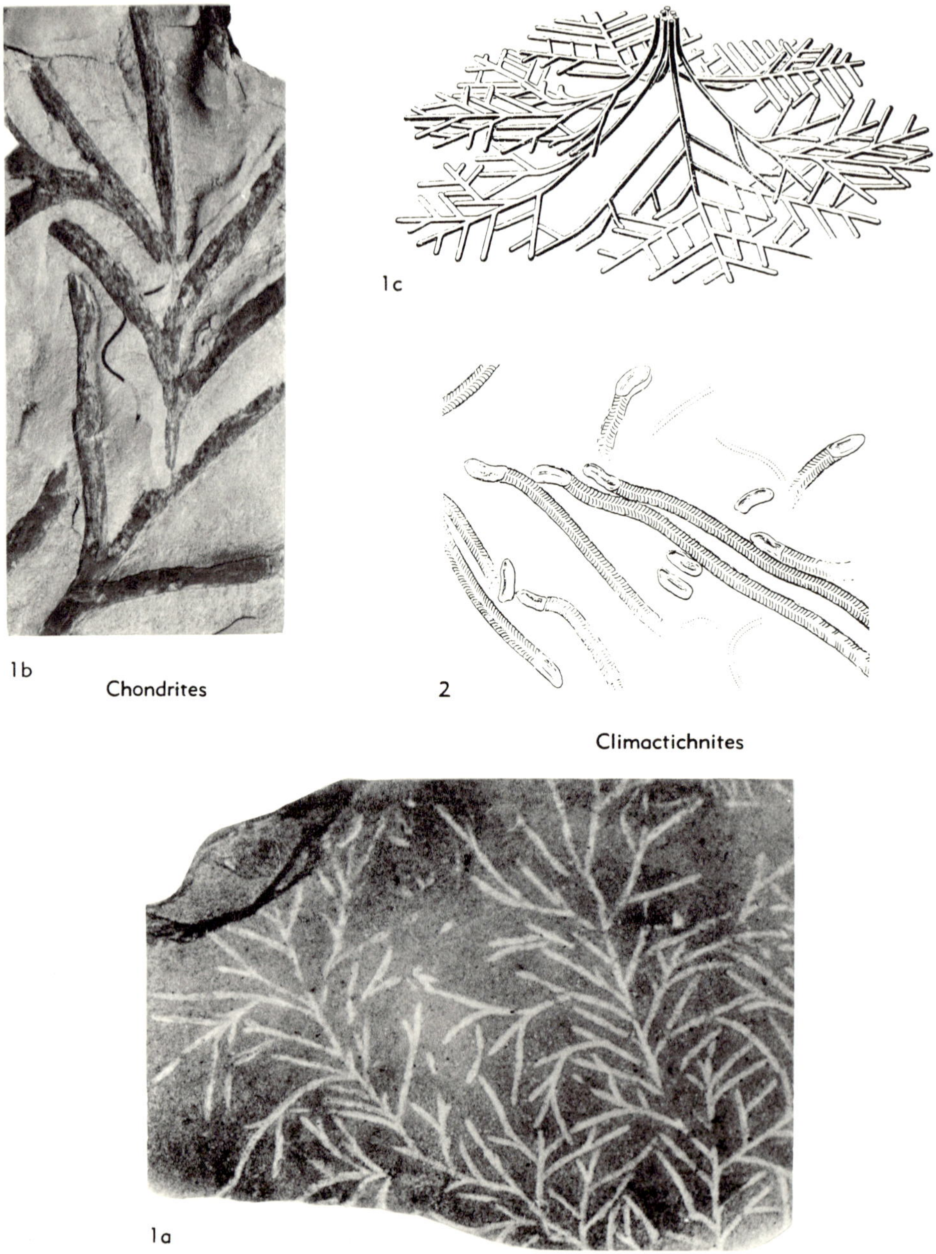

FIG. 115. Trace fossils (p. *W*187-*W*188).

118,*1*. **C. gregarium*, U.Trias.(M.Keuper), Ger.; ×1 (24).

Cylindrites GOEPPERT, 1841 [*non* GESNER, 1758; *nec* GMELIN, 1793; *nec* SOWERBY, 1824] [**C. spongioides;* SD ANDREWS, 1955] [=*Spongites* GEINITZ, 1842 *(partim)* (*non* OKEN, 1815); ?*Astrocladia furcata* GERSTER, 1881; *Goniophycus* SAPORTA, 1884]. Like *Palaeophycus*, used as general term for cylindrical and not vertical fillings of burrows (16). [Junior homonym.] *Mesoz.*, Eu.

Dactylophycus MILLER & DYER, 1878 [**D. tridigitatus*]. Very poorly figured; according to JAMES

(19), fragments of burrows or inorganic; according to SEILACHER (personal communication, 1956), resembling small *Phycodes*. *Ord.*, USA(Ohio).

Daedalus ROUAULT, 1850 [*non* REDTENBACHER, 1891] [**Vexillum desglandi* ROUAULT, 1850, SD HÄNTZSCHEL, herein] [=*Vexillum* ROUAULT, 1850 (*non* BOLTEN, 1798); *Humilis* ROUAULT, 1850; *Vescillum* LEBESCONTE, 1892 *(errore)*]. *Spreiten* structures, J-shaped at beginning, later spirally twisted; *Spreiten* surface may cut through itself, as in *Dictyodora* (1,23). *Ord.-Sil.*, Eu.-N.Am.——FIG. 120,*1*. **D. desglandi* (ROUAULT); *1a,b*, Ord., Fr., ×0.4 (92); *1c*, diagram showing gradation from vertical to spiral, L.Sil., USA (115).

Daimonelix BARBOUR, 1892 [**D. circumaxilis;* SD ANDREWS, 1955] [=*Daemonelix* BARBOUR, 1895; *Helicodaemon* CLAYPOLE, 1895; *Daemonhelix*, AUCTT. (*non Daemonhelix krameri* VON AMMON, 1900)]. Large vertical, open, spiral structures, regular in form, mostly coiled with strict uniformity; with transverse rhizome-like piece at base (1,23). [Explained as fresh-water sponges, or casts of rodent burrows; some forms also resembling concretions.] *Mio.*, N.Am.——FIG. 121,*9*. **D. circumaxilis*, USA(Neb.); side view, ×0.3 (50).

Delesserites STERNBERG, 1833 [**Fucoides lamourouxii* BRONGNIART, 1828; SD ANDREWS, 1955] [=*Delesserites* BRONN, 1853 (*non* RUEDEMANN, 1925) *Delesserella* RUEDEMANN, 1926]. Heterogeneous "genus," including obvious trails (e.g., *D. sinuosus, D. gracilis, D. foliosus* LUDWIG, 1869, from German Paleozoic) and equally obvious plants (e.g., probably *D. lamourouxii*, and, according to PIA, 1927, *D. salicifolia* RUEDEMANN, 1925, Ord., N.Y.); Cenozoic "species" under the name of Recent genus *Delesseria* LAMOUROUX (2).

Dendrotichnium DELLARENA, 1949. Name with neither diagnosis nor species, provisionally given for treelike branched trail. *U.Cret.(Flysch)*, Sp.

Desmograpton FUCHS, 1895. Elevated reliefs, roughly in form of letter *H*, usually lined up in ribbons; form variable; "vertical branches" pressed together or standing apart, ends pointed or club-shaped. *Cret.-L.Tert.(Flysch)*, Eu.——FIG. 120,*2*. *D.* sp., Italy; ×0.6 (Seilacher, n).

Dictyodora WEISS, 1884 [**Dictyophytum? liebeanum* GEINITZ, 1867] [?=*Nemertites* MACLEAY, 1839 (certainly *N. sudeticus* ROEMER, 1870); *Myrianites gracilis* DELGADO, 1910, and very probably several other "new species" of *Myrianites* in DELGADO, 1910]. Complicated spread *(Spreiten)* structure, irregularly conical, with its point toward hanging wall; delicately striated exterior surface (=*Dictyodora s.s.*) intensely "folded" and commonly cutting through itself [sections of this surface=*Palaeochorda marina* (GEINITZ)]; lower, nonhorizontal margin padlike [=*Crossopodia henrici* (GEINITZ)] (41). L.Cam., Asia(Pak.);

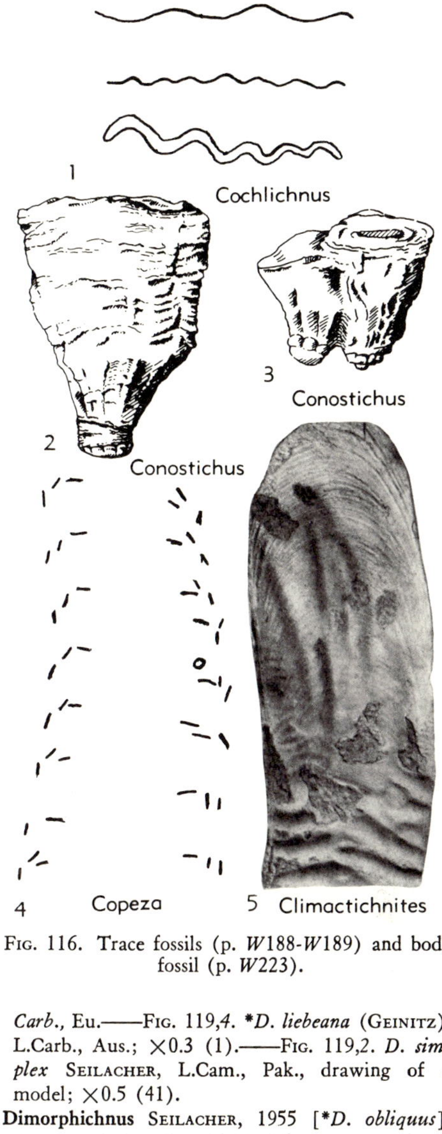

FIG. 116. Trace fossils (p. W188-W189) and body fossil (p. W223).

Carb., Eu.——FIG. 119,*4*. **D. liebeana* (GEINITZ), L.Carb., Aus.; ×0.3 (1).——FIG. 119,*2*. *D. simplex* SEILACHER, L.Cam., Pak., drawing of a model; ×0.5 (41).

Dimorphichnus SEILACHER, 1955 [**D. obliquus*]. Asymmetrical trails with 2 different types of impressions—thin sigmoidal ones, produced by raking movement, and blunt ones, similar to impressions of toes, made by a supporting activity—both types arranged in series oblique to direction of movement (41). [Feeding trail of trilobites.] *L. Cam.*, Swed.-Pak.——FIG. 119,*1*. **D. obliquus*, Magnesian Ss., Pak.; *1a*, ×0.16; *1b*, ×0.8; *1c*, ×0.4 (41).

Diplichnites DAWSON, 1873 [**D. aenigma*] [=?*Acripes* MATTHEW, 1910]. Rather nondescript biserial walking track of arthropods with numerous steps; tracks differently arranged depending on direction of movement (straight or oblique). *L.Cam.*, Asia(Pak.); *?Dev.*, Can.(N.B.); *Carb.*, Can.——Fig. 121,*1*. *D.* sp., L.Cam., Pak.; schematic, ×1.3 (41).

Diplocraterion Torell, 1870 [**D. parallelum;* SD Rudolf Richter, 1926]. U-shaped burrow with *Spreite* similar to *Rhizocorallium*, but always built strictly perpendicular to bedding plane; vertex of U-tube built progressively deeper; tubes ending in large funnels, in small, shallow ones or remaining subcylindrical to surface (23, 33). *L.Cam.*, Eu.-N.Am.——Fig. 120,*3*. **D. parallelum*, *Mickwitzia* Ss., Swed.; ×0.7 (136).

Diplopodichnus Brady, 1947 [**D. biformis*]. Long, continuous arthropod trails, consisting of 2 or 3 parallel grooves, each pair separated by a narrow, low ridge; rarely with faint foot impressions; common in Coconino Sandstone; somewhat similar to *Unisulcus* Hitchcock (16). *M.Perm.*, USA (Ariz.).

Dreginozoum von der Marck, 1894 [**D. nereitiforme*]. Oval, patchlike structures on both sides of narrow median ridge; width about 15 mm. (16). *U.Cret.-Oligo.*, Eu.——Fig. 120,*5*. **D. nereitiforme*, U.Cret.(Campan.), Ger.; ×1.3 (Häntzschel, n).

Durvillides Squinabol, 1887 [**D. eocenicus*]. Meandering trail (2). *Eoc.*, Italy.

Ephemerites Abel, 1935 [*non* Geinitz, 1865]. Horizontal U-shaped burrows produced by larvae of ephemerids; occurring in fresh-water deposits (1). [Shown by Seilacher, 1951, to be *Spreiten* burrows.] *Rec.*

Eugyrichnites Ami, 1905 [**E. minutus*]. Minute tortuous trails, about 1 mm. wide; with fine annulation (25 to 30 closely set parallel lines in 10 mm.); somewhat resembling *Gyrichnites* Whiteaves, 1883; never figured and no specimens

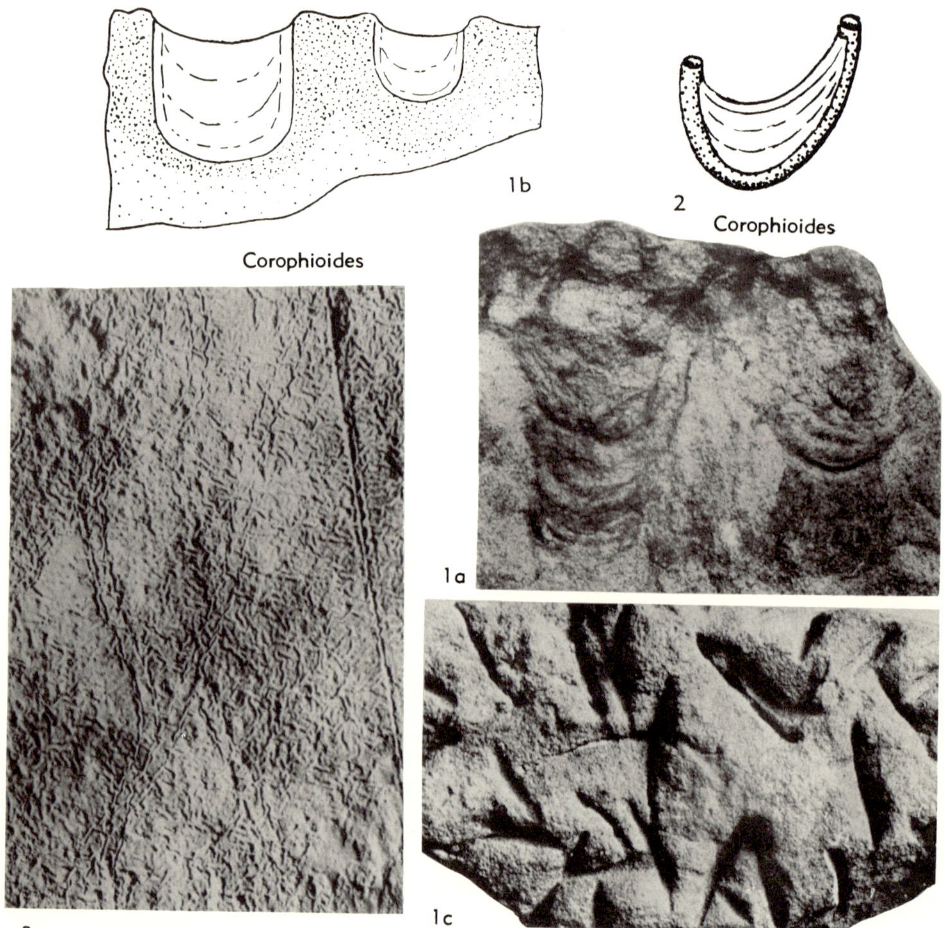

Fig. 117. Trace fossils (p. W188-W189).

could be located in Canadian collections (16). *?Sil.*, Can.(N.B.).

Felixium DELAUBENFELS, 1955 [**Rhizocorallium gläseli* FELIX, 1913]. Erroneously regarded as sponge by DELAUBENFELS (*Treatise*, p. *E*36); undoubtedly a burrow resembling *Spongites saxonicus* GEINITZ. *Cret.(Pleist.* drift), Ger.

Fraena ROUAULT, 1850 [**F. sainthilairei;* SD PÉNEAU, 1946] [=*Froena* PÉNEAU, 1946]. "Genus" at first comprising heterogeneous "species," especially bilobate trails belonging partly to *Cruziana,* later called *Rouaultia* (e.g., *F. lyelli*); on suggestion of TROMELIN & LEBESCONTE, 1875, and BUREAU, 1900, the name *Fraena* has been restricted to nonbilobate, smooth trails and elongate, cylindrical tunnel fillings, usually arranged horizontally; "genus" figured only once (16, 23). *Ord.*, Fr.

Fucoides BRONGNIART, 1823 [**F. orbignyanus;* SD ANDREWS, 1955]. Formerly used as generic name for regularly branching, plantlike tunnel structures; at present only used informally ("fucoid"), because too many widely differing "species" have been recognized. JAMES wrote in 1892-93 "that before many years the genus *(Fucoides)* began to overflow and then, like an overloaded wagon, broke down. . . . Among the debris we find tracks of crustaceans, burrows of worms, trails of mollusks, marks made by trailing tentacles of medusae, markings made by the tide or waves, rills made by running water, and holes formed by burrowing worms." [See also *Chondrites.*]

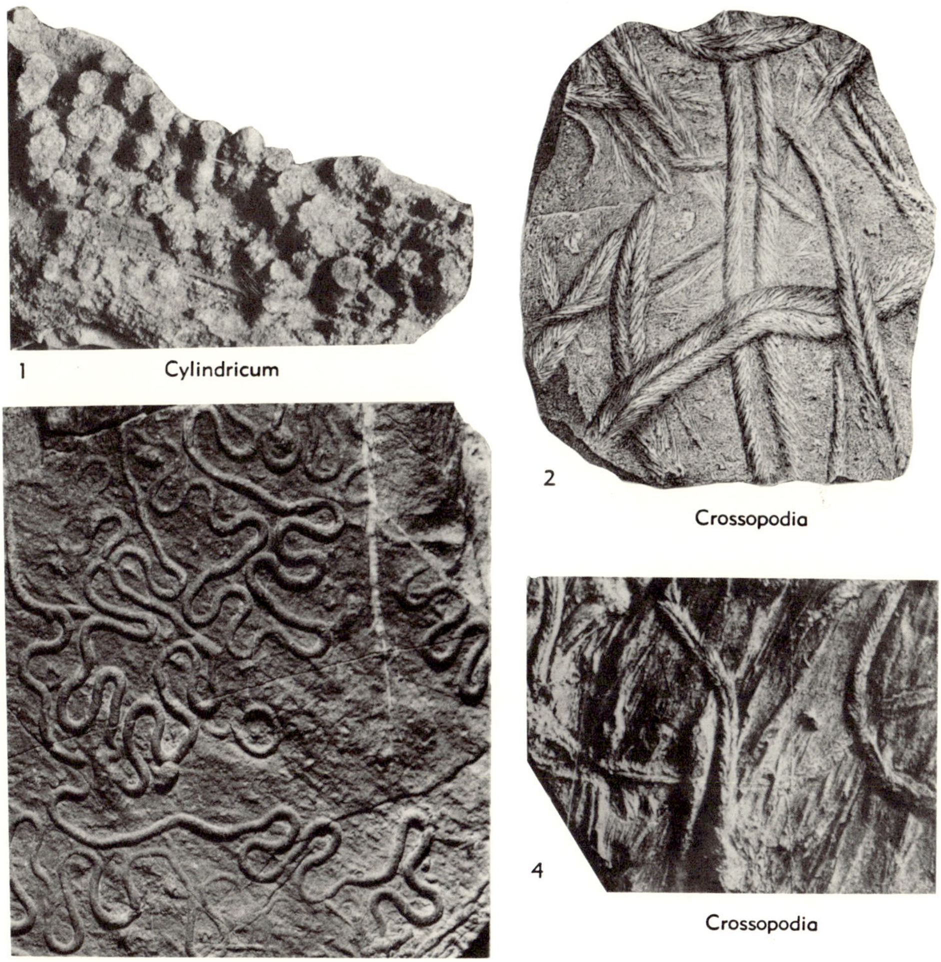

1 Cylindricum

2 Crossopodia

3 Cosmorhaphe

4 Crossopodia

FIG. 118. Trace fossils (p. *W*189).

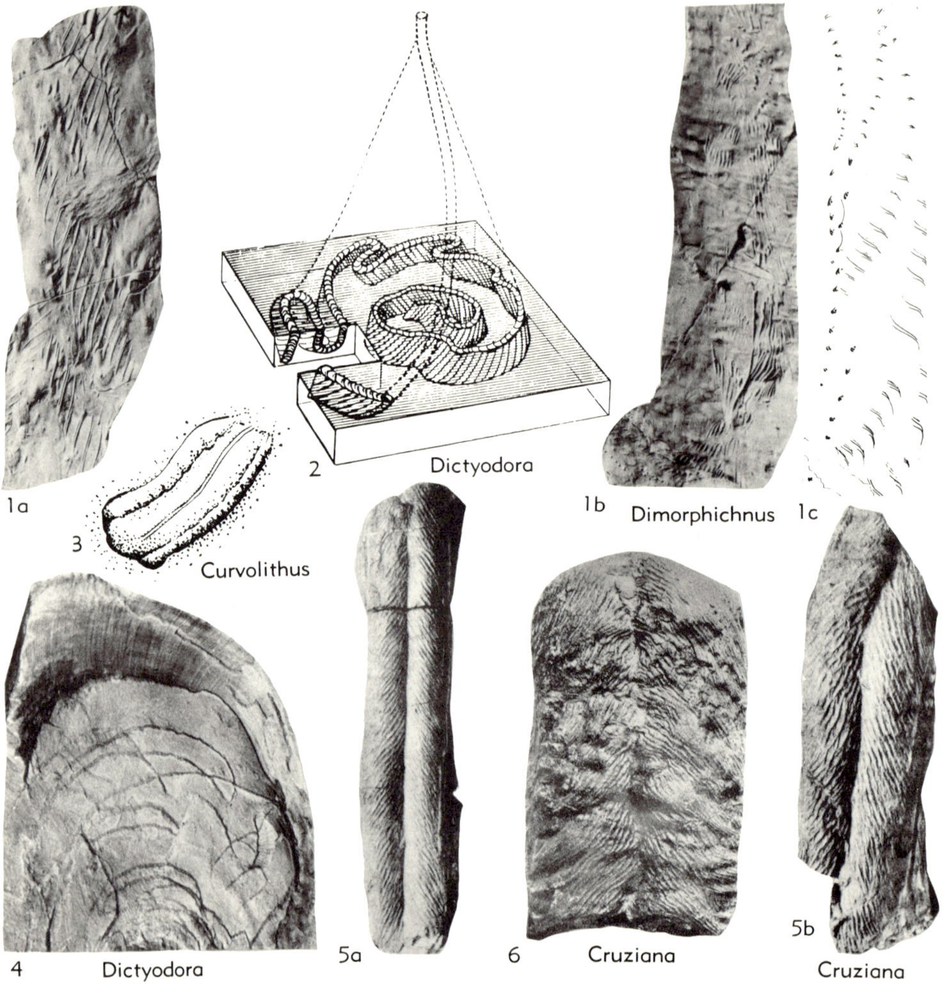

FIG. 119. Trace fossils (p. W189, W191).

Fucusopsis VASSOIEVITCH, 1932 [*F. *angulatus*] [=*Fucopsis* GROSSHEIM, 1946 (obj.)]. Originally described as "hieroglyphs in form of tubes"; according to SEILACHER (1959), stretched burrows with threadlike sculpture (16). *Cret.-Tert.*, Eu.——FIG. 120,*4*. *F. *angulatus*, U.Cret.(Senon.), USSR; ×0.3 (Hecker, n).

Goniadichnites MATTHEW, 1891 [*G. *trichiformis*]. Small, sinuous trails no larger than slender thread, commonly branching, apparently forking dichotomously; resembling trails of Recent *Goniada* as figured by NATHORST (16). *Cam.*, Can.

Gordia EMMONS, 1844 [*non* MELICHAR, 1903] [*G. *marina*] [=*Palaeochorda* M'COY, 1848 (*non P. marina* (EMMONS) *sensu* GEINITZ, 1867; see *Dictyodora* WEISS, 1884); *Palaeochordia* EICHWALD, 1855; *Herpystezoum* HITCHCOCK, 1848; *Helminthoidichnites* FITCH, 1850; *Unisulcus* HITCHCOCK, 1858; *Gordiopsis* HEER, 1865 (*nom. nud.*)]. Long, slender, smooth wormlike trails of uniform thickness throughout; mostly bent but not meandering; resembling hair-worm, *Gordius* (16). *Palaeoz.-Cenoz.*, Eu.-N.Am.——FIG. 121, *2*. *G.* sp.; schematic drawing, ×0.7 (Häntzschel, n).

Grammepus HITCHCOCK, 1858 [*G. *erismatus*; SD HÄNTZSCHEL, herein]. Doubtful insect trail (17). *Trias.*, USA(Mass.).

Granularia POMEL, 1849 [*non* POLETAEVA, ?1936] [*G. *repanda*; ?SD DESAPORTA, 1873] [=?*Alcyonidiopsis* MASSALONGO, 1856]. Sediment-filled tubes; diameter up to about 1 cm.; walls coarsely dressed with clay particles; somewhat similar to *Chondrites granulatus* (SCHLOTHEIM),

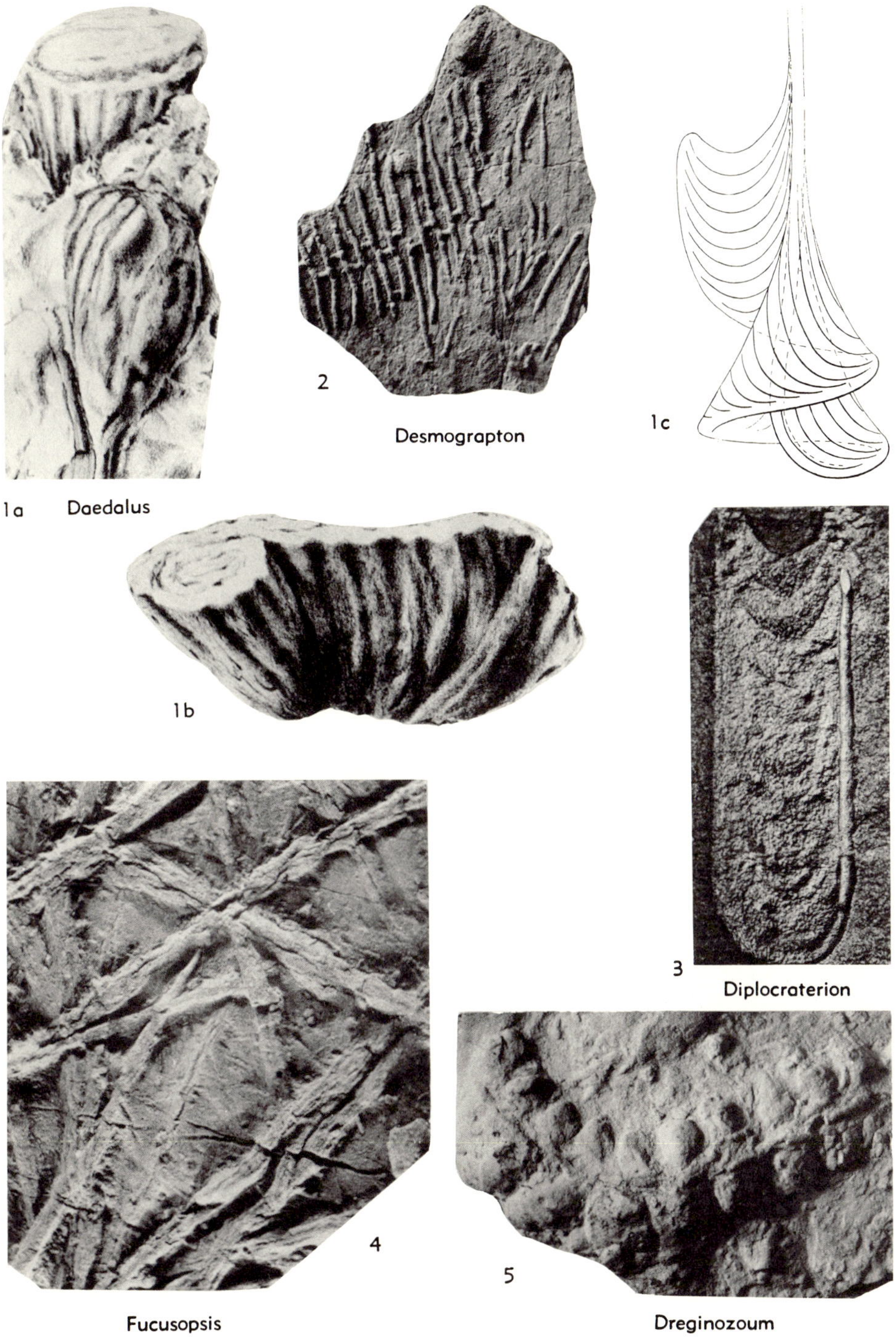

FIG. 120. Trace fossils (p. *W*191-*W*194).

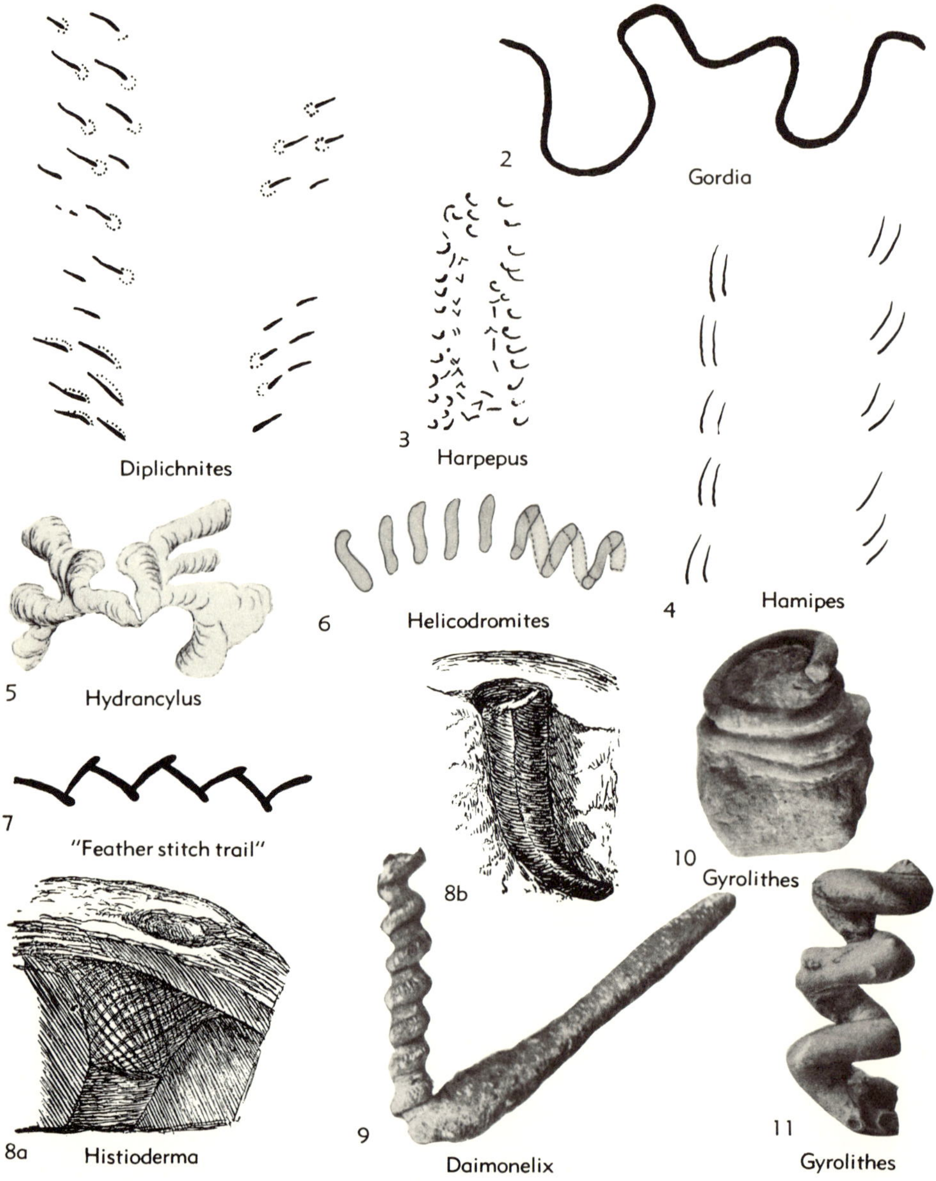

FIG. 121. Trace fossils (p. W191, W194, W196, W200, W220).

but tubes not stuffed (23, 31). *Jur.-Tert.*, Eu.——FIG. 123,*5*. *G. lumbricoides* (HEER), L.Tert.(Alberese), Italy; ×1.25 (109).——FIG. 123,*1*. *G.* sp. cf. *G. arcuata* SCHIMPER, L.Tert.(Alberese), Italy; ×1.25 (109).

Gyrichnites WHITEAVES, 1883 [**G. gaspensis*]. Trails of large size; undulating, slender, rounded furrows of almost equal width throughout and marked transversely by nearly straight, subparallel and subequidistant grooves; name given as "provisional and local" (16). [?Annelid trails.] *?U. Cam.*, USA(N.Y.); *L.Dev.*, Can.——FIG. 124,*4*. **G. gaspensis*, L.Dev., Can.; ×0.3 (139).

Gyrochorte HEER, 1865 [**G. comosa;* SD HÄNTZSCHEL, herein] [=*Gyrochorda* SCHIMPER, 1879; *?Equihenia* MEUNIER, 1886]. *Zopf*-traces of German literature, i.e., ridges on bedding-planes with biserially arranged, obliquely placed transverse

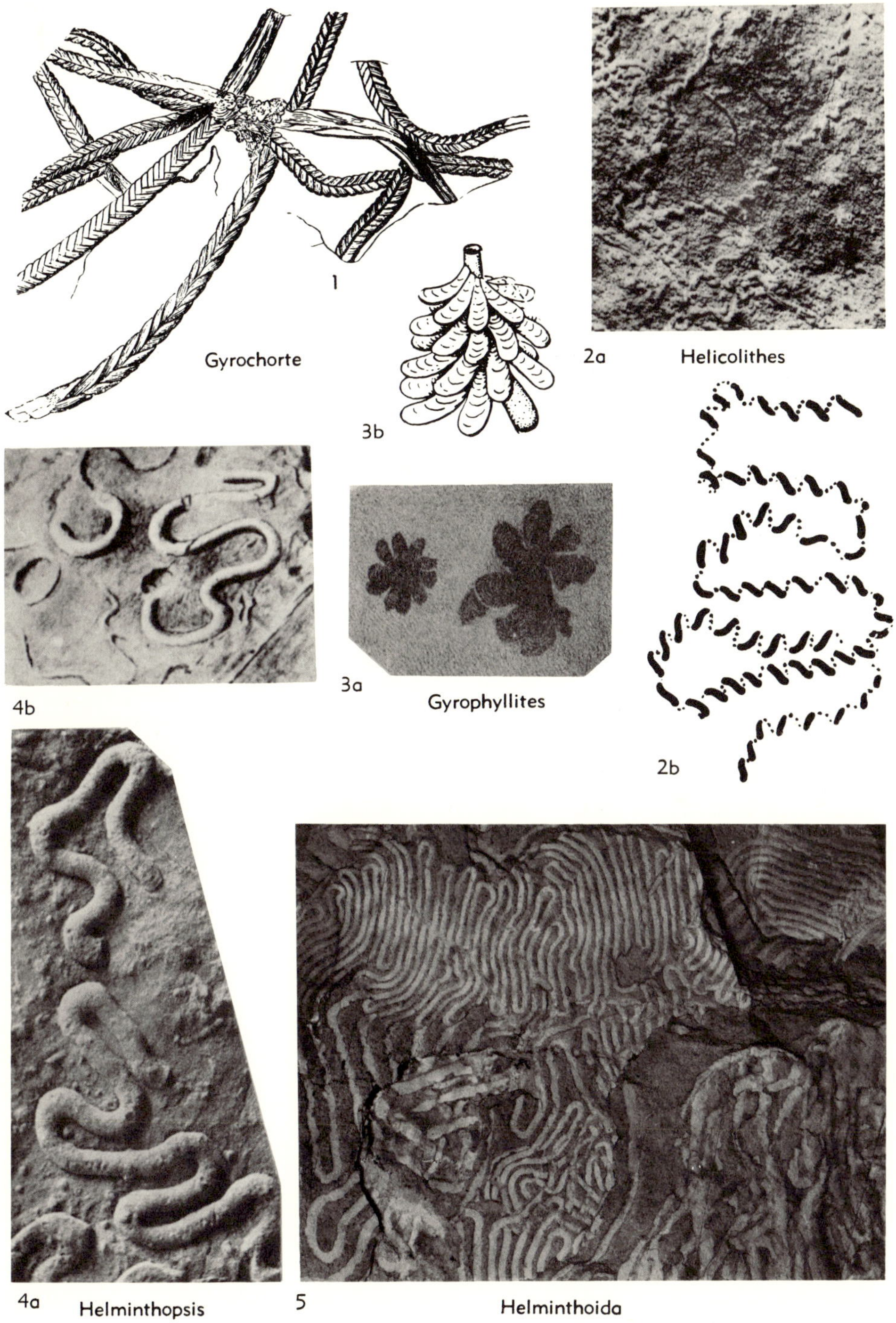

FIG. 122. Trace fossils (p. *W*196, *W*200).

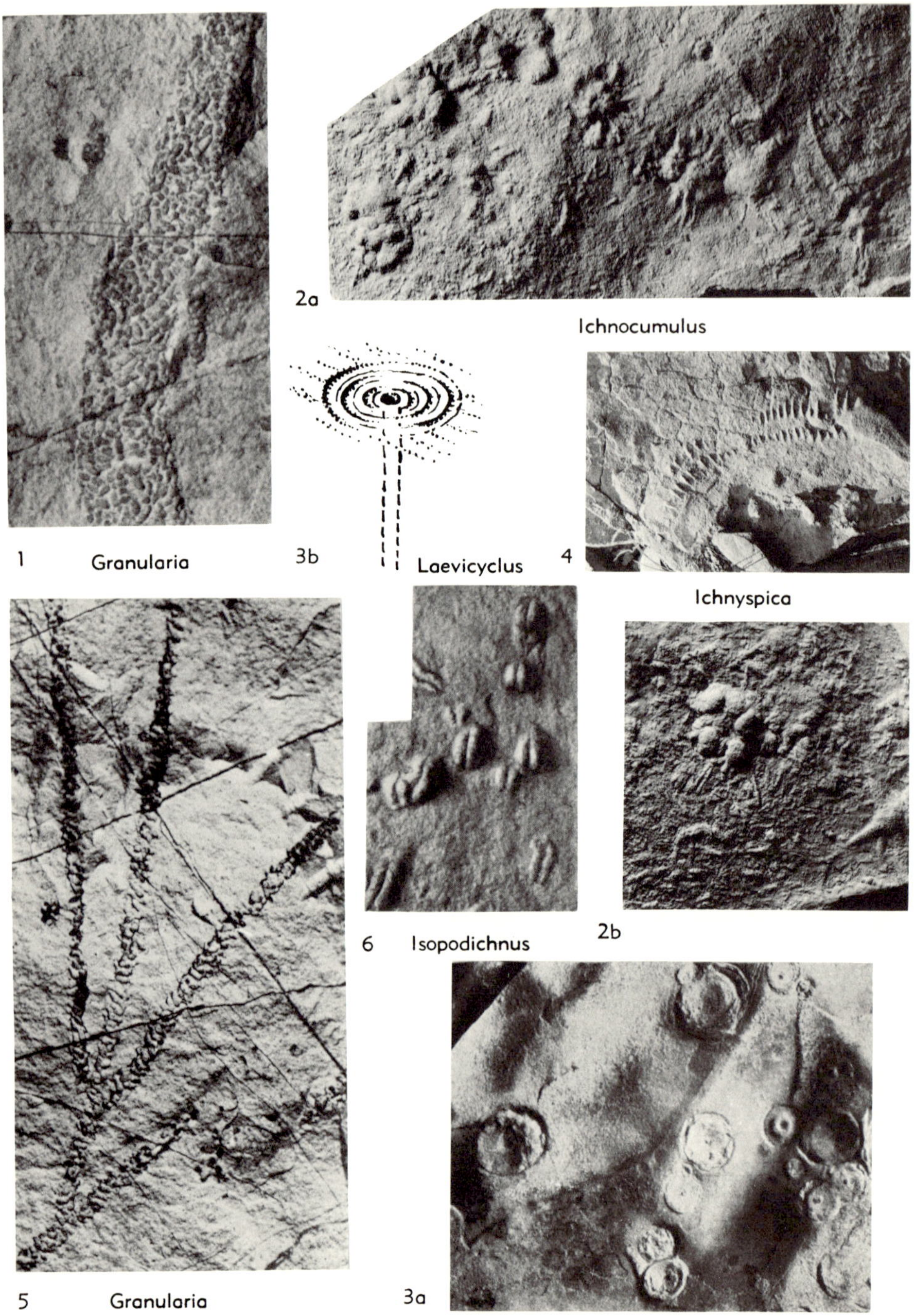

FIG. 123. Trace fossils (p. W194, W200-W201).

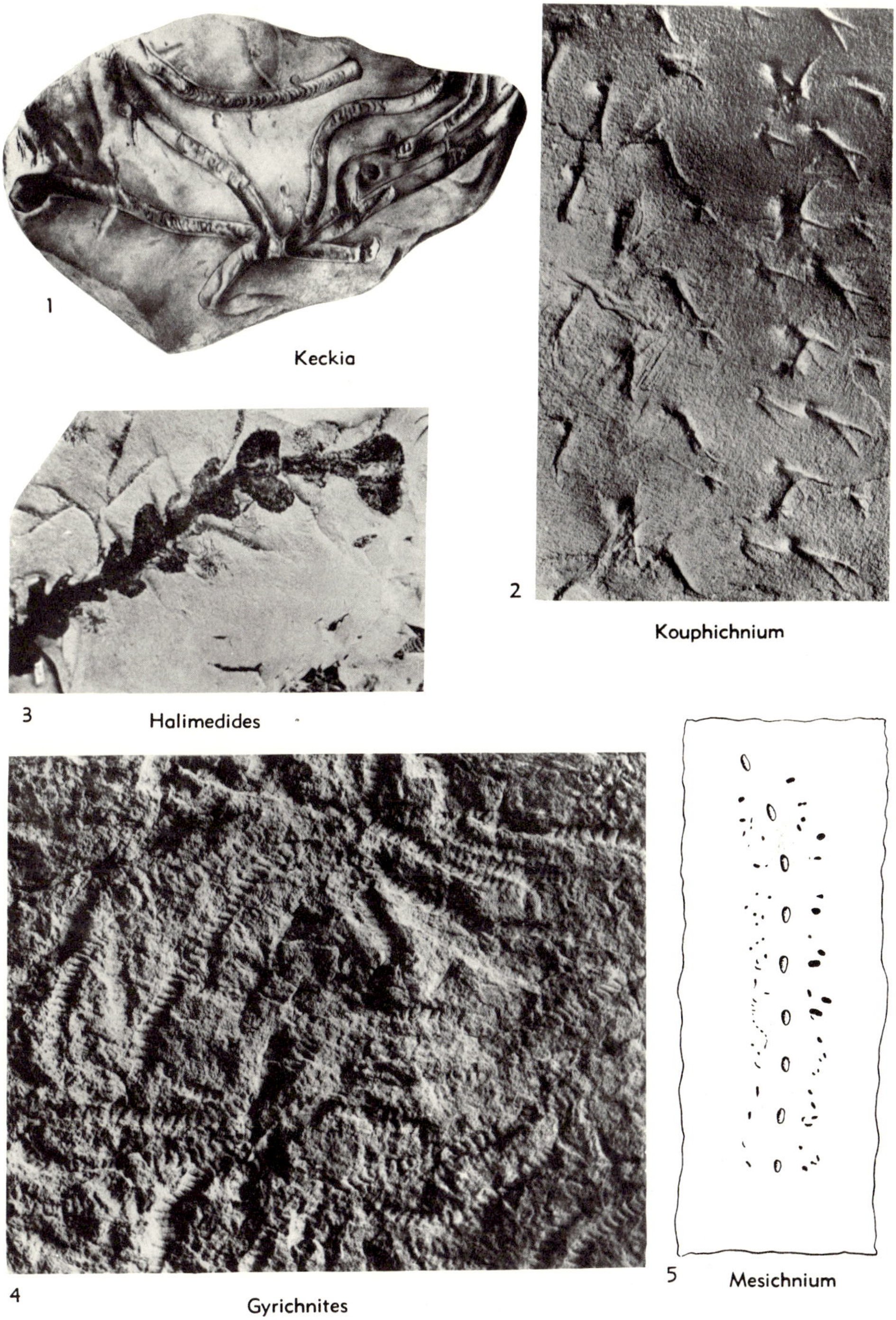

FIG. 124. Trace fossils (p. *W*196,*W*200-*W*201, *W*205).

pads, both series separated by median furrow (46). *Cam.-?Tert.*, Eu.——FIG. 122,*1*. **G. comosa*, M.Jur., Switz.; ×1 (84).

Gyrolithes DESAPORTA, 1884 [**G. davreuxi;* SD HÄNTZSCHEL, herein] [=*"Gyrolithen"* DEBEY, 1849 *(partim); Siphodendron* DESAPORTA, 1884; *Syringodendron* FUCHS, 1895 *(pro Siphodendron); Daemonhelix krameri* VON AMMON, 1900 *(non* BARBOUR, 1892, 1895); *Xenohelix* MANSFIELD, 1927]. Dextrally or sinistrally coiled burrows, upright in deposit; surface with rounded or elongate processes; thin mantle commonly formed by network of small *Chondrites;* diameter of whorls mostly uniform; may branch near upper end (16). [Probably made by Decapoda.] *Jur.-Tert.*, Eu.-USA(Md.-Calif.).——FIG. 121,*11*. *G. marylandicus* (MANSFIELD), ?Mio., Md.; ×? (98).——FIG. 121,*10*. *G. saxonicus* (HÄNTZSCHEL), U.Cret. (Turon.), Ger.; ×0.4 (82).

Gyrophyllites GLOCKER, 1841 [**G. kwassizensis*] [=*?Discophorites* HEER, 1876]. Vertical shaft from which rosettes of short, simple (feeding) tunnels radiate at different levels, as in a mine; "leaves" with sculpture of *Spreiten* burrows; shape of whole structure conical (2). *Dev.-Tert.*, Eu.——FIG. 122,*3*. *G.* sp., U.Cret.(Flysch), Aus.; *3a*, ×1; *3b*, schematic (*3a*, 10; *3b*, 120).

Halimedides LORENZ VON LIBURNAU, 1902 [**Halimeda fuggeri* LORENZ VON LIBURNAU, 1897]. Burrow with bilaterally ("pinnate") arranged, kidney-shaped extensions; morphologically corresponding rather well to Recent alga *Halimeda* LAMOUROUX (2, 40). *Cret.-L.Tert.(Flysch)*, Aus.——FIG. 124,*3*. **H. fuggeri;* ×0.3 (95).

Halopoa TORELL, 1870 [**H. imbricata;* SD ANDREWS, 1955]. Trace fossil, never figured (2). *Cam.*, Swed.

Halysichnus HITCHCOCK, 1858 [**H. laqueatus;* SD HÄNTZSCHEL, herein]. Repeatedly looped, chain-like trail with ridges on each side (17). *Trias.*, USA(Mass.).

Hamipes HITCHCOCK, 1858 [**H. didactylus*]. Two paired, regular, parallel rows of equidistant impressions of steps, curved inward, somewhat hook-shaped; width of trackway 1.6 in.; toes nearly parallel, may be slightly divergent (17). [Arthropod trail.] *Trias.*, USA(Mass.).——FIG. 121,*4*. **H. didactylus;* ×0.7 (17).

Harpepus HITCHCOCK, 1865 [**H. capillaris*]. One or 2 rows of tracks showing slightly curved foot, one end of which forms raised, blunt extremity on track (18). *Trias.*, USA(Mass.).——FIG. 121,*3*. **H. capillaris;* ×0.7 (96).

Helicodromites BERGER, 1957 [**H. mobilis*]. Smooth screw-shaped burrows; horizontal; diameter of tunnels about 2 mm.; interval between spiral turns about 1 cm. (16). *Oligo.(Rupel.)*, S.Ger.——FIG. 121,*6*. **H. mobilis;* ×0.7 (51).

Helicolithus AZPEITIA MOROS, 1933 [**H. sampelayoi,* SD HÄNTZSCHEL, herein]. Small, meandering, screw-shaped burrows; diameter of tunnels 1 mm.; diameter of spiral about 3 mm. (3). *U.Cret.-L.Tert.(Flysch)*, Eu.——FIG. 122,*2*. **H. sampelayoi; 2a*, ×1 (U.Cret., Sp.) (49); *2b*, schematic drawing, ×1.5 (?Cret., Italy) (41).

Helminthoida SCHAFHÄUTL, 1851 [**H. labyrinthica;* SD HÄNTZSCHEL, herein] [=*Helminthoidea* MAILLARD, 1887; *Helminthoides* FUCHS, 1895; *Helmintoidea* VINASSA DEREGNY, 1904 *(non H. mollassica=H. helvetica* HEER, 1865)]. Smooth; numerous, parallel, equidistant concentric furrows, about 2 mm. wide, mostly curved, may be concentric. [According to RUDOLF RICHTER, comprise "guided meanders" (1, 2).] *Cret.-Tert.*, Eu.-Alaska-Chile-Trinidad.——FIG. 122,*5*. **H. labyrinthica* U.Cret.(Flysch), Aus.; ×1 (15).

Helminthopsis HEER, 1877 [*non* GROUVELLE, 1906] [**H. magna;* SD ULRICH, 1904] [=*Helmintopsis* VINASSA DEREGNY, 1904; *Magarikune* MINATO & SUYAMO, 1949; *?Serpentinichnus* MAYER, 1956; *Tosahelminthes* KATTO, 1960]. Simple meandering tracks, but not as strictly developed as *Helminthoida* (s.s., RUDOLF RICHTER, 1928); in part with marginal ridges (46). *Ord.-Tert.;* Eu.-Asia-Alaska-Antarct.——FIG. 122,*4*. *H.* sp.; *4a*, U.Cret., Aus.; ×0.75 (1); *4b*, U.Cret., Alaska; ×1 (44).

Herpetonites VON OTTO, 1855 [**H. holothurioides*]. Superfluous, seldom used name for burrows in sandstones, regarded as bodily preserved holothurians. *U.Cret.*, Ger.(Saxony).

Hexapodichnus HITCHCOCK, 1858 [**H. magnus;* SD HÄNTZSCHEL, herein]. Small trails, very probably made by insects (17). *Trias.*, USA(Mass.).

Himanthalites FISCHER-OOSTER, 1858 [**H. taeniatus*] [=*?Taeniophycus* SCHIMPER, 1869]. Differs from *Chondrites* only in size and fewer ramifications (7). *L.Jur.*, ?Ger.; *Cret.-Tert.*, Switz.-Italy.

Histioderma KINAHAN, 1858 [**H. hibernicum*]. Curved tubes, upper extremity trumpet-shaped, lower turned up at right angle to bedding planes; upper portion of tube marked by several ridges crossing each other at irregular intervals (16). [According to RUDOLF RICHTER (1920), a dwelling-burrow.] *Cam.*, Ire.——FIG. 121,*8*. **H. hibernica; 8a, 8b*, ca. ×0.7 (80).

Hydrancylus FISCHER-OOSTER, 1858 [**Münsteria geniculata* STERNBERG, 1833]. Groups of rounded "leaves" arranged irregularly or lyre-shaped; proposed as subgenus of *Münsteria;* feeding-burrow (7). *Cret.-L.Tert.(Flysch)*, Eu.——FIG. 121,*5*. *H. oosteri* FISCHER-OOSTER, ?U.Cret.(Flysch), Switz.; ×1.5 (7).

Ichnocumulus SEILACHER, 1956 [**I. radiatus*]. Little pustule-shaped heaps with straight, radiate appendages; resting traces made by unknown animals which hid temporarily in sand (16). *L.Jur.-*

M.Jur., Ger.——FIG. 123,*2*. **I. radiatus*, L.Lias. (Angulaten-Schichten); *2a*, holotype, ×1; *2b*, ×1 (120).

Ichnyspica LINCK, 1949 [**I. pectinata*] [=*Ichnispica* LESSERTISSEUR, 1955 *(errore)*]. Double track, each composed of numerous "teeth" as in a comb; teeth straight and ending in very sharp points; rows curved, parallel, and equidistant; "type" of "ear-shaped" tracks (*Ichnia spicea* RUDOLF RICHTER) (24). *U.Trias.(M.Keuper)*, Ger.——FIG. 123,*4*. **I. pectinata*, ×0.3 (24).

Ichthyoidichnites AMI, 1903 [**I. acadiensis*]. Two rows of dashlike impressions with small ridges or monticules at posterior ends; believed to have been made by fin or finlike appendages of acanthodians (AMI, 1903) or by arthropods (ABEL, 1) (23). *L.Dev.*, Can.(Nova Scotia).

Incisifex DAHMER, 1937 [**I. rhenanus*]. Two parallel rows of obliquely arranged notches, stemming from 3-membered extremities of an arthropod *(?Homalonotus)*; between and outside the rows are smooth strips of sediment; made by sliding ventral side of animal (16). *L.Dev.*, Ger.-Belg.; *?Perm.*, S.Afr.——FIG. 125,*6*. **I. rhenanus*, L. Dev.(Seifener beds), Ger.; ×0.7 (62).

Isopodichnus BORNEMANN, 1889 (*emend.* SCHINDEWOLF, 1928) [*non* BRADY, 1947] [**I. problematicus;* SD SCHINDEWOLF, 1928 (=*Ichnium problematicum* SCHINDEWOLF, 1921)]. Dimorphous trace fossils consisting of small, straight or curved double-ribbon trails, 1 to 6 mm. wide, transversely striated by fine furrows; both "ribbons" separated by median ridge; trail may be intermittent; associated with "coffee-bean"-shaped impressions of corresponding size (16). ["Genus" placed by SEILACHER (40) in synonymy with *Rusophycus.*] *U.Sil.(?Downton.)*, Spitz.; *?L.Dev.*, Ger.; *Carb.*, Austral., Can.(Nova Scotia-N.B.); *Trias;* Ger.——FIG. 123,*6*. **I. problematicus*, L.Trias.(Buntsandstein), Ger.; ×0.75 (117).

Keckia GLOCKER, 1841 [**K. annulata*]. Similar to *Taenidium* HEER, 1877, but with flat ribbons (2, 46). *Cret.-Tert.*, Eu.——FIG. 124,*1*. **K. annulata*, U.Cret.(Cenom.), Ger., ×0.18 (75).

Kirklandia CASTER, 1945 [**K. texana*]. Perhaps a trachylinid (*Treatise*, p. *F*70, Fig. 54); perhaps a trace fossil (feeding burrow). *?M.Jur.*, Ger.; *L. Cret.*, USA.

Kouphichnium NOPCSA, 1923 [**Ichnites lithographicus* OPPEL, 1862] [=*Micrichnum* ABEL, 1924; *Micrichnus* ABEL, 1926; *Artiodactylus* ABEL, 1926; *Hypornithes*, *Ornichnites* JAEKEL, 1929; *Protornis* JAEKEL, 1929 (*non* MEYER, 1844); *Paramphibius* WILLARD, 1935; *Limuludichnulus* LINCK, 1943; *Limuludichnus* LINCK, 1949]. Trails of great variability; 2 longitudinal series of digitate tracks with 3 or 4 slender, straight or acuminate, or slightly curved rays; length of rays equal or unequal; some with tail track; interpreted as trails of vertebrates (birds, fishlike amphibians, reptiles or mammals) in first descriptions; now recognized as made by limulids (1, 23). *Dev.-Jur.*, Eu.-N.Am.-Greenl.——FIG. 124,*2*. *K. didactylus* (WILLARD), U.Dev.(Chemung), Penn., ×1.5 (56).——FIG. 125,*1*. *K. gracilis* (LINCK), U.Trias.(Schilfsandstein), Ger.; ×0.7 (24).

Kulindrichnus HALLAM, 1960 [**K. langi*]. Stumpy, cylindrical or conical bodies with apex directed downward; oriented subvertically in bed; up to 130 mm. in length and 75 mm. in diameter; composed of shell aggregates, some aligned peripherally to margin; matrix may be phosphatic (16). [Interpreted as burrow (resting trail) produced by cerianthid sea anemone.] *L.Jur.*, Eu.——FIG. 125,*5*. **K. langi*, Blue Lias, Eng.; *5a*, long. sec. with phosphatic sheath; *5b*, long. sec. without phosphatic sheath; *5c*, reconstr. burrow indicating calcite-filled cracks in phosphatic sheath, ±×0.3 (81).

Laevicyclus QUENSTEDT, 1879 [=*Cyclozoon* WURM, 1912 *(partim)*]. Approximately cylindrical bodies standing at right angles to bedding plane; diameter variable in same specimen; perforated by central canal; visible on bedding planes as regular concentric circles with diameter of several cm. (23, 41). [Interpreted by QUENSTEDT (1879) as coral; by PHILIPP (1904) and WURM (1912) as organism of unknown affinities; by M.SCHMIDT (1934) as inorganic, made by gas-exhalations and water under pressure within sediment; by SEILACHER (38, 41) as trace fossil (feeding burrow) comparable with dwelling shaft and scraping circles of Recent annelid worm, *Scolecolepis.*] *L. Cam.*, Pak.; *Trias.-Jur.*, Eu.——FIG. 123,*3*. *L.* sp.; *3a*, U.Trias.(Campiller beds), Italy; ×0.25 (118); *3b*, reconstr., L.Cam., Pak.; ×2.7 (41).

Lapispira LANGE, 1932 [**L. bispiralis*]. U-shaped tunnel with both legs spirally curved in same direction (23). *L.Jur.(L.Lias.)*, Ger.——FIG. 125, *2*. **L. bispiralis;* wire models of burrows, ×0.2 (91).

Lennea KRÄUSEL & WEYLAND, 1932 [**L. schmidti*]. Vertical shaft about 1 cm. in width with numerous narrower lateral tunnels branching off irregularly at right angles along whole length of vertical shaft; lateral branches approximately horizontal, branching dichotomously (16). *Dev.*, Ger.——FIG. 125,*7*. **L. schmidti*, M.Dev.; ×0.3 (104).

Lithographus HITCHCOCK, 1858 [**L. hieroglyphicus;* SD HÄNTZSCHEL, herein]. Insect trail, very similar to or identical with *Copeza* HITCHCOCK but having oblique markings outside longitudinal ones (17). *Trias.*, USA(Mass.).——FIG. 125,*3*. **L. hieroglyphicus;* ×0.7 (96).

Lophoctenium REINHOLD RICHTER, 1850 [**L. comosum* REINHOLD RICHTER, 1851] [=*Criophycus* TOULA, 1906]. Bunches of closely spaced, inwardly bent "twigs" with comblike branches, joining to form main axis; formerly erroneously thought to belong to graptolites, sertularids, or algae (23). [Feeding burrows, according to SEILACHER (41).] *Dev.-Carb.*, Eu. (Ger.-Port.); *U. Cret.-L. Tert. (Flysch)*, Eu.——FIG. 127,7. **L. comosum*, M. Dev. (*Nereites* beds), Ger.; ×1.5 (40).

Lorenzinia GABELLI, 1900 [**L. apenninica*]. Commonly regarded as scyphozoan (*Treatise*, p. *F*43, Fig. 32,*1-3*); more probably a feeding burrow; some forms resembling *Atollites*. *Cret.-Tert. (Flysch)*, Eu.

Lumbricaria MÜNSTER, 1831 [**L. intestinum;* SD HÄNTZSCHEL, herein] [=*Vermiculitae, Vermiculites* PARKINSON, 1811 (neither name intended for genus); *Medusites* GERMAR, 1827 (long unused name seemingly intended for this fossil); *?Lumbricites* AUCTT. (*non L. antiqua* PORTLOCK, 1843); (*?non L.? gregaria* PORTLOCK, 1843); *Cololites*

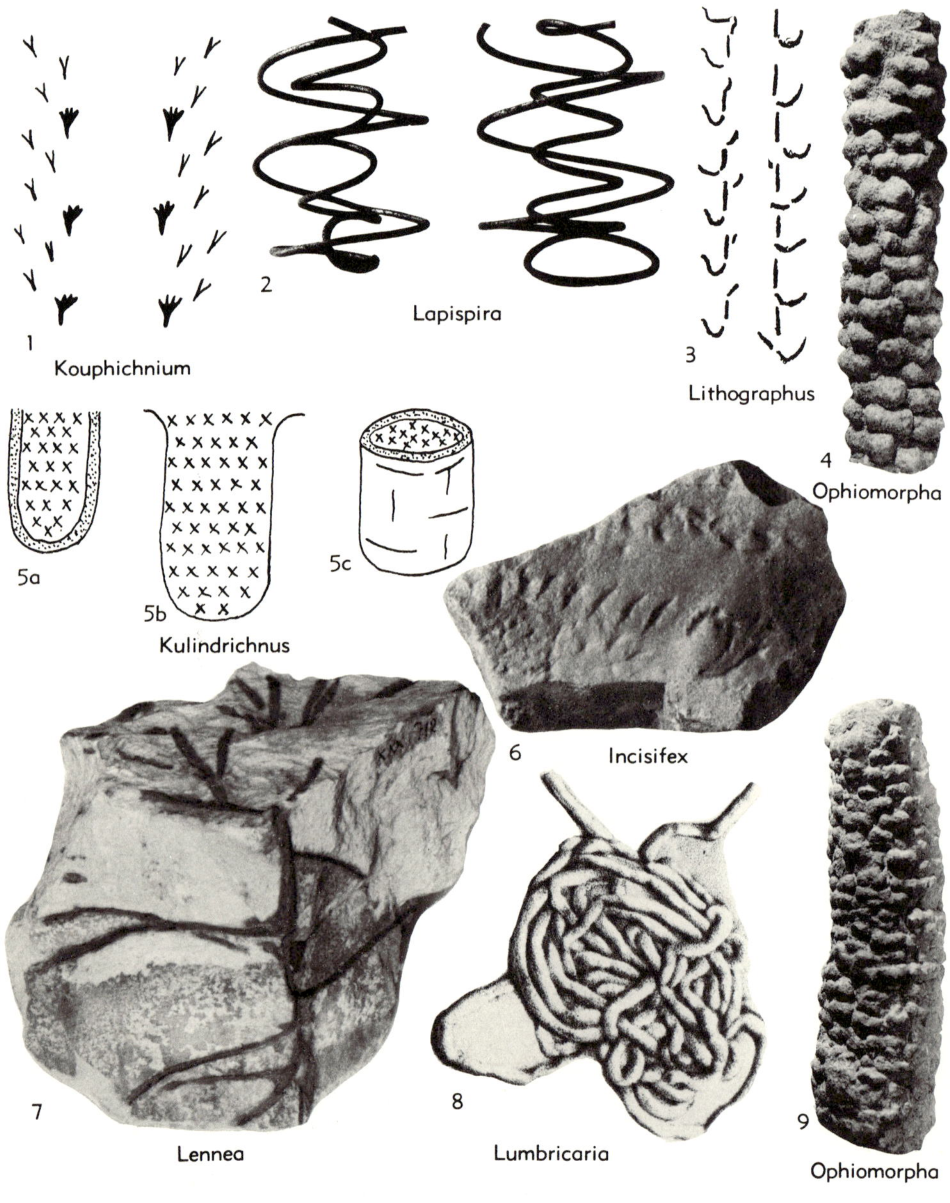

FIG. 125. Trace fossils (p. *W*201-*W*202, *W*205).

Agassiz, 1836 (clearly not intended as generic name)]. Entangled intertwined strings, approximately 3 mm. wide (16). [Interpreted as excrement of fish or ejected entrails of holothurians.] [See also *Treatise,* p. *F*159.] *?U.Trias., Jur.,* Ger. ——Fig. 125,*8*. **L. intestinum,* U.Jur.; ×0.7 (76).

Lunula Hitchcock, 1865 [*non* Koenig, 1825, *nec* Lamarck, 1812] [**L. obscura*]. Trail consisting of narrow axis, with laterally extended lunate im-

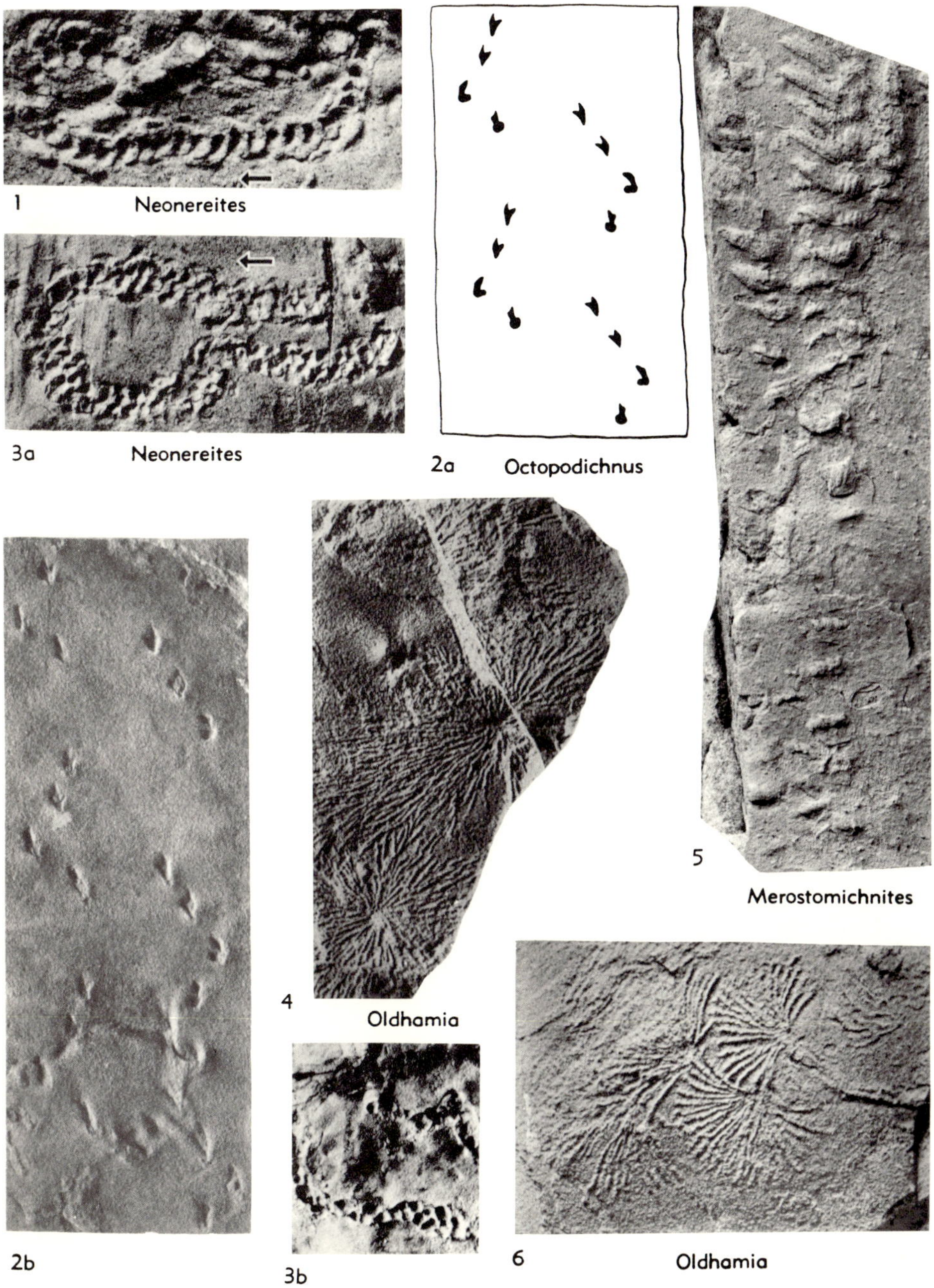

Fig. 126. Trace fossils (p. *W*205).

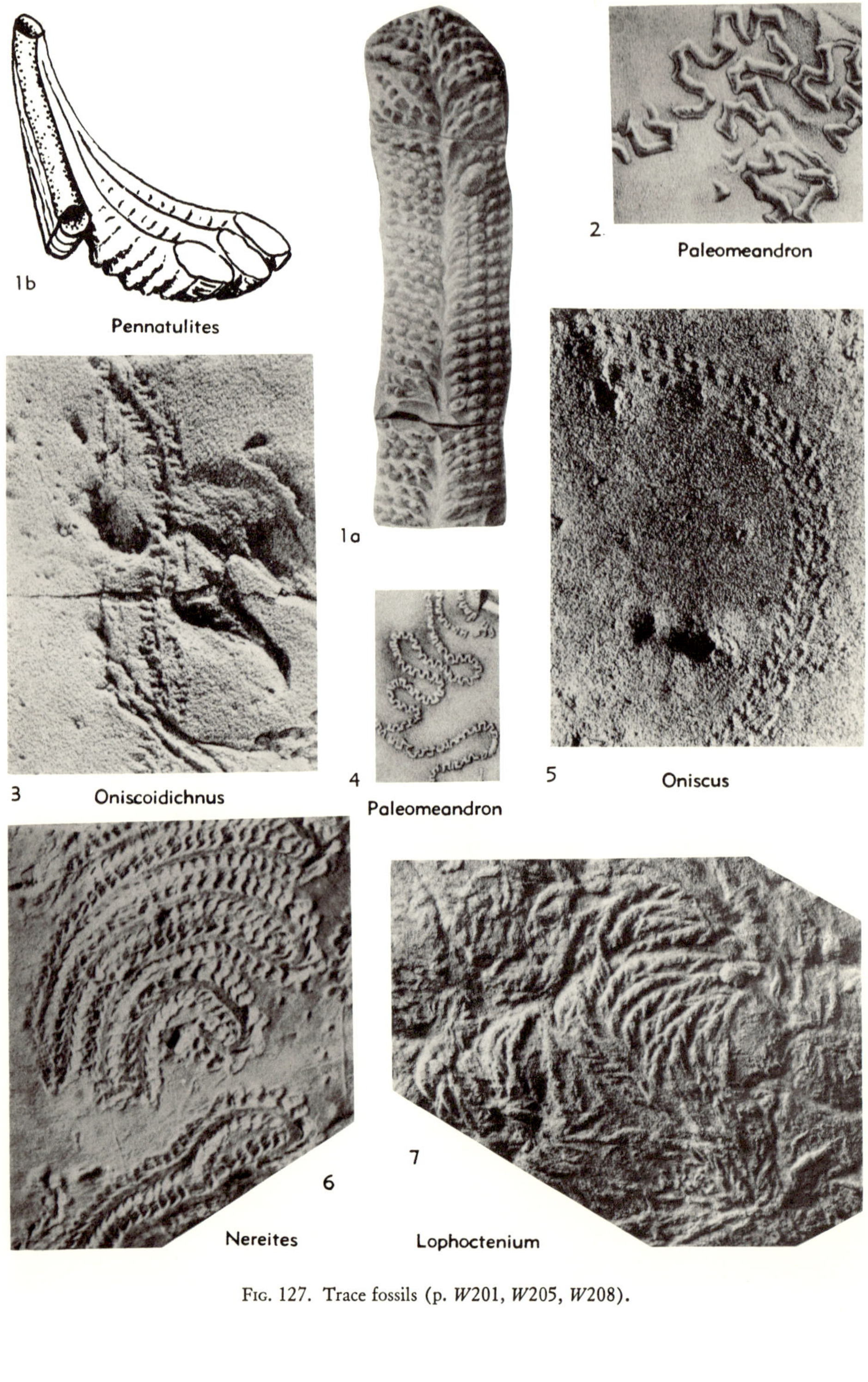

Fig. 127. Trace fossils (p. *W*201, *W*205, *W*208).

pressions on both sides (18). [Possibly made by phyllopod or myriapod; junior homonym.] *Trias.*, USA(Mass.).

Medusina WALCOTT, 1898. "Species" of this "genus" lack valid generic designation. Several species considered to be medusae in the *Treatise* (p. *F*153, *F*156-158), as well as "species" of *"Medusites"* (p. *F*159) and similar forms, probably belong among trace fossils (feeding trails).

Merostomichnites PACKARD, 1900 [**M. beecheri;* SD HÄNTZSCHEL, herein]. Circular, spindle- or bow-shaped, or transversely expanded, rarely ringed; opposite impressions arranged in 2 parallel rows with about 1 cm. interval; Paleozoic forms attributable in part to trilobites, in part to eurypterids, Triassic forms possibly to phyllopods (23). *Cam.-L.Trias.*, Eu.-N.Am.——FIG. 126,*5*. *M. strandi* STØRMER, Downtonian, Nor.(Spitz.); ×1 (126).

Mesichnium GILMORE, 1926 [**M. benjamini*]. Somewhat obscure trail consisting of 2 parallel lines of footprints with median row of suboval regularly spaced depressions; digital formula of producing animal unknown; width about 2 cm.; length of stride (distance between depressions of median row) about 15 mm. (11). *Perm.*, USA (Ariz.).——FIG. 124,*5*. **M. benjamini*, Coconino Ss.; diagram of trackway, ×0.5 (11).

Montfortia LEBESCONTE, 1886 [jr. hom.; *non* RÉCLUZ, 1843]. Small horizontal, oblique, or perpendicular burrows, 1 to 2 mm. wide, occasionally showing annulation; very similar to *Planolites* (42). [Probably worm trails; not a sponge as interpreted by LEBESCONTE.] *Precam.*, Fr.

Muensteria STERNBERG, 1833 [*non* KROGERUS, 1931]. "Genus" based on cylindrical, transversely striped bodies; in part body fossils (?sponges), in part trace fossils (?coprolites, *Taenidium* or other "genera") (7).

Myriapodites MATTHEW, 1903. Two opposite rows of impressions about 6 mm. apart, each row 2 mm. wide; rows consisting of closely set linear prints arranged in double series of elongated scratches or claw-markings, mostly directed from outside to inside of row; tracks commonly only round dots (16). *Carb.*, Can.(Nova Scotia).

Myriodocites MARCOU (before 1880). According to ZITTEL (1880), resembling *Nereites* and like Paleozoic trails; MARCOU's description not found.

Nemapodia EMMONS, 1844 [**N. tenuissima*]. According to RICHTER (1924), very probably track of Recent gastropod feeding in meanders on surface of slabs of slate (as shown by RICHTER for *N. tenuissima*, described by GEINITZ (1854) from L.Carb. slates of Saxony) (32).

Neonereites SEILACHER, 1960 [**N. biserialis*]. Typically (as negative epireliefs) consisting of irregularly curved chains of deep, smooth-walled dimples on upper surface of thin sandstone layers; chains restricted in length, some bordered laterally by flabby structures caused by burrowing; corresponding hyporeliefs (on lower surface of sandstone beds) forming indistinct rows of pustules (one row, *N. uniserialis;* two rows *N. biserialis*); interior with some clay pills corresponding to dimples of epirelief (16). *L.Jur.-M. Jur.*, Ger.——FIG. 126,*3*. **N. biserialis*, M.Jur. (Dogger β); *3a, 3b*, ×0.6 (120).——FIG. 126,*1*. *N. uniserialis* SEILACHER, L.Jur.(Lias α_2); ×0.9 (120).

Nereites MACLEAY, 1839 [*non* EMMONS, 1846] [**N. cambrensis;* SD HÄNTZSCHEL, herein] [=*Myrianites* MACLEAY, 1839 *(partim); Nereograpsus* GEINITZ, 1852; *Myrianytes* MASSALONGO, 1855; *Nereograptus* HALL, 1865]. Meandering feeding trails with narrow central axis and regularly spaced lateral, leaf-shaped or lobelike, commonly with finely striated elevations; width of trail about 1 to 2 cm.; trails highly variable in width and length of lobelike projections (23). [Made by annelids (RICHTER, 35) or gastropods (ABEL, 1).] *Paleoz.*, cosmop.——FIG. 127,*6*. *N. loomisi* EMMONS, M.Dev., Ger.; ×0.3 (35).

Octopodichnus GILMORE, 1927 [**O. didactylus*]. Trail of apparently 8-footed animal; tracks arranged in 4 groups; alternating; 2 anterior impressions of each group didactyle, 2 posterior, unidactyle; probably made by arachnid (ABEL, 1) or scorpionid (BRADY, 1947) (11). *Perm.*, USA (Ariz.).——FIG. 126,*2*. **O. didactylus*, Coconino Ss.; *2a*, ×0.8; *2b*, diagram of trackway, ×0.25 (11).

Oldhamia FORBES, 1849 [**O. antiqua;* SD ANDREWS, 1955] [=*Murchisonites* GOEPPERT, 1860]. Bunches of fine rills, radiating from joints of a sympodial axis; variously explained as actual remains of organisms, small burrows of worms, or inorganic; very probably trace fossils (16). *Cam.*, Eu.-N.Am. ——FIG. 126,*6*. **O. antiqua*, Ire.; ×1.3 (123). ——FIG. 126,*4*. *O. radiata* FORBES, Ire.; ×1.3 (123).

Oncophorus GLOCKER, 1850 [*non* RUDOW, 1870; *nec* EPPELSHEIM, 1885] [**O. beskidensis*]. Sinuous trail resembling *Gyrochorte bisulcata* GEINITZ, 1883 (16). *?U.Cret.*, Czech.

Oniscoidichnus BRADY, 1949 [**Isopodichnus filiciformis* BRADY, 1947] [=*Isopodichnus* BRADY, 1947 (*non* BORNEMANN, 1889)]. Trail with low, sinuous median ridge and forward-pointing bractlike footprints on each side at intervals of about 1 mm.; width of entire trail about 1 cm.; resembles track of Recent isopod, *Oniscus* (16). *M. Perm.*, USA(Ariz.).——FIG. 127,*3*. **O. filiformis* (BRADY), Coconino Ss.; ×0.5 (52).——FIG. 127, *5*. *Oniscus* sp.; track of Recent isopod, ×0.5 (52).

Ophiomorpha LUNDGREN, 1891 [*non* SZEPLIGETI, 1905] [**O. nodosa*] [=*Ophiomorpha* NILSSON,

1836 *(nom. nud.)*; *Cylindrites spongioides* GOEPPERT, 1841; *Spongites saxonicus* GEINITZ, 1842; *Cylindrites tuberosus* EICHWALD, 1865; *Phymatoderma dienvalii* WATELET, 1866; *Halymenites major* LESQUEREUX, 1873; *?Broeckia* CARTER, 1877; *?Halymenidium* SCHIMPER, 1879; *Astrophora* DEECKE, 1895]. Tunnel trails with tuberclelike or wartlike ornamentation of outer wall but smooth inside; width 1 to 2 cm.; may be branched with place of ramification widened in blistered or pear-shaped way (23). *U.Cret.-Tert.*, Eu.-N.Am.-Japan.——FIG. 125,9. **O. nodosa*, ?U.Cret. or L. Tert., S.Swed.(Scania); ×0.4 (82).——FIG. 125, 4. *O. major* (LESQUEREUX), U.Cret., USA (N.Dak.), ×0.5 (82).

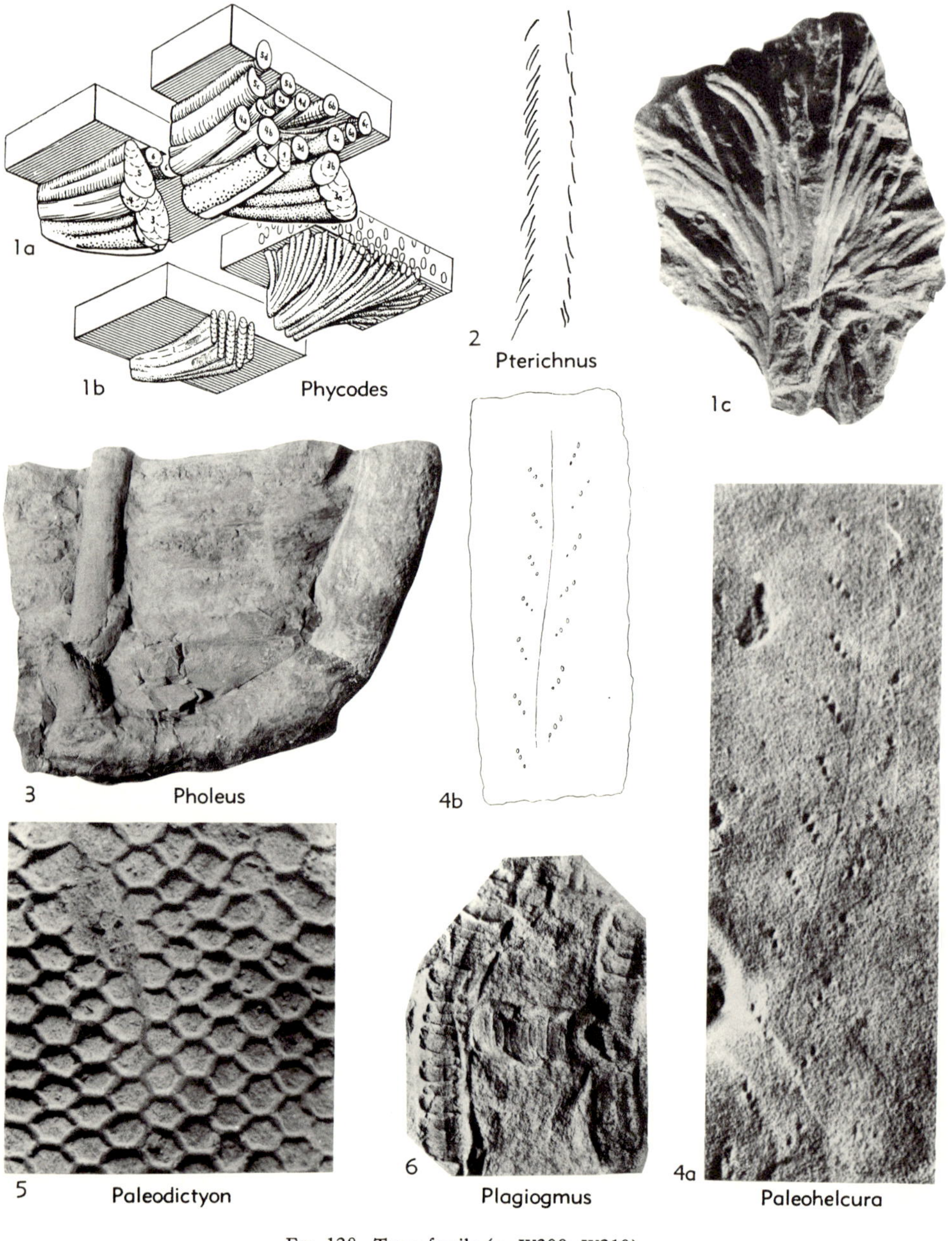

FIG. 128. Trace fossils (p. *W*208, *W*210).

Ormathichnus Miller, 1880 [**O. moniliformis*]. Single, continuous, beaded trail, resembling impression of small column of *Heterocrinus* (16). *Ord.*, USA(Ohio).

Ostrakichnites Packard, 1900. Designation for trails described insufficiently by Dawson (1873) as *Protichnites carbonarius*, according to Packard, they do not belong to *Protichnites*, nor were they made by limulids (16). *Carb.*, N.Am.(Nova Scotia).

Palaeocrista Hundt, 1941 [*nom. nud.;* diagnosis and designation of type-species missing]. (16). *L.Ord.*, Ger.

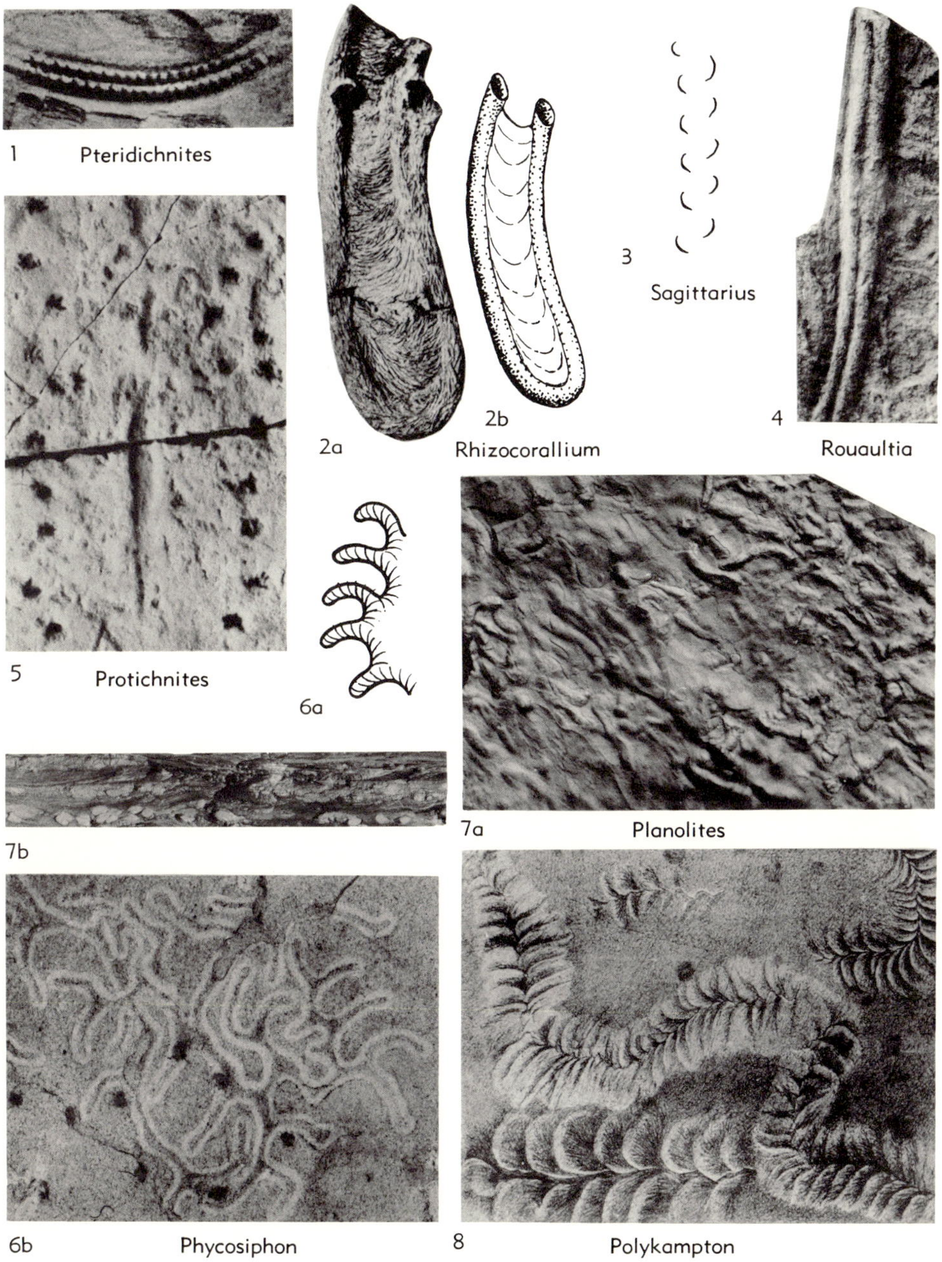

Fig. 129. Trace fossils (p. *W*208, *W*210, *W*212, *W*215).

Palaeophycus Hall, 1847 [**P. tubularis;* SD Wilson, 1948] [=*?Palaeospongia* d'Orbigny, 1849 *(partim); Foralites* Rouault, 1850; *?Aulacophycus* Massalongo, 1859; *?Aulacophycus* Eichwald, 1860; *Spongillopsis* Geinitz, 1862; *?Scolecites* Salter, 1873; *?Aulacophycus* Heer, 1877; *Paleophycus* James, 1879; *?Scoyenia* White, 1929]. Fillings of cylindrical or subcylindrical horizontal galleries, branched and irregularly winding (46). *Paleoz.-Mesoz.*, Eu.-N.Am.

Palaeosemaeostoma Rüger, 1925 [**Medusina geryonides* von Huene, 1901 (=*Medusa gorgonoides* G. Wagner, 1932]. (See *Treatise,* p. *F*76, Fig. 61,*2*.) Very probably not belonging to the Trachylinida; according to Fuchs (1901) and Seilacher (41) a feeding burrow resembling *Gyrophyllites* Glocker, 1841. *M.Jur.*, Ger.; *Cret.*, Czech.

Paleodictyon Meneghini, 1850 [**P. strozii*] [=*Palaeodictyon, Palaeodictyum* auctt. (*non* Heer, 1877); *Scyphia maxima* Eichwald, ?1846; *Reticulipora buzzoni, R. villae, R. quadrata, R. briantea* Stoppani, 1857; *Glenodictyum* von der Marck, 1863; *Cephalites maximus* Eichwald, 1865; *Paretodictyon* Mayer, 1878; *Palaeodyction* de Stefani, 1879; *Retiofucus* Keeping, 1882; *Retiphycus* Ulrich, 1904]. Honeycomb-like networks, mostly regularly 6-sided, but also 5- to 8-sided meshes which are commonly open on one side; size variable; in relief on lower surface of beds (1, 3). [Interpreted by some authors as organic, by others as inorganic; according to Seilacher (40), these structures represent feeding trails.] *Ord.-Tert.*, Eu.-Asia-N.Am.——Fig. 128,*5*. *P. regulare* Sacco, L.Tert.(Flysch), Italy; ×0.4 (Seilacher, n).

Paleohelcura Gilmore, 1926 [**P. tridactyla*]. Two parallel rows of tracks with drag of caudal appendage between them; clusters of 3 imprints made by tridactyl, pointed extremities; their longer axis about 45 degrees to line of direction; clusters alternating on 2 sides; greatest width 22 mm.; probably made by scorpionid (11). *Perm.*, USA (Ariz.).——Fig. 128,*4*. **P. tridactyla,* Coconino Ss.; *4a,* ×0.5; *4b,* diagram of trackway, ×0.3 (11).

Paleomeandron Peruzzi, 1881 [**P. rude;* SD Andrews, 1955]. [=*Palaeomaeandron* Fuchs, 1895 *(errore)*]. Irregular, wide meanders made up of smaller, rather angular meanders or H-shaped figures (10). *Tert.*, Italy, Sp.——Fig. 127, *2*. **P. rude,* Italy; ca. ×2 (105).——Fig. 127,*4*. *P. elegans* Peruzzi, Italy; ca. ×0.5 (105).

Palmichnium Rudolf Richter, 1954 [**P. palmatum*]. Large, plantlike arthropod track; opposed symmetrical rows of leg impressions; median keel; divided at regular pace intervals; bordered by longitudinally directed club-shaped impressions distinctly set off toward interior, but indistinctly toward exterior (16). *L.Dev.*, Ger.——Fig. 130,*1*. **P. palmatum;* ×0.25 (111).

Parinassa Hundt, 1941 [**P. pennaeformis*] [*nom. nud.;* diagnosis missing]. (16). *L.Ord.*, Ger.

Pelecypodichnus Seilacher, 1953 [**P. amygdaloides*] [=*?Lockeia* James, 1879]. Small podlike fossils, tapering to sharp and obtuse points at each end; resting trails of pelecypods (39). *Ord.-Tert.*, Eu.-N.Am.——Fig. 130,*4*. **P. amygdaloides,* M. Jur.(Dogger β, Donzdorfer Ss.), Ger.; (shown with *Asteriacites quinquefolius* (Quenstedt), ×0.75 (39).

Pennatulites de Stefani, 1885 [**P. longespicata*] [=*Paleosceptron* de Stafani, 1885; *Virgularia presbytes* Bayer, 1955 (Tert. forms only)]. Thick cylindrical stalk, followed by a club- or ear-shaped part consisting of biserially arranged overlapping "sawed" leaves divided by deep median furrow (10). [According to Seilacher (40, 41), a feeding burrow.] *Cret.-L.Tert.(Flysch);* Eu.-W.Indies (Trinidad).——Fig. 127,*1*. **P. longespicata,* U. Cret., Italy; *1a,* ×0.5; *1b,* model, ×0.2 (*1a,* 125; *1b,* 41).

Permichnium Guthörl, 1934 [**P. völckeri*]. Two parallel, equal, and equidistant rows of V-shaped impressions, open to exterior; indicative of equal walking feet with 2 claws each; similar to *Bifurculapes* Hitchcock, 1858 (16). [Running trail of insect (?blattoid).]. *L.Perm.(Rotl.)*, Ger.——Fig. 130,*3*. **P. voelckeri;* holotype, ×1.4 (78).

Petalichnus Miller, 1880 [**P. multipartitus*]. Wide trail, consisting of numerous transversely elongated depressions arranged without distinct order; apparently equivalent to 3 or more interlocking rows (16). [Made by trilobites.] *Ord.*, USA (Ohio).

Pholeus Fiege, 1944 [**P. abomasoformis*]. Large compactly cylindrical dwelling burrows; longitudinal axis horizontal; anterior and posterior end closed and rounded; 2 or more rounded tubes, running obliquely or vertically, leading to surface; walls lined with flakes (16). [Probably made by decapods.] *M.Trias.(L.Muschelkalk)*, Ger.——Fig. 128,*3*. **P. abomasoformis,* ×0.4 (68).

Phycodes Richter, 1850 [*non* Guenee, 1852; *nec* Milne-Edwards, 1869] [**P. circinatum*] [=*Licrophycus* Billings, 1865; *Vexillum rouvillei* Saporta, 1884; *Lycrophycus* Twenhofel, 1928]. Bundled cylindrical fillings of tunnels on lower surface of quartzite beds, may show faint regular transverse fluting (41). [According to Seilacher, 1955, feeding burrow; see also *Arthrophycus*.] *L. Cam.*, Pak.; *Ord.*, Eu.-Am.-AsiaM.——Fig. 128 *1b-c*. **P. circinatum; 1b,* model, L.Ord., Ger.; ×0.3 (41); *1c,* Ord., Galena F., USA(Minn.); ×0.7 (97).——Fig. 128,*1a*. *P.* cf. *palmatum* (Hall), L.Cam., Pak.; model, ×0.7 (41).

Phycosiphon Fischer-Ooster, 1858 [**P. incertum*] [=*Palaeodictyon* Heer, 1865 (*non* Meneghini,

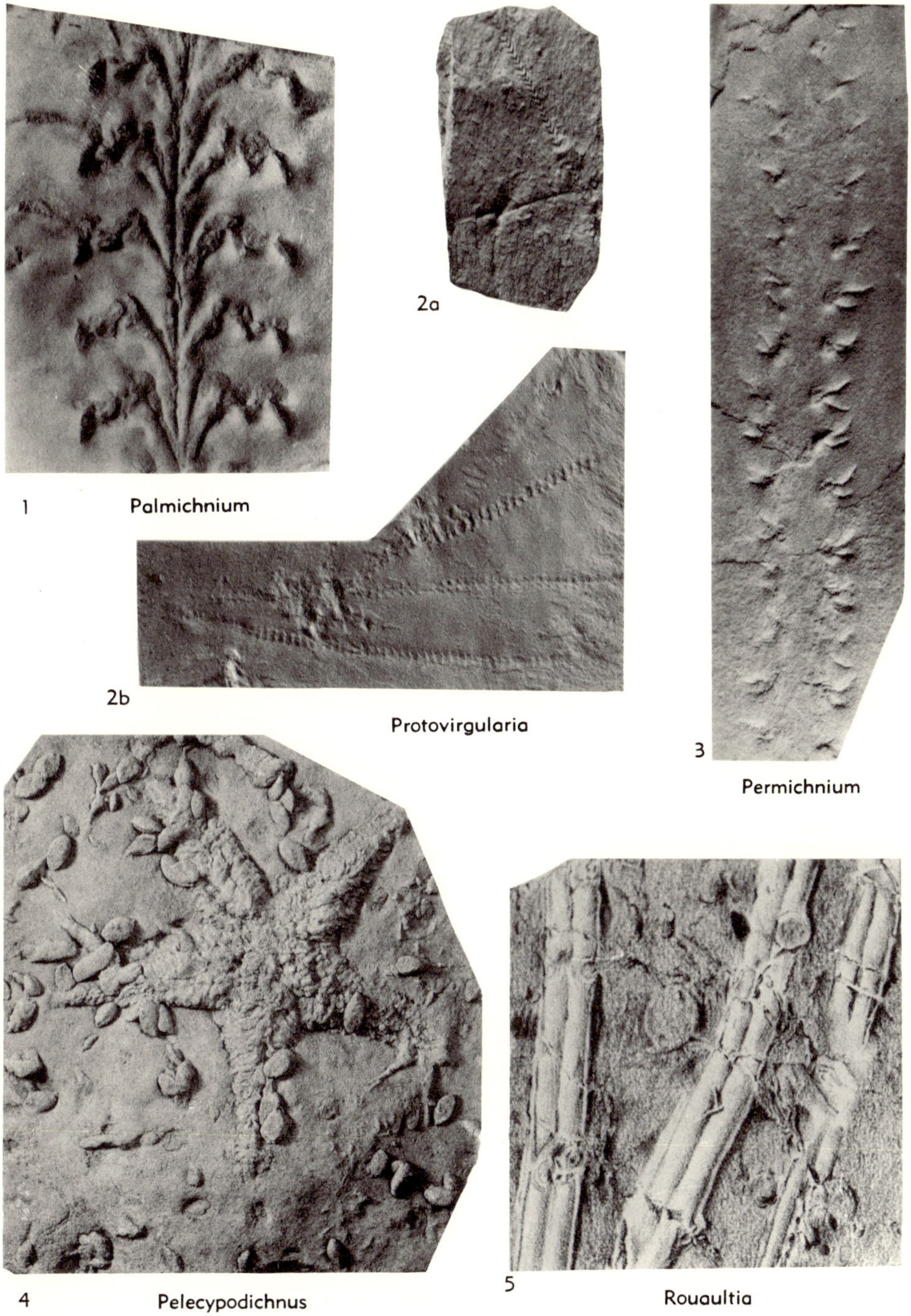

FIG. 130. Trace fossils (p. *W*208, *W*210, *W*212).

1850); *Reticulum* DE STEFANI, 1879; *Eterodictyon* PERUZZI, 1881)]. Elongate, U-shaped loops connected by constructed *Spreite;* frequently branched in large numbers to form antler-shaped systems; similar to *Rhizocorallium* ZENKER, but much smaller and asymmetrical; oblique or parallel to bedding planes (7). *Sil.-L.Dev., ?L.Carb., Jur.-Tert.,* Eu.-?N.Am.——FIG. 129,6. **P. incertum; 6a,* Eoc., Italy; ×2 (Seilacher, n); *6b,* U.Cret., Aus.; ×1 (41).

Phyllodocites GEINITZ, 1867 [**Crossopodia thuringiaca* GEINITZ, 1864]. Sinuous trails consisting of overlapping, somewhat irregularly placed oval depressions on either side of median narrow furrow; width of trail about 1 cm.; very similar to *Nereites* (23). *Paleoz.,* Eu.-N.Am.

Phyllonia HUNDT, 1941 [*nom. nud.;* diagnosis and designation of type-species missing.] (16). *L.Ord.,* Ger.

Phytopsis HALL, 1847 [*non* TOWNSEND, 1915] [**Fucoides demissus* EMMONS, 1842]. Inosculating straight or flexuous tubes, nearly circular in section (about 1 cm. in diam.), with diverging and anastomosing branches; pseudonym "birdseye" (2). [Burrows, according to RAYMOND (1931).] *Ord.,* USA(Ky.-Tenn.-N.Y.).

Plagiogmus ROEDEL, 1929 [**P. arcuatus;* SD HÄNTZSCHEL, herein]. Smooth, flat, concave ribbons, 1.5 to 2 cm. wide, slightly curved, with pronounced transverse bulges at irregular intervals or closely crowded; slight longitudinal furrow (16). *?L. Cam., M.Cam.,* Swed.; *Cam.(Pleist.* drift), Ger. ——FIG. 128,6. **P. arcuatus,* Cam.(Pleist. drift), Ger.; ×0.4 (112).

Planolites NICHOLSON, 1873 [**P. vulgaris;* SD NICHOLSON & ETHERIDGE, 1875]. Fillings of burrows, about 1 cm. wide, penetrating sediment in irregular course and direction; sand and silt of filling brought in by animals (presumably worms) after passing alimentary canal (16). *Precam.-Mesoz.,* cosmop.——FIG. 129,7. *P. montanus,* U. Carb., Ger.; *7a,* ×1; *7b,* transv. sec., ×0.7 (111).

Polykampton OOSTER, 1869 [**P. alpinum*] [*Polycampton* FUCHS, 1895 *(errore)*]. Zigzag-shaped stalk, at angles of which feather-like bunches grow out at both sides with backward directed curvature; externally similar to *Sertularia* (10). [Interpreted by FUCHS as spawn ribbons of prosobranchs; according to SEILACHER (1959), feeding burrow with alternating fanlike feeding fields.] *U.Cret.-L.Tert.,* Eu.——FIG. 129,8. **P. alpinum,* Trias., Switz.; ca. ×0.3 (102).

Protichnites OWEN, 1852 [**P. septemnotatus;* SD HÄNTZSCHEL, herein] [=*Protichnides* CHAPMAN, 1878]. Trackways consisting of trifid impressions, with or without median furrow (1, 23, 41). [Some made by trilobites.] *L.Cam.,* Pak.; *U.Cam.,* N. Am.——FIG. 129,5. **P. septemnotatus,* U.Cam. (Potsdam Ss.), Can.(Que.); track, ×0.3 (134). ——FIG. 131,4. *P. logananus,* U.Cam., USA (N.Y.), trail, ×0.1 (134).

Protopalaeodictyon KSIAZKIEWICZ, 1958 [*nom. nud.,* provisionally published with neither diagnosis nor named species] [=*Protopalaeodictyum* NOWAK, 1959]. Name proposed for initial forms of *Palaeodictyon,* resembling *Belorhaphe* (16). *L. Tert.(Flysch),* Eu.(Pol.).

Protovirgularia M'COY, 1850 [**P. dichotoma* (=?*Cladograpsus nereitarum* REINHOLD RICHTER, 1853)] [*Triplograptus* REINHOLD RICHTER, 1871; *Triplograpsus* GÜMBEL, 1879; *Provirgularia* GÜMBEL, 1879 *(errore)*]. Formerly believed to be octocoral owing to similarity with Recent *Virgularia,* or a graptolite; very probably an earlike trail (group *Ichnia spicea,* RUDOLF RICHTER) (16). [According to NATHORST (28), a trail belonging to the Chordophyceae of SCHIMPER (see *Treatise,* p. V95, p. F231).] *Ord.,* Eng.; *M.Dev.,* Ger.——FIG. 130,2. **P. dichotoma,* U.Ord.(L. Bala beds), Scot.; *2a, 2b,* ×1 (Brighton, n).

Pterichnus HITCHCOCK, 1865 [**Acanthichnus tardigradus* HITCHCOCK, 1858] [=*Acanthichnus* HITCHCOCK, 1858 *(partim)*]. Two rows of tracks, usually quite numerous, turned outward at angle of 15 to 20 degrees from median line; feet lineate, acuminate (18). [?Myriapod trail.] *Trias.,* USA (Mass.).——FIG. 128,2. **P. tardigradus* (HITCHCOCK); ×0.7 (17).

Pteridichnites CLARKE & SWARTZ, 1913 [**P. biseriatus*]. Two rows of small pits bordered by narrow elevated margin; width about 4 mm.; median ridge crenulated; pits nearly equidimensional, alternating in position (16). [Explained as crawling trail of arthropod or annelid.] *U.Dev.,* USA(Md.). ——FIG. 129,1. **P. biseriatus,* Jennings F., ×1 (59).

Pyrophyllites. Starlike ?trace-fossil; cited by DAWSON (5); description not found.

Rhizocorallium ZENKER, 1836 [**R. jenense*] [=*Spongia rhizocorallium* GEINITZ, 1846; ?*Lithochela* GÜMBEL, 1861; *Glossifungites* LOMNICKI, 1886; *Lissonites* DOUVILLÉ, 1908 *(nom. nud.); Cavernaecola* BENTZ, 1929; *Upsiloides* BYRNE & BRANSON, 1941; *Rhizocorallum* SULLIVAN & ÖPIK, 1951 (obj.)]. U-shaped tubes with transverse packing (Ger., *Spreite;* Fr., *traverse*) and nearly parallel legs which are distinctly set off from the vertex arch; tubes relatively thick; mostly oblique or parallel to bedding; preponderantly feeding burrows; prolongation of dwelling tube by removal and redeposition of sediment in vertex; may show lateral flaps and at outer side of tubes scratch markings (indicative of crustaceans); ellipsoidal excrement pills may be incorporated (1, 23, 33). *Cam.-Tert.,* cosmop.——FIG. 129,2. *R.* sp.; *2a,* U.Cret., Fr.; ×0.5 (1); *2b,* model, L.Cam., Pak.; ×0.4 (41).

Rosselia DAHMER, 1937 [**R. socialis*]. Cylindrical, pencil-thick tubes, oblique to bedding, widening cuplike; opening traversed by peg of equal thickness and filled out by rock lamellae fitted into each other concentrically. [According to DAHMER, dwelling burrows; according to SEILACHER (1955), feeding burrows.] *L.Cam., Perm.,* Asia(Pak.); *L.Dev., ?Jur.,* Ger.——FIG. 131,2. **R. socialis,*

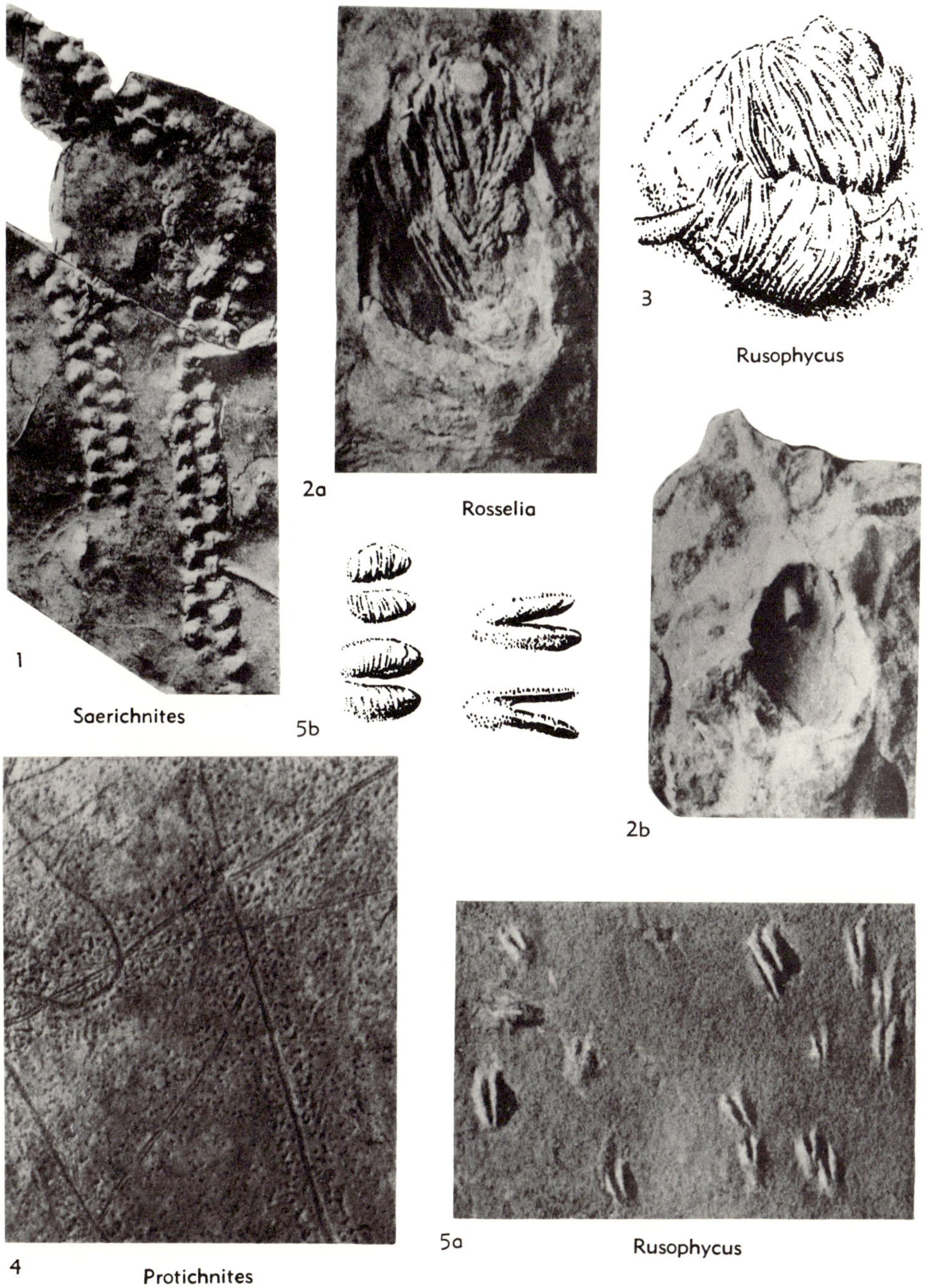

FIG. 131. Trace fossils (p. *W*210-*W*212, *W*215).

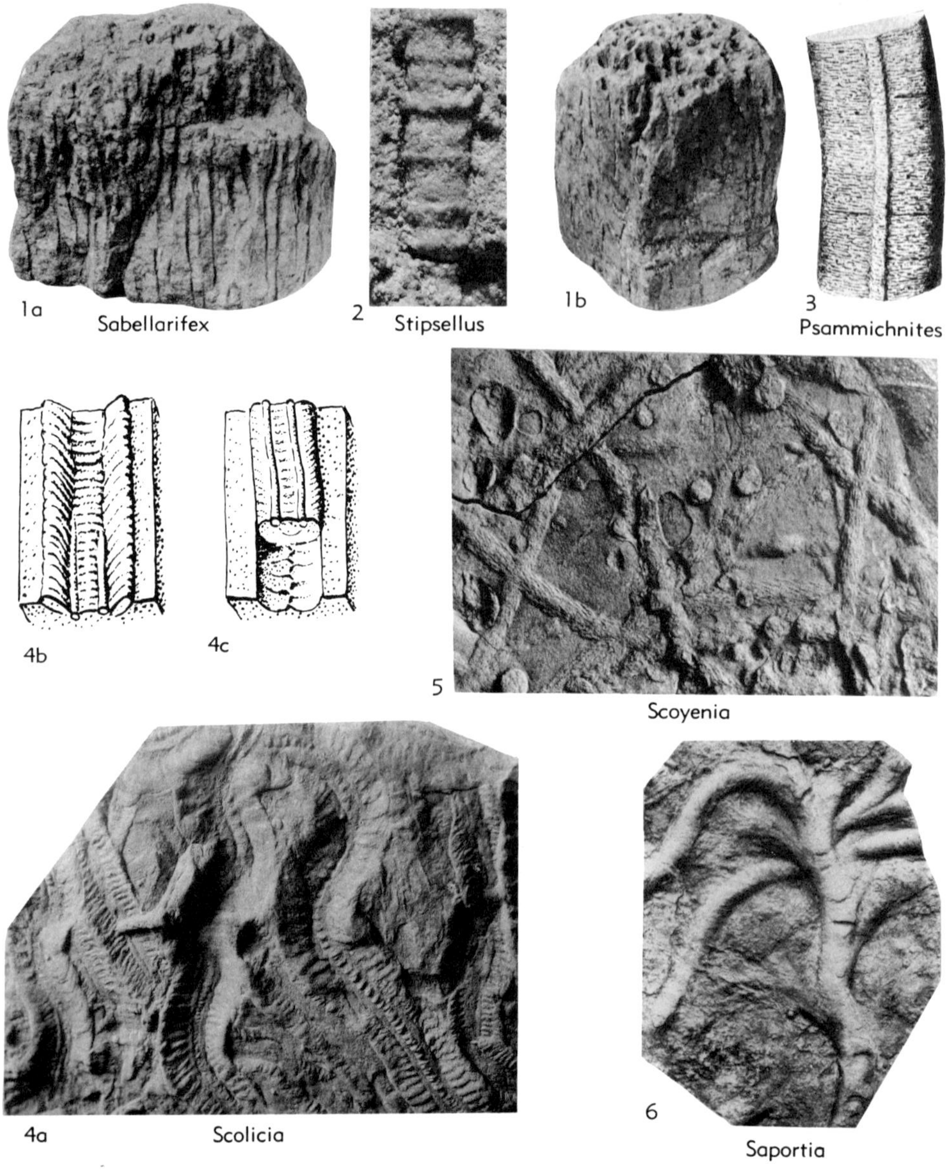

FIG. 132. Trace fossils (p. *W*214-*W*215, *W*218).

L.Dev.(L.Taunus quartzite), Ger.; *2a,* opening, ×0.5; *2b,* upper end of dwelling burrow with opening, ×0.5 (62).

Rouaulita DETROMELIN, 1877 [*non* BELLARDI, 1878] [**Fraena lyelli* ROUAULT, 1850]. Smooth, bilobate crawling trails, up to 1 cm. wide; some with 2 indistinct lateral furrows in addition to median furrow; mostly very long (16, 23). *Ord.,* Fr.-Port.; *?Sil.,* N.Afr.——FIG. 129,*4*. **R. lyelli* (ROUAULT), Ord., Port.; ×0.7 (63).——FIG. 130, *5*. *R. rouaulti* (LEBESCONTE), L.Ord.(Arenig.), Fr.; ×0.75 (92).

Rusophycus HALL, 1852 [**R. bilobatus*] [=*Rhysophycus* EICHWALD, 1860; *Rusichnites* DAWSON, 1864; *Rysophycus* DETROMELIN & LEBESCONTE, 1876; *Rhysophycus* SCHIMPER, 1879; *Rhizophycus* PENEAU, 1946]. Bilobate forms, resembling shape of coffee beans; transversely wrinkled, with deep median groove; some forming beaded rows of ribbons by horizontal repetition (39). [Typical rest-

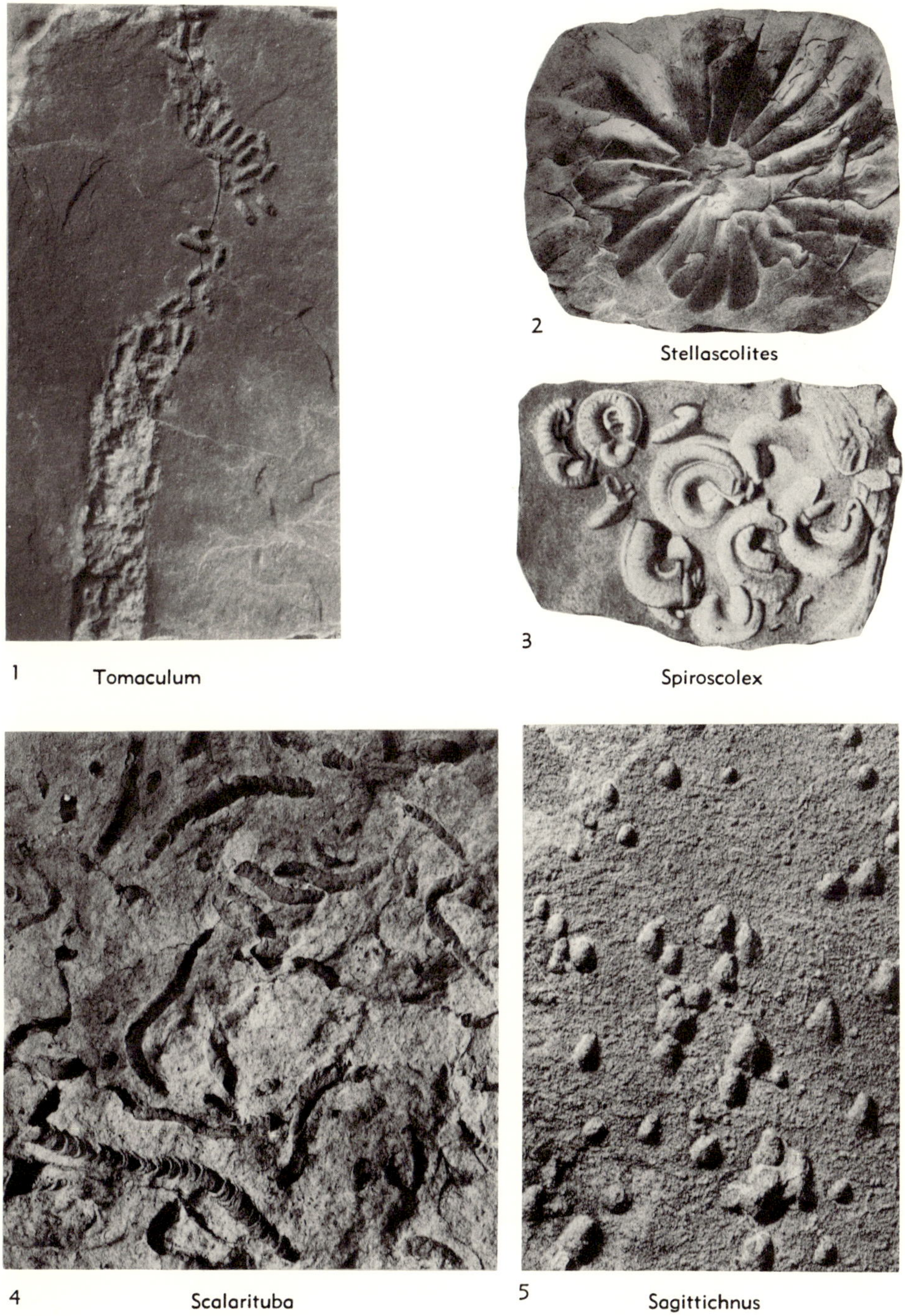

FIG. 133. Trace fossils (p. *W*215-*W*216, *W*218).

ing trail made by trilobites; see also *Isopodichnus* BORNEMANN, 1889.] *Paleoz.*, Eu.-N.Am.-N.Afr.-Asia(Pak.).——FIG. 131,*3*. **R. bilobatus* (VANUXEM), *L.Cam.*, Pak.; ×0.5 (41).——FIG. 131,*5*. *R. didymus* (SALTER), L.Cam., Pak.; *5a*, ×0.5; *5b*, ×1 (41).

Sabellarifex RUDOLF RICHTER, 1921 [**S. eifliensis*] [=*Skolithos* HALDEMAN, 1840; *Sabellarites*

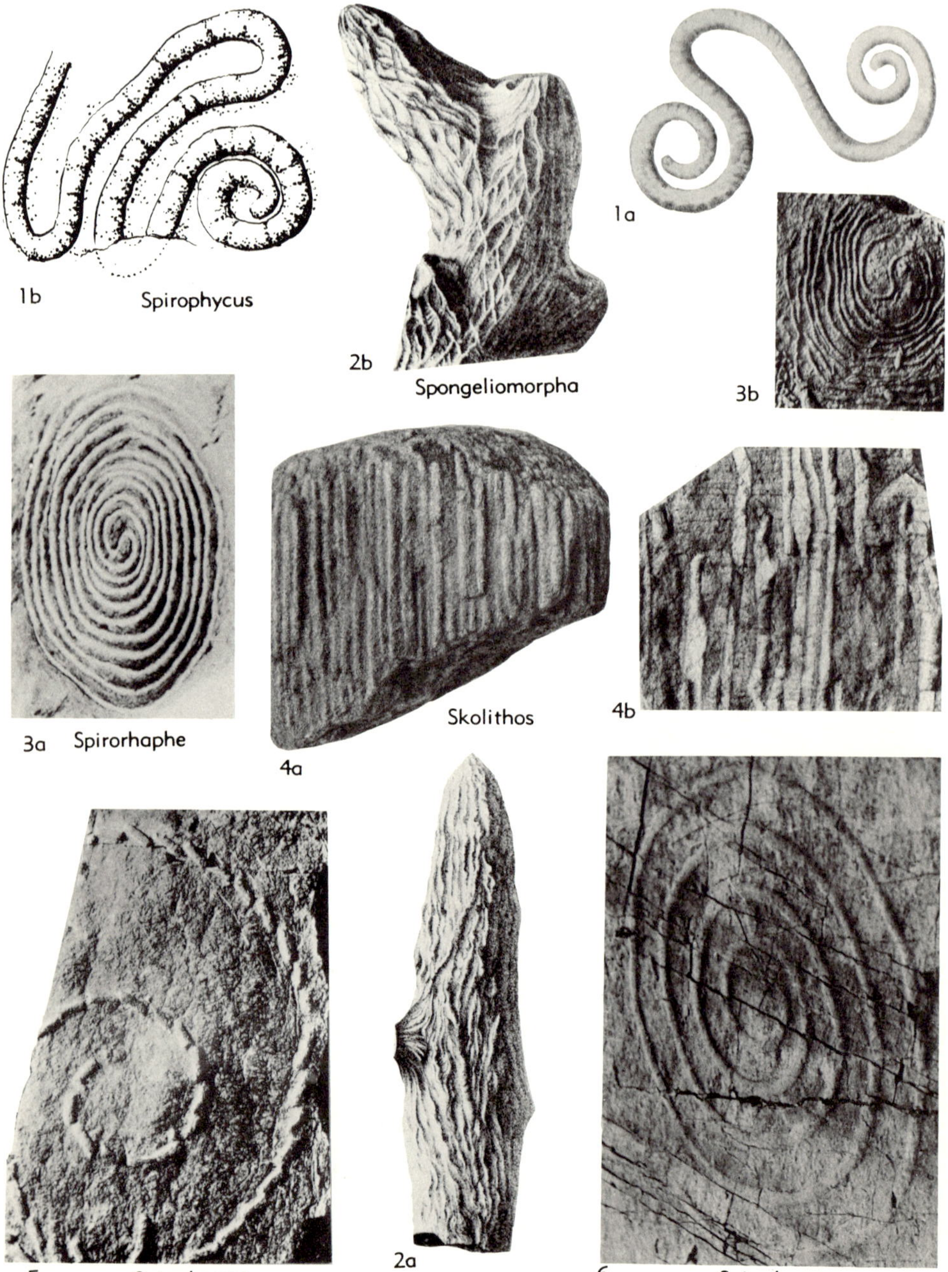

FIG. 134. Trace fossils (p. W215-W216).

RICHTER, 1920 (*non* DAWSON, 1890)]. Like *Skolithos*, but individual tubes less straight and not as crowded (23). *Cam.-L.Dev.*, Ger.-Swed.-?N.Am.——FIG. 132,*1*. **S. eifliensis*, L.Dev., Ger.; *1a*, ×0.65; *1b*, ×0.6 (111).

Sabellarites DAWSON, 1890 [*non* RUDOLF RICHTER, 1920] [**S. trentonensis;* SD HÄNTZSCHEL, herein]. Tortuous tubes, 1 to 3 mm. in diameter; up to 3 cm. in length; walls thick and composed of fragmental material cemented by organic substance; some in groups of 2 or more attached together (5). *Ord.*, Can.

Saerichnites BILLINGS, 1866 [**S. abruptus*]. Two parallel rows of semicircular or subquadrate pits 0.5 in. in diameter, alternating with each other uniformly, somewhat curved in outline on outer margin; anterior and posterior margins nearly straight; bottom nearly flat (16). *Ord.*, Can. (Anticosti).——FIG. 131,*1*. **S. abruptus*, English Head F.; ×0.14 (131).

Sagittarius HITCHCOCK, 1865 [jr. hom.; *non* VOSMAER, 1767; *nec* HERMANN, 1783] [**S. alternans*]. Two parallel rows of delicately curved tracks, with concave sides toward each other, resembling many small bows alternating with one another (18). [Insect trail.] *Trias.*, USA(Mass.).——FIG. 129, *3*. **S. alternans;* ×0.7 (18).

Sagittichnus SEILACHER, 1953 [**S. lincki*]. Resting trails suggestive of arrowheads with median keel; up to 5 mm. long; occurring in masses and equally oriented rheotactically (39). [Producer unknown, belonging to epipsammonts.] *U.Trias.*, Ger.——FIG. 133,*5*. **S. lincki*, M.Keuper; ×2 (39).

Saportia SQUINABOL, 1891 [**Zonarides striatus* SQUINABOL, 1887]. Large passageways, commonly branching dichotomously; surface with rhombic pattern produced by delicate arched parallel striations in 2 systems (2, 46). *Tert.*, Italy.——FIG. 132,*6*. **S. striata*, Flysch; ×0.3 (124).

Scalarituba WELLER, 1899 [**S. missouriensis*]. Subcylindrical burrows, 2 to 4 mm. in diameter, curving in all directions, marked by transverse ridges situated at distances of 1 to 2 mm. (16). *L.Miss.*, USA(Mo.).——FIG. 133,*4*. **S. missouriensis*, Kinderhook.; ×0.8 (Häntzschel, n).

Scolicia DEQUATREFAGES, 1849 [**S. prisca*]. Used for various trails presumably made by gastropods; typical is flattened ribbon-like shape with peculiar "gill-like" transverse structures produced by repeated displacement of sediment; longitudinal furrows in varied arrangement may occur; some differences may have surface or subsurface origin, as shown by GÖTZINGER & BECKER (1932) (41). [The following "genera" belong to this group but are not classifiable as synonyms: *Nemertilites* MENEGHINI, 1850; *Nereiserpula* STOPPANI, 1857; *Psammichnites* TORELL, 1870; *Cymaderma* DUNS, 1877; *?Phyllochorda* SCHIMPER, 1879; *Bolonia* MEUNIER, 1886; *Tetraichnites* DESTEFANI, 1895; *Curvolithus* FRITSCH, 1908; *Scolithia* KINDELAN, 1919 *(errore); Palaeobullia* GÖTZINGER & BECKER, 1932; *Olivellites* FENTON & FENTON, 1937.] *Cam.-Tert.*, Eu.-N.Afr.-N.Am.-Asia(Pak.).——FIG. 132, *4*. **S. prisca*, Eoc.(Flysch), Aus., Italy; *4a*, ×0.3 (1); *4b*, *4c*, models, ×0.4 (41); *4a*, *4b*=upper side forms=*Palaeobullia* GÖTZINGER & BECKER; *4c*=lower side form=*Subphyllochorda* GÖTZINGER & BECKER).——FIG. 135,*3*. *Olivellites plummeri* FENTON & FENTON, Penn. (Cisco), Tex.; ×0.6 (Bur. Econ. Geol., Austin).——FIG. 132,*3*. *Psammichnites gigas* (TORELL), L.Cam., S.Swed.; ×0.7 (129).

Scoyenia WHITE, 1929 [**S. gracilis*]. Slender rope-like remains; 2 to 5 mm. in diameter; in half relief or flattened; linear and commonly curved; densely clothed with closely appressed, tapering, acute, bract- or leaflike appendages; resembling lycopod such as *Selaginella;* obviously a trail (16). *Perm.*, Ger.-Fr.-USA(Ariz.).——FIG. 132,*5*. **S. gracilis*, Hermit Sh., Ariz., ca. ×0.7 (138).

Siphonites SAPORTA, 1872 [**S. heberti*]. Tubes about 1 cm. in diameter with sandy lining, mostly washed out and collapsed on bedding planes (2). *U.Trias.(Rhaet.)*, Fr.——FIG. 135,*4*. **S. heberti*, L.Lias.; ×0.35 (Laugier, n).

Skolithos HALDEMAN, 1840 [**Fucoides ?linearis*] [=*Tubulites* ROGERS, 1838 *(nom. nud.)* (*non* GESNER, 1758); *Scolithus* HALL, 1847 (and most later authors dealing with this "genus"); *Scolecolithus* ROEMER, 1848; *Scolecolithus* GOEPPERT, 1852; *Scolites* SALTER, 1857; *?Haughtonia* KINAHAN, 1858; *?Scolecites* SALTER, 1873 *(partim)* *(?pro Scolites)*]. Tubes or tube fillings standing vertically in sandstones; diameter about 0.2 to 1 cm.; usually straight, never branched; commonly but not always closely crowded; rarely with fine annulations (1, 23). [Made by worms or phoronids.] *Cam.-Ord.*, Eu.-Am.-Greenl.-Tasm.——FIG. 134,*4*. **S. linearis*, L.Cam.; *4a*, Swed.(Öland), ×0.6; *4b*, Swed., ×0.5 (136).

Spirochorda SCHIMPER, 1879 [**Dictyota spiralis* LUDWIG, 1869]. Possibly braided trail (2). *U.Dev.*, Ger.

Spirodesmos ANDRÉE, 1920 [**S. interruptus*]. Large spiral consisting of individual parts which usually are closely packed; individual parts 2 to 3 cm. long and 7 to 8 mm. wide; in outer coils parts are displaced toward interior with respect to each other; possibly part of large double spiral such as *Spirophycus* (16). *L.Carb.*, Ger.——FIG. 134,*5*. **S. interruptus*, Culm; ×0.17 (48).——FIG. 134, *6*. *S. archimedeus* HUCKRIEDE, Culm; ×0.2 (86).

Spirophycus HÄNTZSCHEL, *nom. nov.* [*pro Ceratophycus* SCHIMPER, 1879[1] (*non* FISCHER DE WALD-

---

[1] SCHIMPER, W. P. in ZITTEL, K. A. VON, 1890: *Handbuch der Palaeontologie*, pt. 2 (Palaeophytologie), p. 59 (München, Leipzig). The section of this work dealing with the Thallophyta was first issued in 1879.

HEIM, 1824)] [*Münsteria bicornis* HEER, 1877; SD HÄNTZSCHEL, herein (=*Münsteria caprina* HEER, 1877; *M. involutissima* SACCO, 1888)]. Transversely folded or rugose cylindrical bulges, curved like horns or bent spirally at the ends; 1 to 2 cm. thick (2, 16). *Cret.-L.Tert.(Flysch)*, Eu.——FIG. 134,*1*. **S. bicornis; 1a,* Switz., ca. ×0.3; *1b,* Aus., ca. ×0.4 (*1a,* 84; *1b,* 41).

Spirorhaphe FUCHS, 1895 [=*Gilbertina* ULRICH, 1904 (*non* MORLET, 1888; *nec* JORDAN & STARKS, 1895); *Spiroraphe* ABEL, 1935 (*non* PERNER, 1907)]. Spirally coiled threads, turning in center with loop and running back between primary coils (10). *Cret.-Tert.,* Eu.-Alaska.——FIG. 134,*3*. *S.* sp.; *3a, "Gilbertina,"* U.Cret.(Yakutat F.), Alaska; ×0.7 (44); *3b,* Flysch, Aus.; ca. ×0.3 (1).

Spiroscolex TORELL, 1870 [**Arenicolites spiralis* TORELL, 1868]. Transversely ribbed, strongly curved, spiral structures 2 cm. in diameter; transverse ribs slightly elevated (16). *Cam.;* Swed.-Est.——FIG. 133,*3*. **S. spiralis* (TORELL), Swed.; ×0.5 (134).

Spongeliomorpha DE SAPORTA, 1887 [**S. iberica*] [=*Spongiliomorpha* DARDER, 1945 *(errore)*.]

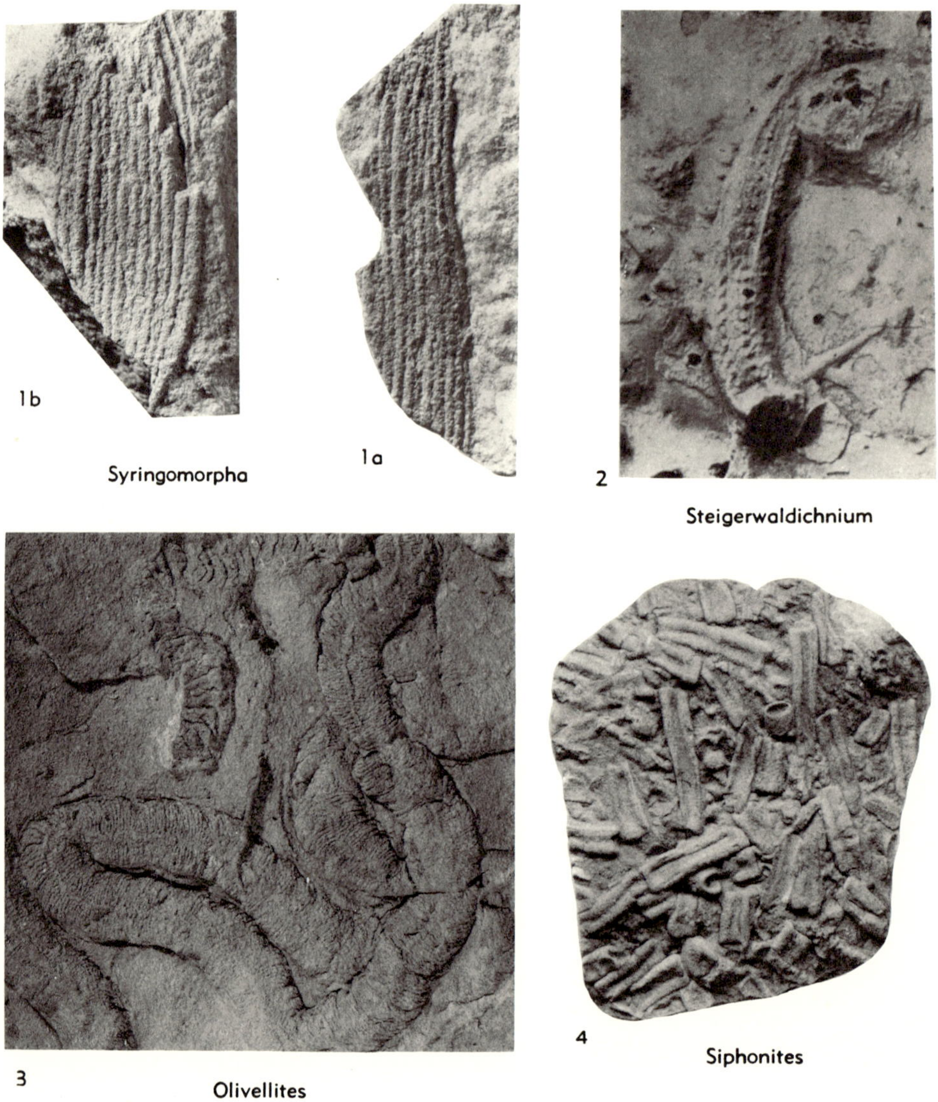

FIG. 135. Trace fossils (p. W215, W218).

Thick, elongate bodies suggestive of antlers; with ramifications and lateral tapering offshoots; surface with network of scratching traces crossing each other at acute angles (2). [Commonly regarded as sponges (*Treatise,* p. *E*36); burrows according to REIS (1922).] *Trias.-Tert.,* Eu.-?N. Am.——FIG. 134,*2*. **S. iberica,* ?Tert., Sp.; *2a, 2b,* ×0.7 (114).

"Spongia ottoi" GEINITZ, 1849. Starlike trails with elevated center; about 5 cm. in diameter; gen-

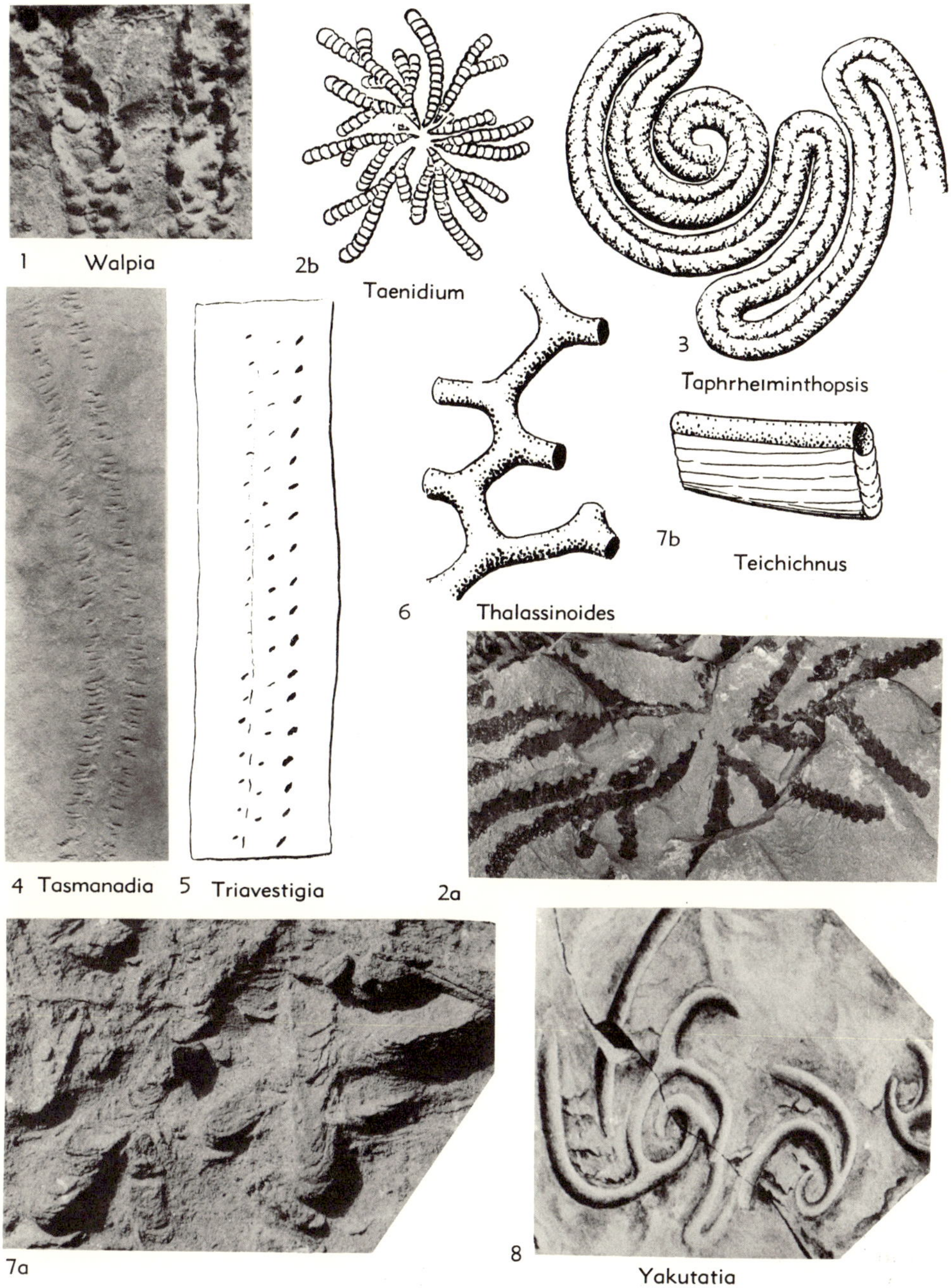

FIG. 136. Trace fossils (p. *W*218-*W*220).

erally 6 to 10 radiating grooves, rather irregularly and commonly only unilaterally developed; formerly described as sponge similar to *Peronidella furcata* (GOLDFUSS) (16). [Surely not belonging to Recent genus *Spongia* LINNÉ; probably feeding burrows made by crustaceans or worms.] *U.Cret. (Cenom.)*, Ger.(Saxony).

Steigerwaldichnium KUHN, 1937 [**S. heimi*]. Straight, rarely curved tunnel traces parallel to bedding with distinct longitudinal rows of tiny projections and impressions from doubtful parapodia (16). [Made by a polychaete.] *U.Trias.*, Ger.——FIG. 135,*2*. **S. heimi*, M.Keuper; ca. ×1.5 (90).

Stellascolites ETHERIDGE, 1876 [**S. radiatus*]. Radiate or stellate disclike impression with 16 rays of nearly equal length radiating from central round space, becoming broader at their extremities which are not clearly defined; diameter 20 to 25 cm. (16). *Ord.*, Eng.——FIG. 133,2. **S. radiatus;* ×0.17 (67).

Stipsellus HOWELL, 1957 [**S. annulatus*] [=*Stripsellus* HOWELL, 1957 *(errore)*]. Perpendicular, cylindrical burrows, spaced about 2 cm. apart in sediment; differing from *Skolithos* by distinct ringlike expanded belts regularly distributed throughout their length; diameter about 1 cm. (16). [Perhaps identical with *Trachyderma serrata* SALTER, 1864.] *Cam.*, USA(Ariz.); *?Penn.*, USA(Md.)-?Arabia.——FIG. 132,2. **S. annulatus*, Cam.(Tapeats Ss.), Ariz.; ×1 (85).

Syringomorpha NATHORST, 1886 [**Cordaites? nilssoni* TORELL, 1868; SD NATHORST, 1886]. Rollerlike sticks several cm. in length and 1 to 2 mm. in thickness lying close together; slightly arched; touching each other along whole length and forming complete slab; occurring in large numbers independent of bedding (2, 23). [Possibly seaweed; work of gregarious worms on flat substratum, according to RUDOLF RICHTER (1927).] *L. Cam.*, Swed.; *Pleist.* (drift), Ger.——FIG. 135,*1*. **S. nilssoni* (TORELL); *1a, 1b*, ×1 (111).

Taenidium HEER, 1877 [**T. serpentinum;* SD HÄNTZSCHEL, herein] [=*Münsteria* STERNBERG, 1833 *(partim); Caulinites catuli* MASSALONGO, 1858; *Eione* TATE, 1859 (*non* RAFINESQUE, 1814; *nec* RISSO, 1826); *?Volubilites* VON LIBURNAU, 1901; *Pseudocrinus* ANELLI, 1935 (*non* GEINITZ, 1846); *Notaculites* KOBAYASHI, 1945; *Scolecocoprus* BRADY, 1947; *Tebagacolites* MATHIEU, 1949; *?Rhizocorallites* MÜLLER, 1955; *Toenidium* LESSERTISSEUR, 1955 *(errore)*]. Cylindrical tunnel fillings with segmentation reminiscent of *Orthoceras;* segmentation may also be indicated on outside by annular constrictions (46). [Interpreted as periodic filling of tunnel in backward direction.] *?Carb., Perm.-Tert.*, Eu.-N.Am.-Jap.-Antarct.——FIG. 136,*2*. *T.* sp., U.Cret. (Flysch), Aus.; *2a*, ×0.7; *2b*, ×0.27 (*2a*, Papp, n; *2b*, 41).

Taphrhelminthopsis SACCO, 1888 [**T. auricularis;* SD HÄNTZSCHEL, herein]. Meandering bilobate trails, similar to *Scolicia* DE QUATREFAGES (41), but in tightly coiled spirals or meanders (2). *Tert.*, Eu.——FIG. 136,*3*. **T. auricularis*, Flysch, Italy; ×0.2 (41).

Tasmanadia CHAPMAN, 1929 [**T. twelvetreesi*]. Double row of very sharp transverse imprints, mostly single but some joined internally or rarely externally to form bifid impression (16). [Sets of imprints explained by CHAPMAN as bristles of polychaete worm, but GLAESSNER (1957) conclusively proved them to be arthropod trails.] *Cam.*, Austral. (Tasm.).——FIG. 136,*4*. **T. twelvetreesi;* part of holotype, ×0.8 (74).

Teichichnus SEILACHER, 1955 [**T. rectus*]. *Spreiten-Bauten* formed by vertical (mostly upward) displacement of horizontal burrows; somewhat flexuous; rarely branched; feeding burrows (41). *L. Cam.*, Asia(Pak.); *M.Trias.-L.Jur.*, Ger.; *Tert.*, Belg.——FIG. 136,*7*. **T. rectus*, L.Cam.(*Neobolus* Ss.), Pak. (Salt Range); *7a*, ×0.7; *7b*, model, ×0.4 (41).

Teratichnus MILLER, 1880 [**T. confertus*]. Trackway of numerous narrow, somewhat rarely bifurcated impressions, arranged in oblique, asymmetrical sets (23). [Probably individual variation of trilobite track.] *Ord.*, USA(Ohio).

Thalassinoides EHRENBERG, 1944 [**T. callianassae*] [=*Spongites saxonicus* GEINITZ, 1842 *(nom. nud.); Cylindrites spongioides* GOEPPERT, 1841 *(nom. nud.); ?Aschemonia* DETTMER, 1915; *Vomacispongites* DE LAUBENFELS, 1955]. Branched burrows and tunnel systems, forkings mostly Y-shaped, without special surface ornamentation, commonly widened to form pear-shaped cavities (16). [Produced by decapod crustaceans.] *Trias.-Tert.*, Eu.-Asia.——FIG. 136,*6*. *T.* sp., Mio. (Meeres-Molasse), Switz.; ca. ×0.07 (41).

Tigillites ROUAULT, 1850 [**T. dufrenoyi;* SD HÄNTZSCHEL, herein] [=*Foralites* ROUAULT, 1850; *Monocraterion* TORELL, 1870; *Lepocraterion* STEHMANN, 1934; *non Tigillites habichi* LISSON, 1904]. Simple vertical burrows without special lining, smooth or regularly annulated; openings may be funnel-shaped, not crowded like *Skolithos* (23). *Cam.-Jur.*, Eu.-N.Am.-Arabia.

Tisoa DE SERRES, 1840 [**T. siphonalis*] [=*?Tissoa* REYNES, 1868 *(errore)*]. Two vertical cylindrical tubes 2 to 30 mm. in diameter, lying about 1 to 15 mm. apart, forming axis of long nodules reaching 1 m. or more in length; not branched; uncertain whether lower ends unite as 2 limbs of U-shaped burrow (23). *Jur.(Lias.)*, Fr.-Madagascar; *L.Cret.*, USSR; *Oligo.*, Afr.(Tunisia).——FIG. 137, *4*. **T. siphonalis*, L.Jur.(Lias), Fr.; ca. ×0.7 (121).

Tomaculum GROOM, 1902 [**T. problematicum*] [=*Syncoprulus* RICHTER & RICHTER, 1939].

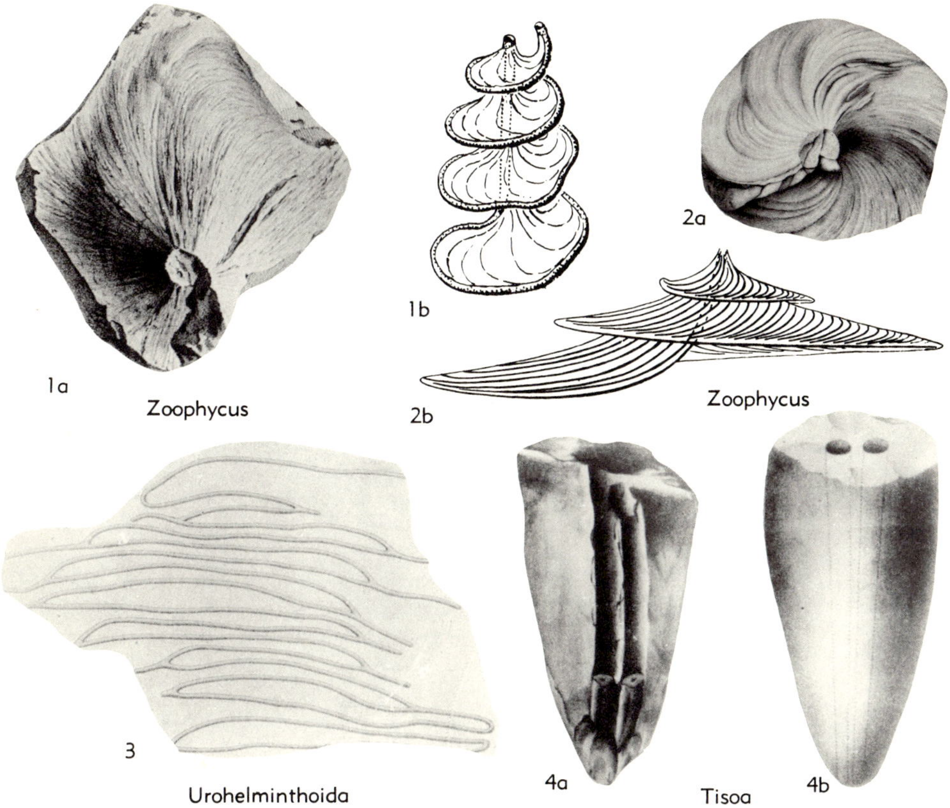

FIG. 137. Trace fossils (p. *W*218-*W*220).

Strands of elliptical fecal pellets (=*Coprulus*) up to 10 cm. long and 1 to 2 cm. broad; lying on bedding planes; within strands pellets commonly lumped together in clusters (16). *Ord.*, Eu.——FIG. 133,*1*. *T. problematicum*, Herscheid slates, Ger.; ×2.5 (111).

Trachomatichnus MILLER, 1880 [**T. numerosus*; SD HÄNTZSCHEL, herein]. Trackway consisting of 2 rows of numerous, simple or beaded impressions (16). [Believed by some to have been made by cephalopods, but probably one of numerous variations of trilobite tracks.] *Ord.*, USA(Ohio).

Triadonereites MAYER, 1954 [**T. mesotriadica*]. Believed to be burrows of *Triadonereis* MAYER, 1954 (16). *M.Trias.*, Ger.

Triavestigia GILMORE, 1927 [**T. niningeri*]. Continuous trail of 3 parallel rows of footlike impressions, between 2 of which is faintly impressed tail track; longer axes of foot markings placed slightly diagonal to direction of movement, alternating; feet seemingly unidactyl; somewhat similar to *Bifurculapes* HITCHCOCK (11). *Perm.*, USA(Ariz.).——FIG. 136,*5*. **T. niningeri*, Coconino Ss.; ×.3 (11).

Trichophycus MILLER & DYER, 1878 [**T. lanosus*]. According to JAMES (1884), rill marks identical to *Blastophycus* MILLER & DYER, 1878; probably inorganic, according to NATHORST (1881). SEILACHER (personal communication, 1955) holds that they are burrows with delicately scraped walls, a conclusion with which the writer agrees (19). *Ord.*, USA(Ohio).

Trisulcus HITCHCOCK, 1865 [**T. laqueatus*]. Sinuous track, consisting of 3 continuous grooves with intermediate ridges; sometimes showing slight protuberances like those of *Sphaerapus* (18). *Trias.*, USA(Mass.).

Trypanites MÄGDEFRAU, 1932 [**T. weisei*]. Straight bore tunnels, usually vertical, 1 to 2 mm. wide, without ramifications, closely spaced (up to 12 per square cm.); some with excrement of producer (16). *Sil.*, USSR [*Pleist.* drift, Ger.]; *M. Trias.(Muschelkalk)*, Ger.

Tubulites H. D. ROGERS, 1838 [*nom. nud.*, provided for *Skolithos*, not published; preoccupied by *Tubulites* GESNER, 1758].

Urohelminthoida SACCO, 1888 [**Helminthoida appendiculata* HEER, 1877; SD HÄNTZSCHEL, herein]

[=*Hercorhaphe* FUCHS, 1895]. Threadlike reliefs forming broad meanders with tail-like appendage at each turn (10). *Cret.-L.Tert.(Flysch)*, Eu.——FIG. 137,*3*. **U. appendiculata* (HEER), Switz.; ×0.3 (84).

Walpia WHITE, 1929 [**W. hermitensis*]. Mined tunnels lined with flattened, lenticular, smooth pellicles of rather leathery texture; irregularly crowded or imbricated; probably representing excrement backed against walls of tunnel; somewhat similar to *Ophiomorpha* (16). [Possibly made by crustaceans.] *Perm.*, USA(Ariz.).——FIG. 136,*1*. **W. hermitensis*, Hermit Sh.; ×0.9 (138).

Yakutatia HÄNTZSCHEL *nom. nov.* [*pro Gyrodendron* ULRICH, 1904 (*non* QUENSTEDT, 1880)] [**Gyrodendron emersoni* ULRICH]. Cylindrical bodies, varying in thickness from 2 to 6 mm.; bifurcating 1 to 3 times, forming 1 to 1.7 volutions about acuminate inner extremity; outer end obtuse (44). *U.Cret.*(*Yakutat* F.), Alaska——FIG. 136,*8*. **Y. emersoni* (ULRICH); ×0.5 (44).

Zonarites STERNBERG, 1833 [jr. hom., *non Zonarites* RAFINESQUE, 1831] [**Fucoides flabellaris* BRONGNIART, 1823; SD ANDREWS, 1955] [Probably= *Zonarides striatus* SQUINABOL, 1887 (*Saportia* SQUINABOL, 1891), as well as plants (e.g., *Z. digitatus* STERNBERG, 1833, =*Zonarides* SCHIMPER, 1869)]. "Genus" comprising starlike trace fossils (e.g., *Z. alcicornis* FISCHER-OOSTER, 1858) (2). [According to SEILACHER (1955) branched feeding burrows with faecal pellets stuffed transversely into them.] *?Perm.*, *Tert.*, Eu.

Zoophycos MASSALONGO, 1855 [**Fucoides brianteus*, VILLA, 1844] [=?*Umbellularia longimana* FISCHER DE WALDHEIM, 1811; *Chondrites scoparius* THIOLLIÈRE, 1858; *Taonurus* FISCHER-OOSTER, 1858; *Spirophyton* HALL, 1863; ?*Sagminaria* TRAUTSCHOLD, 1867; *Alectorurus, Physophycus, Zoophycus* SCHIMPER, 1869; *Cancellophycus* SAPORTA, 1873; *Glossophycus* SAPORTA & MARION, 1881; ?*Flabellophycus* SQUINABOL, 1890; ?*Myelophycus* ULRICH, 1904; *Physiophycus* FRITEL, 1925; *Zoophicos* WASSOJEWITSCH, 1953; *Zoophycus* LESSERTISSEUR, 1955]. Variously shaped *Spreiten* structures with thin tube and large but variable radius of curvature; without strict separation of legs and vertex (thus unlike *Rhizocorallium*); *Spreite* comprise thin slab of varied outline, in part screw-shaped (23). [Perhaps made by worms; all forms are feeding burrows.] *Dev.-Tert.*, Eu.-N.Am.-Afr.——FIG. 137,*1*. **Z. brianteus* (VILLA), *1a*, Eoc., Italy; ×0.4 (99); *1b*, schem. drawing after Tert. Ital. specimen (115).——FIG. 137,*2*. *Z. crassus* (HALL) ["*Spirophyton crassum*" HALL], U.Dev., USA; *2a*, ca. ×0.3 (79); *2b*, schem. drawing (41).

DOUBTFULLY DISTINGUISHED TRACE FOSSILS

Algites SEWARD, 1894 [*emend.* STOPES, 1913]. Seldom used, comprehensive generic name given to replace all older generic names of "algae" which suggest relationship with living forms. Generally =*Chondrites* STERNBERG, 1838.

Apodichnites FITCH, 1850 [*non* HITCHCOCK, 1869]. Suggested as "new order" including all sorts of "footless" tracks (e.g., *Helminthoidichnites* FITCH, 1850).

Coprolithus PAREJAS, 1948. Informal name used for coprolites of crustaceans, obviously not thought of as designation of "genus," although several "species" have been erected and described by PAREJAS (16). *U.Jur.*, Eu.(Switz.)-AsiaM.(Turk.)

Coprulus RICHTER & RICHTER, 1939. Mechanical-ecological subsidiary name for excrement in form of isolated, loose pills; not considered generic name, but sometimes used as such (16).

Dipodichnites HITCHCOCK, 1841. Name proposed for an "order" including tracks of biped animals.

"Feather-stitch trail" WILSON, 1948. Straight or curved burrows in form of zigzag feather-stitch pattern (16). *M.Ord.(Trenton)*, Can.; *L.Jur.*, S. Ger.——FIG. 121,*7*. "Feather-stitch trail" WILSON, schem. drawing.

Graphoglypten FUCHS, 1895 [=*Hieroglyphs* s.s., FUCHS, 1895] Trace fossils appearing as reliefs on lower surface of beds (mostly sandstones) and resembling ornaments or letters (e.g., *Paleodictyon, Paleomeandron*), indicative of Flysch sediments. [Explained by FUCHS, 1895, as strings of spawn of gastropods.]

Helminthites SALTER, 1857 [=*Helmintholites* MURCHISON, 1867; *Helmintholithes* ETHERIDGE, 1881]. Name proposed for long, sinuous surface trails or filled-up burrows of marine worms without impressions of lateral appendages; not used as generic name; published without designation of "species" (16).

Ichnites (or **Ichnytes**) VINASSA DA REGNY, 1904. Not a "genus"; general designation for various trails; for instance, applied by OPPEL (1862) and WINKLER (1886) to trails from the Upper Jurassic Solnhofen Limestone (Bavaria).

Ichnium [*non Ichnium* PABST, 1896 (amphib.), *nec* SOLLAS, 1900 (?worm)]. General formal designation of a trail; not a "genus"; used in connection with "species" designations for various vertebrate and invertebrate trails.

Nucleocavia RICHTER & RICHTER, 1930. General name (not generic) for small, usually winding canals which generally occur in form of furrows on surface of *steinkerns*. [Producers are worms, arthropods, and other animal groups.]

Polypodichnites HITCHCOCK, 1841. Name proposed for an "order" including tracks of animals with more than 4 feet.

Rhabdoglyphen FUCHS, 1895. General and informal name for nearly straight bulges, mostly on undersurface of sandstone beds of Flysch and similar sediments; greatest diameter several centimeters; *Rhabdoglyphus* used by WASSOJEWITSCH

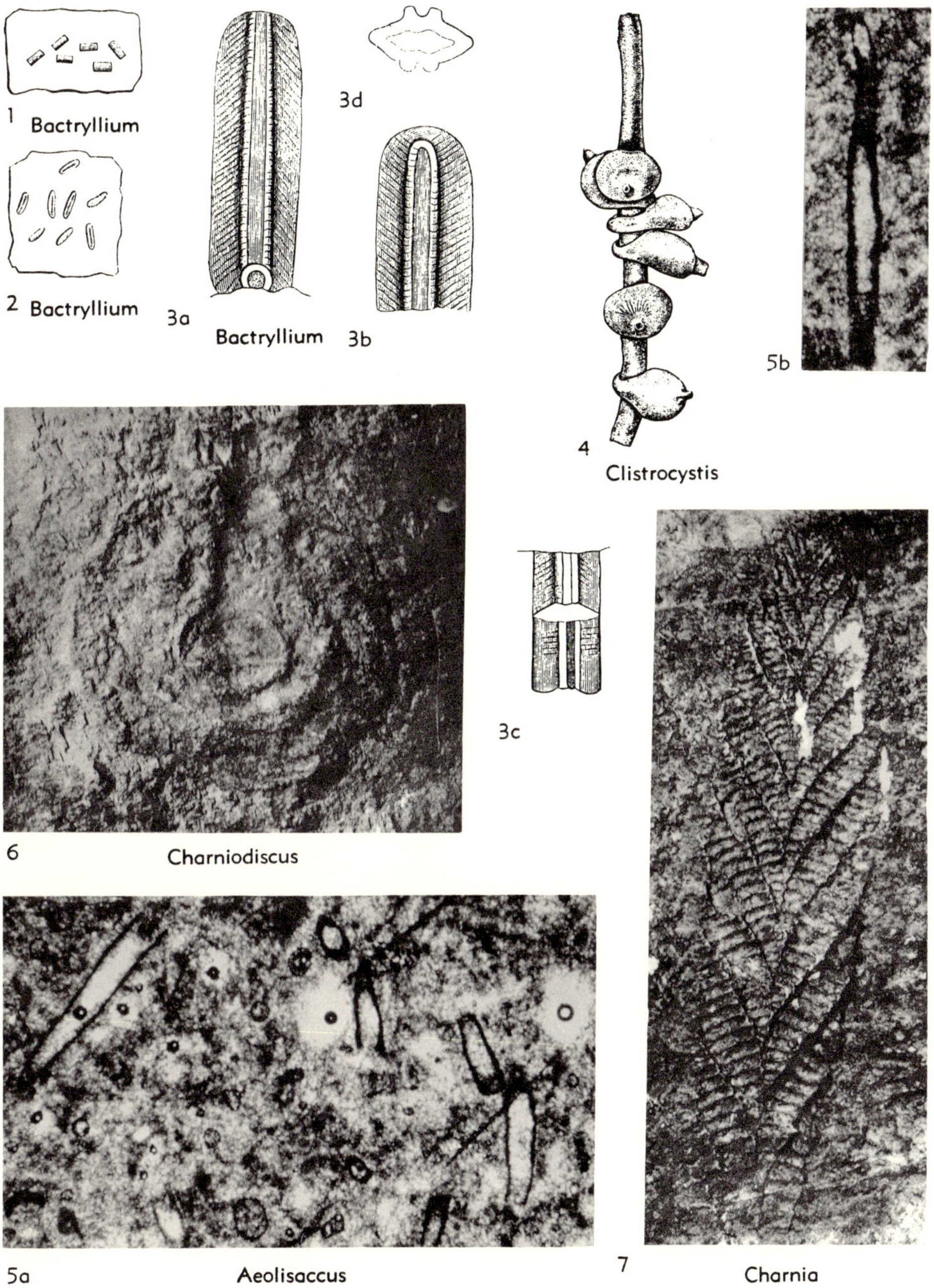

FIG. 138. Body fossils (p. *W*223).

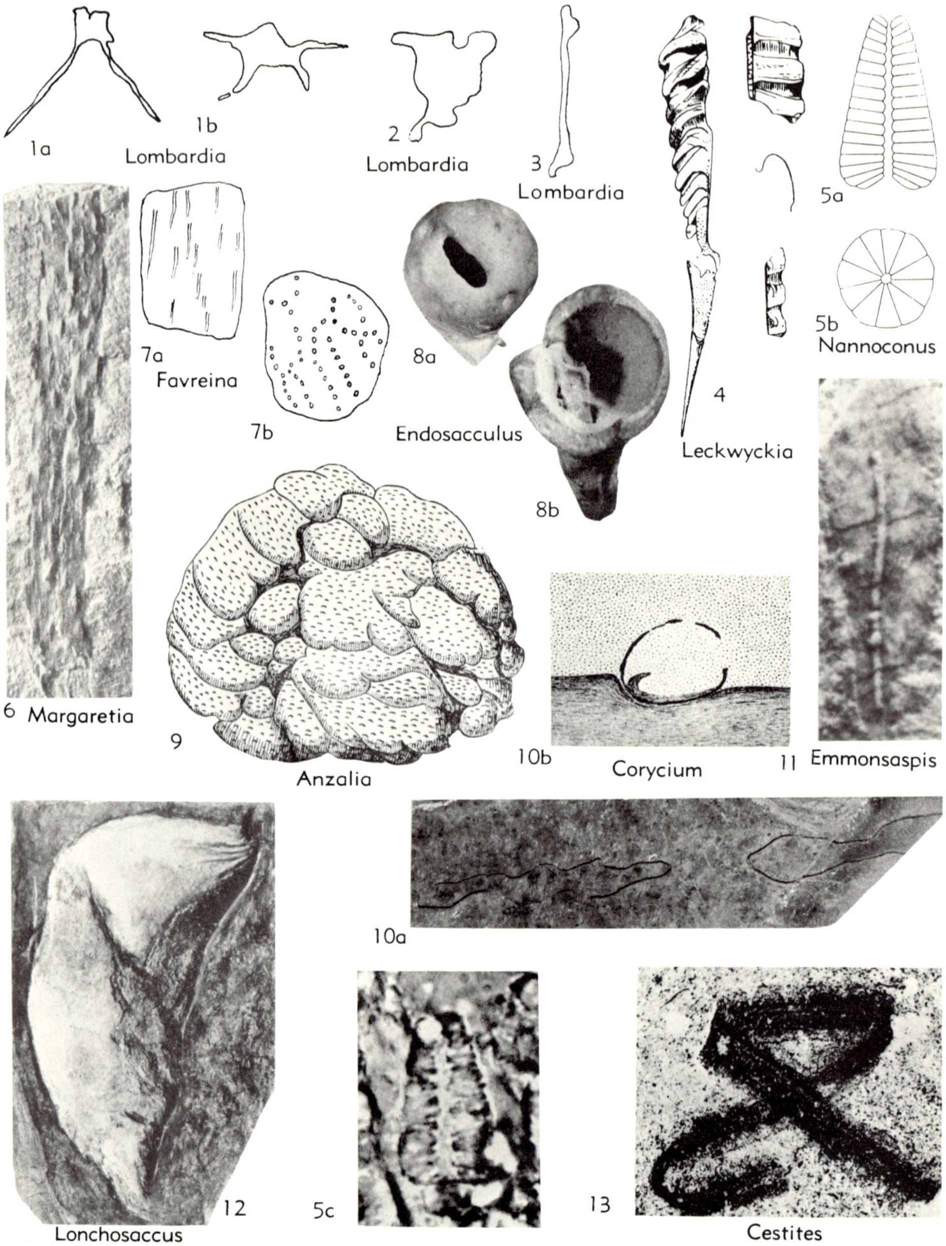

FIG. 139. Body fossils (p. W223-W225).

(1933), with the "species" *R. grossheimi* from the Flysch of USSR, is invalid owing to absence of diagnosis of the "genus."

Tetrapodichnites HITCHCOCK, 1841. Name proposed for "order" containing tracks of quadruped animals.

Vermiglyphen FUCHS, 1895. Collective name for threadlike, straight or variously winding reliefs on undersurface of sandstone beds in Flysch and similar sediments; mostly unbranched; width usually only a few mm.

BODY FOSSILS

This chapter contains "genera" of doubtful or completely uncertain classificatory status. Frequently they have been described only once and have never been discussed again. Additional "genera" of this type may be found in the sections on "unrecognizable genera" in this and in other parts of the *Treatise*. The forms listed in the parts of the *Treatise* published before this manuscript was finished have generally not been listed here again. The larger groups of doubtful affinities such as conodonts and hyolithids, have been dealt with by other authors.

Aeolisaccus Elliott, 1958 [**A. dunningtoni*]. Small thin-walled tubes, gently tapering, open at both ends, maximum length 1.7 mm., diameter 0.1 mm.; wall structure of crystalline calcite, walls irregularly annular (16). [Doubtfully inferred to be shells of small extinct pteropod.] *Perm.-Low. M.Jur.*, Middle East.——Fig. 138,*5*. **A. dunningtoni*, U.Perm., Arabia; *5a*, sec. showing numerous individuals; *5b*, approx. long. sec. of irregular elongate tube; both ×50 (66).

Anzalia Termier & Termier, 1947 [**A. cerebriformis*]. Reef-forming organisms of brainlike aspect, with large central cavity and very numerous small apertures resembling oscula of sponges (16). *Cam.*, Morocco.——Fig. 139,*9*. **A. cerebriformis;* ×.04 (128).

Armelia ?Lebesconte, ?1891. Description missing. [According to Seilacher (personal communication, 1956) problematic body fossil.] *Ord.*, Fr.

Bactryllium Heer, 1853 [**B. canaliculatum;* SD Andrews, 1955]. Small rounded or flat bacilliform bodies, a few mm. to 1 cm. in length, about 0.6 mm. wide; smooth or mostly with delicate transverse striations and 1 or 2 longitudinal furrows; ends rounded; material siliceous (2). [Interpretation as diatoms very improbable; ?pellets.] *Trias.-Jur.*, Eu.——Fig. 138,*1*. **B. canaliculatum*, U.Trias., Switz., Italy; ×1 (116).——Fig. 138, *2*. *B. schmidi* Heer, U.Trias., Italy, Switz.; ×1 (116).——Fig. 138,*3a-d*. *B. striolatum* Heer, U. Trias., Switz., Italy; (*3c*, transv. sec.), ca. ×8 (116).

Bovicornellum Howell, 1934 [**B. vermontense*]. Small horn-shaped tube; about 1.5 cm. long; walls smooth (16). [May be worm or mollusk. Index fossil of Highgate Shale.] *U.Cam.*, USA(Vt.).

Ceramites Liebmann (in Forchhammer), 1845 [*non* Massalongo, 1859] [**C. hisingeri*]. This fossil, described from alum shales (U.Cam.) of Scandinavia as a fucoid, represents a species of *Dictyonema* Hall, 1851, probably *D. flabelliforme* Eichwald.

Cestites Caster & Brooks, 1956 [**C. mirabilis*]. Fringed ribbon reduced to carbonaceous film, with longitudinal lines (16). [Regarded as lobe of fossil cestid ctenophoran, but identification questionable.] *Ord.*, USA(Tenn.).——Fig. 139, *13*. **C. mirabilis;* ×2 (56).

Charnia Ford, 1958 [**C. masoni*] [=?*Rangea* sp. Glaessner, 1959]. Frondlike organisms, 10 to 25 cm. in length, 4 to 5 cm. wide; composed of segmented oblique lobes, diverging alternately on either side of sinuous median axial line, whole fossil tapering to pointed apex at one end and blunt stalk at other; frond possibly grown up from a disc named *Charniodiscus* (16). [Interpreted by Ford (1958) as algal frond or primitive colenterate of unknown affinities, and by Glaessner (1959) as coelenterate related to the Pennatulacea.] *Precam.*, Eng.-S.Austral.——Fig. 138,*7*. **C. masoni*, Woodhouse beds, Eng.; ×0.4 (70).

Charniodiscus Ford, 1958 [**C. concentricus*]. Disclike structures, possibly organic, 5 to 30 cm. in diameter; central area rough-surfaced; smooth flange with or without concentric corrugations; possibly associated with frondlike fossil *Charnia* (16). [Interpreted by Ford (1958) as basal part of the "alga" *Charnia*, and by Glaessner (1959) as medusa-like base of coelenterate related to the Pennatulacea.] *Precam.*, Eng.——Fig. 138,*6*. **C. concentricus*, Woodhouse beds; ×0.7 (70).

Clistrocystis Kozlowski, 1959 [**C. graptolithophilius*]. Padlock-like chitinous forms bearing a very small cone about 0.5 mm. long; individual side by side on stipes of *Mastigograptus* sp. and embracing them; longitudinal axis perpendicular to graptolite stipes (16). [Possibly cysts of aquatic invertebrate; systematic position unknown.] *M. Ord.*(drift), Pol.——Fig. 138,*4*. **C. graptolithophilius*, on a stipe of *Mastigograptus* sp.; ×25 (88).

Coelenteratella Korde, 1959 [**C. antiqua*]. Small cuplike bodies; height about 7 mm., wall thickness about 0.15 mm.; fixed by foot about 8 mm. long (16). [Questionable coelenterate]. *M.Cam.*, USSR(Sib.).

Conostichus Lesquereux, 1876 [**C. ornatus*] [=*Conostychus* Lesquereux, 1880]. Cone with flattened or cup-shaped top; showing a series of several successive layers; diminishing in diameter from base toward top; wrinkled lengthwise; somewhat similar to feeding burrow, *Rosselia* (10). [Branson (1961) has designated *Conostichus* (=*Duodecimedusina* King, 1955) as type of n. fam. Conostichidae of the Scyphozoa, Order Coronatida.] ?*Dev.*, S.Am.(Bol.), *Penn.*, USA.

——FIG. 116,2. *C. *ornatus;* ×0.3 (69).——FIG. 116,3. *C.* sp., Ohio; ×0.3 (69).

Corycium SEDERHOLM, 1911 [**C. enigmaticum*] [=*Corycinium* C. L. FENTON, 1946]. Saclike structures with carbonaceous walls occurring in sandy beds; filling mass commonly shows concentric internal structure (37). [Regarded as alga by SEDERHOLM, but considered to be inorganic by VAN STRAATEN (1949); carbonaceous material proved by isotope investigation to be of organic origin.] *Precam.*, Fin.——FIG. 139,*10*. **C. enigmaticum; 10a,* ×0.2; *10b,* vert. sec., ×0.7 (*10a,* Geol. Survey Finland; *10b,* 119).

Curculionites KOLBE, 1888 [jr. hom.; *non* HEER, 1847; *nec* GIEBEL, 1856] [**C. senonicus*] [=*Curculidium* HANDLIRSCH, 1907]. Name proposed for burrow of curculionid, presumably in wood; recognized by W. QUENSTEDT (1932) as belonging to *Doratoteuthis syriaca* WOODWARD (16). *U. Cret.(Senon.),* AsiaM.(Syria).

Emmonsaspis RESSER & HOWELL, 1938 [**Phyllograptus? cambrensis* WALCOTT, 1890]. Oval shape, blunter at one end than other, with rod beginning about a third of way back and extending almost to posterior end, mostly with ribbing beginning at about center line and extending to outer margins (16). [Possibly a chordate.] *L.Cam.,* USA (Vt.).——FIG. 139,*11*. **E. cambriensis* (WALCOTT), Rome F.(*Olenellus* Z.); mag. unknown (110).

Endosacculus VOIGT, 1959 [**E. moltkiae*]. Globular, gall-like swellings in internodes of octocoral *Moltkia minuta* NIELSEN; diameter about 5 mm.; with narrow ventral slitlike opening, length about 2.5 mm.; interior of "cyst" smooth (16). [Possibly made by barnacles (Ascothoracida).] *U.Cret. (Camp.-U.Maastr.),* Netherl.-Swed.——FIG. 139, *8*. **E. moltkiae,* Maastricht., Neth.; *8a,* cyst with somewhat damaged opening; *8b,* cyst opened, showing the thin walls; both ×3 (132).

Favreina BRÖNNIMANN, 1955 [**F. joukowskyi*] [="Organisme B" JOUKOWSKY & FAVRE, 1913; *Coprolithus salevensis* PAREJAS, 1948; "Characé primitive(?)" CUVILLIER, 1951]. Subtriangular and rounded dark organic remains of apparently homogeneous texture; 0.5 to 1.5 mm. in length, 0.2 to 0.4 mm. in width; longitudinal section showing long, thin, straight and parallel canals distributed in regular but intermittent pattern; transverse section showing minute pores either arranged in 2 or more flattened, oblong rings or distributed irregularly; diameter of pores 12 to 40 microns (16). [Explained by PAREJAS (1948) as coprolites of crustaceans, by CUVILLIER (1951) as primitive Charophyta, by BRÖNNIMANN (1955) as microfossils *incertae sedis,* and by BRÖNNIMANN & NORTON, 1960, as coprolites of crustaceans.] *L.Jur. (Infra.-Lias.)-L.Cret.(Apt.),* Eu.(Switz.-Fr.)-W. Indies(Cuba) - USA(Tex.) - C.Am.(Guatemala)-Trinidad.——FIG. 139,*7*. **F. joukowskyi,* U.Jur. (M.Portland.), Cuba; *7a, 7b,* long. sec., transv. sec., ×22 (54).

Halysium SWIDZINSKI, 1934 [**H. problematicum*]. [=?*Hormosira moniliformis* HEER, 1877; *Halimeda saportae* FUCHS, 1894; *Arthrodendron* ULRICH, 1904 (*non* SEWARD, 1898 *nec* SCOTT, 1900)]. Ovate capsules, commonly flattened, smooth or minutely granulated, with consistency differing from matrix; some specimens with carbonaceous lining; capsules forming branching rows (16, 44). *U.Cret.-L.Tert.,* Eu.-N.Am.(Alaska).——FIG. 141, *4*. **H. problematicum,* Italy; ×0.6 (Seilacher, n).

Hensonella ELLIOTT, 1960 [**H. cylindrica*]. Calcareous tubes, hollow, cylindrical, straight, slightly tapering, length up to 2.5 mm. (?incomplete); diameter 0.1 to 0.5 mm.; walls consisting of very thin dark inner layer and thick outer layer of aragonite with radiate structure (16). [Affinities doubtful; according to ELLIOTT not a dasyclad alga; perhaps a small scaphopod.] *L.Cret.,* AsiaM. (Iraq-Iran)-N.Afr.(Algeria)-E.Indies(Borneo).——FIG. 141,*5*. **H. cylindrica,* NE.Iraq; ×30 (66).

Leckwyckia TERMIER & TERMIER, 1951 [**L. aenigmatica*]. Smooth, sharply pointed, acutely conical tube; upper end widening regularly and showing transverse units separated by constrictions (16). *Ord.,* Morocco.——FIG. 139,*4*. **L. aenigmatica;* ×0.9 (128).

Lenaella KORDE, 1959 [**L. reticulata*]. Cylindrical calcareous organisms, about 1 mm. long and 0.5 mm. wide; wall perforated by very fine holes (16). [Systematic position unknown (?hydrozoan).] *L.Cam.,* USSR(Sib.).

Lombardia BRÖNNIMANN, 1955 [**L. arachnoidea*] [="Formes découpées" LOMBARD, 1938; "Sections de thalles" LOMBARD, 1945]. Free, calcareous, transparent microfossils; spined, broad-branching or angularly bone-shaped; symmetrical; central body of variable size and shape and granular in aspect; extensions with dark median line; diameter up to about 1.5 mm. (16). [Interpreted by LOMBARD (1945) as algae, by PAREJAS (1938) as remains of sponge skeletons, and by BRÖNNIMANN (1955) as sections of microscopic symmetrical holothurian remains or microscopic planktonic crinoids or ophiuroids.] *U.Jur.,* Eu.(Fr.-Switz.)-W.Indies(Cuba).——FIG. 139,*1*. **L. arachnoidea,* Portland., Cuba, *1a, 1b,* ×62 (54).——FIG. 139, *2*. *L. perplexa* BRÖNNIMANN, Portland., Cuba; ×62 (54).——FIG. 139,*3*. *L. angulata* BRÖNNIMANN, Portland., Cuba; ×62 (54).

Lonchosaccus RUEDEMANN, 1925 [**L. uticanus*]. Formed like bent bag, length more than twice width, with thick, substantial wall, now carbonized; 2 "extremities" drawn into apertures (16).

[Systematic position unknown.] *Ord.,* USA(N.Y.). ——Fig. 139,*12*. **L. uticanus,* Utica Sh.; holotype, ×? (113).

Margaretia Walcott, 1931 [**M. dorus*]. Thin membranous sheet with elongate oval perforations arranged on longitudinal and obliquely transverse lines; tegument presumably leathery (16). [Compared with algae and alcyonarians.] *M.Cam.,* Can.(B.C.)-USA(Idaho). —— Fig. 139,*6*. **M. dorus,* Burgess Sh., B.C.; holotype, ×0.7 (134).

Nannoconus Kamptner, 1931 [**Lagena colomi* de Lapparent, 1931; SD Brönnimann, 1955]. Micro-

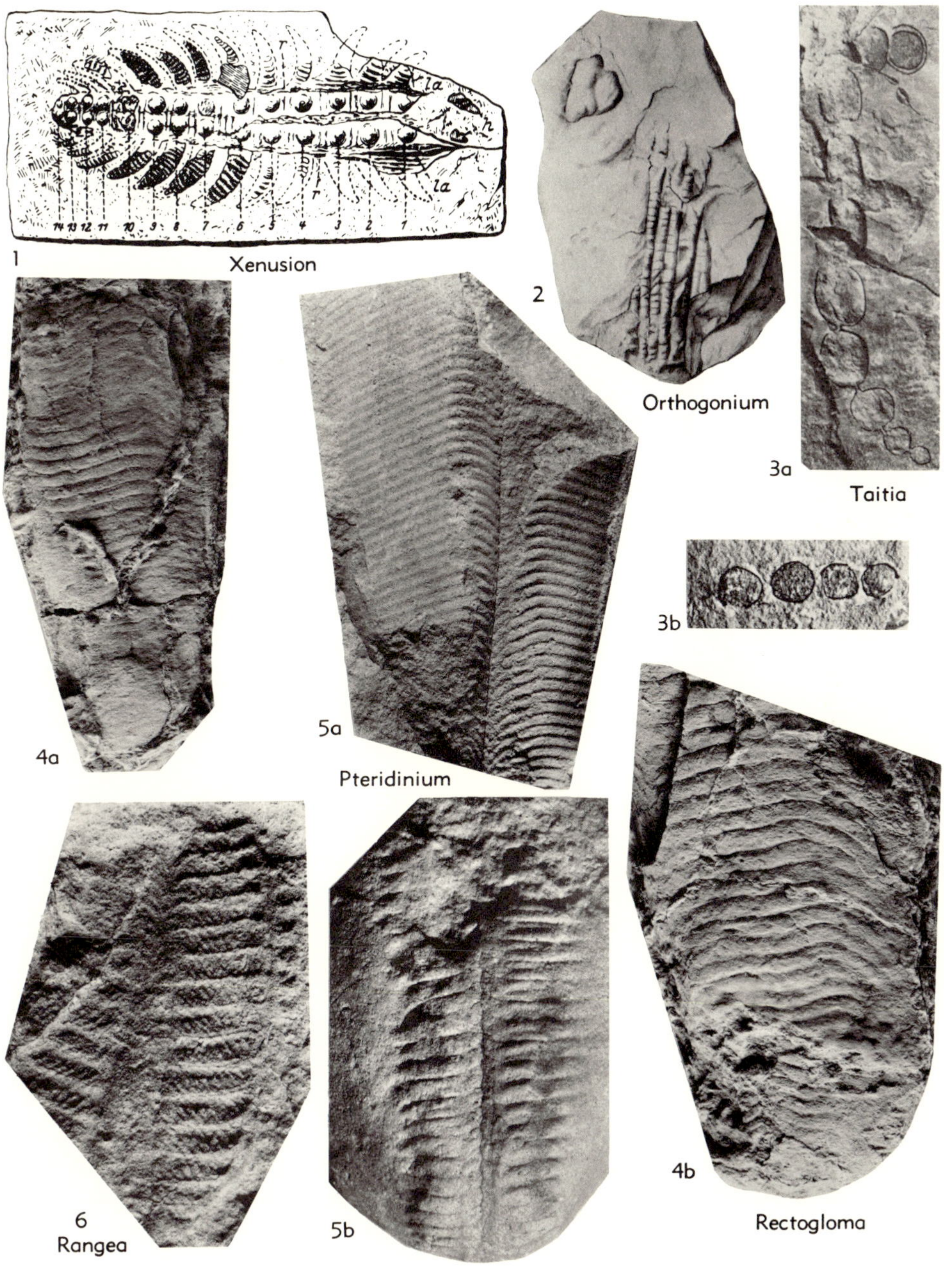

Fig. 140. Body fossils (p. W226, W228).

scopically small, peg-shaped structures with axial canal, 5 to more than 50 microns (mostly 15 to 20 microns) long, 5 to 15 microns wide; composed of numerous wedge-shaped individual elements (16). [?Skeletal remains of planktonic Protozoa (*Treatise*, p. D170-171).] *U.Jur.-L.Cret.*, S.Eu.-N.Afr.-Carib.——FIG. 139,*5*. *N. steinmanni* KAMPTNER, U.Tithon, Italy; *5a*, schem. drawing of "cone," long. sec., ×1333; *5b*, transv. sec., ×2000; *5c*, single "cone," long. sec.; ×1250 (87).

Orthogonium GÜRICH, 1930 [**O. parallelum*]. Problematical body fossil consisting of several articulated rows suggestive of crinoid arms; width of a row 3 or 4 mm., length about 6 cm.; sponge-like body similar to dictyospongiids (16). *?Precam.-?L.Paleoz.* (*Nama* F., *Kuibis Quartzite*), S.Afr.——FIG. 140,*2*. **O. parallelum*, L.Paleoz. (Nama F., Kuibis Quartzite); ×0.4 (77).

Palaeobalanus VON SEEBACH, 1876 [*nom. nud.*] [**P. schmidi*]. Name given for little bodies found on *Lima*, recognized by POHLIG (1888) as *Discina*, which allegedly occupy the opening of little bore holes (16). *M.Trias.(Muschelkalk)*, Ger.

Palaxius BRÖNNIMANN & NORTON, 1960 [**P. habanensis*]. Coprolites of oval to sub-pentagonal or subrectangular shape; width 0.5-2 mm., breadth ±0.5 mm.; pierced by crescent or hook-shaped longitudinal canals (length 45-140 meters, width 15-35 meters), arranged in two symmetric groups; coprolites structurally closely related to those of the Recent thalassinid *Axius stirhynchus* (16). *Eoc.*, C.Am.(Guatemala); *Mio.*, W.Indies(Cuba)-Libya.

Paramedusium GÜRICH, 1930 [**P. africanum*]. ?Medusa; ?inorganica (*Treatise*, p. F154, Fig. 125). *?Precam.-?L.Paleoz.* (*Nama* F., *Kuibis Quartzite*), S.Afr.

Parvancorina GLAESSNER, 1958 [**P. minchami*]. Small shieldlike body with oval outline; length up to 25 mm.; ?front margin curved in low arc and gently tapering to rounded end; center formed by prominent smooth anchor- or T-shaped unsegmented and undivided ridge; this ridge separated from ?anterior rim by distinct furrow; 2 lateral areas divided by 7 or more fine oblique lines (?traces of appendages) (16). [Systematic position and affinities unknown; possibly a larval form.] *Precam.(Ediacara Quartzite, Adelaide Syst.)*, S.Austral.——FIG. 141,*3*. **P. minchami; 3a*, ×1.2; *3b*, ×1 (74).

Porocystis CRAGIN, 1893 [**Siphonia globularis* GIEBEL, 1853 (=*Araucarites? wardi* HILL, 1893)]. Spheroids, generally prolate, with flattened, slightly protuberant area; whole surface covered with ridges and oval or circular depressions; arranged mostly rather irregularly in rows; diameter about 2 cm. (16). [Interpreted by GIEBEL (1853) as alga, by HILL (1889-93) as fruit of *Goniolina*, *Parkeria* or *Araucarites*, by CRAGIN (1893) as cheilostomatous byrozoan, by RAUFF (1895) as calcareous alga, and by JARVIS (1905) as gigantic monothalamnian foraminifer.] *L.Cret.*, USA (Tex.).——FIG. 141,*1*. *P. pruniformis* CRAGIN, L. Alb.(large specimens)-M.Alb.(small specimens); *1a-d*, ×1 (Geol. Staatsinst. Hamburg).

Pteridinium GÜRICH, 1933 [**Pteridium simplex* GÜRICH, 1930] [=*Pteridium* GÜRICH, 1930 (*non* SCOPOLI, 1777)]. Long, thin, bilaterally symmetrical "leaves" with transverse ribs (16). [According to RUDOLF RICHTER (1955) belongs to Gorgonaria (together with *Rangea* GÜRICH, 1930); according to GLAESSNER (1959) (together with *Rangea*) probably closely related to Pennatulacea.] *Precam.(Ediacara Quartzite)*, S.Austral., *?Precam.-?L.Paleoz.*, S.Afr.——FIG. 140,*5*. **P. simplex*, ?L.Paleoz.(Kuibis Quartzite), S.Afr.; *5a*, ×0.8; *5b*, ×0.7 (111).

Rangea GÜRICH, 1930 [**R. schneiderhöhni*]. Leaf-shaped main body with median field (axis); lateral branches separated by transverse lateral furrows and subdivided by secondary furrows arising from their proximal margins (pinnae correspondingly) (16). [Interpretations: GÜRICH (1930, 1933), compared with the Ctenophora; RICHTER (1955), placed (together with *Pteridinium*) in Gorgonacea; GLAESSNER (1959), near Pennatulacea.] *Precam.(Ediacara Quartzite)*, S.Austral.; *?Precam. or L.Paleoz.(Kuibis Quartzite, Nama* F.), S.Afr.——FIG. 140,*6*. **R. schneiderhoehni*, ?L. Paleoz.(Kuibis Quartzite), S.Afr.; holotype, ×0.7 (111).

Rectogloma VAN TUYL & BERCKHEMER, 1914 [**R. problematica*]. Cephalopod-like fossil, elliptical in transverse section; apex terminating in spiral coil, closely placed sinuous sutures on surface which disappear completely on apical coil (16). *U.Dev.*, USA(Pa.).——FIG. 140,*4*. **R. problematica; 4a, 4b*, ×1.2 (Am. Mus. Nat. Hist.).

Stromatolite KALKOWSKY, 1908 [=*Coenoplase* TWENHOFEL, 1919]. General name for variously shaped, finely stratified calcareous crusts and calcareous bodies (also called stromatoliths); obviously formed by lime-precipitating algae; commonly associated with oolites or ooid-grains. Many "genera" belong here (e.g., *Anomalophycus* FENTON & FENTON, 1937; *Aphrostoma* GÜRICH, 1906; *Archaeozoon* MATTHEW, 1890; *Chondrostoma* GÜRICH, 1906; *Codonophycus* FENTON & FENTON, 1939; *Collenia* WALCOTT, 1914; *Cryptozoon* HALL, 1884; *Conophyton* MASLOV, 1937; *Dolatophycus* FENTON & FENTON, 1937; *Gouldina* JOHNSON, 1940; *Gymnosolen* STEINMANN, 1911; *Malacostroma* GÜRICH, 1906; *Osagia* TWENHOFEL, 1919 (only pisoids or ooids?) *Ottonosia* TWENHOFEL, 1919; *Pycnostroma* GÜRICH, 1906; *Spongiostroma* GÜRICH, 1906; *Stylophycus* JOHNSON, 1940; *Tetonophycus* FENTON & FENTON, 1939; *Weedia* WALCOTT,

1914, etc.) (37). [Genera named above are differentiated only on basis of general form. According to Holtedahl (1919) and Schindewolf (37), they are pseudofossils (inorganic structures), and should not be accorded generic and specific names]. *Precam.-Rec.*, cosmop.

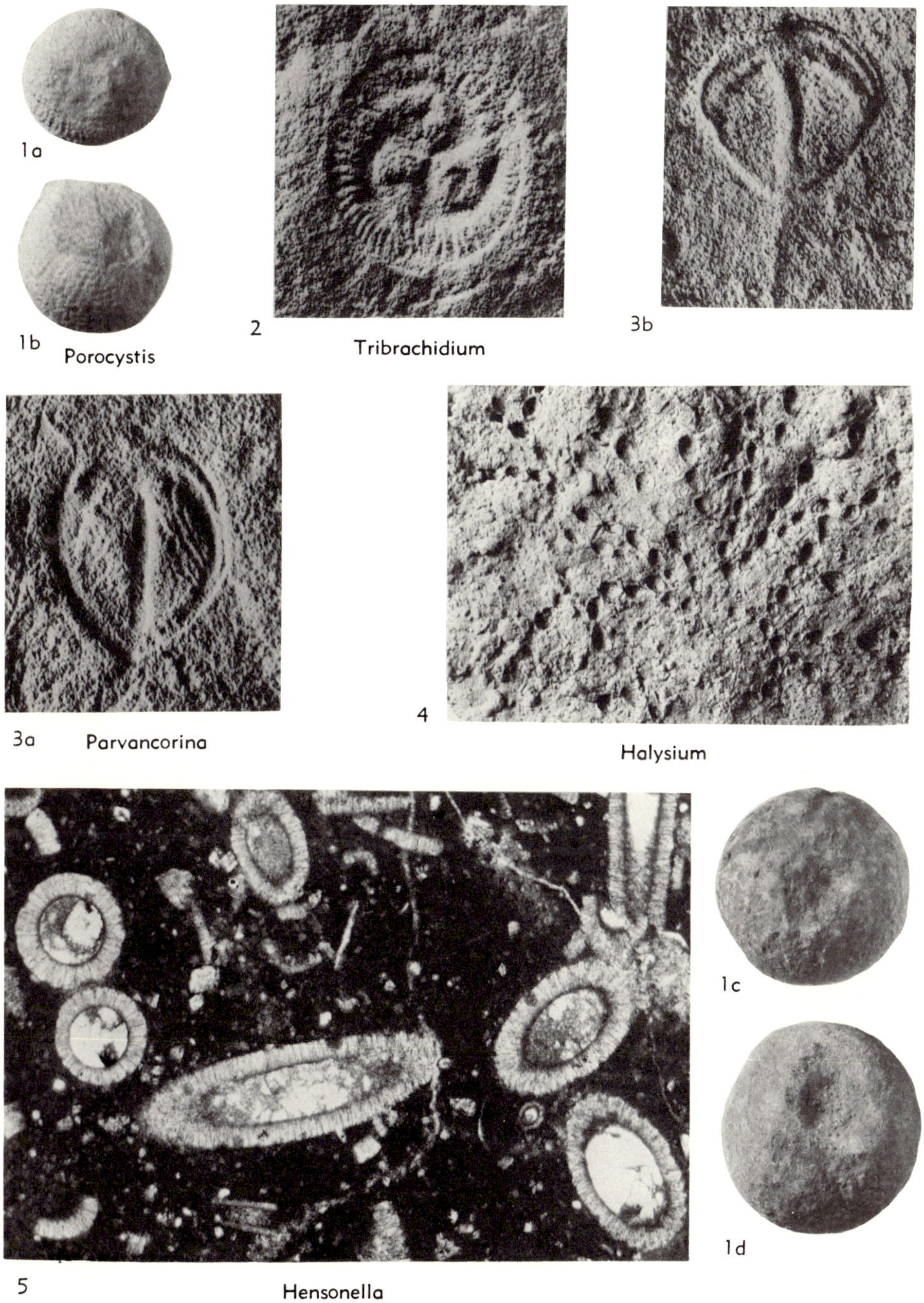

Fig. 141. Body fossils (p. *W*224, *W*226, *W*228).

Taitia CROOKALL, 1931 [*T. catena]. Little chains commonly composed of 6 to 7 (maximum, 11) circular or oval bodies; adjacent bodies united by thin isthmus 1 mm. long and 1 mm. wide; bodies generally constant in size (diameter 1 cm.), some with progressive diminution in size toward extremity; characteristic but problematical body fossil of Scottish Downtonian rocks (2). *U.Sil.*, Scot. ——FIG. 140,*3*. **T. catena; 3a,* ×1; *3b,* ×0.7 (61).

Tribrachidium GLAESSNER, 1959 [*T. heraldicum]. Subcircular impressions up to 26 mm. in diameter; rim sharply impressed, with distinct sculpture consisting of 3 hooked ridges of similar size and shape radiating from the center, ending along periphery in fringe of tentacular projections; all known specimens seemingly external molds; excluded from all known major groups on basis of its 3 tentaculate arms (16). [Perhaps an aberrant coelenterate.] *Precam.(Ediacara Quartzite),* S. Austral.——FIG. 141,*2*. **T. heraldicum;* ×1 (impression) (74).

Xenusion POMPECKJ, 1927 [*X. auerswaldae]. Bilaterally symmetrical; length 8.5 cm.; central field divided into segments, with median longitudinal furrow; segments separated by broad transverse furrows; each segment supporting a bosslike elevation; finely fluted marginal stripes along central field; transversely ringed, slightly curved, delicately striped lateral appendages along both sides (16). [Affinities unknown, although some have regarded it as a ctenophore (*Treatise*, p. F478).] *Precam.* [more probably *Cam.* (drift)], N.Ger. ——FIG. 140,*1*. **X. auerswaldae,* ?L.Cam.(Pleist. drift); ×0.7 (106).

BORINGS

Borings or, more properly, etching traces in shells, bones, or other hard parts of invertebrates and vertebrates occupy a special position among the trace fossils which entitles them to a chapter of their own. Of the few papers on this subject that of MÄGDEFRAU (26) deserves mention. Boring traces are known as far back as Early Paleozoic. They may be produced by plants or by animals. Those produced by plants are made by algae or fungi, but the cavities left by them do not allow conclusions as to a definite producer. Within the animal kingdom certain sponges, worms, bryozoans, and barnacles bore into shells; the last mentioned especially leave very characteristic cavities. The cavities made by the others often leave uncertainty as to the producer. Boring ctenostome Bryozoa (Terebriporidae) have not been dealt with in this chapter, inasmuch as BASSLER has already described them in Part G of the *Treatise.*

Abeliella MÄGDEFRAU, 1937 [*A. riccioides; SD HÄNTZSCHEL, herein]. Dichotomously branching borings in fish scales; width of individual borings 4 to 8 microns, of the whole system 0.25 to 0.5 mm. (26). *U.Cret.-Oligo.*, Eu.(Ger.-Eng.).——FIG. 142,*7*. **A. riccioides,* Oligo., Ger.; (in fish scale), ×110 (26).

Anobichnium LINCK, 1949 [*A. simile]. Smooth cylindrical perforations in fossil wood, 1 to 1.5 mm. in diameter, with numerous openings to each gallery; very similar to the borings of Recent beetles of the genus *Anobium* (16). *U.Trias.*, Ger. ——FIG. 144,*2*. **A. simile,* Keuper; in wood, ×0.7 (94).

Brachyzapfes CODEZ, 1957 [*B. elliptica]. Borings of barnacles; short and broad; cross section elliptical; depth half the length; observed in belemnoids and pelecypods (16). *L.Cret.*, Fr.——FIG. 142,*2*. **B. elliptica;* schem. drawings; *2a,* opening; *2b,* tang. sec. (max.); *2c,* long. sec.; *2d,* chamber (60).

Calcideletrix MÄGDEFRAU, 1937 [*C. flexuosa; SD HÄNTZSCHEL, herein]. Cavity systems in belemnoids; one or more openings, shrublike, ramified; diameter of branches 0.02 to 0.1 mm. (26). *U. Cret.*, Ger.——FIG. 142,*4*. **C. flexuosa;* in *Belemnitella,* ×8 (26).——FIG. 142,*5*. *C. breviramosa* MÄGDEFRAU; in *Actinocamax,* ×8 (26).

Calciroda MAYER, 1952 [*C. kraichgoviae]. Cylindrical boring tunnels up to 1 mm. wide; usually built parallel to outer surface in shells of mollusks or in stalk members of *Encrinus;* may be ramified, cutting through or crossing each other (16). [According to A. H. MÜLLER (1956), probably identical with *Trypanites* MÄGDEFRAU.] *M.Trias. (Trochiten-Kalk),* Ger.

Caulostrepsis CLARKE, 1908 [*C. taeniola] [=*Polydorites* DOUVILLÉ, 1908 (*according to* BATHER, 1910, not intended as an independent generic name)]. U-shaped boring tunnels with constructed *Spreite*, corresponding to a tiny *Rhizocorallium;* up to 2 cm. long and 5 mm. wide; mostly in shells of brachiopods, mollusks, and echinoids (23). *L.Dev.*, Ger.; *U.Trias.-?L.Jur.*, Eng.; *Tert.*, Austral.-Port.——FIG. 142,*3*. **C. taeniola;* L.Dev., Ger.; in shell of *Stropheodonta,* ×0.75 (58).

Chaetophorites PRATJE, 1922 [*C. gomontoides]. Ramifying tunnels in rostra of belemnoids and

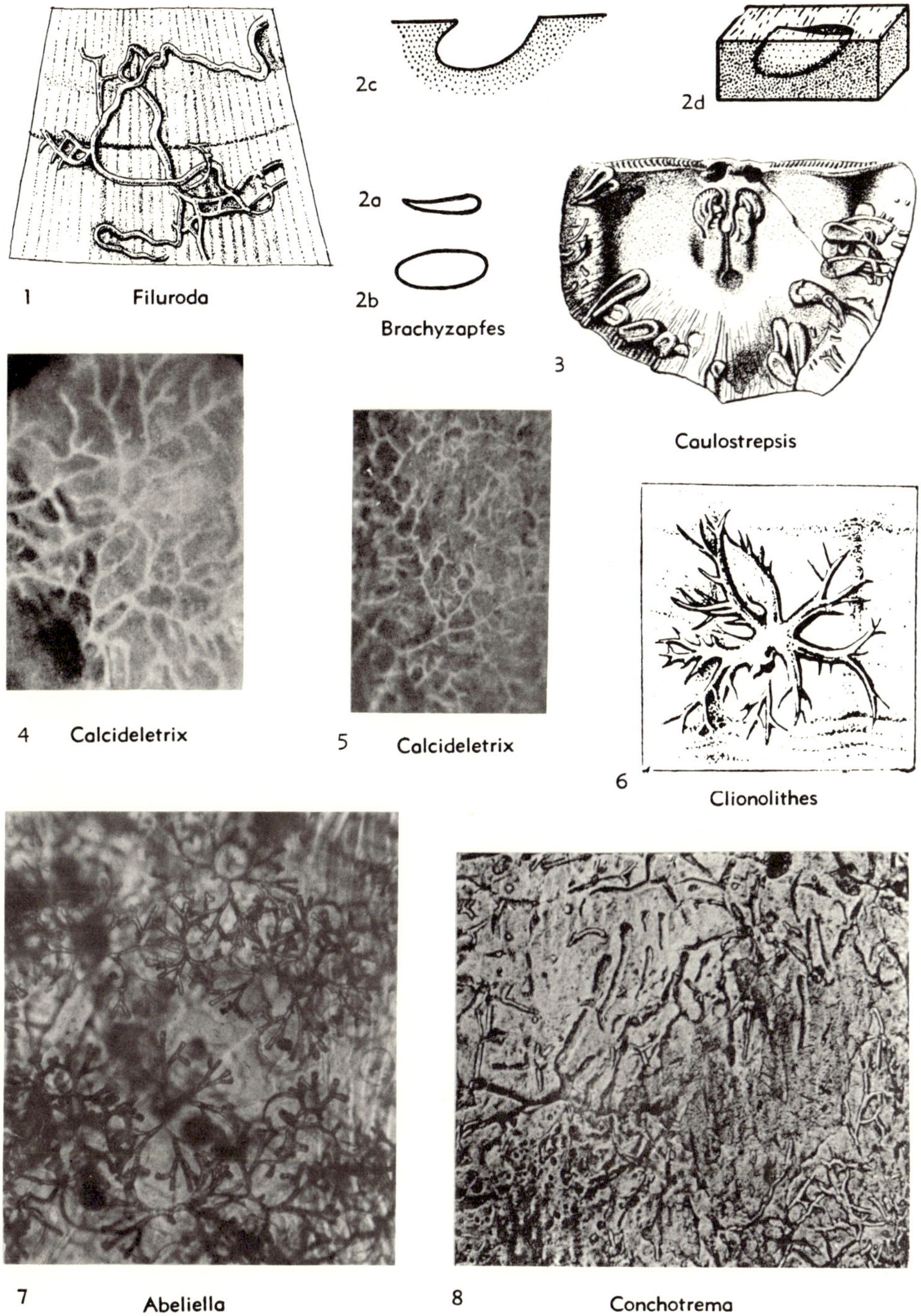

FIG. 142. Borings (p. *W*228, *W*230).

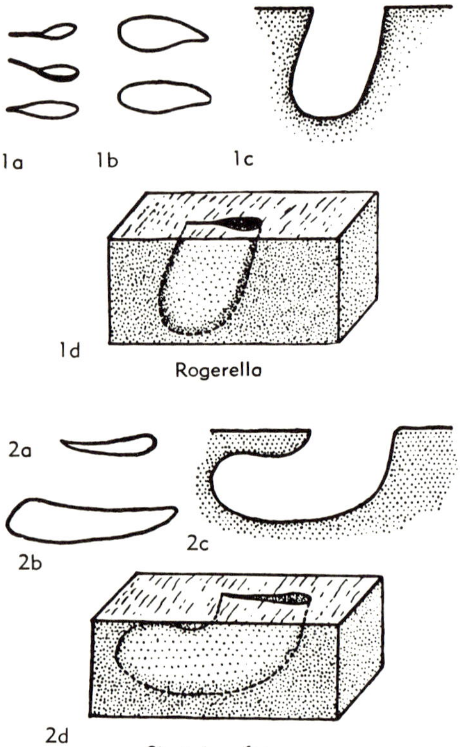

FIG. 143. Borings (p. W231).

shells of brachiopods and mollusks; usually straight; diameter less than 0.02 mm.; located close to surface of shell (26). *Jur.-Plio.*, Eu.——FIG. 144,*4*. **C. gomontoides*, L.Jur.(Lias. δ), Ger.; in pelecypod shell, ×106 (108).

Cliona GRANT, 1826 (see *Treatise*, p. *E*40).

Clionoides FENTON & FENTON, 1932 (see *Treatise*, p. *E*40).

Clionolithes CLARKE, 1908 [**C. radicans;* SD FENTON & FENTON, 1932] [=*Pyritonema? gigas* FRITSCH, 1908 (*non* M'COY, 1850); *Olkenbachia* SOLLE, 1938]. Bent or cracked borings of sponges, generally radiating in one plane to all sides from very small, central cavity; commonly branching dichotomously; diameter several mm.; always etched into shell of some host animal (see *Treatise*, p. *E*40) (16). *Ord.*, Czech.; *Dev.-Carb.*, Ger.-USA-China.——FIG. 142,*6*. **C. radicans*, U.Dev. (Chemung Ss.), USA; in *Atrypa* shell, ×6 (58).

Conchotrema TEICHERT, 1945 [**C. tubulosa*]. Narrow tubular borings in shells (diameter about 0.2 mm.), communicating with the surface, straight or gently curved; branching (16). [Probably made by worms; observed in brachiopods.] *L.Carb.*, Scot.; *Perm.*, W.Austral.——FIG. 142,*8*. **C. tubulosa*, Wandagee Ser.; in *Taeniothaerus* valve, ×2 (127).

Dendrina QUENSTEDT, 1848 [**Talpina dendrina* MORRIS, 1851]. Borings just below surface in brachiopods and in rostra of belemnoids; without aperture; forming rosettes 1.5 to 6 mm. in diameter; ramifying intensely and irregularly; diameter of borings about 0.05 mm. (26). *Ord.-U.Cret.*, Eu.(Ger.-Eng.-USSR.).——FIG. 144,*6*. *D. belemniticola* MÄGDEFRAU, U.Cret., Ger.; in *Belemnitella*, ×5 (26).

Dictyoporus MÄGDEFRAU, 1937 [*non* HOULBERT, 1934] [**D. nodosus*]. Borings in rostra of belemnoids; without exterior aperture; distinctly netlike; width of canals about 0.07 mm. (26). *U. Cret.*, Eu.(Ger.).——FIG. 144,*5*. **D. nodosus;* in *Belemnitella*, ×5 (26).

Entobia BRONN, 1838 [**E. cretacea* PORTLOCK, 1843; SD HÄNTZSCHEL, herein]. Borings consisting of small cavities connected by processes nearly hair-thin; occurring in belemnoids and in shells of pelecypods, particularly *Inoceramus;* similar forms also in trilobites (16). [Possibly made by boring sponges.] *?Sil.*, Ire.; *U.Cret.*, Eng.-Ire.

Filuroda SOLLE, 1938 [**Clionolithes reptans* CLARKE, 1908] [=*Clionolithes* CLARKE, 1908 (*partim*)]. Threadlike, strongly curved borings in shells, running closely below surface of shell (16). [Possibly made by boring sponges (*Treatise*, p. *E*40).] *L.Dev.-M.Dev.*, N.Am.-Eu.(Ger.).——142,*1*. **F. reptans* (CLARKE), L.Dev.(Oriskany Ss.), USA; in *Leptostrophia*, ca. ×2 (58).

Mycelites ROUX, 1887 [**M. ossifragus*]. General name for various irregularly branching tunnels about 2 to 6 microns wide in hard parts (shells, bones, teeth, scales) of invertebrates and vertebrates (26). [Caused by algae and/or fungi.] *?Sil.*, *Rec.*, cosmop.

Nygmites MÄGDEFRAU, 1937 [**Talpina solitaria* VON HAGENOW, 1840; SD HÄNTZSCHEL, herein] [=*Talpina* VON HAGENOW, 1840 (*partim*)]. Simple, unbranched tunnels in rostra of belemnoids; oblique to surface; leading from outside inward (26). *L.Jur.-U.Cret.*, Eu.(Ger.-Fr.-USSR.).

Palaeachlya DUNCAN, 1876 [**P. perforans*]. Small tubes of parasitic algae in fossils; diameter 0.008 inch; usually straight, rarely curved; not varying much in size; running more or less inward at different angles to surface; some branched (2). *Sil.-Dev.*, Eu.-Austral.-N.Am.(Can.).

Palaeosabella CLARKE, 1908. [See *Treatise*, p. *E*41.] (16).

Paleobuprestis WALKER, 1938 [**P. maxima;* SD HÄNTZSCHEL, herein]. Channels under bark of *Araucarioxylon arizonicum;* diameter 2 to 10 mm.; recognizable all around tree; channels resembling work of Recent buprestids (16). *Trias.*, USA(Ariz.).

Paleoipidus WALKER, 1938 [**P. perforatus;* SD HÄNTZSCHEL, herein]. Tunnels and burrows penetrating heart-wood of *Araucarioxylon arizonicum* (see also *Paleobuprestis* and *Paleoscolytus*); diameter 2 to 5 mm.; boring near bark or through wood (16). *Trias.*, USA(Ariz.).

Paleoscolytus Walker, 1938 [**P. divergus*]. Channels under bark of *Araucarioxylon arizonicum;* diameter 5 mm.; running in all directions; not filled with castings; resembling channels of Recent bark beetles or engraver beetles of family Scolytidae (16). *Trias.*, USA(Ariz.).

Rogerella de Saint-Seine, 1951 [**R. lecointrei*]. Very deep borings of barnacles; cross section short and broad; observed in shells of echinoids, pelecypods, belemnoids and corals (16). *M.Jur.-U.Cret.*, Fr.-Eng.; *Mio.*, Fr.; *Plio.*, Morocco.——Fig. 143,*1*. *R. mathieui* de Saint-Seine; schem., *1a, 1b*, various kinds of openings and tang. secs.; *1c*, long. sec., *1d*, chamber (60).

Simonizapfes Codez, 1957 [**S. elongata*]. Long, narrow borings of barnacles; length (max.) 4.5 mm.; width (max.) 1.1 mm.; shallow; observed in shells of oysters, gastropods, belemnoids, crinoids, corals, etc. (16). *Jur.*, Fr.-Eng.——Fig. 143, 2. **S. elongata;* schem.; *2a*, opening; *2b*, tang. sec. (max.); *2c*, long. sec.; *2d*, chamber (60).

Talpina von Hagenow, 1840 [**T. ramosa;* SD Häntzschel, herein]. Tunnels in rostra of belemnoids; width about 0.2 mm.; numerous circular or oval openings toward exterior; commonly branched (26). *U.Cret.*, Eu.(Ger.-Fr.-USSR.).

Tarrichnium Wanner, 1938 [**T. balanocrini*]. Irregularly branched, ribbonlike, sharply entrenched traces on stalks of *Balanocrinus;* surface of ribbons slightly convex, some divided by 1 or 2 very thin longitudinal furrows; with fine bowl-shaped impressions (16). [Made by ?hydrozoan

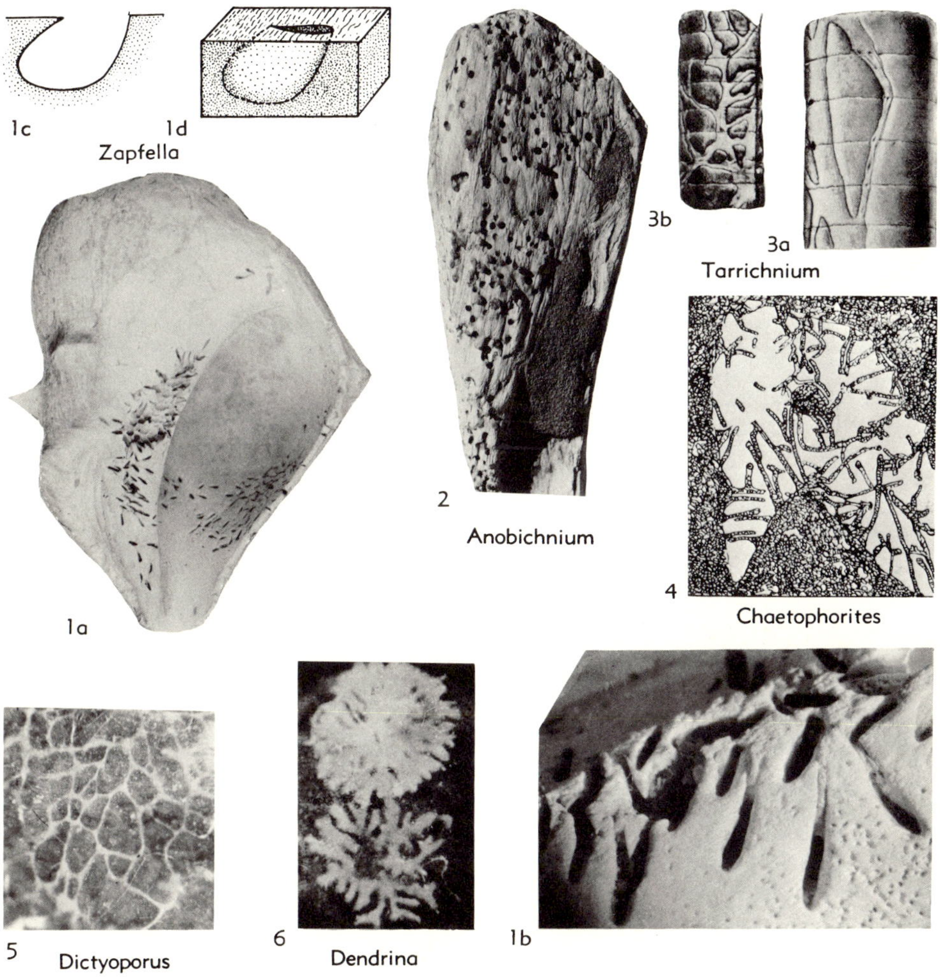

Fig. 144. Borings (p. *W*228-*W*232).

(*Treatise*, p. *F*88).]. *Mio.*, E.Ind.——FIG. 144,*3*. **T. balanocrini*; *3a*, ×2.2; *3b*, ×1.2 (135).

Topsentopsis DELAUBENFELS, 1955 [*pro Topsentia* CLARKE, 1921 (*non* BERG, 1899)] [**Topsentia devonica* CLARKE, 1921]. (See *Treatise*, p. *E*41.)

Zapfella DE SAINT-SEINE, 1955 [**Z. pattei*]. Saclike bore holes, 1 to 4 mm. long, 0.5 to 1 mm. wide and up to 5 mm. deep; slitlike opening (23). [Made by barnacles (*Acrothoracica*); found in mollusks, brachiopods, corals, and solid rock.] *Tert.*, Eu.-N.Afr.——FIG. 144,*1*. Borings of *Zapfella* in *Galeodes (Volema) cornuta* AGASSIZ, Mio., Hung.; *1a*, ×.08; *1b*, ×4; *1c*, schem. long sec.; *1d*, schem. chamber (*1a*, *1b*, 142; *1c*, *1d*, 60).

"FOSSILS" PROBABLY OF INORGANIC ORIGIN

Concretions, clay galls, various trail-like markings and even mud cracks and structures of diagenetic origin have sometimes been described and named as plant or animal fossils. One of the best-known examples is *Eozoon canadense*. Errors of this type occurred frequently when paleontology was a new field, but more recent examples may be found (e.g., markings and structures of tectonic or diagenetic origin described by FUCINI in the Verrucano of Florence, also described and named in voluminous books with many plates). The "arthropod" remains in very old rocks of South Australia, described and named only two decades ago, ought to be mentioned in this connection too. Naturally, all names listed here are worthless and have no right to exist. They are included here at the request of the Editor for their historical interest and for the sake of completeness. Naming of "type-species" is, of course, unnecessary. Nevertheless, in those cases in which "type-species" have been formally designated, they have been cited.

Aenigmichnus HITCHCOCK, 1865 [**A. multiformis*]. Parallel lines, commonly changing to rows of dots or to moniliform lines, covering wide spaces; highly variable; surely inorganic (markings of drifting or rolling bodies) (18). *Trias.*, USA (Mass.).

Antholithina CHOUBERT, TERMIER & TERMIER, 1951 [**A. rosacea*]. Almost circular cross sections with radially disposed structures ("septa"), observed in thin sections; regarded by authors as calcareous algae. [According to SCHINDEWOLF (37) oolitic grains with outer cover of iron-oxyhydrate which in part has penetrated radially into the interior.] *Precam.*, Morocco.

Archaeophyton BRITTON, 1888 [**A. newberryanum*]. Thin films of graphite lying parallel to bedding planes of limestones; at first regarded as "the most ancient plant yet discovered" (16). *Precam.*, USA(N.J.).

Aristophycus MILLER & DYER, 1878 [**A. ramosus*]. ?Rill markings (19). *Ord.*, USA(Ohio).

Aspidella BILLINGS, 1872 [**A. terranovica*]. Small, ovate, narrow ringlike border; having general aspect of small *Patella* flattened by pressure (37). [Regarded by MATTHEW (1898) as slickensided mud concretion striated by pressure; somewhat similar to *Guilielmites* GEINITZ.] *Precam.*, Can. (Newf.).——FIG. 145,*3*. **A. terranovica*, Etchemin Gr.; *3a*, *3b*, mag. unknown (134).

Atikokania WALCOTT, 1912 [**A. lawsoni*]. Compared at first with sponges and Archaeocyathinae; now considered inorganic by RAYMOND (1935) and other authors (*Treatise*, p. *E*20, *E*33 and *E*103). *Precam.*, Can.

Batrachoides HITCHCOCK, 1858 [jr. hom.; *non* LACEPÈDE, 1800] [**B. nidificans*] [=*Batrachioides* WEIGELT, 1927; *Batracoides* ILIE, 1937]. Shallow contiguous pits on bedding planes; about 2.5 cm. in width, depth about 1 cm.; compared with similar Recent excavations made by small fishes and tadpoles (SILLIMAN, 1850; HITCHCOCK, 17). [Reasonably explained by KINDLE (1914) as interference ripples.] *Sil.*, USA(N.Y.); *Trias.*, USA(Mass.).——FIG. 145,*10*. **B. nidificans*, U. Trias. (New Red Ss.), Mass.; ×0.2 (1).

Bisulcus HITCHCOCK, 1865 [**B. undulatus*]. Continuous paired grooves separated by single ridge (18). [According to BROWN (1912), drag markings.] *Trias.*, USA(Mass.).

Camasia WALCOTT, 1914 [**C. spongiosa*]. Cross sections showing spongioid appearance; originally regarded as algae. [According to SCHINDEWOLF (37), probably of inorganic origin (diagenetic structure).] *Precam.*, USA(Mont.).——FIG. 145, *6*. **C. spongiosa*, Belt Ser. (Newland Ls.); vert. sec., ×0.4 (45).

Chloephycus MILLER & DYER, 1878 [**C. plumosum*]. Rill marks or, according to SEILACHER (personal communication, 1955), drag marks (19). *Ord.*, USA(Ohio).

Chuaria WALCOTT, 1899 [**C. circularis*]. Disclike bodies resembling compressed conical shells of discinoid or patelloid shape; 2 to 5 mm. in diameter; concentrically wrinkled; dark bituminous matter covering surface. Certainly inorganic.

[According to SCHINDEWOLF (37), possibly small, wrinkled clay galls or concretions. *C. wimani* BROTZEN (1941) has been variously regarded as trilobite eggs (WIMAN, 1894); ?chitinous foraminifers (BROTZEN); and hystrichosphaerids (EISENACK, 1951). May be inorganic.] *Precam.*, USA-USSR-Swed. —— FIG. 145,5. **C. circularis*, Algonk., Ariz.; *5a, 5b*, ×12 (134).

Copperia WALCOTT, 1914 [**C. tubiformis*] [=*Cooperia* CHOUBERT, TERMIER & TERMIER,

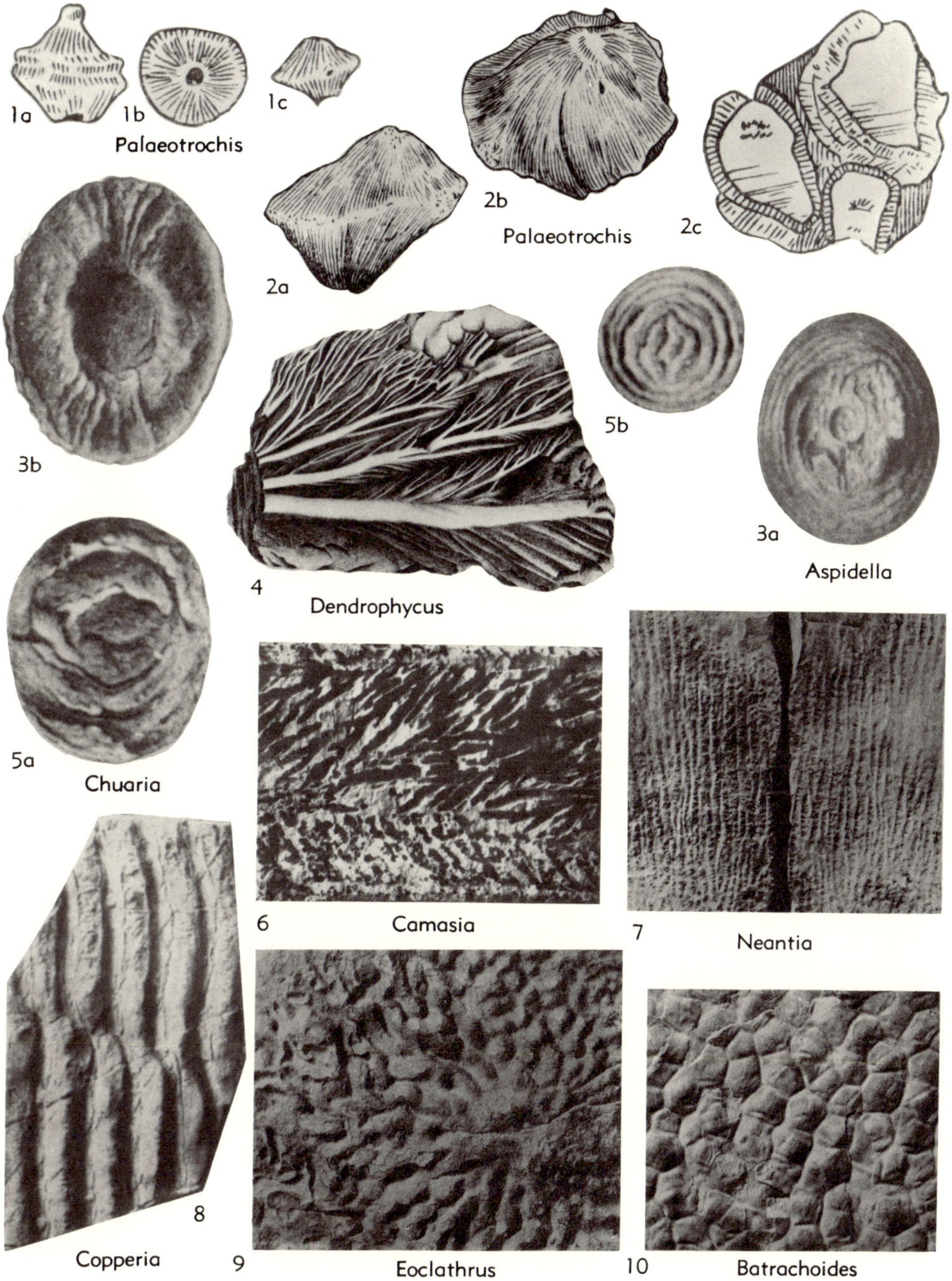

FIG. 145. "Fossils" of inorganic origin (p. *W232-W234, W236*).

1951]. Differs from *Greysonia* WALCOTT, 1914 (45), in greater irregularity of "growth" and more nearly cylindrical nature of tubes. [According to FENTON & FENTON (1936), identical with *Greysonia* and both "genera" of inorganic origin; according to SCHINDEWOLF (37), partly resembling ripplemarks deformed by diagenetic and tectonic processes.] *Precam.*, USA(Mont.)-?N.Afr.——FIG. 145,*8*. **C. tubiformis*, Belt Ser. (Newland Ls.), Mont.; surface of group of tubes formed in horiz. position, ×0.7 (45).

Corticites FUCINI, 1938 [*nom. nud.*] [jr. hom.; *non* ROSSMAESSLER, 1840]. ?Inorganic (16). *?L.Perm. (Verrucano)* Italy.

Cupulicyclus QUENSTEDT, 1879. Pressure cone, recognized as inorganic by QUENSTEDT (16). *Dev.-Tert.;* Ger.

Cyathospongia? eozoica MATTHEW, 1890. "Fossils" interpreted as skeletal fragments of sponges (37). [Recognized by RAUFF (1893) as inorganic in origin.] *Precam.*, Can.(N.B.).

Dendrophycus LESQUEREUX, 1884 [**D. desorii*]. Branched rill marks?, described as algae (2).——FIG. 145,*4*. *D. triassicus* NEWBERRY, U.Trias., Conn.; ?ca. ×0.2 (101).

Dictuolites CONRAD, 1838 [**D. beckii*]. Mud cracks regarded by HALL with some doubt as plants (2). *L.Sil.*, USA(N.Y.).

Dinocochlea WOODWARD, 1922 [**D. ingens*]. Large horizontal bodies, spirally twisted to right or left; erroneously described as gastropod *steinkerns;* now interpreted as concretions (23). *L.Cret.*, Eng.

Eoclathrus SQUINABOL, 1887 [**E. fenestratus*]. Irregular, elongate, ridgelike structures nearly parallel with each other; probably inorganic (6). *L.Dev., Tert.*, Italy-N.Afr.——FIG. 145,*9*. *E. balboi* DESIO, L.Dev., N.Afr.; ×0.3 (6).

Eophyton TORELL, 1868 [**E. linneanum*] [=*Rabdichnites* DAWSON, 1873 *(partim); Taonichnites* MATTHEW, 1890 *(partim); Medusichnites, Eoichnites, Ctenichnites* MATTHEW, 1891; *Aspidiaria* VLCEK, 1902 (*non* PRESL, 1838)]. Straight, parallel or curved drag markings on bedding planes; produced by organisms or comprising inorganic objects (2, 46). *Cam.-Rec.*, cosmop.——FIG. 147, *1*. *E.* sp., Cam.(*Mickwitzia* Ss.), Swed.; ×0.3 Regnéll, n, Paleont. Coll., Paleont. Inst. Lund).

Eopteris DE SAPORTA, 1878 [**E. andegavensis;* SD ANDREWS, 1955]. *Cardiopteris*-like "fossil." According to GOTHAN (1909), *E. morierei* DE SAPORTA comprises ferric sulphide dendritic marking (16). *Ord.*, Fr.

Eospicula DELAUBENFELS, 1955 [**E. cayeuxi*]. Doubtful "fossils" resembling spicules of calcisponge; lumpy and crooked. [Believed by CAYEUX (1895) to be sponge; regarded as inorganic in origin by RAUFF (1896)] (*Treatise*, p. *E*33). *?Precam.*, Fr.

Eozoon DAWSON, 1865 [**E. canadense*]. Banded structures of coarsely crystalline calcite and serpentine; originally interpreted as gigantic Foraminifera (16). *Precam.*, Can.

Flabellaria johnstrupi HEER, 1883. Regarded by SCHENK, 1890 as ripple marks rather than palm leaf (16). *Tert.*, Greenl.

Forchhammera GOEPPERT, 1860 [**F. silurica*]. According to POULSEN and ROSENKRANTZ (personal communication, 1956), inorganic; probably dendritic markings (2). *L.Ord.*, Den.(Bornholm).

Gallatinia WALCOTT, 1914 [**G. pertexa*]. Septarian concretions (RAYMOND, 1935) (30, 45). *Precam.*, USA(Mont.).——FIG. 146,*4*. **G. pertexa*, Belt Ser. (Newland Ls.); upper surface, ×0.3 (45).

Gloeocapsomorpha tazenakhtensis CHOUBERT, TERMIER & TERMIER, 1951. "Organisms" observed in thin sections of limestones; interpreted as calcareous algae (37). [According to MOSEBACH (1956), certainly inorganic structures produced by combination of tectonic movements and metamorphic recrystallization.] *Precam.*, Morocco.

Gothaniella FUCINI, 1936 [**G. sphenophylloides*]. Small rosettes, occurring together with bigger and more pronounced ones called *Sewardiella* FUCINI (16). [Interpreted by FUCINI (1936) as algae; by SACCO (1940) as *?Sphenophyllum*; and by PIA (1937) as probably inorganic.] *?L.Perm., Trias.*, Italy.——FIG. 147,*4*. **G. sphenophylloides*, ?U.Trias.(Verrucano); ×2 (71).

Grammichnus HITCHCOCK, 1865 [**G. alpha*]. [According to BROWN (4) and LULL (25), probably roll or drag markings (18).] *Trias.*, USA(Mass.).

Greysonia WALCOTT, 1914 [**G. basaltica*]. Shrinkage cracks (RAYMOND, 1935) or results of segregation of $CaCO_3$ and dolomite by percolating waters (FENTON & FENTON, 1936). [According to SCHINDEWOLF (37), partly resembling ripple marks transformed by tectonic and diagenetic processes.] *Precam.*, USA(Mont.).——FIG. 146,*2*. **G. basaltica*, Belt Ser. (Newland Ls.); *2a*, sec. of mass of basaltic-like columns, *2b*, view of end of tubes; both ×0.7 (45).

Guilielmites GEINITZ, 1858 [=*Calvasia* sp. STERNBERG, 1820; *Carpolites umbonatus* STERNBERG, 1825; *Cardiocarpum umbonatum* BRONN, 1837; *Carpolites clipeiformis* GEINITZ, 1856; *?Gaussia* CHACHLOF, 1934 *(partim); ?Gaussia* NEUBURG, 1934; *Verrucania* FUCINI, 1936]. Ellipsoidal bodies, 1 or 2 cm. in diameter; originally thought to be seeds; most authors (CARRUTHERS, SCHENK, POTONIÉ, GOTHAN) consider them to be of inorganic origin (concretions or similar diagenetic structures); PRUVOST (1930) interpreted them as burrows of lamellibranchs (2, 23). *Carb.-Perm., ?Jur.*, Eu.-Am.-Asia.——FIG. 147,*2*. *G. umbonatus* (STERNBERG) L.Perm., Ger.; *2a, 2b*, ×1 (72).

Halichondrites graphitiferus MATTHEW, 1890. Long, thin spicules in graphitic shales and graphite lenses (37). [Interpreted as remains of sponges; recognized by RAUFF (1893) as inorganic (?systems of striae on graphite flakes).] *Precam.*, Can., (N.B.).

Halleia Fucini, 1936 [**H. penicillata*]. Probably very slender flow markings; inorganic (16). *?L. Perm.(Verrucano)*, Italy.

Hirmeria Fucini, 1936 [**H. notabilis*]. Small parallel wrinkles, resembling *Eoclathrus* Squinabol, 1887; inorganic (16). *?L.Perm.(Verrucano)*, Italy.

Interconulites Desio, 1941. Suggestion for an international name for cone-in-cone structures.

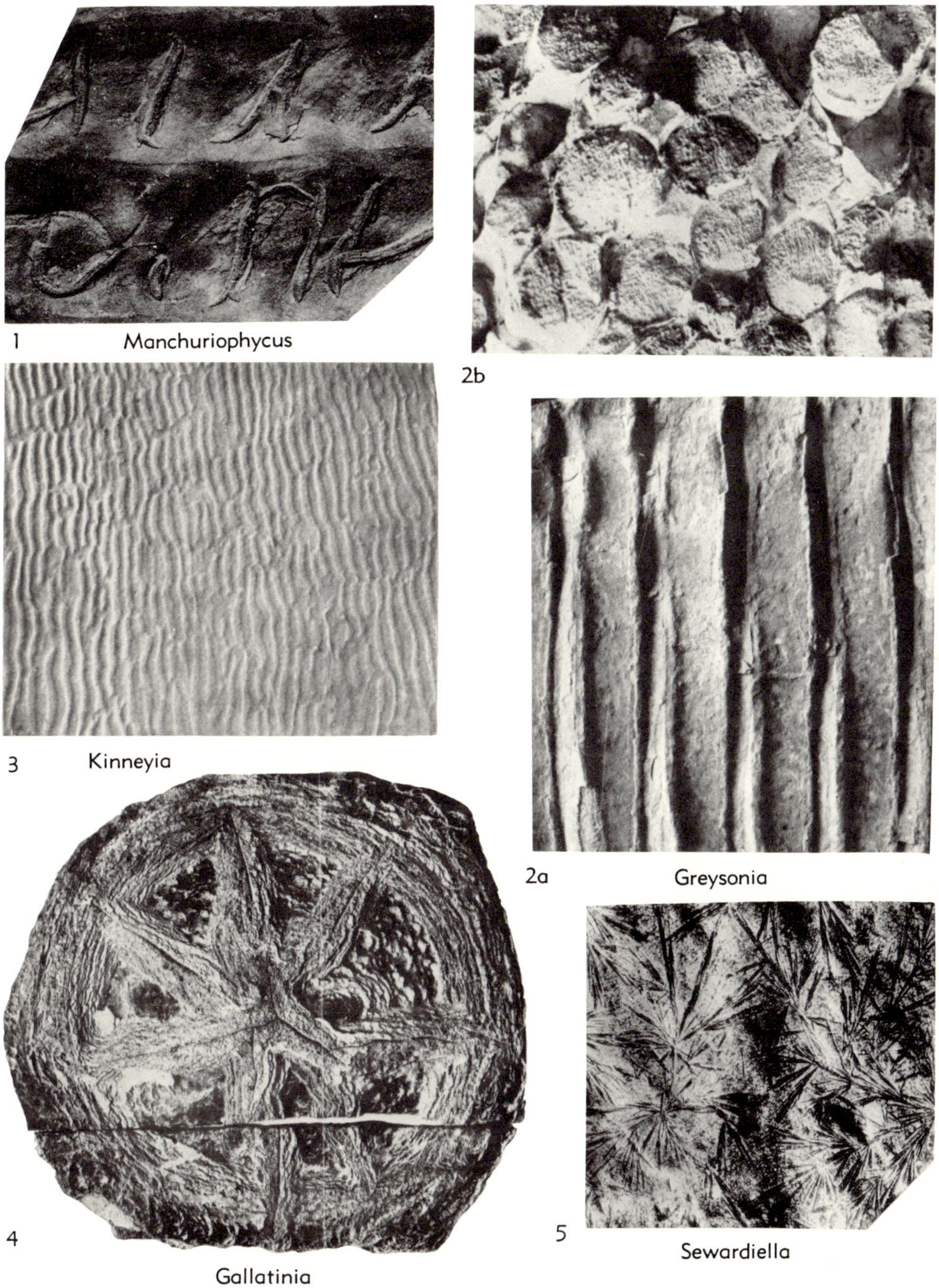

Fig. 146. "Fossils" of inorganic origin (p. *W*234, *W*236).

Kinneyia WALCOTT, 1914 [**K. simulans*]. Reliefs reminiscent of very small ripple marks; 1 to 3 mm. wide, approximately parallel; similar to or identical with *Furchensteine* (furrow-stones) or corroded limestone flags; perhaps inorganic (37, 45). *Precam.*, USA; *Sil.*, N.Afr.——FIG. 146,*3*. **K. simulans*, Precam. Belt. Ser., (Newland Ls.), Mont.; upper surface, ×0.7 (45).

Kräuselia FUCINI, 1936 [**K. verrucana*]. Narrow, long, tapering swellings, apparently screw-shaped, twisted; inorganic (16). *?L.Perm.(Verrucano);* Italy.

Lithodictuon CONRAD, 1837 [**L. beckii*]. Very probably mud cracks (2). *Sil.*, USA(N.Y.).

Manchuriophycus ENDO, 1933 [**M. yamamotoi*]. Mud cracks, in part in normal form of polygons *(M. yamamotoi)*, in part curved in valleys of simple or interference ripples *(M. sawadai)* (37). [Interpreted by ENDO as algae; by LEE (1939) as worm burrows; recognized by HÄNTZSCHEL (1949) as inorganic.] *Precam.-Trias.*, Eu.-Asia-Can.-Greenl.——FIG. 146,*1*. *M. sawadai* YABE, Precam., Asia; ×0.4 (141).

Membranites FUCINI, 1938. Very probably inorganic (16). *?L.Perm.(Verrucano)*, Italy.

Neantia LEBESCONTE, 1886. Wrinkle-like structures very closely resembling rill marks and other markings (42, 46). [Certainly not sponges, as LEBESCONTE thought.] *Precam.*, Fr.——FIG. 145,*7*. *N. rhedonensis* LEBESCONTE*;* ca. ×0.7 (92).

Newlandia WALCOTT, 1914 [**N. frondosa*]. Irregular hemispherical or frondlike bodies; diameter up to 80 cm.; built of concentric, subparallel, subequidistant layers; similar to *Collenia* or *Cryptozoon* (45). [Very probably inorganic.] *Precam.*, USA(Mont.).——FIG. 147,*5*. **N. frondosa*, Belt Ser. (Newland Ls.); upper surface, large frond, ×0.5 (45).

Osagia TWENHOFEL, 1919 [**O. incrustata*]. Resembling *Fusulina* in size and shape; with thin concentric lamellae and with nucleus comprising fragment of rock or shell; forming thin beds of limestone. [Regarded by TWENHOFEL as algal "coenoplases" of small size; according to SCHINDEWOLF (37) perhaps only simple ooids or pisoliths.] *Penn.*, USA(Kans.-Okla.).

Palaeotrochis EMMONS, 1856. Double cone, with grooved surface; cones juxtaposed base to base (37). [Formerly regarded by EMMONS as coral; determined by HALL (1857), MARSH (1868), HOLMES & DILLER (1899) as inorganic; possibly concretions or cone-in-cone structures; according to WALCOTT (1899) spherolite of an acidic magmatic rock. *Precam.*, USA(N.Car.).——FIG. 145,*1*. *P. minor; 1a-c,* mag. unknown (137).——FIG. 145,*2*. *P. major; 2a-c,* mag. unknown (137).

Palmacites martii HEER, 1855 [=*Palmanthium martii* SCHIMPER, 1870]. "Fossil" interpreted as flower or fruit of a palm; according to SCHENK, possibly inorganic (16). *U.Tert.(Molasse)*, Switz.

Panescorsea DE SAPORTA, 1882 [=*Panescorea* ANDREWS, 1955 *(errore)*]. Long parallel ridges on bedding planes. [Erroneously explained by DE SAPORTA as seaweed.] Resembling ripplemarks or mud flow markings (2). *Cret.-Tert.*, Fr.-Italy.

Phyllitites FUCINI, 1936 [**P. rugosus*]. Inorganic (16). *?L.Perm.(Verrucano)*, Italy.

Phytocalyx BORNEMANN, 1886 [**P. antiquus*]. Structureless conical or hemispherical bodies originally regarded as algae (2). [Probably concretionary bodies.] *Cam.*, Italy(Sardinia).

Piaella FUCINI, 1936 [**P. biformis*]. Inorganic (16). *?L.Perm.(Verrucano)*, Italy.

Polygonolites DESIO, 1941. Suggested as international designation for mud cracks.

Protadelaidea TILLYARD, 1936 [**P. howchini*]. Fragments in form of ochreous to black crusts in quartzites, with rather regular angular outlines (37). [Erroneously believed to represent body segments of giant arthropods. According to GLAESSNER (1959), possibly formed by pyritized soft plant tissue. Very similar forms described by HUPÉ (1952) from the Precambrian of Morocco as inorganic (mud flakes or flattened clay pellets).] *Precam.(Adelaide System)*, S.Austral.

Pseudopolyporus HOLLICK, 1910 [**P. carbonicus*]. Concretion, closely resembling fungus (especially *Polyporus*) and originally described as such (16). *Carb.*, USA(W.Va.).

Reynella DAVID, 1928 [**R. howchini*]. Problematical small fragments of exceedingly irregular shape (16). [Erroneously explained by DAVID as belonging to the crustaceans; according to GLAESSNER (personal communication, 1956, 1959) not recognizable as animal remains, perhaps inorganic (16). *Precam.(Brighton Ls., Sturtian)*, S.Austral.

Rivularites FLICHE, 1905 [**R. repertus*]. Explained by FLICHE and D. WHITE (1929) as algal colonies (?Cyanophyceae). American "species," *R. permiensis* WHITE, is very similar to mud flow markings on bedding planes; compared by C. L. FENTON (1946) with small ripple marks (2). *Perm.*, Ariz.; *U.Trias.*, Fr.——FIG. 147,*3*. *R. permiensis* WHITE, Perm., (Hermit Sh.), Ariz., ×0.4 (138).

Schafferia FUCINI, 1938 [**S. verrucana*]. Inorganic (16). *?L.Perm.(Verrucano)*, Italy.

Sewardiella FUCINI, 1936 [**S. verrucana*] [=*Baieropsis* FUCINI, 1928 (*non* FONTAINE, 1889)]. Sharply stamped impressions resembling palm branches, fans, or rosettes on bedding planes (16). [Evidently molds of radiate crystal aggregates (gypsum or ice), not algae as believed by FUCINI.] *?L.Perm.(Verrucano)*, Italy.——FIG. 146,*5*. **S. verrucana;* ca. ×0.7 (71).

Sickleria MÜLLER, 1846 [**S. labyrinthiformis*]. Shrinkage cracks in sandstone; originally regarded as plants (16). *L.Trias.*, Ger.

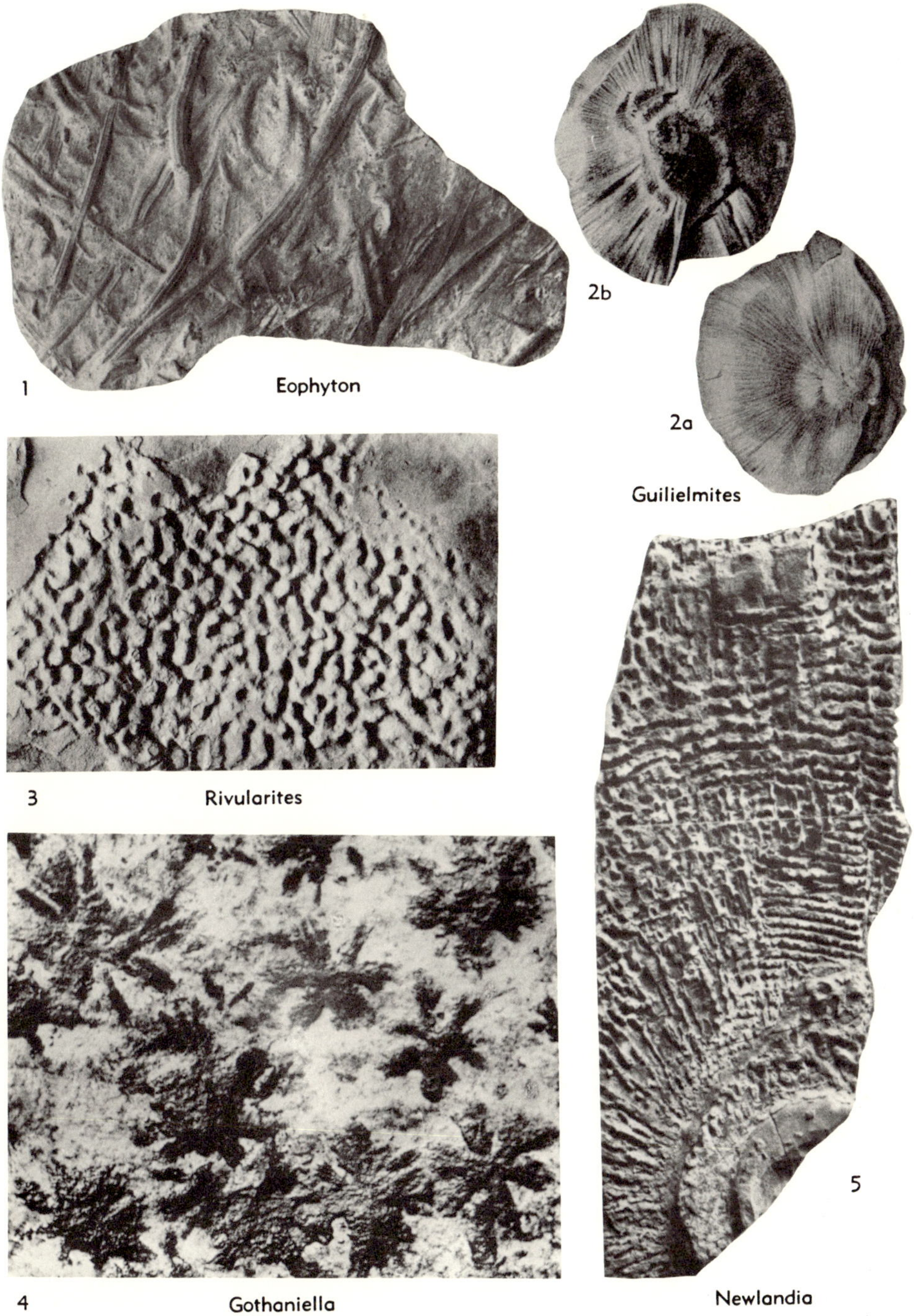

Fig. 147. "Fossils" of inorganic origin (p. *W*234, *W*236).

Sidneyia groenlandica CLEAVES, 1935. Not abdominal segments of poorly preserved arachnid, as originally believed according to EHA (1953), more probably group of ripple marks partly removed by erosion (37). *Precam.*, Greenl.

Squamopsis FUCINI, 1938 [**S. modesta*]. Inorganic (16). *?L.Perm.(Verrucano),* Italy.

Stylolithes KLÖDEN, 1828 [**S. sulcatus*]. Regarded by KLÖDEN as problematical fossil; actually stylolites (16). *M.Trias.(Muschelkalk),* Ger.

Tazenakhtia CHOUBERT, TERMIER & TERMIER, 1951 [**T. aenigmatica*]. "Organisms" observed in thin sections of limestones, interpreted as of animal origin (?Foraminifera), but also compared with calcareous algae (*Nubecularites* MASLOV); according to MOSEBACH (1956) certainly inorganic structures due to combination of tectonic movements and metamorphic recrystallization (37). *Precam.*, Morocco.

Tubiphyton CHOUBERT, TERMIER & TERMIER, 1951 [**T. taghdoutensis*]. "Organisms" observed in thin sections of limestones, interpreted as calcareous algae; according to MOSEBACH (1956), certainly inorganic structures due to combination of tectonic movement and metamorphic recrystallization (37). *Precam.*, Morocco.

Vesicolithus FRITSCH, 1908 [*nom. nud.*] [**V. guttalis*]. Very probably inorganic (?raindrop impressions) (9). *Ord.*, Czech.

UNRECOGNIZED AND UNRECOGNIZABLE "GENERA"

Numerous "genera," mostly of badly preserved fossils, are included in this group, because of insufficient descriptions and inadequate illustrations. The majority of them are so nondescript that they do not deserve to be named, and under no circumstances should their names be revived. Many of these fossils will remain unexplainable for a long time. In only a few cases are investigations of new and better material likely to clarify their systematic position.

Acanthus GROSSHEIM, 1946 [*non* BLOCH, 1795; *nec* DUMONT, 1816; *nec* GISTL, 1834; *nec* LOCKINGTON, 1876)] [**A. dodecimanus*] (16). *L.Tert.*, USSR.

Aequorfossa NEVIANI, 1925. See *Treatise*, p. *F*159.

Agarites DE SAPORTA, 1890 [*non* AGASSIZ, 1841] [**A. fenestratus*] (16). *U.Jur.*, Fr.

Amanlisia LEBESCONTE, 1891 [**A. simplex*]. Nondescript (?) trail; somewhat similar to *Palaeophycus* HALL (16). *Precam.*, Fr.

Amansites BRONGNIART, 1849 [**Fucoides dentatus* BRONGNIART, 1828]. ?Graptolites (16). *?Ord.*, Can.

Ampelichnus HITCHCOCK, 1865 [**Grammepus uniordinatus* HITCHCOCK, 1858] [=*Grammepus* HITCHCOCK, 1858 *(partim)*] (18). *Trias.*, USA (Mass.).

Archaeorrhiza TORRELL, 1870 [**A. tuberosa*]. "Plant" *("Radicibus similis");* never figured (16). *Cam.*, Swed.

Archaeoscolex MATTHEW, 1889 [**A. corneus*]. Dubious fossil interpreted as insect larva; no specimens could be located in Canadian collections (16). *U.Carb.*, Can.(N.B.).

Beaumontia DAVID, 1928 [*non* EDWARDS & HAIME, 1851; *nec* EUDES-DESLONGCHAMPS, 1856] [**B. eckersleyi*] [=*Beaumontella* DAVID, 1928]. Nodular bodies (16). [Interpreted by DAVID as various parts of eurypterids; according to GLAESSNER (personal communication, 1956, 1959) not recognizable as animal remains, perhaps inorganic.] *Precam.(Beaumont Dol., Adelaide System),* S. Austral.

Beltina WALCOTT, 1899 [**B. danai*]. Angular fragments of thin, commonly much distorted and compressed tests without distinctive surface ornamentation (42). [Regarded by WALCOTT as fragmentary remains of Merostomata, and by WHITE (1929) and FENTON & FENTON (1937) as probably noncalcareous algae, if not inorganic (37, 42). *Precam.*, USA(Mont.).——FIG. 148,2. **B. danai,* Belt Ser.(Greyson Sh.); *2a,* body segment, ×?; *2b,* portion of jointed appendage, ×3; *2c,* unidentified fragment with terminal curved spine, ×4; *2d,* appendage with two large basal? joints and two smaller terminal joints, ×2 (134).

Bipezia MATTHEW, 1910 (?1909) [**B. bilobata*] [=*Bipesia* MATTHEW, 1910 *(errore)*]. Spindle-shaped "footprints," pointed at both ends, in pairs opposite each other, coalescing laterally; length 10 mm., width 3 mm. (23). [Interpretation very doubtful, but certainly not of vertebrate origin, as MATTHEW believed; according to GLAESSNER (1957), possibly synonymous with *Isopodichnus* BORNEMANN, 1889.] *Dev.*, Can.(N.B.).

Bitubulites BLUMENBACH, 1803 [**B. problematicus*]. This "genus" (especially the "species" *B. irregularis* VON SCHLOTHEIM, 1820) possibly synonymous with *Rhizocorallium;* name not used for more than 100 years (16). *M.Trias.*, Ger.

Boliviana SALTER, 1861 [**B. melocactus;* SD ANDREWS, 1955]. (16). *Ord.*, Bol.

Bucinella FUCINI, 1936 [**B. verrucana*] (16). *?L. Perm.(Verrucano),* Italy.

Carelozoon Metzger, 1924 [**C. jatulicum*]. Irregularly ramifying and branching, irregularly shaped structures about 0.5 mm. in diameter; circular in cross section, forming network in rock; with crustal layer and possible tabulae; reminiscent of stromatoporoids; affinities unknown; possibly coelenterate, calcareous alga or inorganic (42). *Precam.*, Finl.——Fig. 149,*1*. **C. jatulicum;* cross sec., ×1.1 (Geol. Survey Finland).

Chauviniopsis de Saporta, 1872 [**C. pellati*] (16). *U.Jur.*, Fr.

Chordophyllites Tate, 1876 [or Young & Bird,

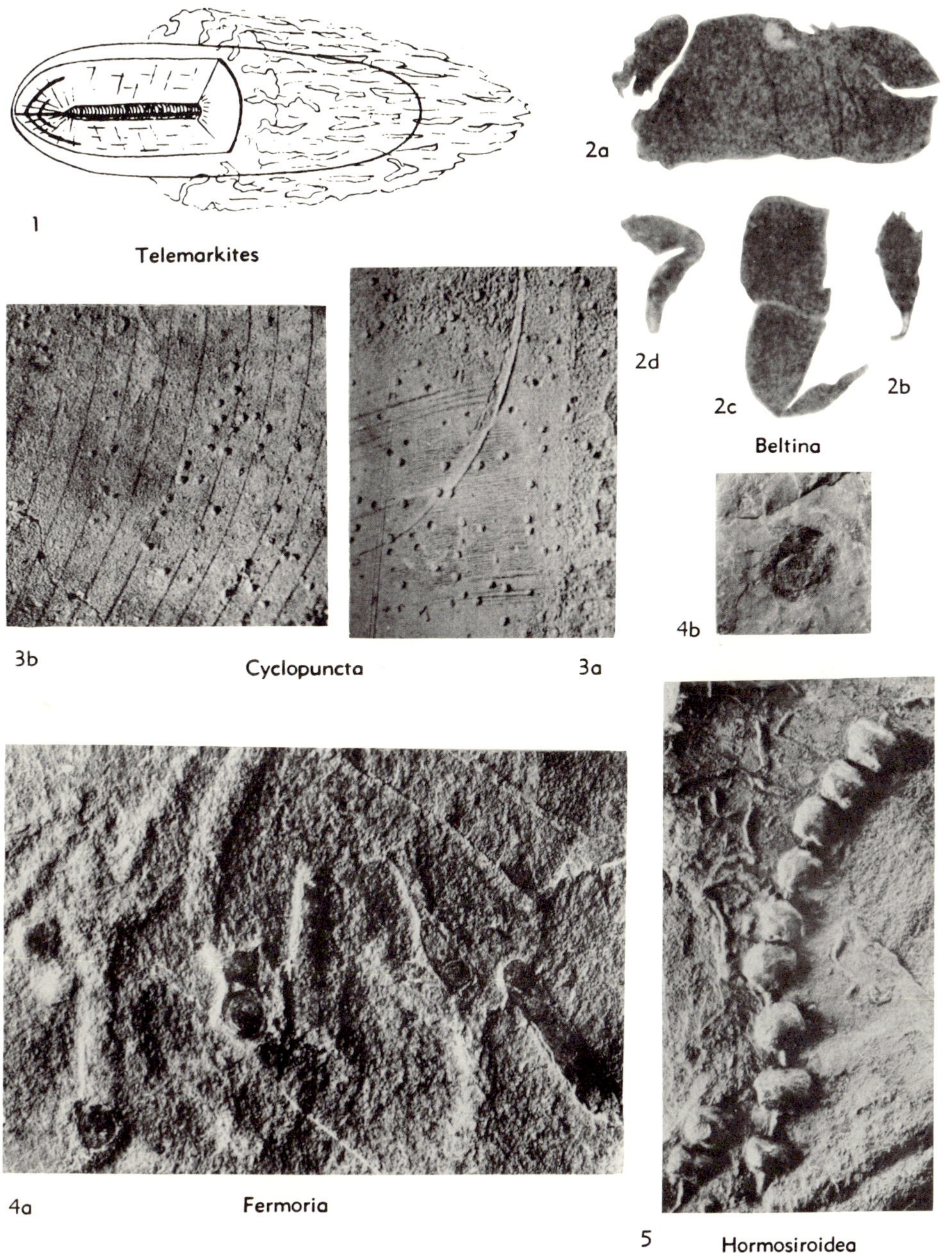

Fig. 148. Unrecognizable "genera" (p. *W*238, *W*240-*W*242).

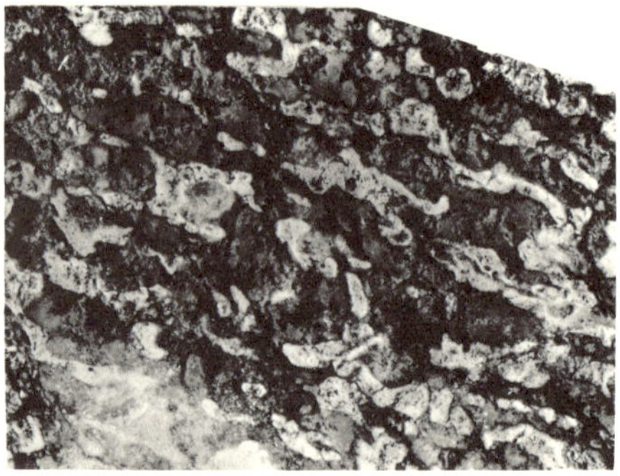

Carelozoon

FIG. 149. Unrecognizable "genera" (p. W239).

1822]. Cylindrical "stems" of great length on bedding planes; "fucoid" (16). *L.Jur.*, Eng.

Codites STERNBERG, 1833 [**C. serpentinus;* SD ANDREWS, 1955]. ?Sponge (16). *U.Jur.*, Fr.

Conchyophycus DE SAPORTA, 1872 [**C. marcignyanus*] (16). *U.Trias.*, Fr.

Cophinus KOENIG, 1839 [**C. dubius*]. Problematical structure resembling an inverted 4-sided pyramid with column-like rounding at each corner; always found in vertical position; tentatively explained by SOWERBY and SALTER (Murchison, 1859) as impressions of rooted crinoid stems which produced observed pattern by wavy and somewhat rotatory motion; possibly inorganic (16). *U.Sil.(Ludlov.)*, Eng.

Crenobaculus FRITSCH, 1908 [**C. draboviensis*] (9). *Ord.*, Czech.

Cyclopuncta ELIAS, 1958 [**C. girtyi*]. Shallow subhemispherical holes; diameter 0.1 to 0.3 mm.; generally irregularly scattered on cephalopod shells, in some specimens tending to follow growth lines (16). [Explained by GIRTY (1909) as perforations in shells, probably made by small gregarious animals (e.g., the lorica-secreting infusorian *Folliculina*), the scars being produced by a prolonged passive attachment.] *Miss.*, USA(Okla.).——FIG. 148,*3*. **C. girtyi; 3a,* on *Bactrites? smithianus*, ×4.6; *3b*, on *Cravenoceras* sp., ×8 (*3a*, 73; *3b*, 65).

Dactyloidites HALL, 1886 [**D. bulbosus* (=*Buthotrephis? asteroides* FITCH, 1850)]. Probably an alga, a medusoid, or rosette-shaped burrows (*Treatise*, p. F159, Fig. 130,*3*) (16). *L.Cam.*, USA(N.Y.); ?M.Cam., USA(Vt.).

Dasycladites FUCINI, 1936 [**D. subclavaeformis*]. (16). *?L.Perm.(Verrucano)*, Italy.

Digitolithus FRITSCH, 1908 [**D. rugatus*]. (9). *Ord.*, Czech.

Discophycus WALCOTT, 1883 [**D. typicalis*]. Discoid, slightly convex bodies; diameter 4 to 12 cm.; outline varying from circular to orbicular, substance ?coriaceous (2). [Interpreted by JAMES (19) as inorganic (fossil mud bubbles); by RUEDEMANN (1925) as fossils of very diverse kinds (seaweeds, sponges, eurypterids).] *Ord.*, N.Am.

Dryalus BARRANDE, 1872 [**D. obscurus*] (9). *Sil.*, Czech.

Duovestigia BUTTS, 1891 [**D. scala*]. Described as amphibian footprint, but obviously of invertebrate origin (16). *U.Carb.*, USA(Mo.).

Dystactophycus MILLER & DYER, 1878 [**D. mamillanum*]. According to JAMES (19), impression of coral base that left its mark in concentric rings. *Ord.*, USA(Ohio).

Eocladophora FUCINI, 1936 [**E. fibrata*]. Long, narrow, threadlike pads or ridges (16). *?L.Perm. (Verrucano)*, Italy.

Eurypterella MATTHEW, 1889 [**E. ornata*]. Dubious fossil interpreted as peculiar crustacean; no specimens could be located in Canadian collections (16). *U.Carb.*, Can.(N.B.).

Fengtienia ENDO & RESSER, 1937 [**F. peculiaris*]. According to ÖPIK (1959) perhaps "only a *Rusophycus*" (p. W212); see also *Treatise*, p. O102 (16). *M.Cam.*, Manchur.

Fermoria CHAPMAN, 1935 [**F. minima*] [=*Protobolella* CHAPMAN, 1935]. Small disc-shaped carbonaceous structures, 2 to 4 mm. in diameter, concentrically wrinkled (37). [Interpreted as ?atremate brachiopod (CHAPMAN, 1935); compared with algae (SAHNI & SHRIVASTAVA, 1954, HOWELL, 1956, and others); probably inorganic, according to MISRA & DUBE (1952).] *Precam.* or *Cam.(Vindhyan F.)*, India.——FIG. 148,*4*. *F.* sp.; *4a*, attached to filament-like bodies, ×2; *4b*, ×4 (*4a*, Sahni, n; *4b*, Howell, n).

Fruticristatum WEBSTER, 1920 [**F. iowense;* SD ANDREWS, 1955]. Never figured; first described as alga (2). [Obviously represents fillings of nondescript burrows.] *Dev.*, USA(Iowa).

Furca FRITSCH, 1909 [BARRANDE MS] [**F. bohemica*]. Possibly pluteus larva of crinoid (9). *Ord.*, Czech.

Gleichenophycus MASSALONGO, ?date [*?*G. granulosus* (=*?Caulerpa lehmanni* HEER, 1877)] (16). *Cret.*, Italy-Switz.

Gordioides FRITSCH, 1908 [*nom. nud.*] [**G. spiralis*] (9). *Sil.*, Czech.

Gracilerectus WEBSTER, 1920 [**G. hackberryensis*]. Nondescript straight or curved, cylindrical "stems" (2). [Originally regarded as algae ("fucoids"), but obviously burrows.] *U.Dev.*, USA(Iowa).

Granifer FRITSCH, 1908 [**G. stolatus*] (9). *Ord.*, Czech.

Guttolithus FRITSCH, 1908 [**G. strasseri*] (9). *Ord.*, Czech.

Harpagopus HITCHCOCK, 1848 [**H. dubius;* SD HÄNTZSCHEL herein]. Obliquely placed elliptical impressions (25). *?M.Dev.*, USA(N.Y.); *Trias.*, USA(Mass.).

Helviensia DE LIMA, 1895 [**H. delgadoi*] (16). *Cam.,* Port.

Hippodophycus HALL & WHITFIELD, 1872 [**H. cowlesi*]. Described as marine plants with swelling roots, which are laterally expanded in form of subcircular disc, with one edge truncate; known only from impressions; holotype (single described specimen) probably lost; perhaps inorganic (2). *U.Dev.,* USA(N.Y.).

Hoplichnus HITCHCOCK, 1848 [**H. quadrupedans* (?=*Chelichnus gigas* JARDINE, 1850)]. Hoof-shaped, semioval reliefs resembling impressions of horseshoes; diameter about 2 inches; perhaps markings or (particularly the "species" *H. equus* HITCHCOCK, 1858) belonging to *Spreiten-bauten* (17). *Penn.-Trias.,* N.Am.-?Eng.-?Ger.

Hormosiroidea SCHAFFER, 1928 [**H. florentina*] [=?*Corallinites rosarium* MASSALONGO, 1851; *C. tuna* MASSALONGO, 1855; ?*Halimedopsis* MASSALONGO, 1859]. Spherical or hemispherical bodies arranged on thin strings like pearls; diameter of hemispheres about 0.5 to 1 cm.; surface of some specimens coarsely granulose (16). [Regarded by HEER (1877) and SCHAFFER (1928) as an alga, by SEILACHER (1959) as a rosary-like trail of unknown origin.] *Cret.-L.Tert.(Flysch),* Eu.——FIG. 148,*5*. **H. florentina,* U.Cret., Italy; ×0.7 (Naturhist. Mus. Wien).

Hydrocytium (?) **silicula** MATTHEW, 1890. Minute oval bodies; length 0.5 mm., width 0.25 mm.; with strong cuticle and pedicle-like knob at one end (2). *Cam.,* Can.(Nova Scotia).

Hylopus? variabilis MATTHEW, 1910. Very doubtful "footprints," referred to vertebrates; according to ABEL (1) not a vertebrate track, but an unrecognizable form. *Dev.,* Can.(N.B.).

Ichnophycus HALL, 1852 [**I. tridactylus*]. Doubtful tridactyl impressions (16). *Sil.,* USA(N.Y.).

Itieria DE SAPORTA, 1872 [jr. hom.; *non* MATHERON, 1842]. (2). *U.Jur.,* Fr.

Krishnania SAHNI & SHRIVASTAVA, 1954 [**K. acuminata*]. Acutely ovate discs, superficially resembling *Lingula,* with deep marginal furrow; longest axis 7.5 mm., maximum width 4 mm.; probably belonging to the Fermoriidae SAHNI (16). *Precam.* or *Cam.*?(*Vindhyan* F.), India.

Laminarites STERNBERG, 1833 [**Fucoides tuberculosus* BRONGNIART, 1828; SD ANDREWS, 1955]. "Genus" comprising very heterogenous "species," similar to *Laminaria;* straight and parallel structures on bedding planes (2). [Seemingly in part of plant origin (e.g., *L. antiquissimus* EICHWALD), in part certainly ripples or flow casts (e.g., *L. langrangei* DE SAPORTA & MARION), in part (e.g., *L. pseudoichnites,* according to MESCHINELLI & SQUINABOL, 1892) also trails.] *Cret.,* Fr.

Laminopsis FUCINI, 1938 [**L. insignis*] (16). ?*L.Perm.(Verrucano),* Italy.

Lepidotruncus FRITSCH, 1908 [**L. fortis*] (9). *Ord.,* Czech.

Leptophycus FRITSCH, 1908 [**L. venosus,* SD ANDREWS, 1955] (9). *Ord.,* Czech.

Leuconoe BOGATSCHEW, 1930 [jr. hom.; *non* BOIE, 1830] [**L. paradoxa*]. Larva of arthropod of unknown systematic position (16). *L.Eoc.(Flysch),* USSR.

Lithodictyon TORELL, 1870 [**L. fistulosum*]. Possibly small mud cracks (16). *Cam.,* Swed.

Lithostachys FISCHER-OOSTER, 1858 [**L. alpina*]. ?Plant (7). *Jur.,* Switz.

Macrocystites FUCINI, 1936 [**M. similis*]. Trail or inorganic (16). ?*L.Perm.(Verrucano),* Italy.

Mastocarpites ZIGNO, 1956 [*non* TREVISAN, 1849 *(nom. nud.)*] [**Algacites crucaeformis* STERNBERG, 1833]. ?Coprolite (2). *U.Jur.,* Ger.

Micrapium TORELL, 1870 [**M. erectum*]. Never figured (16). *Cam.,* Swed.

Naites GEINITZ, 1867 [**N. priscus*]. Rather valueless name for a trail somewhat resembling that of the Recent genus *Nais* (16). [Interpreted by GEINITZ as a bodily preserved annelid.] *L.Carb.,* Ger.

Nanopus? vetustus MATTHEW, 1910. Very doubtful "footprints," referred to vertebrates; according to ABEL (1) not a vertebrate track, but an unrecognizable form. *Dev.,* Can.(N.B.).

Nematolites KEEPING, 1882 [**N. edwardsii;* SD ANDREWS, 1955]. Poorly preserved, "curious irregular branching structures" (2). *Sil.,* Eng.

Nereitopsis GREEN, 1899 [**N. cornubicus*]. ?Trail, somewhat similar to *Nereites;* poorly described and figured (16). *L.Dev.,* Eng.

Nimbus BOGATSCHEW, 1930 [jr. hom.; *non* MULSANT & REY, 1870] [**N. helianthoides*]. Starlike? trace fossil, somewhat resembling *Atollites* or similar forms (16). [Explained as belonging to Trachymedusae or Narcomedusae.] *L.Eoc.(Flysch),* USSR.

Nisea DE SERRES, 1840 [*non* RAFINESQUE, 1815 *(nom. nud.)*] [=*Nemausina* DUMAS, 1876]. Irregularly shaped globular or ellipsoidal bodies which give off 2 or more long, transversely striped or slightly segmented tubes (16). [Interpreted as annelids, mollusks, or coelenterates.] *L.Cret.,* S.Fr.

Orthocaris FRITSCH, 1908 [**O. splendens*] (9). *Ord.,* Czech.

Palaeonereis EICHWALD, 1856 [*non* HUNDT, 1940] [**P. prisca*]. Poorly based "genus" interpreted as a bodily preserved polychaete worm, but obviously a trail, possibly related to *Nereites*; in one paper of EICHWALD (1856) described as "*P.* mihi"; in a later one (1860) as "*Palaeonereis* SOWERBY"; no description given by SOWERBY was to be found (16). ?*Ord.,* USSR.

Papilionata SPRIGG, 1947 [**P. eyrei*]. According to GLAESSNER (1959), possibly a poorly and incompletely preserved specimen of *Dickinsonia* SPRIGG, (1947), and perhaps belonging to the Annelida. *Precam.,* Austral.

Phycoidella MATTHEW, 1890 [**P. stichidifera*]. Strap-shaped "fronds" showing irregular rows of dark spots or granules transversely arranged on "stem"; some "fronds" with enlarged extremity "like a *stichida*"; according to MATTHEW, related to *"Fucoides circinnatus"* BRONGNIART and belonging to algae; perhaps trace fossil (2). *Cam.*, Can.

Platyrhynchus GLOCKER, 1850 [jr. hom.; *non* LEUCKART, 1816; *nec* SWAINSON, 1820; *nec* CUVIER, 1826; *nec* WAGLER, 1830; *nec* AGASSIZ, 1846; *nec* VAN BENEDEN, 1876; *nec* CHEVROLÂT, 1882] [**P. problematicus*] (16). *?U.Cret.*, Ger.

Portelia BOURSAULT, 1889 [jr. hom.; *non* DE QUATREFAGES, 1850] [**P. meunieri*]. Nondescript, branched cylindrical fillings of tunnels; very poorly figured (2). *U.Jur.*, Fr.

Ptilichnus HITCHCOCK, 1858. Finlike impressions, arranged in rows; others consisting of parallel, slightly curved grooves (17). [According to HITCHCOCK, swim trails of fishes; more probably markings of rolling or dragging objects.] *Trias.*, USA(Mass.).

Ptychoplasma FENTON & FENTON, 1937 [**P. excelsum*] (23). *Penn.*, USA(Tex.).

Pucksia SOLLAS, 1895 [**P. machenryi*]. Long, narrow, threadlike markings in slate (16). *Cam.*, Ire.

Punctatumvestigium BUTTS, 1891 [**P. circuliformis*]. Described as amphibian footprint, but obviously of invertebrate origin (16). *U.Carb.*, USA (Mo.).

Quallites FRITSCH, 1908 [**Q. graptolitarum* (=*Q. problematicus* FRITSCH, 1908)] (9). *Sil.*, Czech.

Radicites FRITSCH, 1908 [**R. rugosus*]. Name used only in explanation of figure; in text FRITSCH calls same fossil *Radix corrugatus* (9). *Ord.*, Czech.

Radicopsis FUCINI, 1938. Probably inorganic (16). *?L.Perm.(Verrucano)*, Italy.

Radiophyton MEUNIER, 1887 [**R. sixii*]. Tetraradiate, probably accidental structure (2). *U.Jur.*, Fr.

Radix FRITSCH, 1908 [jr. hom.; *non* DE MONTFORT, 1810] [**R. corrugatus*]. In explanation of figure same fossil named *Radicites rugosus* (9). *Ord.*, Czech.

Rauffella ULRICH, 1889 [**R. filosa*] [=*Raufella* SARDESON, 1896 *(errore)*]. Possibly trace fossils (16). [See *Treatise*, p. *E*107]. *Ord.*, USA.

Rhizomorpha HERNANDEZ-PACHECO, 1908 [**R. calderoni*]. Bulging structures on bedding planes; 3 to 12 mm. in diameter; irregularly branched; curved (16). *Sil.*, Sp.

Saccophycus JAMES, 1879 [**S. inortus*]. Possibly burrows, smooth or striated longitudinally; never figured (2, 19). *Ord.*, USA(Ohio).

Saltator HITCHCOCK, 1858 [jr. hom.; *non* VIEILLOT, 1816]. Inorganic markings or tracks made by animals moving by leaps; 2 "species" having little in common (17). *Trias.*, USA(Mass.).

Scotolithus LINNARSSON, 1871 [**S. mirabilis*] (16). *Cam.*, Swed.

Solicyclus QUENSTEDT, 1879. Elliptical reliefs, smooth internally; marginal seam divided by numerous radial rays (16). *L.Jur.*, Ger.

Sphaerapus HITCHCOCK, 1858. ?Trackway consisting of 2 rows of small (diameter 3 to 5 mm.) hemispherical impressions (17). *Trias.*, USA(Mass.).

Sphenopus FRITSCH, 1908 [jr. hom.; *non* STEENSTRUP, 1856] [**S. pectinatus*] (9). *Ord.*, Czech.

Spongia paradoxica WOODWARD, 1833 [=*Siphonia paradoxica*, AUCTT.]. "Ramifying zoophyte, resembling the roots of trees, about an inch thick, branching and interweaving in every direction, fragments not unlike the horns of a stag" (16). [According to MCKENNY HUGHES (1884), not sponges, but of inorganic (concretionary) origin.] *L.Cret.*, Eng.

Spongolithus FRITSCH, 1908. Very heterogenous group of ridgelike and tracklike structures (9). *Ord.*, Czech.

Squamularia ROTHPLETZ, 1896 [*non* GEMMELLARO, 1899] [**Caulerpa cicatricosa* HEER, 1877; SD HÄNTZSCHEL, herein]. Possibly small fucoids (2, 23). *Tert.*, Eu.

Staurophyton MEUNIER, 1891 [**S. bagnolensis*]. Similar to *Radiophyton* MEUNIER, 1887 (16). [See *Treatise*, p. *F*23, Fig. 12,2.] *Ord.*, Fr.

Stratipes HITCHCOCK, 1858 [**S. latus*]. Very large trail, about 20 inches between 2 rows of impressions, doubtful if made by invertebrate (17). *Trias.*, USA(Mass.).

Striocyclus QUENSTEDT, 1879. Reliefs on bedding planes with radial, wormlike ornament and central hollow (16). *L.Jur.*, Ger.

Telemarkites DONS, 1959 [**T. enigmaticus*]. Ellipsoidal nodules; long axis parallel to bedding planes; 2 to 4 cm. in length; 1 to 2 cm. across; composed of fine-grained quartz, with central tube filled with coarse-grained quartz, lying parallel to the long axis (16). [According to DONS, organic or organic-controlled origin (primitive sponges or concretions formed by intervention of algae).] *Precam.*, S.Norway.——FIG. 148,*1*. **T. enigmaticus*, simpl. reconstr. showing internal structures, ×1.5 (64).

Thinopus antiquus MARSH, 1896. Single "footprint" with 3 "toe-impressions," described as earliest record of a terrestrial vertebrate; according to ABEL (1) and others, not a vertebrate footprint; in ABEL's opinion a "fossil" allowing of various explanations; possibly only a fish coprolite (1). *U.Dev.*, USA(Pa.).

Trianisites RAFINESQUE, 1821 [**T. cliffordi*]. See *Treatise*, p. *F*159.

Trichoides HARKNESS, 1855 [**T. ambiguus*]. Hair like bodies; generally straight, some slightly curved; length irregular, about 1 inch; never figured (2). *Ord.(Llandeil.)*, Scot.

Tropidaulus FENTON & FENTON, 1937 [**T. magnus*] (23). *Penn.*, USA(Tex.).

Truncus FRITSCH, 1908 [**T. ramifer*] (9). *Ord.*, Czech.

Valonites SORDELLI, 1873 [**V. utriculosus*] (16). *Plio.*, Italy.

Vermiculites ROUAULT, 1850 [jr. hom.; *non* BRONN, 1848] [**V. panderi*]. Poorly described and never figured (2). *Ord.*, Fr.

Walcottia MILLER & DYER, 1878 [**W. rugosa*]. "Genus" including 3 different "species" of long, tapering, rugose, flexuous impressions of worm-like shape (16). [The 3 "species" explained by JAMES (1886) as a burrow, an impression of a crinoid, and an imprint of a starfish arm, respectively.] *Ord.*, USA (Ohio).

Zearamosus WEBSTER, 1920 [**Z. elleria*]. (2). *Dev.*, USA (Iowa).

ACKNOWLEDGMENTS

The number of paleontologists in all parts of the world who have helped me in such friendly ways in the often difficult task of finding and procuring type specimens and literature is so large that it is impossible to name them individually. My sincere thanks are due to all who aided and promoted my work by frequent painstaking, time-consuming search for identification of often-neglected Problematica. Special thanks are due to my colleague Dr. ULRICH LEHMANN (Hamburg) for the difficult translation of the German text.

REFERENCES

The included papers represent only a very small selection from the widely scattered literature on palichnology. As far as could be ascertained, at least 900 publications dealing exclusively with the fossil *Lebensspuren* of invertebrates and with the Problematica had been published up to 1957. Commonly, however, the descriptions and interpretations of these sorts of fossils are contained in purely stratigraphic or faunistic texts where they easily can be overlooked. Altogether more than 1,100 publications have been consulted.

The best and most complete bibliographies on palichnology are found in WILCKENS (46) and especially LESSERTISSEUR (23). Those of NATHORST (28) and WINKLER (47) are older. A bibliographical index of all "genera" treated here (including synonyms), to be published by the author in 1962 (16), will make it possible to find their original descriptions.

Abel, Othenio

(1) 1935, *Vorzeitliche Lebensspuren:* xv+644 p., 530 fig., G. Fischer (Jena).

Andrews, H. N., Jr.

(2) 1955, *Index of generic names of fossil plants, 1820-1950:* U. S. Geol. Survey, Bull. 1013, p. 1-262.

Azpeitia Moros, Florentino

(3) 1933, *Datos para el estudio paleontológico del Flysch de la Costa Cantàbrica y de algunos otros puntos de España:* Inst. Geol. & Minero España, Bull., v. 53, p. 1-65, pl. 1-19.

Brown, A. P.

(4) 1912, *The formation of ripple-marks, tracks and trails:* Acad. Nat. Sci. Philadelphia, Proc., v. 63, p. 536-547, pl. 41-42.

Dawson, J. W.

(5) 1890, *On burrows and tracks of invertebrate animals in Palaeozoic rocks, and other markings:* Geol. Soc. London, Quart. Jour., v. 46, p. 595-617, fig. 1-19.

Desio, Ardito

(6) 1940, *Vestigia problematiche Paleozoiche della Libia:* Museo Libico Storia Nat., Ann., v. 2, p. 47-92, pl. 4-13.

Fischer-Ooster, Carl von

(7) 1858, *Die fossilen Fucoiden der Schweitzer-Alpen nebst Eröterungen über deren geologisches Alter:* 74 p., 18 pl., Huber (Bern).

Fraipont, Charles

(8) 1915, *Essais de paléontologie expérimentale:* Geol. Fören. Stockholm, Förhandl., v. 37, p. 435-451, 1 fig.

Fritsch, Anton

(9) 1908, *Problematica Silurica:* 28 p., 12 pl. (Praha).

Fuchs, Theodor

(10) 1895, *Studien über Fucoiden und Hieroglyphen:* Akad. Wiss. Wien, math.-naturwiss. Kl., Denkschr., Bd. 62, p. 369-448, pl. 1-9.

Gilmore, C. W.

(11) 1926-28, *Fossil footprints from the Grand Canyon:* Smithsonian Misc. Coll., v. 77, no. 9, 41 p., 12 pl. (1926); v. 80, no. 3, 78 p., 21 pl. (1927); v. 80, no. 8, 16 p., 5 pl., 7 fig. (1928).

Gómez de Llarena, Joaquín

(12) 1946, *Revision de algunos datos paleontologi-*

cos del Flysch Cretaceo y Numulitico de Guipuzcoa: Inst. Geol. & Minero España, Notas & Comun., no. 15, p. 113-165, pl. 1-8, fig. 1-5.

Häntzschel, Walter

(13) 1937-39, *Fährten und Problematika (Reviews):* Fortschr. Paläont., Bd. 1, p. 52-57 (1937); Bd. 2, p. 58-65 (1939).

(14) 1955, *Lebensspuren als Kennzeichen des Sedimentationsraumes:* Geol. Rundschau, Bd. 43, p. 551-562, fig. 1-2.

(15) 1955, *Rezente und fossile Lebensspuren, ihre Deutung und geologische Auswertung:* Experientia, v. 11, no. 10, p. 373-382, fig. 1-12.

(16) 1962, *Vestigia invertebratorum et Problematica:* Fossilium Catalogus, I. Animalia, pt. (in press), Junk (s'Gravenhage).

Hitchcock, Edward

(17) 1858, *Ichnology of New England; a report on the sandstone of the Connecticut Valley, especially its footmarks:* xii+220 p., 60 pl., White (Boston).

(18) 1865, *Supplement to the ichnology of New England:* 96 p., 20 pl., Wright & Potter (Boston).

James, J. F.

(19) 1884, *The fucoids of the Cincinnati Group:* Cincinnati Soc. Nat. History, Jour., v. 7, p. 124-132, 151-166, pl. 5-6, 8-9.

(20) 1892, *Studies in problematic organisms—the genus Scolithus:* Geol. Soc. America, Bull., v. 3, p. 32-44, 15 fig.

(21) 1894, *Studies in problematic organisms. No. II. The genus Fucoides:* Cincinnati Soc. Nat. History, Jour., v. 16, p. 62-81.

Krejci-Graf, Karl

(22) 1932, *Definition der Begriffe Marken, Spuren, Fährten, Bauten, Hieroglyphen und Fucoiden:* Senckenbergiana, v. 14, p. 19-39.

Lessertisseur, Jacques

(23) 1955, *Traces fossiles d'activité animale et leur signification paléobiologique:* Soc. géol. France, Mém., new ser., no. 74, p. 150, pl. 1-11.

Linck, Otto

(24) 1949, *Lebens-Spuren aus dem Schilfsandstein (Mittl. Keuper km 2) NW-Württembergs und ihre Bedeutung für die Bildungsgeschichte der Stufe:* Ver. vaterländ. Naturk. Württemberg, Jahresh., v. 97-101, p. 1-100, pl. 1-8, fig. 1-5.

Lull, R. S.

(25) 1953, *Triassic life of the Connecticut Valley:* Connecticut Geol. & Nat. History Survey, Bull., no. 81, p. 1-336, pl. 1-12, fig. 1-168.

Mägdefrau, Karl

(26) 1937, *Lebensspuren fossiler "Bohr"-Organismen:* Beiträge naturkundl. Forsch. Südwestdeutschland, Bd. 2, p. 54-67, pl. 4-6, fig. 1-2.

Maillard, Gustave

(27) 1887, *Considérations sur les fossiles décrits comme Algues:* Soc. Paléont. Suisse, Mém., v. 14, no. 4, p. 1-40, pl. 1-5.

Nathorst, A. G.

(28) 1881, *Om spar af nagra evertebrerade djur M.M. och deras paleontologiska betydelse:* K. Svenska Vetenskapsakad., Handl., v. 18, no. 7, p. 1-104, pl. 1-11.

(29) 1886, *Nouvelles observations sur des traces d'animaux et autres phénomènes d'origine purement mécanique décrits comme "algues fossiles":* K. Svenska Vetenskapsakad., Handl., v. 21, no. 14, p. 1-58, pl. 1-5, fig. 1-24.

Raymond, P. E.

(30) 1935, *Precambrian life:* Geol. Soc. America, Bull., v. 46, p. 375-392.

Reis, O. M.

(31) 1910, *Zur Fukoidenfrage:* K.K. Geol. Reichsanst. Wien., Jahrb., Bd. 59, p. 615-638, pl. 17.

Richter, Rudolf

(32) 1924, *Flachseebeobachtungen zur Paläontologie und Geologie VII-XI:* Senckenbergiana, Bd. 6, p. 119-165, fig. 1-8.

(33) 1926, *Flachseebeobachtungen zur Paläontologie und Geologie XII-XIV:* Senckenbergiana, Bd. 8, p. 200-224, pl. 3.

(34) 1927, *Die fossilen Fährten und Bauten der Würmer, ein Überblick über ihre biologischen Grundformen und deren geologische Bedeutung:* Paläont. Zeitschr., Bd. 9, p. 193-240, pl. 1-4, fig. 1-11.

(35) 1928, *Psychische Reaktionen fossiler Tiere:* Palaeobiologica, v. 1, p. 225-244, pl. 23, fig. 1-4.

(36) 1941, *Fährten als Zeugnisse des Lebens auf dem Meeres-Grunde:* Senckenbergiana, Bd. 23, p. 218-260, fig. 1-17.

Schindewolf, O. H.

(37) 1956, *Über präkambrische Fossilien: Geotektonisches Symposium zu Ehren von H. Stille, herausgegeben von F. Lotze:* p. 455-480, pl. 31-34, Enke (Stuttgart).

Seilacher, Adolf

(38) 1953, *Studien zur Palichnologie. I. Über die Methoden der Palichnologie:* Neues Jahrb. Geol. & Paläont., Abhandl., v. 96, p. 421-452, pl. 14, fig. 1-10.

(39) 1953, *Studien zur Palichnologie. II. Die fossilen Ruhespuren (Cubichnia):* Neues Jahrb. Geol. & Paläont., Abhandl., v. 98, p. 87-124, pl. 7-13, fig. 1-5.

(40) 1954, *Die geologische Bedeutung fossiler Lebensspuren:* Deutsch. geol. Gesell.,

Zeitschr., Bd. 105, p. 214-227, pl. 7-8, fig. 1-3.
(41) 1955, *Spuren und Lebensweise der Trilobiten; Spuren und Fazies im Unterkambrium:* in SCHINDEWOLF, O. H., & SEILACHER, A., Beiträge zur Kenntnis des Kambriums in der Salt Range (Pakistan): Akad. Wiss. & Lit. Mainz. Math.-naturwiss. Kl., Abhandl., Jahrg. 1955, no. 10, p. 86-143, pl. 16-27, fig. 1-11.
(42) 1956, *Der Bebinn des Kambriums als biologische Wende:* Neues Jahrb. Geol. & Paläont., Abhandl., v. 103, p. 155-180, pl. 8-9, fig. 1-2.

Simpson, Scott
(43) 1957, *On the trace-fossil Chondrites:* Geol. Soc. London, Quart. Jour., v. 112, p. 475-499, pl. 21-24, fig. 1-2.

Ulrich, E. O.
(44) 1904, *Fossils and age of the Yakutat Formation: description of collections made chiefly near Kadiak, Alaska:* Smithsonian Inst., Harriman Alaska Ser., v. 4, Geol. & Paleont., p. 125-146, pl. 11-21.

Walcott, C. D.
(45) 1914, *Cambrian geology and paleontology. III., No. 2. Pre-Cambrian Algonkian algal flora:* Smithsonian Misc. Coll., v. 64, p. 77-156, pl. 4-23.

Wilckens, Otto
(46) 1947, *Paläontologische und geologische Ergebnisse der Reise von Kohl-Larsen (1928-1929) nach Süd-Georgien:* Senckenberg. Naturf. Gesell., Abhandl., no. 474, p. 1-75, pl. 1-9. [bibliography on trace-fossils, p. 50-61.]

Winkler, T. C.
(47) 1886, *Histoire de l'ichnologie. Etude ichnologique sur les empreintes de pas d'animaux fossiles:* Arch. Musée Teyler, ser. 2, v. 2, p. 241-440, pl. 8-19.

SOURCES OF ILLUSTRATIONS

(48) **Andrée, Karl**
(49) **Azpeitia Moros, Florentino**
(50) **Barbour, E. H.**
(51) **Berger, Walter**
(52) **Brady, L. F.**
(53) **Brighton, A. G.**
(54) **Brönnimann, Paul**
(55) **Brown, Barnum, & Vokes, H. E.**
(56) **Caster, K. E.**
(57) **———, & Brooks, H. K.**
(58) **Clarke, J. M.**
(59) **———, & Swartz, H. K.**
(60) **Codez, Jean, & Saint-Seine, Roseline de**
(61) **Crookall, R.**
(62) **Dahmer, Georg**
(63) **Delgado, J. F. N.**
(64) **Dons, J. A.**
(65) **Elias, M. K.**
(66) **Elliott, G. F.**
(67) **Etheridge, Robert**
(68) **Fiege, Kurt**
(69) **Fisher Marple, Mildred, & Stout, Wilber**
(70) **Ford, T. D.**
(71) **Fucini, Alberto**
(72) **Geinitz, H. B.**
(73) **Girty, G. H.**
(74) **Glaessner, M. F.**
(75) **Glocker, F. E.**
(76) **Goldfuss, A. G.**
(77) **Gürich, Georg**
(78) **Guthörl, Paul**
(79) **Hall, James**
(80) **Hallissy, T.**
(81) **Hallam, A.**
(82) **Häntzschel, Walter**
(83) **Haug, Emile**
(84) **Heer, Oswald**
(85) **Howell, B. F.**
(86) **Huckriede, Reinhold**
(87) **Kamptner, Erwin**
(88) **Kozlowski, Roman**
(89) **Ksiazkiewicz, Marian**
(90) **Kuhn, Oskar**
(91) **Lange, Werner**
(92) **Lebesconte, Paul**
(93) **Lesquereux, Leo**
(94) **Linck, Otto**
(95) **Lorenz von Liburnau, J.R.**
(96) **Lull, R. S.**
(97) **Mägdefrau, Karl**
(98) **Mansfield, W. C.**
(99) **Massalongo, A. B.**
(100) **Miller, S. A.**
(101) **Newberry, J. S.**
(102) **Ooster, W. A.**
(103) **Otto, Ernst von**
(104) **Paulus, Bruno**
(105) **Peruzzi, D. G.**
(106) **Pompeckj, J. F.**
(107) **Prantl, Ferdinand**
(108) **Pratje, Otto**
(109) **Reis, O. M.**
(110) **Resser, C. E., & Howell, B. F.**
(111) **Richter, Rudolf**
(112) **Roedel, Hugo**
(113) **Ruedemann, Rudolf**
(114) **Saporta, Gaston de**
(115) **Sarle, C. J.**
(116) **Schimper, W. P.**
(117) **Schindewolf, O. H.**
(118) **Schmidt, Martin**
(119) **Sederholm, J. J.**
(120) **Seilacher, Adolf**
(121) **Serres, Marcel de**
(122) *See* Glocker (75)
(123) **Sollas, W. F.**
(124) **Squinabol, S.**
(125) **Stefani, Carlo de**
(126) **Størmer, Leif**
(127) **Teichert, Curt**
(128) **Termier, Henri, & Termier, Geneviève**
(129) **Torell, O. M.**
(130) **Trusheim, Ferdinand**
(131) **Twenhofel, W. H.**
(132) **Voigt, Ehrhard**
(133) **Volk, Max**
(134) **Walcott, C. D.**
(135) **Wanner, Johannes**
(136) **Westergård, A. H.**
(137) **White, C. H.**
(138) **White, David**
(139) **Whiteaves, J. F.**
(140) **Williamson, W. C.**
(141) **Yabe, Hisakatsu**
(142) **Zapfe, Helmuth**

SUPPLEMENT TO SYSTEMATICS OF CONODONTS

By Klaus J. Müller

Dr. W. H. Hass, author of the main *Treatise* article on conodonts, including systematic descriptions, unfortunately died prior to publication of this volume. Shortly before his death he made revisions and last additions to his typescript, which therefore was well up to date until the middle part of 1959. This chapter provides information for the period extending from mid-1959 to August, 1961.

CONODONT TAXA CLASSED AS VALID

Belodina Ethington, 1959 [**Belodus grandis* Stauffer, 1935]. Family Belodontidae Huddle, 1934. Anteriorly directed horizontal cusp with series of prominent denticles and upward-bowed main denticle; unit longitudinally asymmetrical, with prominent heel posterior to first denticle and 2 conical basal cavities. *M.Ord.-U.Ord.*, N.Am. (USA-Can.).——Fig. 150,*1*. **B. grandis* (Stauffer), U.Ord.(Galena F.), USA(Iowa); lat., ×60 (2).

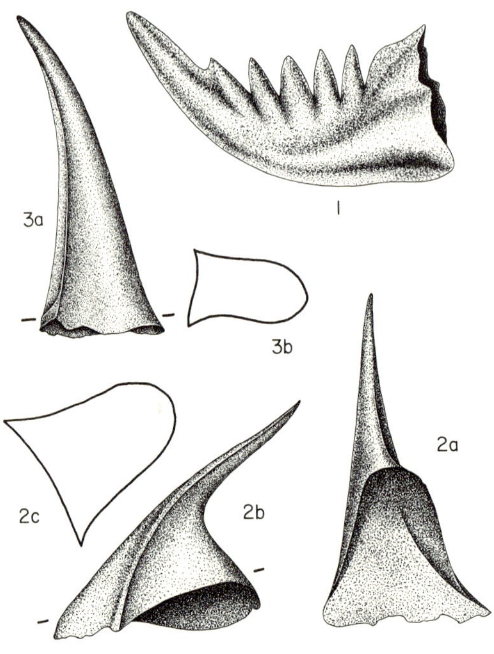

Fig. 150. Conodonts (p. W246).——*1. Belodina. 2. Furnishina. 3. Hertzina.*

Furnishina Müller, 1959 [**F. furnishi*]. Family Distacodontidae Ulrich & Bassler, in Bassler, 1925. Unsymmetrical, with very large, triangular-pentagonal basal cavity; anterior side with broad flat area, corresponding to greatest width of fossil. *U.Cam.*, Eu.(Swed.-N.Ger.)- N.Am.(S.Dak.-Wyo.-Utah-Nev.).——Fig. 150,*2*. **F. furnishi,* Gallatin Ls., USA(Wyo.); *2a-c,* post., lat., transv. sec., ×80 (7).

Hertzina Müller, 1959 [**H. americana*]. Family Distacodontidae Ulrich & Bassler, in Bassler, 1925. Subsymmetrically elongated, with narrow, extremely long basal cavity, with flat area on posterior side limited laterally by carinae. *U.Cam.*, Eu.(Swed.-N.Ger.)-N.Am.(Nev.).——Fig. 150,*3*. **H. americana,* Elvinia Z., USA(Nev.); *3a,b,* lat., transv. sec., ×80 (7).

Plegagnathus Ethington & Furnish, 1959 [**P. nelsoni*]. Family Belodontidae Huddle, 1934. Laterally compressed, with cusp; posterior denticulated bar, and aboral process; deep thin-walled cavity opens posteriorly and tapers to sharp point beneath cusp. *U.Ord.*, N.Am.(N.Man.-Wyo.).——Fig. 151,*1*. **P. nelsoni,* Shamattawa Ls., Man.; lat., ca. ×110 (3).

Proacodus Müller, 1959 [**P. obliquus*]. Family Distacodontidae Ulrich & Bassler, in Bassler, 1925. Asymmetrical, with large basal cavity which is rounded anteriorly and posteriorly and bears stretched-out carina on one side. *U.Cam.*, Eu. (Swed.-N.Ger.).——Fig. 151,*2*. **P. obliquus,* 5d Zone, N.Ger.; *2a,* ant. view with transv. secs.; *2b,* post.; all ×80 (7).

Scaphignathus Ziegler, 1960 [**S. velifera*]. Family Polygnathidae Ulrich & Bassler, in Bassler, 1925. Long, narrow platform with peg-shaped or rounded lower surface, small basal cavity at anterior end; blade short and high laterally, as in *Cavusgnathus*. *U. Dev.(Prolobites - Platyclymenia Z.),* Eu.(Ger.).——Fig. 152,*1*. **S. velifera; 1a,b,* upper and outer lat. views, ×35 (12).

Family WESTERGAARDODINIDAE Müller, 1959

Lamellar, with 2 lateral denticles produced by upwardly bowed flanges of main denticle; lower side smooth and undifferentiated. *M.Cam.-L.Ord.*

Westergaardodina Müller, 1959 [**W. bicuspidata*]. Bilaterally symmetrical; lateral denticles may be larger than middle one; large basal cavity, in some forms divided into 2 lateral cavities. [Little balls associated with the type species prob-

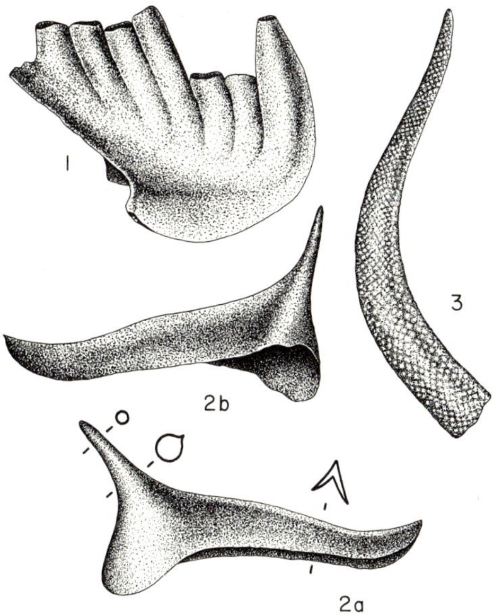

FIG. 151. Conodonts and paraconodont (p. *W*246, *W*248).——*1. Plegagnathus. 2. Proacodus. 3. Rhombocorniculum.*

ably belong to the same conodont taxon (?static organ)]. *M.Cam.-L.Ord.,* Eu.(Swed.-N.Ger.)-N. Am.(S.Dak.-Wyo.).——FIG. 152,2. **W. bicuspidata,* U.Cam. (5b Zone), N.Ger.; ca. ×55 (7).

Conodont Assemblage Classification UNCERTAIN

Westfalicus SCHMIDT in MOORE & SYLVESTER-BRADLEY, 1957 (p. 21) [*pro Gnathodus* SCHMIDT, 1934 (*non* PANDER, 1856)] [**Gnathodus integer* SCHMIDT, 1934]. This generic name was proposed for designation of conodont assemblages named *Gnathodus integer* by SCHMIDT, the gnathodid element of which was (and still is) considered to be congeneric with *G. mosquensis* PANDER, 1856, type-species of *Gnathodus.* MÜLLER, supported by SCHMIDT (personal communication), judges that *Westfalicus* should be cited as a junior subjective synonym of *Gnathodus,* despite the fact that *Westfalicus* is a name based on conodont assemblages and *Gnathodus* was defined on the basis of discrete conodonts.

[*Editorial Note.* It should be observed that because *Westfalicus* as a generic designation for conodont assemblages was premised on acceptance of *Gnathodus* and other names based on discrete conodonts as parataxa, which the XVth Zoölogical Congress (London, 1958) denied, the standing of *Westfalicus* as a zoological name is not affected. The Rules (new Art. 6, Sec. 5c) provide: "The fact that the name of a taxon published prior to 1958 was originally established conditionally does not prevent its name from being available." Also, as is well recognized (Art. 6, Sec. 5e), "A name is not made unavailable because it is based on any part of an animal. . . ." Thus, *Gnathodus* is available for application to assemblages of conodonts, as well as to discrete conodonts, and *Westfalicus* (by no means rejectable on the ground of its provisional first-published status) is also available for assemblages, subject only to considering the name to be a junior subjective synonym of *Gnathodus.*—MOORE]

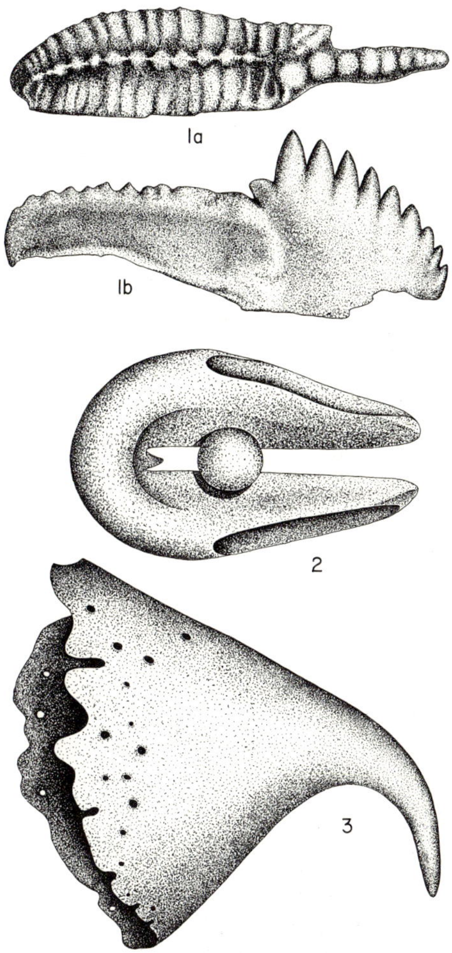

FIG. 152. Conodonts and paraconodont (p. *W*246, *W*248).——*1. Scaphignathus. 2. Westergaardodina. 3. Problematoconites.*

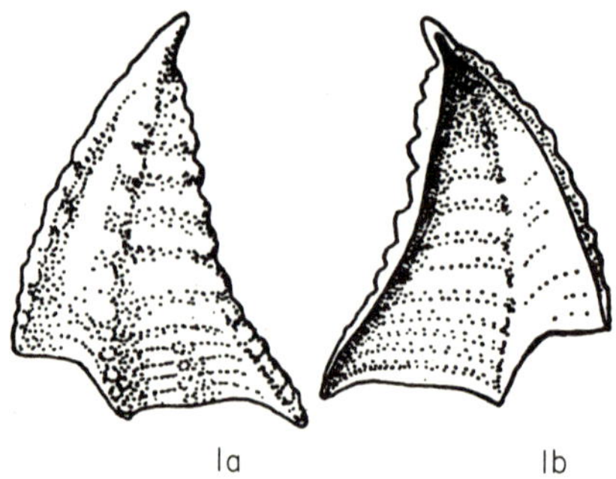

FIG. 153. Paraconodontida (p. W248).——*1. Pygodus.*

CONODONT TAXA CONSIDERED INVALID

SYNONYMS

Ancyrolepis ZIEGLER, 1959 [**A. cruciformis*]. Junior subjective synonym of *Ancyroides* MILLER & YOUNGQUIST, 1947.

Cornudina HIRSCHMANN, 1959 (Dec.) [**Ozarkodina breviramulis* TATGE, 1956]. Junior subjective synonym of *Parachirognathus* CLARK, 1959 (March).

Ctenognathodus FAY, 1959 [*pro Ctenognathus* PANDER, 1856 (*non* FAIRMAIRE, 1843)] [**Ctenognathus murchisoni* PANDER, 1856]. Junior subjective synonym of *Spathognathodus* BRANSON & MEHL, 1941.

Eobelodina SWEET, TURCO, WARNER & WILKIE, 1959 [**Oistodus fornicalis* STAUFFER, 1935]. Junior subjective synonym of *Oistodus* PANDER, 1856.

Gnamptognathus ZIEGLER, 1958 [**G. walliseri*]. Junior subjective synonym of *Scutula* SANNEMANN, 1955.

Order PARACONODONTIDA Müller, n. order

Phosphatic microfossils, which may contain organic matter and which, though resembling conodonts in size range and partly in general appearance, grow by accretion of the lower rim in a manner clearly distinct from the mode of growth in conodontophorids (*see* Fig. 48), are grouped under the name Paraconodontida. It is believed that the paraconodontids, now represented by a few described forms and several undescribed ones, will be useful in stratigraphic studies of early Paleozoic rocks. The presently recorded genera come from pre-Silurian strata. *L.Cam.-M.Ord.*

Problematoconites MÜLLER, 1959 [**P. perforata*]. Conical toothlike forms with large basal cavity; flanges perforated by numerous small oval holes, which demonstrate mode of growth unlike that of conodonts. *U.Cam.*, Eu.(Swed.-N.Ger.).——FIG. 152,*3*. **P. perforata*, 5d Zone, N.Ger.; lat., ×80 (7).

Pygodus LAMONT & LINDSTRÖM, 1957 [**P. anserinus*]. Sheathlike, triangular, flat, with point on one extremity; on reverse side sheath folded down like doublure of trilobite pygidium; accretion on rim opposite to point. *M.Ord.(Llandeil.)*, Eu. (Swed.-Scot.).——FIG. 153,*1*. **P. anserinus* LAMONT & LINDSTRÖM, *1a,b*, lat. obverse and reverse sides, Swed.; ×60 (6).

Rhombocorniculum WALLISER, 1958 [**R. comleyense*]. Conical, slender, laterally compressed, bowed or twisted, with deep basal cavity; surface with regular facet-like sculpture. *L.Cam.*, Eng.——FIG. 151,*3*. **R. comleyense*, Shrops.; lat., ×35 (9).

REFERENCES

Ash, S. R.

(1) 1961, *Bibliography and index of conodonts, 1949-1958:* Micropaleontology, v. 7, p. 213-244, 1 table.

Ethington, R. L.

(2) 1959, *Conodonts of the Ordovician Galena Formation:* Jour. Paleontology, v. 33, p. 257-292, pl. 39-41.

———, & Furnish, W. M.

(3) 1959, *Ordovician conodonts from northern Manitoba:* Jour. Paleontology, v. 33, p. 540-546, pl. 73.

Fay, R. O.

(4) 1959, *Generic and subgeneric homonyms of conodonts:* Jour. Paleontology, v. 33, p. 195-196.

Hirschmann, Christa

(5) 1959, *Über Conodonten aus dem oberen Muschelkalk des Thüringer Beckens:* Freiberger Forschungshefte, v. C 76, p. 33-86, pl. 3-5.

Lamont, Archie, & Lindström, Maurits

(6) 1957, *Arenigian and Llandeilian cherts identified in the southern uplands of Scotland by means of conodonts, etc.:* Edinburgh Geol. Soc., Trans., v. 17, pt. 1, p. 60-70, pl. 5.

Müller, K. J.

(7) 1959, *Kambrische Conodonten:* Deutsch. geol. Gesell., Zeitschr., v. 111, pt. 2, p. 434-485, pl. 11-15.

Sweet, W. C., Turco, C. A., Warner, Earl, Jr., & Wilkie, L. C.

(8) 1959, *The American Upper Ordovician standard. I. Eden conodonts from the Cincinnati region of Ohio and Kentucky:* Jour. Paleontology, v. 33, p. 1029-1068, pl. 130-133.

Walliser, O. H.

(9) 1958, *Rhombocorniculum comleyense n.gen., n.sp. (Incertae sedis, Unterkambrium, Shropshire):* Paläont. Zeitschr., v. 32, p. 176-180, pl. 15.

Ziegler, Willi

(10) 1958, *Conodontenfeinstratigraphische Untersuchungen an der Grenze Mitteldevon/Oberdevon und in der Adorfstufe:* Hess. Landesamt Bodenf., Notizbl., v. 87, p. 7-77, pl. 1-12.

(11) 1959, *Ancyrolepis n.gen. (Conodonta) aus dem höchsten Teil der Manticoceras-Stufe:* Neues Jahrb. Geol. & Paläont., v. 108, p. 75-80, pl. 7.

(12) 1960, *Die Conodonten aus den Geröllen des Zechsteinkonglomerates von Rossenray (südwestlich Rheinberg/Neiderrhein) mit der Beschreibung einiger neuer Conodontenformen:* Fortschr. Geol. Rheinland & Westfalen, v. 6, p. 1-15, pl. 1-4.

INDEX

Names included in the following index are classified typographically as follows: (1) Roman capital letters are used for suprafamilial taxonomic units which are recognized as valid in classification; (2) italic capital letters are employed for suprafamilial categories which are considered to be junior synonyms of valid names; (3) morphological terms and generic family names accepted as valid are printed in roman type; and (4) generic and family names classed as invalid, including junior homonyms and synonyms, are printed in italics. Page numbers printed in boldface type (as **W134**) indicate the location of systematic descriptions or definitions of morphological terms.